Aerospace Science and Engineering
III Aerospace PhD-Days

III Aerospace PhD-Days
International Congress of PhD Students in
Aerospace Science and Engineering
16-19 April 2023, Bertinoro, Italy

Editors
Sara Bagassi[1], Dario Modenini[1], Marco Petrolo[2]

[1]Università di Bologna, Italy
[2]Politecnico di Torino, Italy

Peer review statement

All papers published in this volume of "Materials Research Proceedings" have been peer reviewed. The process of peer review was initiated and overseen by the above proceedings editors. All reviews were conducted by expert referees in accordance to Materials Research Forum LLC high standards.

Published under License by **Materials Research Forum LLC**
Millersville, PA 17551, USA

Published as part of the proceedings series
Materials Research Proceedings
Volume 33 (2023)

ISSN 2474-3941 (Print)
ISSN 2474-395X (Online)

ISBN 978-1-64490-266-0 (Print)
ISBN 978-1-64490-267-7 (eBook)

This book contains information obtained from authentic and highly regarded sources. Reasonable efforts have been made to publish reliable data and information, but the author and publisher cannot assume responsibility for the validity of all materials or the consequences of their use. The authors and publishers have attempted to trace the copyright holders of all material reproduced in this publication and apologize to copyright holders if permission to publish in this form has not been obtained. If any copyright material has not been acknowledged please write and let us know so we may rectify in any future reprint.

Distributed worldwide by

Materials Research Forum LLC
105 Springdale Lane
Millersville, PA 17551
USA
https://www.mrforum.com

Manufactured in the United State of America
10 9 8 7 6 5 4 3 2 1

Table of Contents

Keyword index

Preface

The Aerospace PhD Days are organized by the Italian Association of Aeronautics and Astronautics, AIDAA, and are open to PhD students working on Aerospace Science and Engineering topics. The first two editions were held in Pisa under the supervision of Professor Aldo Frediani. The event has no parallel sessions to allow students to follow all presentations; furthermore, most of the organization is carried out by graduate students.

The 2023 edition had 63 presentations, with students from more than ten institutions, including delegates from China, France, and Spain. Many aerospace disciplines and topics were covered, i.e., dynamics and control, navigation, aeroacoustics, fluid dynamics, human-machine interaction, structures, maintenance and operations, sustainability of aeronautics and space, space economy, propulsion, additive manufacturing, sensors, aerospace systems, aeroelasticity, artificial intelligence, and UAV.

Four invited speakers delivered lectures on opportunities, challenges, and perspectives of the aerospace field: Angelo D'Agostino (Head of Research career and NCPs coordination Unit, APRE), ERC Starting Grant & MSCA-Postdoctordal Fellowships opportunities; Franco Ongaro (Chief Technology and Innovation Officer – Leonardo), Aerospace perspectives of Leonardo Company; Tatjana Bolić (Chair of the SESAR 3 Joint Undertaking's Scientific Committee - University of Westminster), Innovations in ATM; Maria Antonietta Perino (Director Space Economy Exploration and International Network - Thales Alenia Space), Exploration: from LEO to Moon and Mars. The closing ceremony included a speech from Anthea Comellini, a Member of the ESA astronaut reserve.

Aerospace Science and Engineering - III Aerospace PhD-Days　　　　Materials Research Forum LLC
Materials Research Proceedings 33 (2023) 1-8　　　　https://doi.org/10.21741/9781644902677-1

Autonomous navigation methods for spacecraft formation flying in cislunar space

Sergio Bonaccorsi[1,a] *

[1]Politecnico di Milano, Milan, Italy, 20156

[a]sergio.bonaccorsi@polimi.it

Keywords: Three Body Problem, Autonomous Navigation, Formation Flying

Abstract – Due to the increasing number of deep space missions in the following years, the focus on methods and strategies that enable a spacecraft to perform critical operations autonomously is becoming crucial. In the context of cislunar space, this research aims to investigate state estimation methods for such a highly nonlinear dynamics environment in order to identify the best operational scenarios and constraints that enable accurate performance for autonomous spacecraft navigation. In particular, the LiAISON navigation method is considered in the context of a formation of satellites moving in cislunar space exploiting the presence of other cooperative satellites to exchange inter-satellite signals to get range measurements.

Introduction

In the context of current and future lunar missions, the number of operating satellites in the cislunar environment is expected to grow. At the same time, spacecraft formation flying (SFF) is receiving increasing attention for its concept of distributing the functionality of a single spacecraft between several close-flying satellites, and for its higher adaptability and flexibility to different mission concepts. For these missions, the baseline option for orbit determination is, in general, ground-based radiometric navigation. Through technological advancements in autonomous Guidance, Navigation, and Control (GNC) capabilities for formation flying two main benefits would be accomplished: reduced costs and ground operations, and superior performance in terms of control accuracy and mission adaptability. For these reasons, autonomous navigation methods for SFF operating in highly nonlinear dynamics environments are investigated.

Space missions of formation of satellites in regions far from Earth pose different challenges for navigation and communication due to large distances. Nowadays, spacecraft navigation techniques for deep-space missions largely rely on radiometric measurements collected on the ground by the Deep Space Network (DSN). All data are processed on the ground to estimate trajectories and plan maneuvers. Newman et al. [1] have assessed the extent to which the DSN should be utilized to meet the tracking requirements for orbit determination. As a reference, for the Lunar Gateway, in the absence of any onboard sensors, DSN measurements should be used at least three contacts per orbital period, lasting 6 hours each time, to meet the navigation performance requirements [2].

There are several challenges to navigating in cislunar space that the current ground-based navigation paradigm does not readily address, such as their limited scheduling resources due to the increasing number of deep-space missions, constrained observation times, processing time and communication delays [3]. For these reasons, the concept of autonomous navigation is gaining interest as the most favorable approach for lunar missions.

Current Trends for Navigation

In the last decade, the first idea to alleviate the load of ground stations was still to assume the DSN as the primary resource for providing inertial measurements, but with the reduction of its overall work through the utilization of different onboard sensors. Secondly, with the ultimate goal of a completely autonomous system, the problem where a satellite attempts to simultaneously estimate

Aerospace Science and Engineering - III Aerospace PhD-Days   Materials Research Forum LLC
Materials Research Proceedings 33 (2023) 1-8   https://doi.org/10.21741/9781644902677-1

its own inertial state, as well as the inertial state of a target with only relative measurements, was considered. This so-called co-estimation problem was first solved using cross-linked range measurements within the Linked Autonomous Interplanetary Satellite Orbit Navigation (LiAISON) method [4]. Then, newer works expanded LiAISON results to optical measurements [5], and different sensors have been considered within the concept of data fusion [3,6,7,8]. Among them, X-ray pulsar navigation and, most importantly, cameras for optical navigation have been considered to get measurements from observations of asteroids, other satellites, Moon centers, and landmarks.

Regarding fully autonomous navigation strategies, current research interests focus on two main trends. The first relies solely on optical navigation, showing that the most efficacious target observations are images of artificial satellites and observations of the Moon (landmarks and/or the center of the figure). The second trend relies solely on satellite-to-satellite measurements, such as range or range-rate. These two systems differ in the algorithms used to process the sensor output and in the power consumption and weight of the overall equipment. Both of them rely on the concepts of the LiAISON navigation method.

LiAISON navigation is based on the notion of uniqueness. If there exists an orbit that is unique, then its absolute orientation is spatially unambiguous, and therefore any other orbit measured with respect to that orbit will, likewise, also be unambiguous. The characteristics of the acceleration function determine whether a unique trajectory exists, and also the degree to which the absolute position of the spacecraft can be estimated using inter-satellite measurements. Building on this assertion, satellites that are in sufficiently distinguishable orbits, are most attractive for this type of navigation. Fortunately, for spacecraft orbiting in the Earth-Moon neighborhood, the three-body dynamics provide the appropriate conditions for unique trajectories and for the subsequent use of such an autonomous navigation system. The LiAISON technique cannot determine the absolute states of both satellites from a single epoch measurement; rather, it requires a time series of measurements that are long enough for the asymmetric dynamic effects to show up in the data.

First Results

Simulations of hybrid strategies (implementing autonomous navigation methods but still relying on DSN) have shown, as a first remarkable result, the possibility of reducing the workload of DSN and ground operations in general, such as data processing and subsequent satellite uplink. In their work, Yun et al. [7] have shown how the combined use of optical navigation, antennas for GPS, and X-ray pulsar navigation could reduce the dependency on DSN to one hour per pass, a six-fold reduction from the nominal DSN-only approach.

Bradley et al. [3] support the feasibility of autonomous navigation in cislunar space with current technology and the application of LiAISON. They have demonstrated the possibility of meeting different navigation performance requirements based on the ability of three optical cameras with different resolutions to extract information from various target types at different times. The results show average uncertainty levels ranging from tens of kilometers to single kilometers. The more targets in visibility and the more measurements that can be provided to the navigation filter, the more accurate the state estimation will be. The most useful target types turn out to be artificial satellites and lunar centers. In particular, different mission scenarios, including different data cadences and correction maneuvers within the time windows of the estimation process, have been analyzed to also consider the robustness of the filtering process with respect to these aspects.

Following this work, Greaves and Scheeres [9] underline that all the previous work assumed a well-known target state so that only a relative state needed to be estimated. Therefore, they state that performing autonomous navigation using just optical sensors is possible, but they also put in evidence that the range estimate generated using bearing measurements is often only weakly observable and highly dependent on the nominal trajectory. For this reason, they have explored new guidance policies, which seek to maximize range information in the cislunar co-estimation

Aerospace Science and Engineering - III Aerospace PhD-Days   Materials Research Forum LLC
Materials Research Proceedings 33 (2023) 1-8   https://doi.org/10.21741/9781644902677-1

problem given relative optical measurements. This has been investigated for the idea of performing autonomous navigation in cislunar space by exploiting targets whose state has to be estimated as well.

Based on the concept of co-estimation with a target satellite, autonomous navigation performed using crosslink radiometric measurements such as range and range-rate is investigated in recent works [10,11,12,13]. Firstly, the authors have shown that range measurements in general provide better state estimation than range-rate observations for cislunar satellites performing autonomous navigation. Then, they have shown that the LiAISON technique could be a possible approach for the LUMIO mission, based on the existing inter-satellite link with the Lunar Pathfinder satellite without using any ground-based measurements. By means of an observation effectiveness analysis, they have also highlighted the presence of optimal tracking windows, which is a crucial remark considering the goal of optimizing the onboard measurement schedule. The same authors have then focused on the analysis of radiometric navigation for cislunar satellite networks involving three satellites moving in centralized or distributed configuration by exploiting the existing communication links between them. The distributed topology has been found to provide better state estimation and quicker convergent navigation solutions.

Regarding these feasibility analyses involving LiAISON navigation, the authors have underlined the importance to investigate what would happen considering the full operative mission lifetime and the presence of dynamics errors and clock model errors. To consider the practical challenges of implementing LiAISON, Wang et al. [14] developed a high-fidelity model to consider the effects of model errors and time synchronization on navigation performances. The most relevant conclusion drawn from this work is that LiAISON navigation allows to meet quite accurate navigation performances even considering a high-fidelity dynamic model. This has been possible by considering these model errors within the navigation filter algorithm. The authors have finally underlined several times the strong dependence of the estimate accuracy on the measurement geometry, which leads to very different performances to be evaluated according to mission requirements. In other words, the orbital regimes of the satellites involved in the navigation scenario strongly influence the resulting performances.

Summarizing the current trends for autonomous navigation in cislunar space, the Cislunar Autonomous Positioning System Technology, Operations, and Navigation Experiment (CAPSTONE) mission will serve as a pathfinder for navigation and operations in the same near rectilinear halo orbit (NRHO) that will be utilized by NASA's Lunar Gateway [15]. The Cislunar Autonomous Positioning System (CAPS), a navigation product developed by Advanced Space via a NASA Small Business Innovation Research, is intended to allow for the navigation of most small satellites without the requirement of a high-gain antenna, high-sensitivity GNSS receiver, or other specialized onboard hardware [16]. While standard DSN two-way navigation will still be the baseline operational navigation framework for the mission, the ultimate goal is to demonstrate navigation with either no ground-in-the-loop or without a dedicated two-way ground link. Currently, there are two measurement paradigms that can be processed by the CAPS flight software: crosslink measurements between two spacecraft, and one-way uplink measurements made possible by an onboard Microsemi SA.45s chipscale atomic clock (CSAC), which will leverage the enhanced oscillator stability for the estimation of clock bias and drift within the navigation filter.

To demonstrate the expected performance of CAPSTONE, high-fidelity navigation simulations have been performed considering noise values and biases that are expected based on ground tests of mission hardware. According to these preliminary results, one-way measurements or crosslink measurements alone would likely be good enough to support the onboard planning of orbit maintenance maneuvers if desired. Each of these data types can provide onboard solutions comparable to the quality of the ground-based filters given the expected CAPSTONE tracking schedule. Additionally, the utilization of both data types can provide additional improvements to onboard solutions. The scientific community look forward to analyze CAPSTONE's experimental data, and sharing the achieved results with the rest of the world.

Spacecraft Formation Flying Navigation

The aforementioned navigation problem focuses on single spacecraft missions interacting with already present targets in cislunar space, but, considering the presence of close-flying satellites within the concept of SFF, relative navigation issues must be addressed as well. Considering the problem of autonomous relative navigation between an active satellite (chaser) and another space object (target) orbiting in close-proximity, for SFF applications the latter can be included in the category of actively or passively cooperative targets, depending on the strategy adopted. In the actively cooperative target case, both chaser and target have the knowledge of their own state with a certain degree of uncertainty and they exchange information by means of a communication link. In some cases, the target may also be cooperating in a passive way, through artificial markers on the spacecraft body that can be detected and tracked by the chaser spacecraft.

In general, the state estimation problem for SFF has attracted a considerable amount of attention due to its practical and theoretical significance. The most direct and accurate way to solve this problem is to design centralized solutions, in which the chief performs all the computation for the fleet using data collected from the deputies [17]. Unfortunately, as the number of the spacecraft information becomes larger, the computation burden of the chief becomes unbearable due to its requirement of large state covariance and measurement matrix calculation [18]. To overcome this problem, decentralized solutions are designed, which allocates the computational load among all the spacecrafts in the formation by gathering the measurements and estimating the state in a decentralized way (i.e. each spacecraft gathers its own measurements and has its own estimator) [19].

Navigation Filters

Following the guidelines given by Pesce [20], some relevant aspects have to be taken into account while designing a navigation filter. It must be robust to measurements and initialization errors. This implies that an appropriate filter should converge to the desired solution in a reasonable time and with an opportune accuracy, also in the case of measurement noise levels and uncertainty in the dynamical model different from the expected ones. The computational load must be reduced due to the fact that high frequencies can be required but the computational power onboard is limited. As importantly, the choice of the model to describe the state dynamics is crucial for filter design. External disturbances and orbital perturbations could be included, implying an increase of the computational cost, but they are necessary for filters with low frequency updates or long operational time, when the effects of perturbations become significant, like in the case of SFF.

An efficient way to handle the prediction step of a Kalman filter, including an accurate dynamic model to propagate covariance and trajectory, is a relevant topic nowadays with the aim of improving the accuracy of the filter without affecting computational performance. The specific case of navigating in cislunar space turns the attention towards the problem of state estimation in the context of highly nonlinear dynamics. In addition, for autonomous purposes, current researches focus on the use of sequential filtering that allows for the simulation of an actual real-time onboard implementation. In particular, the extended Kalman filter (EKF) and the unscented Kalman filter (UKF) are widely used. Regarding EKF, in some cases, the linear assumption fails to provide an accurate realization of the local trajectory motion due to the low frequency of the estimation process as well as the nature or the limited number of measurements. In such cases, UKF yields superior performance in highly nonlinear situations because it is based on the unscented transformation, which does not contain any linearization. Following the idea of enhancing the filter's ability to manage nonlinear dynamics, in the work of Valli et al. [21], it has been

Aerospace Science and Engineering - III Aerospace PhD-Days   Materials Research Forum LLC
Materials Research Proceedings 33 (2023) 1-8   https://doi.org/10.21741/9781644902677-1

demonstrated that working in the differential algebra framework significantly reduces the complexity and the computational burden related to the standard higher-order approaches.

Additionally, in the context of using the LiAISON strategy, the same nonlinearities that produce observability also cause issues for linearized filters. The filter linearization assumptions are further strained by the inflated uncertainty profiles associated with the co-estimation problem. Thus, it becomes increasingly vital to obtain additional state information and to manage the prediction step better to alleviate nonlinearities, improve observability, and ensure accurate navigation.

Conclusions and Future Perspectives

The main focus of this research concerns the role of the navigation filter for the autonomous navigation of a formation of satellites operating in cislunar space. At the same time, the idea is to evaluate the navigation performance by considering combined metrics that take into account both estimation accuracy and mission objectives in order to consider a practical use of the algorithms and the possible improvements to realistic applications.

The major part of this research regards how to manage state estimation in such highly nonlinear dynamics, understand the behavior of propagated uncertainties, and how to mitigate their accumulation in a computationally efficient way. These points are addressed in the context of autonomous navigation, considering a possible real-time onboard implementation according to appropriate mission limitations and requirements.

To pursue these goals, the first step is the definition of accurate dynamic models and representative measurement models. In a navigation framework including the LiAISON method in the three-body problem of the earth-moon system, the satellites involved can operate in orbital regimes experiencing very different acceleration fields. For a practical implementation of the desired strategy, it is fundamental to consider dynamic model errors and clock errors. For the same reason, the actual sensors to be used onboard must be appropriately modeled considering real parameters and noise values. The most realistic dynamic environment should be reproduced through the implementation of a high-fidelity model of cislunar space, and an appropriate dynamic model must be included for the filter's prediction step considering computational constraints.

The effects of nonlinearity on the curvature of the state distribution after a certain propagation time could lead to an untreatable knowledge of the spacecraft state because of the accumulated uncertainty. For this reason, as a first step, this work aims to apply well-known uncertainty propagation techniques in order to analyze their behavior when used in the three-body dynamics with the ultimate goal of exploring strategies to mitigate their accumulation over time. Considering the peculiarities of such a navigation scenario, different state representations and uncertainty propagation techniques will be used within the prediction step of the estimation filter with the aim of defining a suitable strategy to meet requirements on the accuracy of the solution and computational load.

Additionally, considering the typical sparse observations of these navigation scenarios, and the need to speed up computations for efficient onboard implementation, the use of multi-fidelity propagation methods should be considered as a potential turning point. The choice to investigate the potential of these methods is also motivated by the different orbital regimes covered by the satellites involved in our scenario. This situation leads to the presence of time windows in which the model of the orbital dynamics could be simplified leading to a more performing propagation step.

Finally, considering formations of satellites flying at relatively short distances, the co-estimation problem results difficult to be solved [4]. For this reason, according to decentralized state estimation for SFF, the idea is to define a navigation strategy such that a leader satellite is involved in the solution of a co-estimation problem with appropriate external targets in cislunar space, while for the deputies the problem could be reduced to the sole estimation of the relative state through the use of RF-based or vision-based systems. The main goal is to investigate the effect of performing decentralized relative navigation with respect to a target whose state is retrieved in real-time through the LiAISON strategy.

Once the issues related to state estimation for SFF operating in this environment have been addressed, the idea is to recover the general framework of the desired mission to apply navigation strategies and methods to a practical test case. This research aims to put the navigation system of an SFF mission into the specific framework of a space mission, thus not just evaluating the state estimation performance in an ideal case without considering the actual mission requirements and onboard capabilities. The driven idea is to identify one or more specific surveillance missions in cislunar space for which SFF could provide benefits in terms of mission objectives. This will provide requirements on SFF orbit, sensor, configuration, and the number of satellites. Then, after a first simple evaluation of autonomous navigation performances, the idea is to define comprehensive metrics to combine the needs of guidance and navigation procedures with the objectives of a surveillance mission. The result is therefore to drive optimization procedures to identify guidance strategies to meet favorable navigation accuracy and surveillance performances. Finally, the potential role of SFF for other cislunar missions could be investigated.

Through the evaluation of the aforementioned metrics along one or more orbital periods or the entire mission lifetime, it is possible to highlight the presence of regions of space or time intervals in which the navigation strategy or the mission objective is not performing well. In this way, one can identify possible limitations of the considered mission and navigation strategy, but also use the metrics in order to drive some optimization routines toward more suitable solutions according to the desired requirements.

The final product of this process is to identify proper guidelines to improve the overall mission performance. The evaluation of the defined metrics could be used to identify appropriate solutions in terms of different aspects of the mission, such as orbital regime, target for the LiAISON strategy, type of sensors for surveillance operations, number of spacecraft and configuration for the formation.

To sum up, the following questions can be considered representative of the main objectives pursued by this research:

- Could an autonomous navigation strategy, based on the use of the LiAISON method and a relative navigation measurement system, lead to the required accuracy for SFF cislunar missions?
- Could the use of combined metrics for navigation and mission goals lead to the improvement of SFF mission design?
- Which are improvements resulting from the use of a multi-sensor configuration of satellites to perform surveillance operations of uncooperative targets in cislunar space?

References

[1] Newman, C. P., Davis, D. C., Whitley, R. J., Guinn, J. R., & Ryne, M. S. (2018). Stationkeeping, orbit determination, and attitude control for spacecraft in near rectilinear halo orbits.

[2] Guinn, J. R., Bhaskaran, S., Ely, T. A., Kennedy, B. M., Martin-Mur, T. J., Park, R. S., ... & Vaughan, A. T. (2019). The Deep Space Positioning System (DPS)-navigator concept for the Lunar Gateway.

[3] Bradley, N., Olikara, Z., Bhaskaran, S., & Young, B. (2020). Cislunar navigation accuracy using optical observations of natural and artificial targets. Journal of Spacecraft and Rockets, 57(4), 777-792. https://doi.org/10.2514/1.A34694

Aerospace Science and Engineering - III Aerospace PhD-Days   Materials Research Forum LLC
Materials Research Proceedings 33 (2023) 1-8   https://doi.org/10.21741/9781644902677-1

[4] Hill, K., & Born, G. H. (2007). Autonomous interplanetary orbit determination using satellite-to-satellite tracking. Journal of guidance, control, and dynamics, 30(3), 679-686. https://doi.org/10.2514/1.24574

[5] Greaves, J. A., & Scheeres, D. J. (2021). Relative estimation in the cis-lunar regime using optical sensors. In Proceedings of the Advanced Maui Optical and Space Surveillance Technologies Conference.

[6] Christian, J. A., & Lightsey, E. G. (2009). Review of options for autonomous cislunar navigation. Journal of Spacecraft and Rockets, 46(5), 1023-1036. https://doi.org/10.2514/1.42819

[7] Yun, S., Tuggle, K., Zanetti, R., & D'Souza, C. (2020). Sensor configuration trade study for navigation in near rectilinear halo orbits. The Journal of the astronautical sciences, 67, 1755-1774. https://doi.org/10.1007/s40295-020-00224-1

[8] Winternitz, L. B., Hassouneh, M. A., Long, A. C., Wayne, H. Y., Small, J. L., Price, S. R., & Mitchell, J. W. (2022, February). A High-fidelity Performance and Sensitivity Analysis of X-ray Pulsar Navigation in Near-Earth and Cislunar Orbits. In 44th Annual AAS Guidance, Navigation and Control Conference. The American Astronautical Society (AAS).

[9] Greaves, J., & Scheeres, D. Autonomous Guidance and Navigation with Optical Measurements for Formation Flying in Cislunar Space.

[10] Turan, E., Speretta, S., & Gill, E. K. A. (2022). Performance Analysis of Radiometric Autonomous Navigation for Lunar Satellite Network Topologies. In 11th International Workshop on Satellite Constellations and Formation Flying,.

[11] Turan, E., Speretta, S., & Gill, E. (2022). Autonomous Crosslink Radionavigation for a Lunar CubeSat Mission. arXiv preprint arXiv:2204.14155. https://doi.org/10.3389/frspt.2022.919311

[12] Turan, E., Speretta, S., & Gill, E. (2022, March). Autonomous navigation performance of cislunar orbits considering high crosslink measurement errors. In 2022 IEEE Aerospace Conference (AERO) (pp. 1-11). IEEE. https://doi.org/10.1109/AERO53065.2022.9843772

[13] Turan, E., Speretta, S., & Gill, E. (2022). Radiometric autonomous navigation for cislunar satellite Formations. In NAVITEC 2022.

[14] Wang, W., Shu, L., Liu, J., & Gao, Y. (2019). Joint navigation performance of distant retrograde orbits and cislunar orbits via LiAISON considering dynamic and clock model errors. Navigation, 66(4), 781-802. https://doi.org/10.1002/navi.340

[15] Cheetham, B. (2021). Cislunar autonomous positioning system technology operations and navigation experiment (Capstone). In ASCEND 2021 (p. 4128). https://doi.org/10.2514/6.2021-4128

[16] Thompson, M. R., Forsman, A., Chikine, S., Peters, B. C., Ely, T., Sorensen, D., ... & Cheetham, B. (2022, January). Cislunar Navigation Technology Demonstrations on the CAPSTONE Mission. In Proceedings of the 2022 International Technical Meeting of The Institute of Navigation (pp. 471-484). https://doi.org/10.33012/2022.18208

[17] Viegas, D., Batista, P., Oliveira, P., & Silvestre, C. (2012, June). Decentralized range-based linear motion estimation in acyclic vehicle formations with fixed topologies. In 2012 American Control Conference (ACC) (pp. 6575-6580). IEEE. https://doi.org/10.1109/ACC.2012.6314915

[18] Busse, F., & How, J. (2002). Real-time experimental demonstration of precise decentralized relative navigation for formation flying spacecraft. In AIAA guidance, navigation, and control conference and exhibit (p. 5003). https://doi.org/10.2514/6.2002-5003

[19] Mu, H., Bailey, T., Thompson, P., & Durrant-Whyte, H. (2011). Decentralised solutions to the cooperative multi-platform navigation problem. IEEE Transactions on Aerospace and Electronic Systems, 47(2), 1433-1449. https://doi.org/10.1109/TAES.2011.5751268

[20] Pesce, V. (2019). Autonomous navigation for close proximity operations around uncooperative space objects.

[21] Valli, M., Armellin, R., Di Lizia, P., & Lavagna, M. R. (2014). Nonlinear filtering methods for spacecraft navigation based on differential algebra. Acta Astronautica, 94(1), 363-374. https://doi.org/10.1016/j.actaastro.2013.03.009

Aerospace Science and Engineering - III Aerospace PhD-Days   Materials Research Forum LLC
Materials Research Proceedings 33 (2023) 9-14   https://doi.org/10.21741/9781644902677-2

Relative visual navigation based on CNN in a proximity operation space mission

A. D'Ortona, G. Daddi

Politecnico di Torino, DIMEAS, corso Duca degli Abruzzi 24, Torino

antonio.dortona@polito.it, guglielmo.daddi@polito.it

Keywords: Navigation, Space, CNN

Abstract. This article explores a solution utilizing a convolutional neural network (CNN) to simulate robust monocular visual navigation during proximity operations of a space mission, where a precise determination of relative pose is crucial for mission safety. This operation involves closely observing a spacecraft with a CubeSat under challenging illumination conditions. The methodology involves generating a dataset using Blender software and training a Mask-CNN with a ResNet-50 architecture to identify relevant features representing the target's 3D model. The dataset's ground truth is obtained through an inverse Perspective-n-Point (PnP) problem. Overall, this work provides valuable insights into the potential of deep learning-based visual navigation techniques for enhancing space mission operations.

Introduction

In modern times, digital cameras have become compact, precise, non-invasive, and affordable, which has led to their widespread use in vehicle and robot navigation. Over the years, various techniques have been developed for this purpose, with visual navigation being one of the most accurate ways to estimate position and attitude, also known as *camera pose* estimation.

Visual navigation has been extensively studied in the context of robotic space exploration, including the Mars exploration rovers in 2003. However, there has been a growing interest in applying visual-based navigation techniques for on-orbit servicing missions in recent years. This is especially important for automatic rendezvous operations, which require precise determination of relative pose to ensure safe mission completion.

The traditional approach of pose estimation involves multi-view geometry, which compares two or more consecutive frames finding a set of 2D/2D correspondences to determine the camera's movement as in (Fravolini, 2010).

These methods are mostly applied in conditions where no known target object is observed. On a space mission, during proximity maneuvers, a target object can be observed and, if the geometry is known, a single image is enough to estimate the camera pose. Classical single-image pose estimation methods aim to solve the *Perspective-n-Point (PnP)* problem. PnP is the problem of estimating the pose of a calibrated camera given a set of n 3D points in the world and their corresponding 2D projections in the image. Model-based methods use a wireframe 3D model of the target spacecraft to match with 2D features extracted from the image and estimate the relative pose. Non-model-based methods compare the in-flight image with a pre-stored database of images to estimate the pose without feature extraction (al., 2012). However, these approaches lack robustness due to low signal-to-ratio, extreme illumination conditions, and dynamic Earth background in space imagery.

Recent developments in computer vision have introduced deep learning for pose estimation. Deep learning-based pose estimation methods typically use *convolutional neural networks (CNNs)* to extract features from input images or sensor data and then use these features to estimate the object's pose. By leveraging large amounts of labeled training data, deep learning methods can

Aerospace Science and Engineering - III Aerospace PhD-Days
Materials Research Proceedings 33 (2023) 9-14

Materials Research Forum LLC
https://doi.org/10.21741/9781644902677-2

learn complex relationships between input data and object poses, enabling them to achieve high accuracy even in challenging conditions such as low-light environments or cluttered scenes.

The objective of this paper is to explore a solution based on a CNN for simulating robust monocular visual navigation during a space mission's proximity operation, which involves closely observing a spacecraft with a CubeSat in critical illumination conditions. The optical sensor will be evaluated as a strong option in synergy with GPS data and IMU to calculate the relative position and attitude accurately. The final part of the mission involves a docking maneuver, where visual navigation is crucial, and hence, an accurate study of the technique is necessary.

Methodology

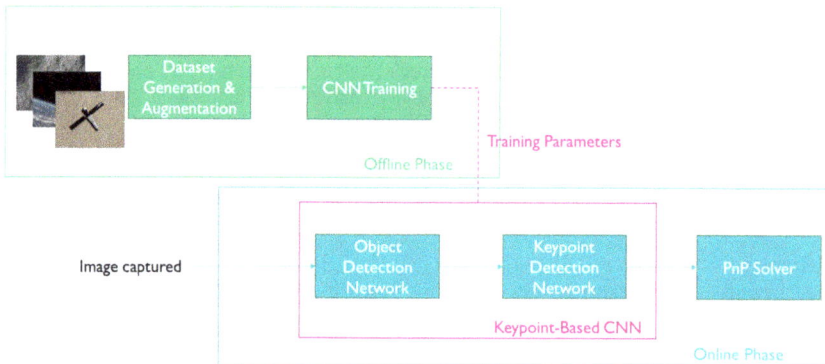

Figure 1 High-level architecture

The high-level architecture first involves the generation of the dataset using the 3D modeling software Blender. We placed in a blender environment a CAD model of SR and used blender's built-in camera model to visualize the camera's field of view, sensor size and resolution. Although this method is not physically or radiometrically accurate, it is a good way to test feature extraction algorithms under ideal circumstances. The environment is built by introducing a realistically scaled Earth, a moving sun, and trajectory import from STK or MATLAB simulations. However, the vast scale differences between Space Rider and Earth introduced core renderer issues that were addressed via workarounds. Three objects - terrain, atmosphere, and clouds - each with their own custom shader, made up the Earth. The terrain was composed of high-resolution satellite imagery, and a layer mask was used to increase roughness and create specular reflections on the water. The atmosphere was modeled to capture Rayleigh scattering and atmospheric pressure decay, which gives the sky its colour during daytime and tinge it red during sunset. The cloud layer imitated volumetric effects through a semi-transparent texture wrapped around the globe, and the sun mesh was placed according to the sun vector, with Blender's lens flare effects overlayed when visible from the camera. Blender's environment can read coordinate files in .csv form taken from STK or MATLAB and use them to procedurally place objects throughout the scene. Overall, this detailed environment provided a good testbed for feature extraction algorithms before introducing complications such as an illuminated moving background, real-world effects such as amplification and radiation noise, motion blur, bloom, or optic-induced blur.

The second component of a visual navigation simulator involves training a CNN to recognize relevant features in the images. In this case, a Mask-CNN with a ResNet-50 architecture is used (He, Gkioxari, Dollar, & Girshick, 2018). The Mask-CNN is a variant of the standard CNN that includes an additional output layer that predicts object masks. This is useful for tasks such as object

Aerospace Science and Engineering - III Aerospace PhD-Days   Materials Research Forum LLC
Materials Research Proceedings 33 (2023) 9-14   https://doi.org/10.21741/9781644902677-2

segmentation, where the goal is to identify the pixels corresponding to specific objects in an image. In our case, the specific objects are 6 keypoint representing the 3D model of our target.

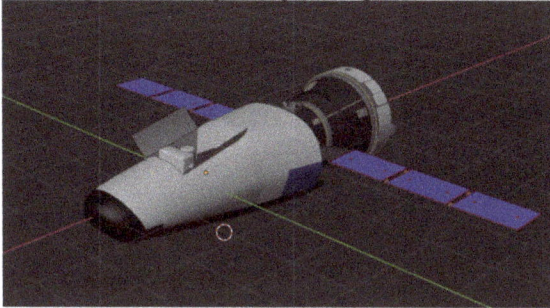

Figure 2 3D Model in Blender environment

The ground truth of the dataset (the 6 keypoints) is calculated through an inverse PnP problem; given the relative attitude and the relative position, the u and v coordinates of the keypoints on the image captured are automatically calculated:

$$(u, v) = \left(\frac{f X_{cam}}{Z_{cam}} + u_0, \frac{f Y_{cam}}{Z_{cam}} + v_0 \right)$$

$$\begin{bmatrix} f X_{cam} \\ f Y_{cam} \\ Z_{cam} \end{bmatrix} = \begin{bmatrix} f & & u_0 \\ & f & v_0 \\ & & 1 \end{bmatrix} \begin{bmatrix} X_{cam} \\ Y_{cam} \\ Z_{cam} \\ 1 \end{bmatrix} = \begin{bmatrix} f & & u_0 & 0 \\ & f & v_0 & 0 \\ & & 1 & 0 \end{bmatrix} \begin{bmatrix} \boldsymbol{R} & \boldsymbol{t} \\ 0 & 1 \end{bmatrix} \begin{bmatrix} X \\ Y \\ Z \\ 1 \end{bmatrix} \qquad (1)$$

In which X_{cam} and Y_{cam} are the coordinates of the keypoints in the camera reference frame, f is the focal length, u_0 and v_0 are the coordinates of the principal point of the camera, X Y and Z are the 3D coordinates in world coordinates frame (we considered a body reference frame).

The CNN architecture was designed with Pytorch framework on Python with a texture-randomized to increase accuracy and robustness. The Resnet 50 architecture is already implemented in the standard Pytorch libraries.

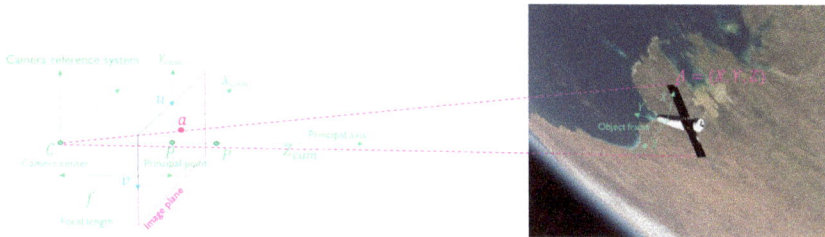

Figure 3 PnP problem depicted

The third component of a visual navigation simulator involves solving the PnP problem to find the relative pose of the target with respect to camera reference. The PnP problem is a classic computer vision problem that involves estimating the position and orientation of an object relative to a camera, given a set of 2D image points and their corresponding 3D points in the object coordinate system. In the context of visual navigation, the PnP problem is used to determine the position and orientation (R and t in Equation 1) of the autonomous agent relative to the environment it is navigating in. This information is critical for the agent to make informed decisions on how to navigate through the environment safely and efficiently. The problem is solved with RANSAC (Random Sample Consensus) (Fischler, 1981) algorithm, which is robust to outliers and can handle noise in the data. The RANSAC algorithm works by randomly selecting a subset of 3D-2D point correspondences to estimate the camera pose.

Overall, the architecture of a visual navigation simulator is designed to enable the creation of realistic environments and provide the necessary tools for training and testing autonomous navigation algorithms. The use of realistic illumination conditions, the training of a CNN, and the solution of the PnP problem are all critical components of a visual navigation simulator that contribute to its effectiveness in training and testing autonomous navigation algorithms.

Results & discussion
The model was trained with 4500 synthetic images generated with Blender as explained in the previous section. The synthetic images are generated in a random position of our chaser in a range from 100 m to 10 m around the target. This dataset replicates the maneuver carried out which is an observation maneuver around the target, and the distance varies as it is a spiral-shaped maneuver with an elliptical-shaped base.

The learning rate is initially set to 0.001 and decays exponentially by a factor of 0.98 after every epoch. The network is trained on an NVIDIA GeForce RTX 3090 for 200 epochs. Our aim was to evaluate the performance of the model in predicting the pose of objects in the scene. We used two metrics to evaluate the error in the predicted pose: E_R for the quaternion error and E_T for the translation error (Sharma & D'Amico, 2019).

$$E_T = |\tilde{t} - t|$$

$$E_R = 2 \arccos|q \cdot \tilde{q}| \tag{2}$$

$$E_{TN} = \frac{|\tilde{t} - t|}{|t|}$$

Where t, q are the predicted unit quaternion and translation vector aligning the target body reference frame and the Camera frame, and t,q are the ground-truth unit quaternion and translation vector. ER corresponds to the angle of the smallest rotation that aligns q and q. ETN is this distance normalized by the ground truth distance between the target and the camera. A final metric combines the two errors:

$EC=ETN+ER$

Table 1 CNN Scores

Metrics	Score
Mean E_T [m]	[0.2978 0.2131 0.3376]
Mean E_R [deg]	4.302
Mean E_C	0.1345

Table 1 reports the CNN's performances (Park, Sharma, & D'Amico, 2019) tested on a validation dataset of 100 synthetic images.

Aerospace Science and Engineering - III Aerospace PhD-Days   Materials Research Forum LLC
Materials Research Proceedings 33 (2023) 9-14   https://doi.org/10.21741/9781644902677-2

Overall, our results demonstrate the effectiveness of the Mask-CNN Resnet 50 model in accurately predicting the pose of objects in synthetic images. Specifically, the mean *ET* is about 50 cm, while the mean *ER* is around 4.3 degrees.

While initial results show promise, they are not as robust as those reported in seminal works in the field (Park, Sharma, & D'Amico, 2019) (Black, 2021). However, it is important to note that this work represents an early iteration in the development of new approaches to pose estimation using CNNs.

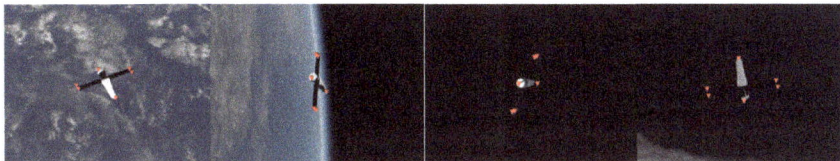

Figure 4 Keypoint detection results

Conclusion

In conclusion, this paper presents a solution based on a convolutional neural network (CNN) for simulating robust monocular visual navigation during a space mission's proximity operation. The proposed approach involves the generation of a dataset using 3D modeling software and training a Mask-CNN with a ResNet-50 architecture to recognize 6 keypoint features in images. The ground truth of the dataset is calculated through an inverse Perspective-n-Point (PnP) problem. The CNN's output is used to estimate the relative pose of a target spacecraft with respect to a CubeSat using a monocular camera in critical illumination conditions.

The proposed approach has several advantages over traditional methods for visual navigation. Deep learning-based pose estimation methods can learn complex relationships between input data and object poses, enabling them to achieve high accuracy even in challenging conditions such as low-light environments or cluttered scenes. The use of a monocular camera in critical illumination conditions reduces the complexity and cost of the system while maintaining high accuracy. The proposed approach can be used for automatic rendezvous operations, which require precise determination of relative pose to ensure safe mission completion.

The present findings underscore the need for further research and refinement of the proposed method to achieve more robust and accurate results. Expanding the dataset, improving its quality evaluating the use of other software and adjusting training parameters may hold promise as solutions to the current limitations of the method.

Another future work involves the integration of the proposed approach with GPS data and Inertial Measurement Unit (IMU) data to calculate the relative position and attitude accurately. Overall, the proposed approach shows promise for improving visual navigation in space missions and could be an essential tool for future on-orbit servicing missions.

References

[1] al., D. A. (2012). Solving the PnP Problem for Visual Odometry – An Evaluation of Methodologies for Mobile Robots. *Conference: Conference Towards Autonomous Robotic Systems*, 451-452. https://doi.org/10.1007/978-3-642-32527-4_54

[2] Black, K. &. (2021). Real-Time, Flight-Ready, Non-Cooperative Spacecraft Pose Estimation Using Monocular Imagery.

[3] Fischler, M. A. (1981). Random sample consensus: a paradigm for model fitting with applications to image analysis and automated cartography. *Communications of the ACM 24.6*, 381-395. https://doi.org/10.1145/358669.358692

[4] Fravolini, S. F. (2010). A Robust Monocular Visual Algorithm for Autonomous Robot Application. *IFAC Proceedings Volumes 43.16*, 551–556. https://doi.org/10.3182/20100906-3-IT-2019.00095

[5] He, K., Gkioxari, G., Dollar, P., & Girshick, R. (2018). Mask R-CNN. *CoRR*. https://doi.org/10.1109/ICCV.2017.322

[6] Park, T. H., Sharma, S., & D'Amico, S. (2019). Towards Robust Learning-Based Pose Estimation of Noncooperative Spacecraft. *ArXiv*.

[7] Sharma, S., & D'Amico, S. (2019). Pose estimation for non-cooperative spacecraft.

Aerospace Science and Engineering - III Aerospace PhD-Days   Materials Research Forum LLC
Materials Research Proceedings 33 (2023) 15-20   https://doi.org/10.21741/9781644902677-3

Enhanced radar detection of small remotely piloted aircraft in U-space scenario

Fausta Mattei[a*]

Department of Industrial Engineering, Università degli Studi di Napoli Federico II, Piazzale Tecchio 80, Naples, Italy

Dept. of Management, Information and Production Engineering, Università di Bergamo, Viale Marconi 5 Dalmine, Italy

[a]fausta.mattei@unina.it

Keywords: Radar Cross Section Enhancement, Unmanned Aerial Systems, Remotely Piloted Aircraft, Passive Surveillance, Active Surveillance, Software Defined Radio, Drones

Abstract. Efficient and safe integration of Unmanned Aerial Systems, both civil and military ones, must be guaranteed in the airspace, which is expected to be threated by problems of collisions, loss of communications and congestion of the air traffic environment. One of the main issues is how to improve the identification of unmanned aircrafts in the low altitude airspace. The identification process, that includes detection, verification, and recognition phases, is affected by different problems such as the difficulty of distinguishing Unmanned Aircraft Vehicles from other small flying objects as birds, because of their similar Radar Cross Section (RCS). To improve this process, an enhancement of the RCS can be a solution. The purpose of my PhD is to find the best passive and active solution for the following assignment.

Introduction

Unmanned Aerial Systems (UASs), i.e. drones, have been recognized as one of the revolutionary advances in the recent technological evolution. The origin of UAS traces back to the military field, for operating in hostile or human-denied areas. They are typically employed for "Dull, Dirty and Dangerous" missions. Today, their technology is improving more and more and making these types of systems more attractive for several applications both in civilian and military domain, including surveillance, search and rescue, traffic and weather monitoring and others [1]. Their main features include improved transportability due to small size, low cost, and speed of implementation. The current level of safety they can provide has made this technology accessible to the general public.

In 2020, the global small UAV market was valued at US$ 411.8 million and is predicted to be more than US$ 1350.1 million by 2027 [2]. Therefore, a large number of projects started in the past years, such as Single European Sky ATM Research (SESAR) [3] and the American "Next Generation Air Transportation System" (NextGen ATS) [4], with the purpose to enable several standards and regulations to fulfil the operational needs of and the basic requirements for UAS Traffic Management [5].

In Europe, a relevant project is the SESAR 3 Joint Undertaking (SJU) [6] which is an institutional European Partnership between private and public sector partners set up to accelerate through research and innovation the delivery of the Digital European Sky. This project started in 2021 and will end in 2031, and it aims to develop all the architectures, services and regulations for the realization of a safe, smart, sustainable and well-connected UTM airspace. More specifically, the European Union has developed a U-Space project [7], that deals with the realization of a set of

Aerospace Science and Engineering - III Aerospace PhD-Days   Materials Research Forum LLC
Materials Research Proceedings 33 (2023) 15-20   https://doi.org/10.21741/9781644902677-3

services to ensure a safe and efficient integration between air traffic of manned aircraft vehicles and air traffic of UAVs.

The "safety and security assessment" aspect is essential both under functional and operational aspects. In general, the challenges that affect the integration of U-Space in Urban Environment are [8]:

o   Public acceptance;

o   Safety metrics for VLL operation;

o   Need for high fidelity updated obstacle data;

o   Hyper-local weather events;

o   Communication and GNSS occlusions;

o   Air rules and airspace organization.

From an operational point of view, loss of communication, congestion problem and traffic collision are the main reasons for the lack of safety and must be reduced for an efficient and smarter UTM.

One of the first steps to enable the integration of unmanned aircraft in Urban Air Mobility (UAM), is to take care of surveillance systems to support UTM. Surveillance is the function that provides a continuous monitoring of trajectories of RPAS flying inside a volume of interest. It must guarantee three main features, such as[9]:

o   Interoperability;

o   Performance;

o   Efficiency.

These three characteristics combined allow cooperative and non-cooperative surveillance systems to deliver a broader range of services and to operate within a wider range of environments.

A big challenge concerning drone surveillance systems is the identification process, which includes three phases: detection, classification, and recognition. Identification means that a small Remotely Piloted Aircraft is properly recognized as soon as a track is initialized after sensor detection. The drone's detection problem is more challenging in urban environments given the number of potential targets, multipath, and often low Signal-to-Noise Ratio (SNR). The problems of identification are multiples [10]:

o   UAVs can be identified after a long time is passed from their initial detection when they fly at very short distances from surveillance sensors;

o   Because of their small Radar Cross Section (RCS), they can be barely distinguished from other small flying objects such as birds;

o   Their low-speed poses challenges when sensors are provided with special processing features such as Moving Target Indicator used to remove ground clutter.

A possible solution to improve the drone detection performance using radar sensors is to do Radar Cross Section RCS enhancement. It is the capability to improve the intensity of radar echo backscattered by a drone. The idea of the proposed PhD project is to evaluate the effect of passive and active solutions for RCS enhancement of a prototypical drone. Both passive and active solutions should be evaluated in light of the flight mission, and the impact on weight, endurance

Aerospace Science and Engineering - III Aerospace PhD-Days   Materials Research Forum LLC
Materials Research Proceedings 33 (2023) 15-20   https://doi.org/10.21741/9781644902677-3

and range, and other parameters related to the safety and efficiency of flight. Passive systems refer to all the methods for radar cross section enhancement which modify the geometry of the drone, for example using corner reflectors on-board, whereas active systems are the ones that do not affect the geometry of the drone, but they can make the enhancement using antenna, sensors or hardware.

Building up a prototypical unit can be easily performed by using Software Defined Radio (SDR). The SDR is a radio communication system that employs reconfigurable software-based components for the processing and conversion of digital signals [11] Flexibility, reduced costs and versatility [12] make it a suitable system for an experimental test of active and passive radar cross section enhancement. It is possible to set the parameter of the hardware through software, also in real-time, on the basis of the available type of radar.

Methodology

The proposed activity can be summarized in a sequence of processing steps, as follows:

I.   Identify and test passive solutions for RCS enhancement of a small drone;
II.   Identify and test active solutions which are able for RCS enhancement of a small drone;
III.   Build a prototype of a small multicopter drone implementing the proposed solution to accomplish a civil flight mission;
IV.   Integrate the proposed active enhanced detectability solutions in the guidance control and navigation system to take advantage of the position, attitude and situational awareness of the drone;
V.   Make a campaign of flight tests with the developed prototype and provide indications on the impact of the proposed solution on flight performance and safety and on the detection rates.

To carry out these tasks, the following procedure is adopted for the research project:

o   *First step*: State-of-the-art analysis. The aim is to identify all interesting projects on the detection of drones in an urban air mobility scenario, to find the most important systems (active and passive) that can fit the proposed objective of research and to study if their application is feasible.

o   *Second step*: set up of requirements needed for the choice of the various system for the drone. The idea is to realize a prototype of a small drone from the beginning, so it is crucial to fix the most important requirements that must be satisfied such as endurance, payload weight, and radar specification. The specifications can be used to size the active and passive systems (*SDR and corner reflectors*).

o   *Third step*: define the system architecture. After the requirement definition, it is possible to proceed with the choice of systems. Table 1 lists all the components needed for the experiments that must be conducted.

Aerospace Science and Engineering - III Aerospace PhD-Days   Materials Research Forum LLC
Materials Research Proceedings 33 (2023) 15-20   https://doi.org/10.21741/9781644902677-3

Table 1. System Architecture.

Drone Configuration		Payload		Ground Control Systems	
Object	*Quantity*	*Object*	*Quantity*	*Object*	*Quantity*
Autopilot [13]	1	Secondary battery	1	Radar [14]	1
Firmware	1	CPU Computer [15]	1	Remote controller	1
GPS receiver	1	SDR transceiver [16]	1	Computer	1
Inertial Measurement Unit (IMU)	1	Antenna	4		
ESC motor controller	6	Corner reflector	1		
Propellers	6				
Motor	1				
Primary battery	1				
Receiver	1				

o   *Fourth step*: define system testing strategy. A set of Flight Test Cards can be filled to describe the proper realization of the experiments. For instance, assuming that 'open category' drones can fly within a maximum altitude of 120 [m] above ground, three different values of altitude are chosen to carry out the same manoeuvre. Then for each altitude, other three different values of speed are imposed, assuming that the maximum velocity allowed in the Urban Air Mobility is 5 [m/s]. For each test, there is a total of 9 acquisitions, as shown in Table 2.

Table 2. Types of tests acquisition.

Altitude [m]	Speed
60	• 60% of V_{MAX} • 75% of V_{MAX} • 90%of V_{MAX}
90	• 60% of V_{MAX} • 75% of V_{MAX} • 90% of V_{MAX}
120	• 60% of V_{MAX} • 75% of V_{MAX} • 90% of V_{MAX}

Aerospace Science and Engineering - III Aerospace PhD-Days   Materials Research Forum LLC
Materials Research Proceedings 33 (2023) 15-20   https://doi.org/10.21741/9781644902677-3

o   *Fifth step*: test execution.

o   *Sixth steps*: data processing and results analysis. The data acquired during the test must be processed using a processing software tool, such as Matlab^{TM}.

Results

As result, it is expected that the signal is amplified when come back to the radar and to find out which are the best conditions for this to happen. In previous work, a preliminary test campaign was carried out and the signal amplification by a commercial SDR system, i.e. HackRF One [17], was validated. It is possible to appreciate in Fig. 1 the peaks which represent the proper amplification of the received tone by the on-board SDR for six different cases, listed as follows:

Test1.   Drone in a steady state mode on the ground, with an active rotor blade;
Test2.   Post take-off phase;
Test3.   Quickly movement up and down;
Test4.   Quickly movement up and down;
Test5.   360° heading rotation in hovering around its vertical axis;
Test6.   Data are acquired subsequentially to the rotation until the leading;

The weaker response is when the drone rotated 360°, because of the presence of the landing gear.

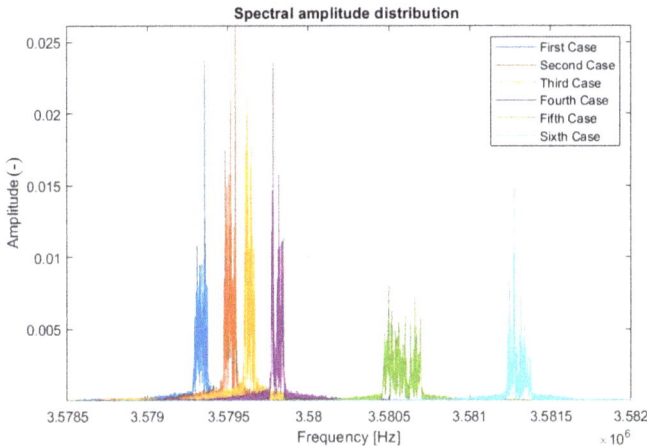

Fig. 1. Peaks of the signal amplified by the HackRF One for six different cases.

Conclusions

An optimization method for the surveillance of drones in the Urban Air Mobility is the Radar Cross Section enhancement. To carry out this task, both active (SDR) and passive (coating, phased antenna array, corner reflector, etc...) systems are investigated. Further activities will be mainly focused on the execution of several tests and simulations in order to assess the best solution to perform RCS enhancement.

References

[1] R. Austin, "UNMANNED AIRCRAFT SYSTEMS UAVS DESIGN, DEVELOPMENT AND DEPLOYMENT". 2011: Wiley. https://doi.org/10.1002/9780470664797

[2] "Small UAV Market Size to Surpass Around US$ 11.31 Bn by 2027."
https://www.precedenceresearch.com/small-uav-market.

[3] "SESAR Joint Undertaking | Delivering the Digital European Sky." https://www.sesarju.eu/

[4] "Next Generation Air Transportation System (NextGen) | Federal Aviation Administration."
https://www.faa.gov/nextgen.

[5] "UTM_ConOps_v2".

[6] "MULTIANNUAL WORK PROGRAMME SESAR 3 Joint Undertaking EUROPEAN
PARTNERSHIP," 2022, doi: 10.2829/156176.

[7] "U-space Blueprint brochure final".

[8] D. G. Castro and E. V. Garcia, "Safety Challenges for Integrating U-Space in Urban
Environments," in *2021 International Conference on Unmanned Aircraft Systems, ICUAS 2021*,
Jun. 2021, pp. 1258–1267. https://doi.org/10.1109/ICUAS51884.2021.9476883

[9] "Communications, navigation and surveillance | EUROCONTROL."
https://www.eurocontrol.int/communications-navigation-and-surveillance.

[10] A. Coluccia, G. Parisi, and A. Fascista, "Detection and classification of multirotor drones in
radar sensor networks: A review," *Sensors (Switzerland)*, vol. 20, no. 15. MDPI AG, pp. 1–22,
Aug. 01, 2020. https://doi.org/10.3390/s20154172

[11] "What is a Software Defined Radio? - everything RF."
https://www.everythingrf.com/community/what-is-a-software-defined-.

[12] "What are the advantages of Software Defined Radio?," 2002.

[13] "Pixhawk | The hardware standard for open-source autopilots." https://pixhawk.org/

[14] "Microwave Sensing, Signals and Systems." http://radar.ewi.tudelft.nl/Facilities/parsax.php.

[15] "ZedBoard Zynq^{TM} Evaluation and Development Hardware User's Guide," 2012.

[16] "AD9361 Datasheet and Product Info | Analog Devices."
https://www.analog.com/en/products/ad9361.html#product-overview.

[17] "HackRF," 2023.

Aerospace Science and Engineering - III Aerospace PhD-Days   Materials Research Forum LLC
Materials Research Proceedings 33 (2023) 21-28   https://doi.org/10.21741/9781644902677-4

Prediction of aeroacoustics of deformable bodies with solid or porous surface through a boundary integral formulation

Beatrice De Rubeis[1,a] *

[1]Roma Tre University, Department of Civil, Computer Science and Aeronautical Technologies Engineering, Via della Vasca Navale 79, Rome, Italy, 00146

[a]beatrice.derubeis@uniroma3.it

Keywords: Deformable Body, Boundary Integral Formulation, Aeroacoustics

Abstract. Novel boundary integral formulations suitable for the radiation of acoustic pressure from deformable, solid or porous, surfaces in arbitrary motion are theoretically/numerically developed. These will first be applied to a translating wing subject to unsteady bending and torsion. The influence of surface deformation on the evaluated perturbation fields will be assessed for different amplitude and frequency values of the bending and torsion modes. Subsequently, pressure will be radiated from a deformable porous sphere to validate the formulations for this type of surface.

Introduction

Nowadays a crucial issue for the aviation industries is the mitigation of the aircraft nuisance to comply with the increasingly demanding constraints in acoustic emissions.

Indeed, the growing interest in Urban Air Mobility (UAM) applications makes the design of more environmentally sustainable vehicles mandatory, representing a key point for the community acceptance of this new mobility concept. At the same time, the enhancement of conventional aircraft sustainability is also of primary importance from the perspective of eco-friendly next-generation aviation. To this aim, several international institutions and committees have established quantitative goals to achieve these common targets [1], in line with the Flightpath 2050 global vision [2].

For these reasons, many industries and research centers are making a great effort to develop new low-noise technologies to improve the sustainability of the current-adopted ones. Thus, to guarantee the sustainability of the aviation described in [2], the need to develop breakthrough concepts able to reduce both noise emissions and fuel consumption [3] has become unquestionable. In this framework, a lot of disruptive design solutions have been proposed like, for instance, Distributed-Propulsion layouts [4] based on the integration of the propulsive system with the aircraft airframe, with the possibility to exploit Boundary–Layer–Ingestion (BLI) technologies [5].

From all the above considerations, it is understood that an accurate description of the vehicle's acoustic emissions is undoubtedly of primary importance in the whole design process since its early stages. Because of this, it is essential to observe that the aeroacoustics of these novel concepts is affected by phenomena that might be negligible in standard configurations. Among these, for instance, the strong aerodynamic interactions occurring in multirotor propulsive systems, which may significantly modify performance and noise radiation of the aircraft more than in a standard configuration [6]. Furthermore, in highly-flexible innovative configurations, particular attention must be paid to the effects produced by the body deformation, which must be accurately taken into account both in aerodynamic and aeroacoustic simulations.

A significant amount of literature concerning the aeroacoustics of rigid bodies is available (see, for instance, the Farassat 1A formulation [7] commonly applied to evaluate noise emitted by roto-translating bodies (like helicopter rotor blades). However, since the distribution of perturbation fields produced by bodies moving in a fluid medium depends on the spatial orientation of the body surface, it can be strongly affected by their deformation (particularly considering slender, flexible structures). Thus, the availability of aerodynamic and aeroacoustic formulations capable of considering such effects can be mandatory. Some acoustic formulations accounting for the body deformation are available in the literature, for instance, the approaches presented in [8] and [9], where the contribution of the blade deformation is included in the Farassat 1A formulation and in its compact-source version, respectively.

Specifically, starting from the most general form of wave equation governing the propagation of perturbations in a fluid medium, the Green function is introduced in order to derive the solution in terms of a boundary integral representation capable to take into account surface deformation effects. Two versions of the integral solution are determined for solid surfaces: one expressed in a reference frame that only observes the deformation of the surface (i.e. the one fixed with the rigid motion of the body) and one expressed in a reference frame fixed with the undisturbed medium. Given the lack of investigation on this topic in the literature, the proposed formulations are validated by cross-comparison of the predictions provided by the integral solutions expressed in the two reference frames. The numerical investigation concerns the effect of surface deformation on the noise emitted by a translating wing subjected to unstable bending and torsional deformations. The effects of the amplitude and frequency of surface deformations on the radiated noise will be analysed by comparison with the acoustic emission of the rigid body, in order to assess the importance of including body deformations in aircraft aeroacoustics. To validate that the written formulations apply to porous surfaces as well, the pressure radiated from a deformable porous sphere containing a source inside was compared with the direct radiation from the source.

Methodology

Let us consider a moving body producing perturbations within a fluid. In a frame of reference $\mathcal{R}(x)$, fixed to the undisturbed fluid (FFR), the potential aerodynamics and noise radiation problems can be described by models that are expressed as subcases of the following general wave equation problem for the generic field, $\Psi(x, t)$

$$\nabla^2 \bar{\Psi} - \frac{1}{c^2} \frac{\partial^2 \bar{\Psi}}{\partial t^2} = \chi + z \cdot \nabla H + \nabla \cdot (Z \nabla H) + k_1 \frac{\partial H}{\partial t} + \frac{\partial}{\partial t} \left(k_2 \frac{\partial H}{\partial t} \right) + \frac{\partial}{\partial t} (w \cdot \nabla H) \quad (1)$$

where $\bar{\Psi} = \Psi H(f)$ is the extension of the field Ψ to the whole domain $\mathbb{R}^3$, with $H(f)$ denoting the Heaviside function with the argument $f(x, t)$ representing the body surface. In addition, c denotes the speed of sound in the undisturbed fluid, while k_1, k_2, z, w and Z are scalar, vector and tensor forcing terms which depend on the specific aerodynamic or aeroacoustic application.

The application of the Green function method provides an integral solution of Eq. (1), which has been extensively used and validated in the past for potential aerodynamics and aeroacoustics of rigid bodies (see, for instance, [10], [11] and [12]). In the following, the extension of the boundary integral solution of Eq. (1) to deformable bodies is briefly described. It derives from the boundary integral formulation for the potential aerodynamics of deformable bodies introduced in [13] which, in turn, is developed as the extension of a boundary integral formulation for rigid bodies expressed in a frame of reference rigidly connected to the body (BFR, $\mathcal{R}(y)$).

Note that, in the case of a rigid body, the introduction of the BFR is particularly useful since, differently from the FFR description, the integration domain (namely, the locus of the emitting points at the instants of emission of the signals that reach the microphone at the observation time) coincides with the real surface of the body.

Following [13], for the development of the integral formulation for deformable bodies, let us introduce a curvilinear coordinate system (ξ^1, ξ^2, ξ^3) that follows the deformation of the body, such that: $\xi^3(y, t) = 0$ identifies the motion of the body surface S with respect to the BFR, and

Aerospace Science and Engineering - III Aerospace PhD-Days   Materials Research Forum LLC
Materials Research Proceedings 33 (2023) 21-28   https://doi.org/10.21741/9781644902677-4

each point $y \in S$ is associated to a point in the plane $(\xi^1, \xi^2) \in \Omega$. Thus, close to the surface of the body one has $\nabla_y \xi^3 = n$ (with n denoting the outward unit normal vector of surface S), and the Heaviside function on the body surface is expressed as $H[\xi^3(y, t)]$.

Noise radiation in the body frame
From the general integral solution of Eq. (1), the integral formulation for sound radiation is derived, as well. To this purpose, for p_0 and ρ_0 denoting pressure and density of the undisturbed fluid, let us introduce the acoustic pressure, $p' = c^2(\rho - \rho_0)$ and the compressive tensor, $P = (p - p_0)I$. Then, for $v_T = v + v_D$, Eq. (1) reduces to the Ffowcs Williams and Hawkings equation for $\bar{\Psi} = \bar{p}'$, $\chi = -\nabla_y \cdot \nabla_y \cdot (HT)$, $Z = P$ and $w = -\rho_0 v_T$, while $z = 0$, $k_1 = 0$ and $k_2 = 0$. Thus, the integral solution of the general differential model in Eq. (1) provides the following BFR representation of the acoustic pressure radiated by deformable bodies [14]

$$
\begin{aligned}
\bar{p}'(y_*, t_*) = & \int_0^\infty \iint_\Omega -\hat{G}_0 \, \nabla_y \cdot \nabla_y \cdot (HT) \, \hat{J} \Big|_{g=0} \, d\xi^1 d\xi^2 d\xi^3 \\
& - \iint_\Omega \left[(Pn) \cdot \nabla_y \hat{G}_0 \, \hat{J} \right]\Big|_{\substack{\xi^3=0 \\ g=0}} d\xi^1 d\xi^2 \\
& + \iint_\Omega \left[\frac{1}{(1+v_D \cdot \nabla_y \hat{\theta})} \frac{\partial}{\partial t}\Big|_\xi \left[(Pn) \cdot \nabla_y \hat{\theta} \, \hat{G}_0 \, \hat{J} \right] \right]\Big|_{\substack{\xi^3=0 \\ g=0}} d\xi^1 d\xi^2 \\
& - \iint_\Omega \left[\rho_0 v_T \cdot n \, v \cdot \nabla_y \hat{G}_0 \, \hat{J} \right]\Big|_{\substack{\xi^3=0 \\ g=0}} d\xi^1 d\xi^2 \\
& - \iint_\Omega \left[\frac{1}{(1+v_D \cdot \nabla_y \hat{\theta})} \frac{\partial}{\partial t}\Big|_\xi \left[\rho_0 v_T \cdot n \left(1 - v \cdot \nabla_y \hat{\theta} \right) \hat{G}_0 \, \hat{J} \right] \right]\Big|_{\substack{\xi^3=0 \\ g=0}} d\xi^1 d\xi^2
\end{aligned}
\tag{2}
$$

Where $\hat{G}_0 = -1/[4\pi r(1 + M_r)]|_{\hat{\theta}}$, with $(...)|_{\hat{\theta}}$ indicating evaluation at time $t_* - \hat{\theta}$ (the symbol $\hat{\theta}$ denotes the time taken by the signal to propagate from a point in the BFR to the observer). In addition, $M_r = M \cdot e_r$, $M = v/c$, with $e_r = r/r$, while v and v_D are rigid and deformation velocity of the body, respectively, and $\hat{J} = |a_1 \times a_2 \cdot a_3|/|1 + v_D \cdot \nabla_y \hat{\theta}|$ is the Jacobian of the curvilinear co-ordinates transformation, with a_β denoting the covariant basis vectors of the curvilinear co-ordinate system. Furthermore, the symbol $(...)|_{\xi^3=0}$ indicates evaluation over the body surface, whereas the symbol $(...)|_{g=0}$ represents evaluation at the instant of emission from the point of the deformed surface of the signal that at the time t_* reaches the observer.

Noise radiation in the air frame
An equivalent integral formulation for the acoustic pressure fields related to moving deformable bodies can be derived in the FFR.

This is readily derived from Eq. (2) by observing that $v = 0$ (the velocity of the fluid reference frame is null), replacing v_D with v_T (in the air frame of reference the velocity of the body is given by the superposition of that related to the rigid body motion with that due to the deformation), replacing $\hat{G}_0$ with $G_0 = -1/(4\pi r)$ (Green's function in the air reference), and replacing $\hat{\theta}$ with $\theta = r/c$ (time taken by a signal to propagate from a fixed point in the FFR to the observer). Thus, the following representation of the acoustic pressure radiated by deformable bodies in the fluid frame of reference is obtained [14]

$$
\begin{aligned}
\bar{p}'(\mathbf{x}_*, t_*) = &\int_0^\infty \iiint_\Omega -G_0 \nabla \cdot \nabla \cdot (HT) \, J \Big|_{g=0} d\xi^1 d\xi^2 d\xi^3 \\
&- \iint_\Omega \left[(\mathbf{Pn}) \cdot \nabla G_0 \, J \right] \Big|_{\substack{\xi^3=0 \\ g=0}} d\xi^1 d\xi^2 \\
&+ \iint_\Omega \left[\frac{1}{(1 + \mathbf{v}_T \cdot \nabla \theta)} \frac{\partial}{\partial t} \Big|_\xi \left[(\mathbf{Pn}) \cdot \nabla \theta \, G_0 \, J \right] \right] \Big|_{\substack{\xi^3=0 \\ g=0}} d\xi^1 d\xi^2 \\
&- \iint_\Omega \left[\frac{1}{(1 + \mathbf{v}_T \cdot \nabla \theta)} \frac{\partial}{\partial t} \Big|_\xi \left[\rho_0 \mathbf{v}_T \cdot \mathbf{n} \, G_0 \, J \right] \right] \Big|_{\substack{\xi^3=0 \\ g=0}} d\xi^1 d\xi^2
\end{aligned}
\tag{3}
$$

where $J = |\boldsymbol{a}_1 \times \boldsymbol{a}_2 \cdot \boldsymbol{a}_3| / |1 + \mathbf{v}_T \cdot \nabla \theta|$ is the Jacobian of the curvilinear co-ordinates transformation in the FFR.

For rigid body applications, the integral formulation in the BFR is convenient because the integration domain coincides with the real surface of the body. This does not occur in the case of deformable bodies. However, as already stated, the BFR formulation is introduced in order to provide analytical/numerical comparison with the FFR integral formulation for cross-validation purposes.

Noise radiation from a porous surface

From the general integral solution of Eq. (1) we also derive the integral formulation for sound radiation from porous surfaces. Making the same assumptions as for the formulations for solid surfaces and defining $\boldsymbol{u}$ as the fluid velocity, Eq. (1) reduces to the Ffowcs Williams and Hawkings equation for porous surfaces for $\bar{\boldsymbol{\Psi}} = \bar{p}'$, $\chi = -\nabla \cdot \nabla \cdot (HT)$, $Z = P + \rho \boldsymbol{u} \otimes (\boldsymbol{u} - \boldsymbol{v}_T)$ and $\boldsymbol{w} = -\rho_0 \boldsymbol{v}_T - \rho(\boldsymbol{u} - \boldsymbol{v}_T)$, while $\boldsymbol{z} = \boldsymbol{0}$, $k_1 = 0$ and $k_2 = 0$. Therefore, the integral solution of the general differential model in Eq. (1) gives the following FFR representation of the acoustic pressure radiated by deformable porous surfaces:

$$
\begin{aligned}
\bar{p}'(\mathbf{x}_*, t_*) = &\int_0^\infty \iiint_\Omega -G_0 \nabla \cdot \nabla \cdot (HT) \, J \Big|_{g=0} d\xi^1 d\xi^2 d\xi^3 \\
&- \iint_\Omega \left[(\mathbf{Pn} + \rho \left[\mathbf{u} \otimes (\mathbf{u} - \mathbf{v}_T) \right] \mathbf{n}) \cdot \nabla G_0 \, J \right] \Big|_{\substack{\xi^3=0 \\ g=0}} d\xi^1 d\xi^2 \\
&+ \iint_\Omega \left[\frac{1}{(1 + \mathbf{v}_T \cdot \nabla \theta)} \frac{\partial}{\partial t} \Big|_\xi \left[(\mathbf{Pn} + \rho \left[\mathbf{u} \otimes (\mathbf{u} - \mathbf{v}_T) \right] \mathbf{n}) \cdot \nabla \theta \, G_0 \, J \right] \right] \Big|_{\substack{\xi^3=0 \\ g=0}} d\xi^1 d\xi^2 \\
&- \iint_\Omega \left[\frac{1}{(1 + \mathbf{v}_T \cdot \nabla \theta)} \frac{\partial}{\partial t} \Big|_\xi \left[(\rho_0 \mathbf{v}_T + \rho(\mathbf{u} - \mathbf{v}_T)) \cdot \mathbf{n} \, G_0 \, J \right] \right] \Big|_{\substack{\xi^3=0 \\ g=0}} d\xi^1 d\xi^2
\end{aligned}
\tag{4}
$$

where G_0, J and θ are exactly the same as defined for the formulation for solid surfaces in the FFR. It is also useful to note that for solid surfaces, the $\boldsymbol{u} - \boldsymbol{v}_T$ term is null, and Eq. (4) reduces to Eq. (3).

Numerical Results

A preliminary numerical investigation on the body deformation effects on the aerodynamic and acoustic fields radiation is here addressed. To this purpose, an upswept, untapered wing in uniform translation, with NACA 0012 airfoil sections, semi-span equal to $\mathbf{1.5} \, \boldsymbol{m}$, and chord equal to $\mathbf{1} \, \boldsymbol{m}$ is considered. In a right-handed BFR with origin at the mid-wing section leading edge, $\boldsymbol{y}_1$-axis

Aerospace Science and Engineering - III Aerospace PhD-Days   Materials Research Forum LLC
Materials Research Proceedings 33 (2023) 21-28   https://doi.org/10.21741/9781644902677-4

coincident with the advancing direction and pointing forward, and y_2-axis directed starboard, the observer is placed at $[0, -2\ m, -0.5\ m]$. A lifting case is examined by setting the wing angle of attack equal to $4°$. The aerodynamic solution over the body surface is evaluated by an aerodynamic integral solver for the rigid body motion, with the deformation only taken into account in the definition of the boundary conditions. The body deformation is expressed as $x_D(y, t) = \Psi_b(y)q_b(t) + \Psi_t(y)q_t(t)$, with the two bending and torsion shape functions given by

$$\psi_b(y) = \begin{Bmatrix} 0 \\ 0 \\ y_2^2 \end{Bmatrix}; \qquad \psi_t(y) = \begin{Bmatrix} y_3 \sin(\Omega_t\ y_2) \\ 0 \\ -y_1 \sin(\Omega_t\ y_2) \end{Bmatrix} \qquad (5)$$

Where $\Omega_t = \pi/b$, with b denoting the semi-span, whereas the lagrangean coordinates are expressed as $q_b(t) = A\cos(\omega_b t)$ and $q_t(t) = B\cos(\omega_t t)$ with ω_b and ω_t representing the pulsations of the bending and torsion deformations, respectively. First, for several Mach numbers, Fig. 1 shows the comparisons between the time signatures at the observer position evaluated through the integral formulations written in the FFR and in the BFR, respectively for the acoustic pressure. In this case the wing deformation is defined with $A = 0.01$ and $B = 0.1$, $\omega_b = 4\pi/T$ and $\omega_t = 6\pi/T$, where T is the observation time interval length. This time interval is assumed to be $T = 0.12$ s, i.e. such that it includes at least two periods of deformation oscillations. These parameters correspond to a maximum bending deflection equal to 1.5% of the wing semi-span, and maximum twist angle equal to $\alpha = 3°$. The results shown in Fig. 1 demonstrate that, regardless the Mach number as expected, the integral formulations expressed in the FFR and BFR perfectly match, thus confirming their full equivalence.

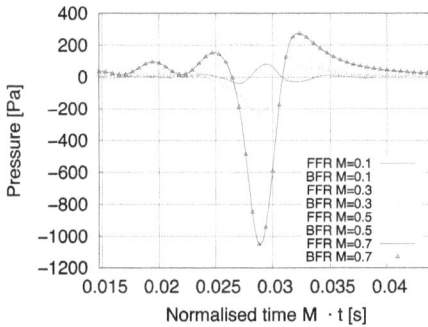

Figure 1. *Comparison between time signatures of acoustic pressure evaluated by the integral formulations written in the FFR and in the BFR, at different Mach numbers.*

In addition, the effect of the deformation parameters on the radiated pressure is investigated. Specifically, three different deformations are analysed: i) "deformation 1": $A = 0.01$, $B = 0.1$, $\omega_b = 4\pi/T$ and $\omega_t = 6\pi/T$ ($w = 0.015$ and $\alpha = 3°$); ii) "deformation 2": $A = 0.02$, $B = 0.2$, and the same frequencies as in "deformation 1" ($w = 0.03$ and $\alpha = 6°$); iii) "deformation 3" same amplitudes as in "deformation 1", $\omega_b = 2\pi/T$ and $\omega_t = 3\pi/T$. For $M = 0.5$ and for the three wing deformations considered, Fig. 2 depicts the comparison between the pressure signatures evaluated through the integral formulations written in FFR and BFR. These results confirm that the predictions provided by the integral formulations expressed in the fixed and moving frames of reference perfectly coincide, regardless amplitude and frequency of deformations.

Aerospace Science and Engineering - III Aerospace PhD-Days   Materials Research Forum LLC
Materials Research Proceedings 33 (2023) 21-28   https://doi.org/10.21741/9781644902677-4

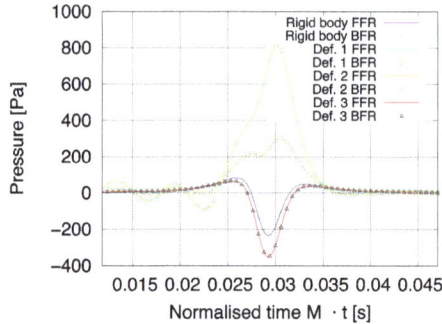

Figure 2. Acoustic pressure signature evaluated for different body deformations.

To validate the formulation for porous surfaces, a porous sphere with a radius of $R = 1\ m$ was used. A point source emitting a potential $\varphi(t) = \sin{(\omega t)}$, with $\omega = 5\ rad/s$, was placed at the center of the sphere (which was located at the origin of the y_1, y_2, y_3 axes). The sphere and source moved at a constant speed of $100\ m/s$ in the opposite direction along the y_1 axis. The deformation of the sphere was described by the bending shape function $\Psi_b(y)$ and the Lagrangean coordinate $q_b(t)$ illustrated earlier, with $A = 0.5$ and $\omega_b = 4\ rad/s$. The observer was positioned at at $[-150\ m, 10\ m, 0\ m]$. Fig. 3 shows the trends in acoustic pressure reaching the observer, calculated in two ways: directly from the source and from the porous sphere. To calculate the pressure reaching the sphere emitted by the source, Bernoulli's theorem was used. From Fig. 3, it is evident that the two trends are similar enough to confirm the correctness of the formulation for porous surfaces.

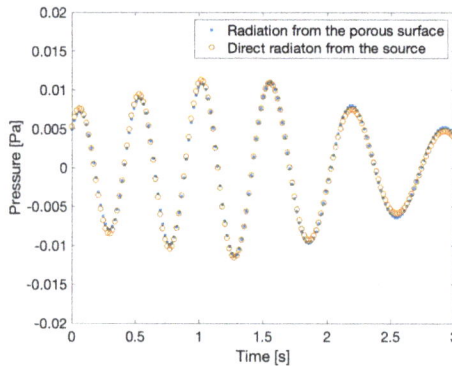

Figure 3. Comparison of acoustic pressure radiation from a point source and from a porous surface, where the porous surface always radiates the signal emitted by the point source.

The objective in the near future is to perform similar analyses to those conducted for the translating wing but for a rotor in forward flight. The numerical investigation will have a twofold purpose: to evaluate the significance of body deformation on the assessment of noise radiated by the rotor, and to discuss the advantages and disadvantages of the BFR and FFR approaches in terms of numerical performance.

Conclusions

In conclusion, two novel boundary integral formulations for the aerodynamics and aerocoustics of deformable bodies, with both porous and solid boundaries, have been presented. Through numerical investigations of a translating wing subject to bending and torsion deflections, and by validating the formulations using two reference frames, one fixed with undisturbed air and the other fixed with the undeformed body, we have confirmed the full equivalence of the formulations, regardless of the body velocity and magnitude and frequency of the deformation. Additionally, we have shown that the effect of wing deformation on acoustic radiated field is not negligible, and that the perturbation amplitude increases with the deformation amplitude. The developed formulation for porous deformable surfaces has also been validated through a comparison of direct radiation from a point source and radiation from the porous surface. The novel formulations presented in this article offer valuable insights and tools for accurately modelling the aerodynamics and acoustics of complex deformable bodies, and have the potential to enhance the design and performance of a wide range of engineering applications.

References

[1] ICAO, "Environmental Report 2016," Tech. rep., International Civil Aviation Organization, 2016. URL https://www.icao.int/environmental-protection/Documents/ICAO%20Environmental%20Report%202016.pdf .

[2] EC, "Flightpath 2050, Europe's Vision for Aviation," Tech. rep., European Commission, 2011.

[3] Nesbitt, E., "Current engine noise and reduction technology," CEAS Aeronautical Journal, Vol. 10, No. 1, 2019, pp. 93-100. https://doi.org/10.1007/s13272-019-00381-6

[4] De Vries, R., Brown, M., and Vos, R., "Preliminary sizing method for hybrid-electric distributed-propulsion aircraft," Journal of Aircraft, Vol. 56, No. 6, 2019, pp. 2172-2188.https://doi.org/10.2514/6.2018-4228

[5] Felder, J., Kim, H., Brown, G., and Kummer, J., "An examination of the effect of boundary layer ingestion on turboelectric distributed propulsion systems," 49th AIAA aerospace sciences meeting including the new horizons forum and aerospace exposition, 2011, p. 300. https://doi.org/10.2514/6.2011-300

[6] Poggi, C., Bernardini, G., and Gennaretti, M., "Aeroacoustic analysis of wing-mounted propeller arrays," AIAA AVIATION 2021 FORUM, 2021, p. 2236. https://doi.org/10.2514/6.2021-2236

[7] Farassat, F., "Derivation of Formulations 1 and 1A of Farassat," Tech. rep., 2007.

[8] Hennes, C. C., and Brentner, K. S., "The effect of blade deformation on rotorcraft acoustics," Journal of the American Helicopter Society, Vol. 53, No. 4, 2008, pp. 398-411. https://doi.org/10.4050/JAHS.53.398

[9] Lopes, L. V., "Compact assumption applied to monopole term of Farassat's formulations," Journal of Aircraft, Vol. 54, No. 5, 2017, pp. 1649-1663. https://doi.org/10.2514/1.C034048

[10] Morino, L., and Gennaretti, M., "Boundary integral equation methods for aerodynamics," Progress in Astronautics and Aeronautics, Vol. 146, 1992, pp. 279-320. https://doi.org/10.2514/5.9781600866180.0279.0320

[11] Gennaretti, M., Luceri, L., and Morino, L., "A unified boundary integral methodology for aerodynamics and aeroacoustics of rotors," Journal of Sound and Vibration, Vol. 200, No. 4, 1997, pp. 467-489. https://doi.org/10.1006/jsvi.1996.0713

Aerospace Science and Engineering - III Aerospace PhD-Days   Materials Research Forum LLC
Materials Research Proceedings 33 (2023) 21-28   https://doi.org/10.21741/9781644902677-4

[12] Testa, C., Poggi, C., Bernardini, G., and Gennaretti, M., "Pressure-field permeable-surface integral formulations for sound scattered by moving bodies," Journal of Sound and Vibration, Vol. 459, 2019, p. 114860. https://doi.org/10.1016/j.jsv.2019.114860

[13] Gennaretti, M., "Una formulazione integrale di contorno per la trattazione unificata di flussi aeronautici viscosi e potenziali," Ph.D. thesis, PhD thesis, Department of Applied Mechanics, University of Roma "La Sapienza, 1993 (in Italian).

[14] Gennaretti, M., De Rubeis, B., Poggi, C., and Bernardini, G., "Prediction of Aerodynamics and Aeroacoustics of Deformable Bodies Through a Boundary Integral Formulation," Atti del XXV Convegno AIMETA, 2022. 5

Aerospace Science and Engineering - III Aerospace PhD-Days   Materials Research Forum LLC
Materials Research Proceedings 33 (2023) 29-35   https://doi.org/10.21741/9781644902677-5

Numerical prediction of plasma formation on a sphere in hypersonic sub-orbital flight regime

Salvatore Esposito [*]

Politecnico di Torino, DET, Corso Duca degli Abruzzi, 24, 10129 TORINO, Italy

salvatore_esposito@polito.it

Keywords: Hypersonics, Plasma

Abstract. Hypersonic flight is challenging for vehicle design and operation due to the intense heating generated by kinetic energy transfer from the vehicle to the gas surrounding it. As a result, plasma is produced, which can interfere with radar tracking and communication, particularly upon re-entry into the Earth's atmosphere. Plasma affects wave propagation, and if the electron density is high enough, waves may lose intensity as they propagate, distorting radar traces. The objective of this research is to predict plasma formation during suborbital hypersonic flight, with a specific focus on determining the Mach number and altitude conditions that generate critical levels of plasma density. To achieve this, Computational Fluid Dynamics is employed to solve the Navier-Stokes equations, and a multi-temperature thermochemical model is adopted to accurately predict plasma behavior. The model is applied in a simplified scenario involving a sphere exposed to hypersonic flow.

Introduction

The hypersonic regime is a flight condition that poses significant difficulties to flight vehicle design and operation. In hypersonic flight, objects traveling at speeds much higher than the speed of sound transfer a significant amount of kinetic energy to the gas surrounding them. This leads to an increase in the gas's internal energy and generates a region of high temperature around the body. The resulting gas flow characteristics are highly complex and differ from those observed in subsonic and supersonic flows. The transfer of kinetic energy from the object to the gas primarily occurs through an intense bow shock wave, which envelops the flying object. This phenomenon triggers chemical reactions and air ionization, leading to the formation of a plasma, which is a state of matter consisting of charged particles (i.e., ions and electrons) that interact strongly with electromagnetic fields. The plasma formed around the hypersonic vehicle can affect the vehicle's radio communication capabilities, produce electromagnetic interference, and significantly disrupt radar tracking, leading to the "blackout" phenomenon. In fact, the presence of a non-negligible number of ions and electrons around an object in its hypersonic motion can cause a significant reduction in electromagnetic wave transmission and reflection efficiency, leading to a complete loss of communication signals. This phenomenon can occur at any altitude but is most severe during re-entry into the Earth's atmosphere, where plasma formation around the hypersonic vehicle is the most pronounced. Therefore, understanding and predicting plasma formation around hypersonic vehicles is crucial. This would help evaluate their radio communication capabilities, obtain reliable data for radiation heating simulations, and accurately track such objects.

To go into further detail, if the density of charges is high enough in plasma, the wave can undergo complete reflection. The properties that govern wave propagation in a medium are defined by the relative electric permittivity ϵ_r or, alternatively, the refractive index (provided that the medium is non-magnetic). According to the standard theory [1], the refractive index n is related to the frequency f of the EM wave (e.g. one generated by radar) and the electron density n_e of the plasma according to the following relationship:

Aerospace Science and Engineering - III Aerospace PhD-Days   Materials Research Forum LLC
Materials Research Proceedings 33 (2023) 29-35   https://doi.org/10.21741/9781644902677-5

$$n^2 = \varepsilon_r = 1 - \frac{f_{pe}^2}{f(f+if_c)} = 1 - \frac{f_{pe}^2}{f^2+f_c^2} + i\,\frac{f_{pe}^2 f_c}{f(f^2+f_c^2)}. \tag{1}$$

the electron number density n_e is related through the plasma frequency:

$$f_{pe} = \sqrt{e^2 n_e / \varepsilon_0 m_e}. \tag{2}$$

where n_e is the number density of electrons, e is the electric charge, m_e is the effective mass of the electron, and ϵ the permittivity of free space.

f_c is the collision frequency (between electrons and neutral particles) defined as [2]:

$$f_c = \sum_i n_i \sigma_{i,e} \sqrt{\frac{8k_b T}{\pi m_e}}. \tag{3}$$

with n_i being the neutral density, $\sigma_{i,e}$ the neutral-electron cross section for the neutral species i, T the temperature and k_b the Boltzmann constant.

The real part of the refractive index becomes zero or negative when the frequency of the wave is lower or equal to the plasma frequency (f_{pe}), which occurs when the electron density is sufficiently high. In regions where the square of the refractive index is less than or equal to zero, the wave is classified as evanescent and gradually loses intensity as it propagates, leading to the reflection of the radiation by the plasma surface. This results in the replacement of the body surface by the plasma surface, causing a distortion in the radar trace. If the plasma density fails to reach the cut-off threshold of $\frac{f_{pe}}{f} > 1$, the phenomenon of refraction or absorption can still occur, leading to a redistribution of the electromagnetic waves and a decrease in re-radiation.

The real part of ϵ_r governs wave propagation, while the imaginary part determines collisional absorption (i.e. the transfer of energy from electrons to neutral species).

This research aims to create a reliable numerical model capable of predicting plasma formation in the context of suborbital hypersonic flight. More specifically, the aim is to determine the Mach number and altitude conditions that could generate regions surrounding the vehicle, where the value of $\frac{f_{pe}}{f}$ and $\frac{f_c}{f}$ reach critical levels.

Plasma model and methodology

Since the considered flight altitude is low, the fluid is approximated as continuous and is described by the Navier-Stokes equations. The Computational Fluid Dynamics approach is used as a means of solving these equations and predicting plasma behavior. In this study, air is a chemically reactive and compressible mixture comprised of seven chemical species: oxygen (O2), nitrogen (N2), nitric oxide (NO), nitrogen atoms (N), oxygen atoms (O), nitrous oxide ions (NO+), and electrons (e-). Given the presence of high temperatures, the modes of rotational, translational, vibrational, and electronic energy must be taken into account. For this reason, a multi-temperature model is adopted to describe energetic non-equilibrium phenomena. The thermochemical model utilized in this study is described in [2,3], and the thermodynamic properties of each chemical species are the ones specified in [4]. Additionally, diffusion coefficients for each chemical species are taken from [5].

Given the large computational cost of solving this complex problem, the model is applied to a simple axial-symmetric case of a sphere with a radius of $152.4\ mm$, exposed to a hypersonic flow.

Aerospace Science and Engineering - III Aerospace PhD-Days   Materials Research Forum LLC
Materials Research Proceedings 33 (2023) 29-35   https://doi.org/10.21741/9781644902677-5

The Mach number and altitude conditions correspond to the trajectories typical of a hypersonic Glider available in [6]. A radiative adiabatic wall condition has been applied at the wall

It should be noted that numerical and experimental results related to this type of simulation are very rare in the literature. Therefore, the model's validation becomes even more important. Specifically, experimental data obtained for the RAM-C II [7] will be presented to validate the developed model.

Results

The following results are reported for an interesting case, related to a flight altitude of 40 kilometres and a Mach number equal to 15. The plasma and collision frequencies have been scaled with respect to a reference frequency of 1 GHz. The compressive shock effect on the air generates high temperatures downstream, as observed in Fig.1. Additionally, the adiabatic radiative wall boundary condition causes the temperature at the wall to decrease. However, the most critical zone is where the shock is normal to the direction of the air, leading to considerable high-temperature effects. This observation is supported by Fig.2, which shows the largest concentration of electrons in this region. When examining the area near the axis of symmetry, numerical instabilities cause the shock to become slightly distorted.

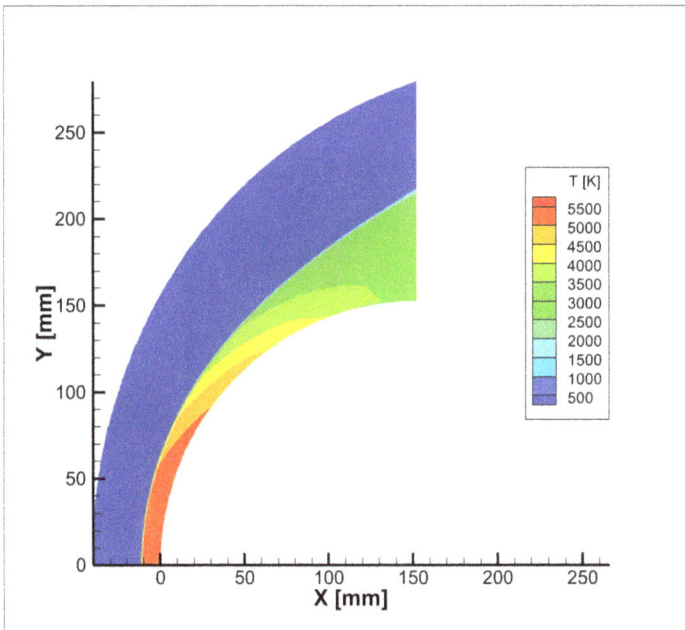

Fig. 1-Temperature around a sphere at an altitude of 40 kilometres at a Mach number of 15.

Aerospace Science and Engineering - III Aerospace PhD-Days   Materials Research Forum LLC
Materials Research Proceedings 33 (2023) 29-35   https://doi.org/10.21741/9781644902677-5

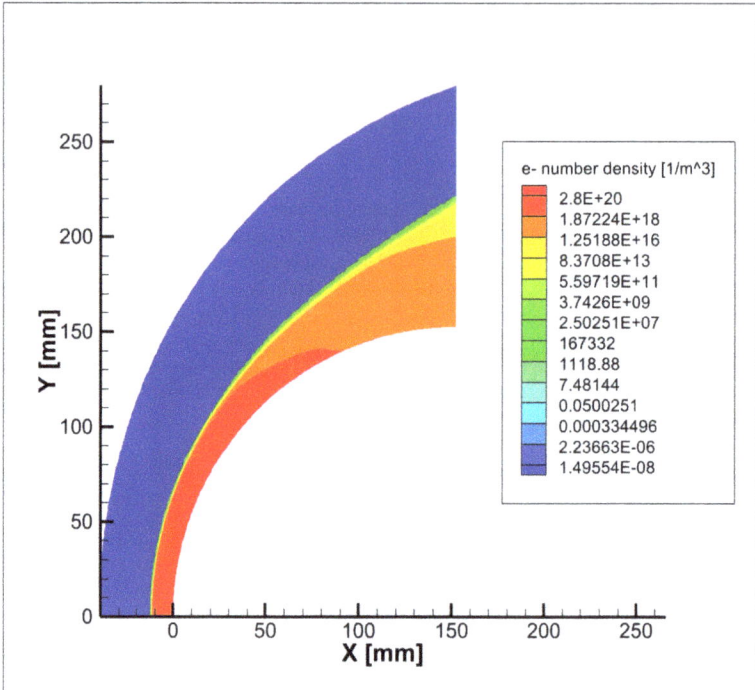

Fig. 2- Electron number density around a sphere at an altitude of 40 kilometres at a Mach number of 15.

Moreover, plasma frequencies in the flow field, as seen in Fig.3, reach extremely high values, up to two orders of magnitude higher than the reference frequency. This implies that any incident wave incident with that frequency will be evanescent downstream of the shock. In Fig.4, the collision frequency also shows interesting behaviour. Due to the sub-orbital altitude, the shock layer contains a collisional region that cannot be ignored, where the frequencies reach up to 45 GHz near the wall and 35 GHz in the normal shock zone.

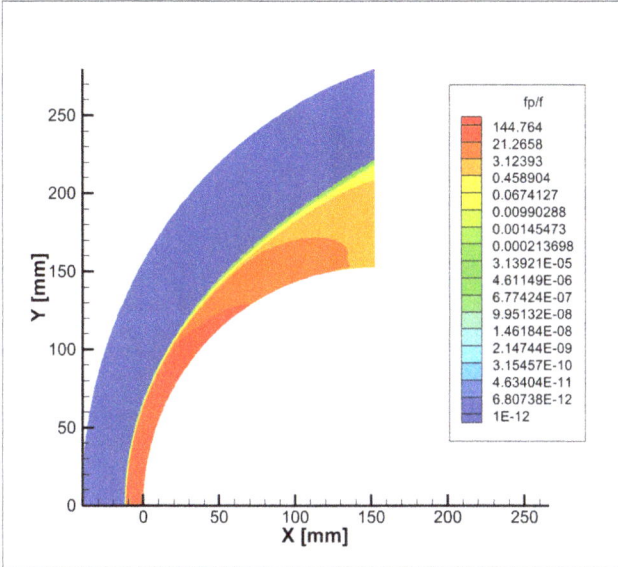

Fig. 3- Plasma Frequency around a sphere at an altitude of 40 kilometres at a Mach number of 15.

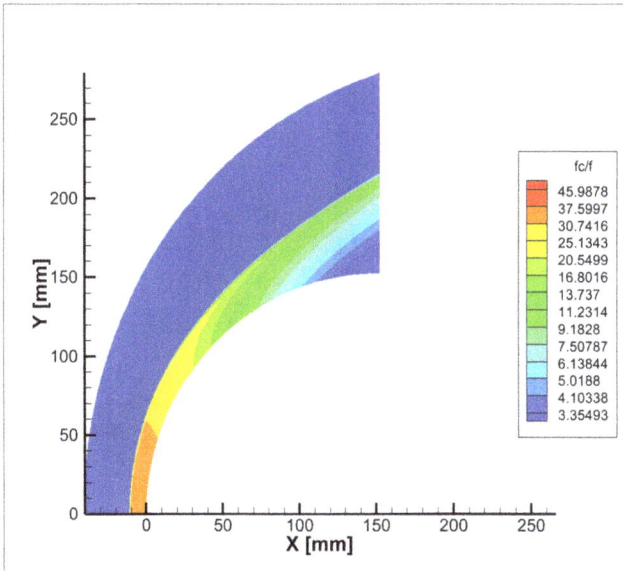

Fig. 4- Collision Frequency around a sphere at an altitude of 40 kilometres at a Mach number of 15.

Aerospace Science and Engineering - III Aerospace PhD-Days   Materials Research Forum LLC
Materials Research Proceedings 33 (2023) 29-35   https://doi.org/10.21741/9781644902677-5

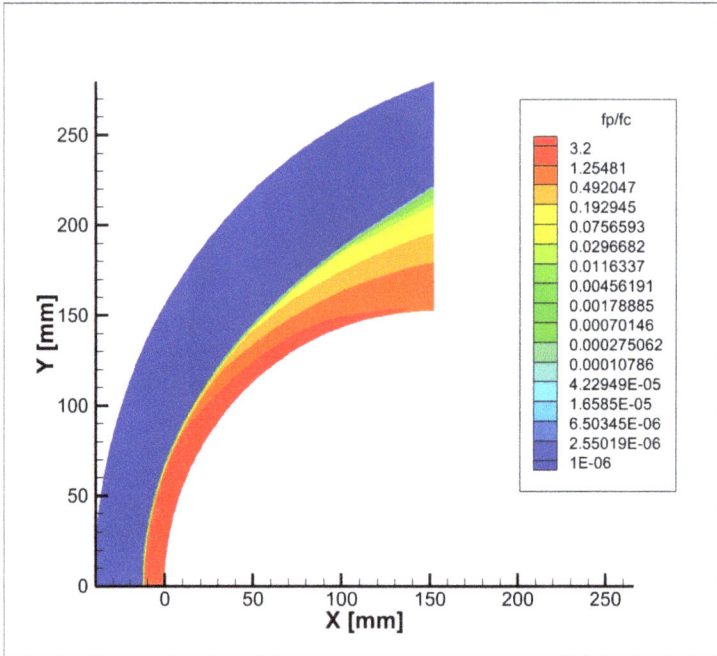

Fig. 5—Collision Frequency and Plasma Frequency ratio around a sphere at an altitude of 40 kilometres at a Mach number of 15.

Understanding the propagation of electromagnetic waves in plasma is crucially dependent on the ratio between two parameters, as illustrated in Fig.5. In the specific flight conditions examined, the results demonstrate that the propagation of electromagnetic waves is evanescent downstream of the shockwave due to the high plasma frequencies. It is worth noting that collisional absorption appears to be less dominant than the aforementioned phenomenon.

Conclusions

This study presents a numerical model for predicting plasma formation in suborbital hypersonic flight conditions. the analysis reveals that at a flight altitude of 40 kilometers and a Mach number of 15, the shockwave induces significant thermal effects downstream, resulting in a critical zone with high-temperature effects. The investigation shows that the plasma and collision frequencies in the flow field reach extremely high values and that the propagation of electromagnetic waves is mainly evanescent for high plasma frequencies. Moreover, collisional absorption is less preponderant than the cut-off phenomenon. Our research provides valuable insights into the behavior of plasma in hypersonic flight conditions and has the potential to aid in improving the radio communication capabilities of hypersonic vehicles.

References

[1] T. H. Stix, Waves in Plasmas. Springer, 1992

[2] L. D. Landau, E. M. Lifshitz, and L. P. Pitaevskii, Statistical Physics, vol. 1. Oxford: Pergamon Press, 1980. https://doi.org/10.1016/B978-0-08-057046-4.50008-7

[3] Park, Chul. "Review of chemical-kinetic problems of future NASA missions. I-Earth entries." Journal of Thermophysics and Heat transfer 7.3 (1993): 385-398. https://doi.org/10.2514/3.431

[4] Park, Chul, Richard L. Jaffe, and Harry Partridge. "Chemical-kinetic parameters of hyperbolic earth entry." Journal of Thermophysics and Heat transfer 15.1 (2001): 76-90. https://doi.org/10.2514/2.6582

[5] McBride, Bonnie J. Coefficients for calculating thermodynamic and transport properties of individual species. Vol. 4513. National Aeronautics and Space Administration, Office of Management, Scientific and Technical Information Program, 1993.

[6] Gupta, Roop N., et al. "A review of reaction rates and thermodynamic and transport properties for an 11-species air model for chemical and thermal nonequilibrium calculations to 30000 K." (1990).

[7] Tracy, Cameron L., and David Wright. "Modeling the performance of hypersonic boost-glide missiles." Science & Global Security 28.3 (2020): 135-170. https://doi.org/10.1080/08929882.2020.1864945

[8] Jones, W. Linwood, and Aubrey E. Cross. Electrostatic-probe measurements of plasma parameters for two reentry flight experiments at 25000 feet per second. Vol. 6617. National Aeronautics and Space Administration, 1972.

Aerospace Science and Engineering - III Aerospace PhD-Days   Materials Research Forum LLC
Materials Research Proceedings 33 (2023) 36-40   https://doi.org/10.21741/9781644902677-6

Digital technologies and human-machine interaction in air traffic control operations

Marzia Corsi[1,a]

[1]Department of Industrial Engineering, Università di Bologna, Forlì, Italy

[a]marzia.corsi2@unibo.it

Keywords: Digital Technologies, Air Traffic Control Operations, ATM, UTM

Abstract. The aim of this study is to exploit digital technologies to develop and validate an innovative *Human Machine Interface for Air Traffic Control Operations* tailored based on actual visibility condition, specific phase of flight and different user working positions. The model for the interface prototype exploits multimodal interaction features and implements technologies and functions such as Extended Reality, aircraft identification and tracking labels, and safety net visualization. Moreover, it includes the future possible scenario of the Air Traffic Management comprising the integration of autonomous drones and Urban Air Mobility.

Introduction

In recent years, innovative and advanced visualization tools for Air Traffic Controllers (ATCOs) such as movement maps, conformance monitoring and conflict detection, have been designed to improve the operational safety of airports. However, the increased number of information required and displayed on the workstation terminals such as Aerodrome Traffic Zone and position and tracking of aircrafts and vehicles, lead to more time in head-down position watching at the screens rather than in a head-up one focusing on the out-of-the-window view with the consequent risk of not detecting unpredictable and potentially dangerous events [1]. In addition, continuously switching from one position to another focusing on two different perspectives of the same environment would lead to a reduction in the situational awareness of controllers [2,3,4,5]. To address this problem, the idea of using Augmented Reality (AR) in the control tower to overlay auxiliary information over the real external view was introduced almost 30 years ago when this technology was still in the very early stages of its industrialization [6]. Over the years, a few experiments with modern hardware have been performed to test the soundness of AR technology in air traffic control towers, confirming its positive potential to benefit control tower operations [7,8,9].

Extended Reality technology in Airport Control Towers

Resilient Synthetic Vision for Advanced Control Tower Air Navigation Service Provision
The *RETINA-The Resilient Synthetic Vision for Advanced Control Tower Air Navigation Service Provision* Single European Sky ATM Research (SESAR) project concept has filled the gap of using the latest advances in Augmented Reality to further investigate the application of synthetic vision tools in a real airport control tower environment (Figure 1) [10].

Aerospace Science and Engineering - III Aerospace PhD-Days   Materials Research Forum LLC
Materials Research Proceedings 33 (2023) 36-40   https://doi.org/10.21741/9781644902677-6

Figure 1: Retina Concept: an ATCO with the ST-HMD visualises the Aircraft Tracking Label through the window of the Control Tower.

The project consortium has investigated the potential and applicability of Virtual/Augmented Reality (V/AR) technologies for the provision of Air Traffic Control (ATC) service in conventional airport control tower [11,12]. RETINA has assessed whether and how the concepts that stand behind tools such as Head-up Displays in the aircraft cockpit, Enhanced Vision Systems and Synthetic Vision Systems could be transferred to ATC with relatively low effort and considerable benefits for controllers' Situational Awareness. In doing so, two different AR systems have been investigated: Conformal-Head-Up Displays (C-HUDs) - which, potentially, can be made to coincide with the tower windows - and See-Through Head-Mounted Displays (ST-HMD). A proof-of-concept of both systems has been implemented and validated by means of human-in-the-loop real-time simulations in a laboratory environment. The external view was provided to the user through a high fidelity 4D model in an immersive environment that replicates the out-of-the tower view and additional information such as airport layout, flight tags, wind velocity and direction and warning detection were placed over the actual out of the window view. RETINA concept, therefore, has enabled the controllers to have a head-up view of the airport traffic in any visibility condition. The outcomes of the project validation, obtained through both subjective qualitative information and objective quantitative data, have proven how this concept could lead to an improvement of the human performance in the control tower, preserving safety and increasing resiliency at airports to low visibility [13]. The results obtained by RETINA consortium have been exploited to feed SESAR's *Digital Technologies for Tower (PJ05-W2 DTT) project* [14,15].

Digital Technologies for Tower
The DTT project aims to contribute to Air Traffic Management (ATM) digitalisation objectives maturing the concept of an Augmented Reality-based interface for Air Traffic Controllers in conventional and remote airport control towers. The project is composed by three different sub-projects, each of which is focusing on specific purposes to be validated and progressively matured for the benefit of the Air Traffic Management (ATM) network in terms of safety, capacity, efficiency and flexibility. The first sub-project, *Multiple Remote Tower and Remote Tower Centre*, proposes the development of a remote aerodrome air traffic service in which services from various aerodromes are combined in a centralised control room independent of airport location.

The second and third sub-projects, *ASR at the TWR CWP supported by AI and Machine Learning* and *Virtual/Augmented Reality applications for Tower*, address the development of new human machine interface (HMI) interaction modes and technologies at the Controller Working

Aerospace Science and Engineering - III Aerospace PhD-Days
Materials Research Proceedings 33 (2023) 36-40

Materials Research Forum LLC
https://doi.org/10.21741/9781644902677-6

Position. They deal with both the current operating airport environment and future environments focusing the activities towards two main areas: Automatic speech recognition (ASR) and Virtual and Augmented Reality. ASR supported with Artificial Intelligence/Machine Learning techniques, is investigated to enable the recognition and translation of spoken language into the system reducing their workload and improving safety whilst, V/AR is expected to allow tower ATCOs to conduct safe operations under any meteorological conditions while maintaining a high taxiway and runway throughput. Within this last area, specific aspects of other features, such as Tracking Labels (TL), multimodal interaction and Attention Guidance/Safety Nets, are investigated by three of the different DTT project partners in different simulation scenarios [16,17,18]. In particular, one of the validation exercises has been carried out as a real-time human-in-the-loop simulation in the *Virtual and Simulation Laboratory* of the University of Bologna (Figure 2) utilizing the more technically advanced version of the platform exploited to validate RETINA concept. The exercise has assessed, at different maturity levels, the introduction of an adaptive HMI encompassing different functions, namely, Virtual/Augmented Reality TL and airport layout overlays, multimodal interaction (voice and Air Gestures) to deliver not time critical clearances and Safety Nets visualisation to guide the attention of the operators towards hazardous situations.

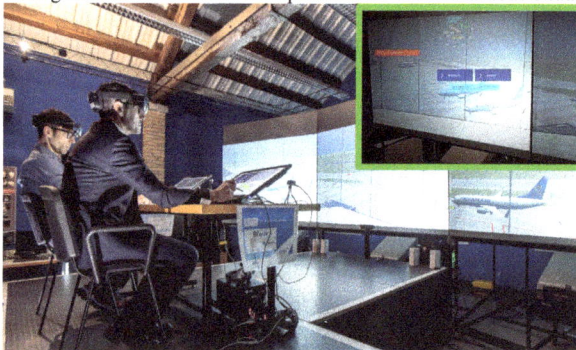

Figure 2: University of Bologna simulation and validation platform at the Virtual and Simulation Laboratory in Forlì premises. The ATCOs are provided with a realistic and consistent scenario of the out-of-the-tower view of the aerodrome onto a CAVE-like virtual environment, by wearing the ST-HMD devices the users can simultaneously see the both the AR overlays and the out-of-the-window view (personal view of the specific ATCO in the green square).

As expected, the results of Bologna validation exercises confirmed that the proposed solution of a V/AR HMI interface in conventional control towers can support the ATC operations. The introduction of synthetic overlays stimulates the controllers in working in a head-up position and reducing the number of switching head-up/head-down with a consequent positive impact on human performances and situational awareness. Nevertheless, further improvements can still be considered to provide an even more effective interaction in airport control towers.

Human Machine Interface for Air Traffic Control Operations
Starting form RETINA and DTT technologies, devices and results, and considering the recent advancement in the aeronautical sector, the aim of this study is to exploit digital technologies to develop and validate an innovative *Human Machine Interface for Air Traffic Control Operations* tailored based on actual visibility condition, specific phase of flight and different user working positions. The interface should enable a more natural and effective interaction in control tower,

Aerospace Science and Engineering - III Aerospace PhD-Days   Materials Research Forum LLC
Materials Research Proceedings 33 (2023) 36-40   https://doi.org/10.21741/9781644902677-6

improving, on one hand, the performance and, on the other hand, the situational awareness of the ATCOs.

The model for the interface prototype exploits multimodal interaction features and implements technologies and functions such as Extended Reality (XR), aircraft identification and tracking labels, safety net visualisation, Conflicting ATC clearances alerts, in-air gesture interaction and speech recognition. Moreover, the future possible scenario of the Air Traffic Management comprising the integration of autonomous drones and Urban Air Mobility (UAM) that requires a closer integration between vehicle and infrastructure capabilities [19] are considered to be included in the proposed project. Within the project, different simulation scenarios integrating ATM and Unmanned Aircraft System Traffic Management (UTM) are planned to be implemented and assessed in a laboratory environment to explore a wide set of possible solutions which are not yet available in the real world. More specifically, the model focuses, on one hand, on the representation of different designs and locations of UAM infrastructures (e.g. vertiports) foreseeing UAM dedicated and integrated airspace and, on the other hand, on the supply to the controllers of the necessary tools to conduct manned and unmanned air traffic control operations.

Conclusions

The proposed research can contribute to lead to a significant benefit for the future aviation system, including, but not limited to, increased safety for passengers, financial savings for carriers and Air Navigation Service Providers, better point-to-point connections and shorter travel times, and improved resilience and efficacy for the control tower IT systems. The development of such interface, indeed, could reduce the workload and increase the productivity and the situational awareness of the ATCOs. Through the use of XR technologies and multimodal interaction, controllers will be provided with high-quality information to operate in any condition of traffic, weather, airport complexity, etc., without endangering safety. Moreover, airports could use the results in planning the infrastructure of the future. As a matter of fact, the integrated traffic forecast given by the increase of autonomous advanced systems, will lead, not only to a whole new batch of actors in the aeronautical panorama, but also to change the perspective and the role of the air traffic controllers.

References

[1] R. Reisman, D. Brown, Design of augmented reality tools for air traffic control towers, in: 6th AIAA Aviation Technology, Integration and Operations Conference (ATIO), 2006, p. 7713. https://doi.org/10.2514/6.2006-7713

[2] E. Pinska, An investigation of the head-up time at tower and ground control positions, EUROCONTROL Experimental Centre, Bretigny (2006).

[3] E. Pinska, C. Tijus, Augmented reality technology for control tower analysis of applicability based on the field study, in: Proceeding of 1st CEAS European Air and Space Conference, Citeseer, 2007, pp. 573-580.

[4] A. Papenfuss, M. Friedrich, C. Möhlenbrink, M. Rudolph, S. Schier, M. Schmidt, N. Fürstenau, Assessing operational validity of remote tower control in high-fidelity tower simulation, IFAC Proceedings Volumes 43 (2010) 117-122. https://doi.org/10.3182/20100831-4-FR-2021.00022

[5] B. Hilburn, Head-down time in aerodrome operations: A scope study, Center for Human Performance Research (2004).

[6] Daniel J Weintraub. Human factors issues in head-up display design: The book of HUD,volume 92. CSERIAC, 1992.

[7] R. J. Reisman, D. M. Brown, Augmented Reality Tower Technology Assessment, Technical Report, 2009.

[8] R. J. Reisman, S. K. Feiner, D. M. Brown, Augmented reality tower technology flight test, in: Proceedings of the International Conference on Human-Computer Interaction in Aerospace, 2014, pp. 1-8. https://doi.org/10.1145/2669592.2669651

[9] M. Hagl, M. Friedrich, A. Papenfuss, N. Scherer-Negenborn, J. Jakobi, T. Rambau, M. Schmidt, Augmented reality in a remote tower environment based on vs/ir fusion and optical tracking, in: International Conference on Engineering Psychology and Cognitive Ergonomics,Springer, 2018, pp. 558-571. https://doi.org/10.1007/978-3-319-91122-9_45

[10] Resilient synthetic vision for advanced control tower air navigation service provision, 2018. URL: http://www.retina-atm.eu/index.html.

[11] S. Bagassi, F. De Crescenzio, F. Lucchi, N. Masotti, Augmented and virtual reality in the airport control tower (2016).

[12] N. Masotti, S. Bagassi, F. D. Crescenzio, Augmented reality for the control tower: The retina concept, in: International Conference on Augmented Reality, Virtual Reality and Computer Graphics, Springer, 2016, pp. 444-452. https://doi.org/10.1007/978-3-319-40621-3_32

[13] S. Bagassi, F. De Crescenzio, S. Piastra, C. A. Persiani, M. Ellejmi, A. R. Groskreutz, J. Higuera, Human-in-the-loop evaluation of an augmented reality based interface for the airport control tower, Computers in Industry 123 (2020) 103291. https://doi.org/10.1016/j.compind.2020.103291

[14] Digital technologies for tower, 2020. URL: https://www.sesarju.eu/projects/DTT.

[15] Project pj05-w2 dtt - digital technologies for tower, 2023. URL: https://www.remote-tower.eu/wp/project-pj05-w2/.

[16] J. Teutsch, T. Bos, M. van Apeldoorn, L. Camara, Attention guidance for tower atc using augmented reality devices, in: 2022 Integrated Communication, Navigation and Surveillance Conference (ICNS), IEEE, 2022, pp. 1-12. https://doi.org/10.1109/ICNS54818.2022.9771479

[17] R. Santarelli, S. Bagassi, M.Corsi, J. Teutsch, R. Garcia Lasheras, M. Angel Amaro Carmona, A. R. Groskreutz, Towards a digital control tower: the use of augmented reality tools to innovate interaction modes, in: Sesar Innovation Days 2022, 2022.

[18] S.Bagassi, M. Corsi, F. Galuppi, V/A-R Air Gestures HMI Interaction in Airport Control Towers, in: ICAS proceedings 33th Congress of the International Council of the Aeronautical Sciences, 2022, ISSN 2958-4647.

[19] Schuchardt, B. I., Geister, D., Lüken, T., Knabe, F., Metz, I. C., Peinecke, N., & Schweiger, K. (2023). Air Traffic Management as a Vital Part of Urban Air Mobility-A Review of DLR's Research Work from 1995 to 2022. Aerospace, 10(1), 81. https://doi.org/10.3390/aerospace10010081

Aerospace Science and Engineering - III Aerospace PhD-Days   Materials Research Forum LLC
Materials Research Proceedings 33 (2023) 41-48   https://doi.org/10.21741/9781644902677-7

DNS of momentum and heat transfer inside rough pipes

Mariangela De Maio[1,a] *

[1]Sapienza Università di Roma, Dipartimento di Ingegneria Meccanica e Aerospaziale, via Eudossiana 18, 00184 Roma, Italy

[a]mariangela.demaio@uniroma1.it

Keywords: Turbulent Flows, Pipe Flows, Roughness, DNS, Heat Transfer

Abstract. We carry out direct numerical simulation (DNS) of turbulent flow in pipes with a grit-blasted surface, to investigate both on momentum and heat transfer. A wide range of Reynolds numbers are considered, while maintaining a constant molecular Prandtl number of 0.7. The large relative roughness influences both the velocity and the temperature fields, indicating that the Reynolds analogy does not hold at high Reynolds bulk number.

Introduction

Pressure-driven flow in ducts is a subject of utmost relevance in mechanical and aerospace engineering applications.

Most simulations of forced convection in circular pipes have been carried out for the canonical case with smooth walls [1]. The case of rough walls is however at least as important, but so far it has mainly received attention through experimental studies and for surfaces with relatively low roughness. However, recent technological advances in the field of additive manufacturing have further prompted investigations of flow over surfaces with large relative roughness [2], which significantly affects both frictional drag and heat transfer. Understanding the behavior of flows over irregular rough surfaces with higher relative roughness than the ones studied in most experiments is thus certainly of great practical interest.

The systematic experimental investigation carried out by Nikuradse [3] is regarded as the starting point for the study of turbulent flows over rough walls. Nikuradse generated an extensive database for fully developed flow in circular pipes whose walls were covered with sieved sand grains. He identified three flow regimes: hydraulically smooth, transitionally rough, and fully-rough. In the first regime the height of the roughness is of the order of the viscous sublayer, hence roughness does not affect the flow. In the transitionally rough regime, the behavior of the flow instead depends strongly on the geometrical parameters of the roughness. Finally, in the fully-rough regime the friction coefficient is nearly unaffected by the Reynolds number. Generally, the friction factor increases with the relative roughness height, except for the laminar regime, where the friction factor of all the tested rough surfaces collapses to the smooth pipe case.

The presence of roughness affects both the mean flow and the turbulent motion of a fluid, which entails an increase of friction with respect to the smooth wall case. This is linked to the downward shift in the inner-scaled profile of the mean streamwise velocity, which can be expressed through the roughness function [4],

$$\Delta U^+ = U_S^+ - U_R^+, \tag{1}$$

where U_S^+ is the value of the mean streamwise velocity of the smooth case scaled in inner-units and U_R^+ is the one for the rough-wall case at the same friction Reynolds. The roughness function is computed in the logarithmic region of the velocity profiles, so that it can be unambiguously defined.

Aerospace Science and Engineering - III Aerospace PhD-Days   Materials Research Forum LLC
Materials Research Proceedings 33 (2023) 41-48   https://doi.org/10.21741/9781644902677-7

Nikuradse's investigation and more recent studies [5] [6] performed experiments with other types of rough surfaces and found that the velocity profiles of the rough-wall case collapse to the smooth-wall case if they are scaled in outer units. This result supports the validity of Townsend's outer-layer similarity hypothesis [7]. According to that hypothesis, smooth and rough wall turbulence behave similarly away from the wall at sufficiently high Reynolds number and at a sufficient scale separation between the typical roughness height (k) and the outer length scale of the flow (δ, e.g., the pipe diameter). Jiménez [8] stated that scale separation requires $k/\delta \lesssim 1/40$. However, several studies have shown that outer-layer similarity still holds in the wake region of the wall layer for surfaces with higher relative roughness [9] [10].

While the effect of wall roughness on momentum transfer is being extensively investigated, less attention has been given to the effect of wall roughness on turbulent heat transfer. Dipprey and Sabersky [11] performed some experiments over a granular surface and found that roughness augments momentum transfer more than heat transfer. The same result was found by Bons [12] who examined a number of realistic roughness geometries of gas turbine blades. On the other hand, so far direct numerical simulations (DNS) have been mainly focused on structured roughness [13]. Lately, more realistic surfaces have been investigated by Peeters et al. [10], who found an outer-layer similarity also for the temperature field. However, they also noticed a decrease in heat transfer efficiency at high Reynolds since momentum transfer is less increased than heat transfer with respect to the smooth-wall case.

The latter result is directly linked to the fact that the Reynolds Analogy hypothesis does not hold for a rough surface. According to this hypothesis the mean temperature field is similar to the velocity field. The discrepancy between the two fields can be quantified by the Reynolds Analogy Factor (RAF), which decreases as the efficiency in heat transfer diminishes.

Analogously to the velocity field, we can define a temperature roughness function,

$$\Delta \Theta^+ = \Theta_S^+ - \Theta_R^+ \tag{2}$$

In this case, the subtraction must be done both at the same friction Reynolds and at the same Prandtl number.

To sum up, the state of the art is mainly focused on relatively low roughness, and most of the DNS performed up to now consist of flows inside channels with infinite extensions in the spanwise direction. The present project aims at investigating the influence of relatively large roughness on the flow inside a circular pipe, with the goal of characterizing the differences between momentum and heat transfer. A relatively wide range of Reynolds numbers is investigated, up to the fully rough regime.

Numerical methodology

We use second-order finite-difference discretization of the incompressible Navier-Stokes equations in Cartesian coordinates, based on the classical marker-and-cell method [14] [15], with staggered arrangement of the flow variables to remove odd-even decoupling.

The divergence-free condition is enforced through the Poisson equation, which is solved by double trigonometric expansion in the streamwise and spanwise directions, and inversion of tridiagonal matrices in the third direction [16]. We use a hybrid third-order Runge-Kutta algorithm to perform the time integration. Furthermore, the convective terms are treated explicitly, whereas the diffusive terms are handled implicitly to alleviate the time step limitation. The mass flow rate is kept constant in time through a time-varying pressure gradient Π, which represents a uniform volumetric forcing applied to the streamwise momentum equation.

Similarly to the streamwise velocity field, the passive scalar equation is also forced with a time-varying, spatially homogeneous forcing term Q, in such a way that the integral of the temperature Θ over the pipe is strictly constant in time [17]. In the present simulations the constant passive scalar value at the wall $\theta_w = 0$ is assigned as boundary condition and a Prandtl number $Pr = 0.7$ is assumed in all the simulations.

The computational domain is a rectangular box of size $L_x \times L_y \times L_z$, covered with a uniform Cartesian mesh of $N_x \times N_y \times N_z$ grid points. The rough pipe with mean radius R and cross-sectional area $A = \pi R^2$ is embedded in it. The no-slip boundary conditions and the temperature at the wall are approximately enforced through the immersed-boundary method.

As a preliminary step, the pipe geometry is generated in the standard Stereo-LiThography format, and a preprocessor based on the ray-tracing algorithm is used to discriminate grid points belonging to the fluid and to the solid [18]. Near the fluid-solid interface the viscous terms dominate the nonlinear and pressure terms, hence the boundary conditions can be enforced by locally changing the finite-difference weights for the approximation of the second derivatives [19].

The controlling parameter of the flow is the bulk Reynolds number $Re_b = 2Ru_b/\nu$, where u_b is the bulk velocity.

The resulting friction Reynolds number is $Re_\tau = Ru_\tau/\nu$, where the friction velocity is evaluated based on the measured pressure gradient, $u_\tau = \sqrt{\tau_w} = \sqrt{R/2 < \Pi >}$.

For normalization of the temperature field, we will consider the friction temperature, $\Theta_\tau = \frac{<q_w>}{u_\tau} = \frac{R}{2}\frac{<Q>}{u_\tau}$, where $< q_w >$ is the heat flux through the walls of the pipe.

The resistance that the fluid encounters inside the pipe is quantified through the friction coefficient, defined as

$$C_f = \frac{2\,\tau_w}{\rho\,u_b^2} \tag{3}$$

Whereas the overall heat transfer performance of the duct is quantified in terms of the Stanton number,

$$St = \frac{<q_w>}{\rho C_p u_b (\theta_w - \Theta_b)} \tag{4}$$

The bulk temperature can be computed as $\Theta_b = \frac{1}{A}\int_A U\theta dA$. In this case, the non-dimensional form for the previous equation is adopted, thus the density ρ and the specific heat C_p are equal 1.

Results

We consider a grit-blasted surface downloaded from the University of Southampton Institutional Repository [20]. The geometry has been scanned and post-processed, more information is reported in [21].

The pipe, shown in figure 1, is obtained by doubling the baseline samples in the spanwise direction and wrapping the obtained surfaces around the mean pipe geometry. In this study we define the roughness height to be the mean-peak-trough height (k), as obtained by partitioning the surface into 5×5 tiles of equal size, and then computing the average of the difference between the maximum and minimum height for each tile [22]. In the case of a rough surface another important parameter that needs to be defined is the roughness Reynolds number $k^+ = k\,u_\tau/\nu$.

Materials Research Forum LLC
https://doi.org/10.21741/9781644902677-7

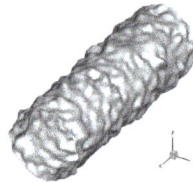

Figure 1 Pipe geometry obtained by wrapping the original rough surfaces around the pipe.

The results and the test conditions of the present DNS are listed in table 1. Since C_f is almost constant moving from $Re_b = 9800$ to 30000, we can state that the fully-rough regime is achieved.

Table 1 The definition of the first parameters is given in the text. $\Delta x^+, \Delta y^+$ and Δz^+ are the grid spacings in the streamwise and cross-stream directions, given in wall units. T is the time interval used to collect the flow statistics, and τ_t is the eddy turnover time.

Re_b	Re_τ	k^+	$C_f \times 10^2$	$St \times 10^3$	N_x	$N_y = N_z$	Δx^+	$\Delta z^+ = \Delta y^+$	T/τ_t	Symbol
500	32.75	6.08	3.43	17.09	384	192	0.53	0.42	104.77	
1000	46.69	8.70	1.75	8.58	384	192	0.76	0.60	74.70	
1500	58.41	10.89	1.21	5.83	384	192	0.96	0.75	62.29	
1750	64.78	12.03	1.10	5.16	384	192	1.06	0.83	59.40	
2500	123.73	23.07	1.95	9.21	384	192	2.01	1.58	79.05	
3000	149.88	27.94	2.00	9.26	384	192	2.45	1.92	79.94	●
4400	224.75	41.90	2.09	9.31	384	192	3.68	1.88	81.73	●
9800	521.67	97.25	2.27	9.01	960	480	3.41	2.67	85.17	●
30000	1613.49	300.02	2.30	8.14	1024	512	9.88	7.72	85.76	●

Figure 2 depicts the mean streamwise temperature distribution averaged over the cross section of the pipe, limited to the volume occupied by the fluid. The velocity isolines are superimposed on the temperature contours.

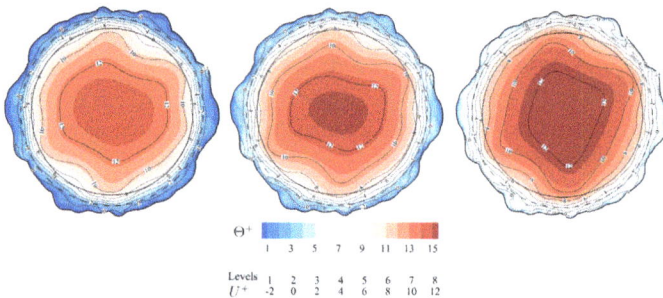

Figure 2 Contours of mean streamwise velocity for pipe at $Re_b = 4400$ (left), $Re_b = 9800$ (centre) and $Re_b = 30000$ (right). The black lines represent the velocity levels. The dashed circular line marks the position of the mean pipe surface, and the solid circular line marks the position of the plane of the crests.

Owing to the large relative roughness, the fields do not show any symmetry, since the effect of the wall is felt throughout the wall layer. Furthermore, the fields iso-lines show clear sensitivity to Reynolds number variations and to the roughness geometry. As expected, temperature and velocity

Aerospace Science and Engineering - III Aerospace PhD-Days   Materials Research Forum LLC
Materials Research Proceedings 33 (2023) 41-48   https://doi.org/10.21741/9781644902677-7

contours present the same shape far from the wall, however they show some differences in the vicinity of the wall.

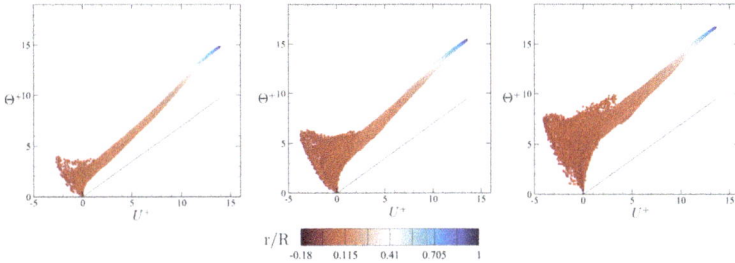

r/R

Figure 3 Scatter plot of mean temperature (Θ^+) versus mean velocity (U^+) for pipe at Re_b = 4400 (left), Re_b = 9800 (centre) and Re_b = 30000 (right). The black line corresponds to the ideal behaviour for the near-wall region in the smooth case, $\Theta^+ = Pr\ U^+$.

These discrepancies are further investigated in figure 3, where we show a scatter plot of mean velocity and temperature scaled in inner units. The colors represent the distance from the mean radius R. In the smooth case a near equality of the two distributions occurs and a linear relationship between the velocity and temperature fields is expected near the wall, $\Theta^+ = Pr\ U^+$. The same does not happen for the rough wall case, where a large scatter between Θ^+ and U^+ occurs near the walls. This is owed to the presence of the recirculation zones where the velocity reaches negative values, whereas the temperature present always positive values. Moreover, the scatter increases with Reynolds bulk since the strength of the recirculation zone increases and the velocity reaches lower mean velocity values. Furthermore, hotter fluid can flow from the bulk region towards the wall, which leads to higher mean temperatures inside the roughness layer [10]. Far from the wall the scatter decreases, also reflecting the similarity of the shapes of the contours in figure 2. However, the temperature presents much higher values than the mean streamwise velocity.

Figure 4 shows the defect velocity profiles of the mean streamwise velocity and temperature. Some scatter is observed near the walls, but both the temperature and the velocity profiles tend to collapse far from it. Despite large relative roughness, our DNS provides evidence that the outer layer similarity is achieved even if $k/R \approx 1/5$.

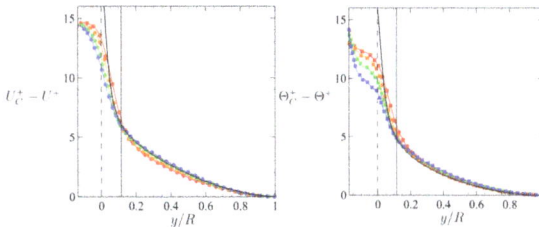

Figure 4 Velocity (left) defect profiles and temperature (right) defect profiles at various Re_b, with colors defined in table 1. The black line corresponds to the smooth wall case. The vertical solid line corresponds to the plane of the crests and the dashed like marks the mean pipe surface $y/R = 0$.

The validity of the outer-layer similarity permits to identify a logarithmic region, even if with narrow extent. This is necessary to properly define both velocity and temperature roughness functions.

Aerospace Science and Engineering - III Aerospace PhD-Days   Materials Research Forum LLC
Materials Research Proceedings 33 (2023) 41-48   https://doi.org/10.21741/9781644902677-7

Figure 5 shows the roughness functions for our DNS as function of k^+, the data from Nikuradse's experiment [3] and the fully-rough asymptote. Velocity roughness function (red circles) has a trend quite similar to that observed in Nikuradse's experiments (black circles). It can be noticed that the velocity roughness function is always increasing with k^+, whereas the temperature roughness function (red squares) has a gradual increase, and it is believed to reach a plateau for high roughness Reynolds numbers [10]. This is directly correlated to the fact that at high Reynolds number the increase in friction with respect to the smooth case is higher than heat transfer augmentation.

As customary we proceed to determine values of the equivalent sand-grain roughness height, by enforcing universality of the roughness function to the fully-rough asymptote of Nikuradse. Data fitting of our DNS results yields $k_s^+ = 0.76\,k^+$ for the surface under consideration. For completeness, we recall that there is no universal behavior in the fully rough regime for the temperature roughness function since the results of previous studies present a wide scatter.

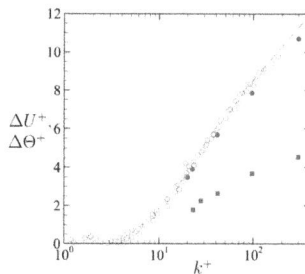

*Figure 5 Variation of the velocity roughness function (red circles)
and the temperature roughness function (blue squares) with inner-scale roughness height.
Empty circles represent Nikuradse's results and the dashed line is the fully-rough asymptote.*

Left panel of figure 6 shows the friction coefficient and the Stanton number obtained from the present DNS as functions of the Reynolds numbers. Referring to the laminar flow region in the low-Re end of the graph, in Nikuradse's experiments the friction factor for all surfaces collapsed to the Hagen-Poiseuille prediction $f = 4C_f = 64/Re_b$, thus suggesting that friction is not affected by the change of the wall geometry. In the present DNS we notice that the computed friction factors are higher than the expected theoretical values by about 7% for the present simulations. This indicates that in our case the larger relative roughness has the role of altering the geometry of the pipe, thus producing modification also in the laminar flow regime. The same result was found by Huang et al. [23], who noticed that as the relative roughness becomes larger, the friction coefficient increases also in the laminar region.

After the transition from laminar to turbulent flow, both the friction factor and the Stanton number start increasing. At the highest Reynolds numbers, C_f attains almost to a constant value, and the fully-rough regime is reached. Since the Stanton number is the thermal counterpart of the friction coefficient, one would expect a constant value of the Stanton number in the fully-rough regime. However, the plot shows that it decreases at the highest Reynolds numbers. Furthermore, in the transitional rough regime the Stanton number is almost constant, as if the fully-rough regime is reached at lower Reynolds number for the temperature field. Similar results have been found by Forooghi et al. [24], who addressed this behavior to the thinner boundary layer of the temperature with respect to the velocity at Prandtl 0.7.

Right panel of figure 6 shows the trend of the ratio between the Reynolds Analogy Factor of the rough (RAF) and smooth case ($\boldsymbol{RAF_S}$) as a function of Reynolds bulk for the flow in the turbulent regime. In accordance with previous studies [9] [12], the Reynolds Analogy Factor decreases with respect to the smooth pipe, this is due to the fact that roughness augments momentum transfer more than heat transfer. As the Reynolds number increases, the efficiency in heat transfer decreases and so does RAF.

Indeed, at low Re_b the viscous transport is still important, and hence, the value of RAF is closer to that of a smooth wall. Whereas, in the fully-rough regime the resistance of the flow is mainly due to the pressure or form drag and not to the viscous drag. However, there is no enhancement mechanism for heat transfer comparable to the pressure, this explains why heat transfer is not increased by roughness as much as skin friction, especially for high Reynolds numbers.

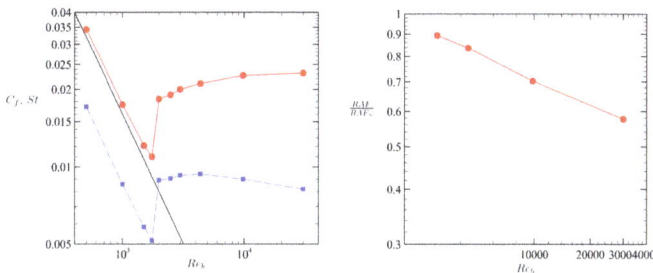

Figure 6 Right: Variation of St (blue squares) and C_f (red circles) with the Reynolds bulk. The black lines represent Hagen-Poiseuille prediction $C_f = 16/Re_b$. Left: Variation of the RAF/RAF_S coefficient with the Reynolds bulk, for $Re_b \geq 3000$.

Summary

To conclude, it is evident that the presence of rough walls influences both the velocity and the temperature fields. Moreover, the large relative roughness influences the flow also in the laminar region, as if the roughness modified the shape of the cross-section of the pipe. Last, it is evident that the Reynolds analogy does not hold, as the Reynolds bulk increases. The latter result is a very important since the industrial design of the cooling channels is based on the use of the Reynolds-averaged Navier-Stokes equations (RANS), which could rely on the Reynolds analogy, thus yielding an overprediction of the heat transfer.

Follow-up studies should include DNS of flows over new rough surfaces. The resulting database could then be used to develop improved predictive correlations for heat transfer and for the tuning of the RANS simulations.

References

[1] S. Pirozzoli, J. Romero, M. Fatica, R. Verzicco e P. Orlandi, «DNS of passive scalars in turbulent pipe flow,» *J. Fluid Mech.*, 940, A45, 2022. https://doi.org/10.1017/jfm.2022.265

[2] C. Stimpson, J. Snyder, K. Thole e D. Mongillo, «Roughness effects on flow and heat transfer for additively manufactured channels.,» *J. Turbomach.*, 138(5):051008, 2016. https://doi.org/10.1115/1.4032167

[3] J. Nikuradse, Strömungsgesetze in rauhen rohren, 1933.

[4] F. Hama, « Boundary layer characteristics for smooth and rough surfaces,» *Trans. Soc. Nav. Arch. Marine Engrs.*, 62:333–358, 1954.

[5] M. Shockling, J. Allen e A. Smits, «Roughness effects in turbulent pipe flow,» *J. Fluid Mech.*, 564:267–285, 2006. https://doi.org/10.1017/S0022112006001467

Aerospace Science and Engineering - III Aerospace PhD-Days   Materials Research Forum LLC
Materials Research Proceedings 33 (2023) 41-48   https://doi.org/10.21741/9781644902677-7

[6] M. Flack e K. Schultz, «The rough-wall turbulent boundary layer from the hydraulically smooth to the fully rough regime», *J. Fluid Mech.,* 580:381–405, 2007. https://doi.org/10.1017/S0022112007005502

[7] A. Townsend, «The structure of turbulent shear flow», *Cambridge University press,* 1976.

[8] J. Jimenez, «Turbulent flows over rough walls», *Annu. Rev. Fluid Mech.,* 36:173–196, 2004. https://doi/10.1146/annurev.fluid.36.050802.122103

[9] P. Forooghi, M. Stripf e B. Frohnapfel, «A systematic study of turbulent heat transfer over rough walls», *Int. J. Heat Mass Transf.,* 127, 1157-1168, 2018. https://doi.org/10.1016/j.ijheatmasstransfer.2018.08.013

[10] J. W. R. Peeters e N. D. Sandham, «Turbulent heat transfer in channels with irregular roughness», *Int. J. Heatm Mass Transf.,* 138:454-467, 2019. https://doi.org/10.1016/j.ijheatmasstransfer.2019.04.013

[11] D. Dipprey e R. Sabersky, «Heat and momentum transfer in smooth and rough tubes at various Prandtl numbers», *Int. J. Heat Mass Transf.,* 6:329-353, 1963. https://doi.org/10.1016/0017-9310(63)90097-8

[12] J. Bons, «A Critical Assessment of Reynolds Analogy for Turbine Flows», *J. Heat Transf.,* 127(5): 472–485, 2005. https://doi.org/10.1115/1.1861919

[13] P. Orlandi, D. Sassun e S. Leonardi, «DNS of conjugate heat transfer in presence of rough surfaces», *Int. J. Heat Mass Transf.,* 100: 250-266., 2016. https://doi.org/10.1016/j.ijheatmasstransfer.2016.04.035

[14] J. Welch e F. Harlow, «Numerical calculation of time-dependent viscous incompressible flow of fluid with free surface», *Phys. Fluids,* 8(12):2182, 1956. https://doi.org/10.1063/1.1761178

[15] P. Orlandi, «Fluid flow phenomena: a numerical toolkit», in *Springer Science & Business Media,* 2000, 55.

[16] P. Moin e J. Kim, «Application of a fractional-step method to incompressible Navier-Stokes equations», *J. Comput. Phys.,* 59:308–323, 1985. https://doi.org/10.1016/0021-9991(85)90148-2

[17] S. Pirozzoli, M. Bernardini e P. Orlandi, «Passive scalars in turbulent channel flow at high Reynolds number», *J. Fluid Mech.,* 788, 614-639., 2016. https://doi.org/10.1017/jfm.2015.711

[18] G. Iaccarino e R. Verzicco, «Immersed boundary technique for turbulent flow simulations», *Appl. Mech. Rev.,* 56(3):331–347, 2003. https://doi.org/10.1115/1.1563627

[19] S. Leonardi e P. Orlandi, «DNS of turbulent channel flows with two-and three-dimensional roughness», *J. Turbul.,* 7:N73, 2006. https://doi.org/10.1080/14685240600827526

[20] M. Thakkar, A. Busse e N. D. Sandham, «Dataset for surface correlations of hydrodynamic drag for transitionally rough engineering surfaces», 2016. [Online]. Available: https://eprints.soton.ac.uk/392562/.

[21] A. Busse, M. Lützner e N. Sandham, «Direct numerical simulation of turbulent flow over a rough surface based on a surface scan», *Comp. Fluids,* 116, 129-147, 2015. https://doi.org/10.1016/j.compfluid.2015.04.008

[22] M. Thakkar, A. Busse e N. D. Sandham, «Surface correlations of hydrodynamic drag for transitionally rough», *J. Turbul.,* 8(2):138–169, 2017. https://doi.org/10.1080/14685248.2016.1258119

[23] K. Huang, J. Wan, C. Chen, Y. Li, D. Mao e Z. M.Y., «Experimental investigation on friction factor in pipes with large roughness», *Exp. Therm. Fluid Sci.,* 50:147–153, 2013. https://doi.org/10.1016/j.expthermflusci.2013.06.002

[24] P. Forooghi, A. Weidenlener, F. Magagnato, B. Böhm, H. Kubach, T. Koch e B. Frohnapfel, «DNS of momentum and heat transfer over rough surfaces based on realistic combustion chamber deposit geometries», *Int. J. Heat Fluid,* 2018. https://doi.org/10.1016/j.ijheatfluidflow.2017.12.002

Aerospace Science and Engineering - III Aerospace PhD-Days   Materials Research Forum LLC
Materials Research Proceedings 33 (2023) 49-56   https://doi.org/10.21741/9781644902677-8

Exact solutions for free vibration analysis of train body by Carrera unified formulation (CUF) and dynamic stiffness method (DSM)

Xiao Liu[1,2,3,4, a], Alfonso Pagani[4*], Dalun Tang[1,2,3], Xiang Liu[1,2,3*]

[1]Key Laboratory of Traffic Safety on Track, Central South University, Ministry of Education, Changsha 410075, China

[2]Joint International Research Laboratory of Key Technologies for Rail Traffic Safety, Changsha 410075, China

[3]National and Local Joint Engineering Research Center of Safety Technology for Rail Vehicle, Changsha 410075, China

[4]Mul2 Lab, Politecnico di Torino, Italy

[a]xiaoliu11@csu.edu.cn

Keywords: Vibration Analysis, Train Body, Parallel Axis Theorem, Carrera Unified Formulation, Dynamic Stiffness Method

Abstract. A novel approach for free vibration analysis of train body structures is introduced by using the Carrera Unified Formulation (CUF) and Dynamic Stiffness Method (DSM). Higher-order kinematic fields are developed using the Carrera Unified Formulation, which allows for straightforward implementation of any-order theory without the need for ad hoc formulations, in the case of beam theories. In particular, the parallel axis theorem is introduced on the basis of the Taylor expansion cross-sectional displacement variables, which unifies the different shape subsections of the train into the same coordinate system. The Principle of Virtual Displacements is used to derive the governing differential equations and the associated natural boundary conditions. An exact dynamic stiffness matrix is then developed by relating the amplitudes of harmonically varying loads to those of the responses. Finally, the Wittrick–Williams (WW) algorithm was used to carry out the free vibration analysis of the train body and the natural frequencies and corresponding modal shapes are presented.

Introduction

Train body frame is an important part of the train as a load-carrying system. The operating environment of high-speed trains is complex, with the increase of speed, the vibration of train body becomes more and more obvious, which has a great impact on the stability, comfort and safety of train operation. As the first step in the optimal design of train body structures, free vibration analysis is an important part of the analysis of train dynamic characteristics. At present, high-speed trains adopt the concept of modular design, and analyze the structure of different parts of the frame separately. The dynamic analysis of the train body frame is still in its infancy [1], so it is necessary to establish an efficient and accurate modeling method and analysis method for the body frame.

As a typical frame structure in engineering, the shape of the train body structure is extremely complex. In order to be able to calculate, certain assumptions and simplifications are often adopted [2]. Among many approximate analysis methods, the finite element method (FEM) is undoubtedly the most extensive and effective numerical calculation method. The finite element (FE) modeling method of the train body can choose refined modeling and equivalent modeling. Refined modeling is to appropriately simplify the train body structure under the premise of retaining the main shape, and reduce the number of units and calculation time of the model while reflecting the actual structural characteristics of the train body as much as possible. Sung-Cheol Yoon [3], H.Kurtaran [4], Wang Wei [5] and others studied the refined modeling of train body structure respectively,

Aerospace Science and Engineering - III Aerospace PhD-Days   Materials Research Forum LLC
Materials Research Proceedings 33 (2023) 49-56   https://doi.org/10.21741/9781644902677-8

and on the basis of the model, the analysis and calculation of strength and mode were carried out. The equivalent modeling is to perform equivalent processing on the structure of the train body. There is a certain difference between the shape and the actual model. The mechanical parameters are derived and calculated according to the actual structural materials. The equivalent model has advantages in terms of the number of units compared to the refined modelling. D.Ribeiro[1] and others established the equivalent single-layer plate model of the BBN train body in the Alpha train. Shen Zhenhong [6] used different methods to establish the train sandwich plate structure. The finite element model of different modeling methods is compared with the deviation of the calculation results of the solid element method, and the feasibility of the lamination theory modeling method is verified. However, the FEM essentially obtains approximate solutions by dividing units. The above two methods still require a certain number of units to meet the accuracy requirements for complex frame structures, and their high calculation costs are hard to accept for the optimal design of train body structures.

The present work is intended to provide a more powerful approach for the free vibration analysis of train body as a beam structure through the application of the Carrera Unified Formulation (CUF) and dynamic stiffness method (DSM) in a much broader context by allowing for the cross-sectional deformation. CUF is a hierarchical formulation that considers the order of ofthe model, N, as a free-parameter (i.e. as an input) of the analysis or in other words, refined models are obtained without having the need for any ad hoc formulation [7-9]. On this basis of the Taylor expansion (TE), we introduce the parallel axis theorem (PAT) for train body structure, which broadens the applicable field of TE. On the other hand, the DSM is appealing in dynamic analysis because unlike the FEM, it provides exact solution of the equations of motion of a structure once the initial assumptions on the displacements field have been made. This essentially means that, unlike the FEM and other approximate methods, the model accuracy is not unduly compromised when a small number of elements are used in the analysis.

In this work, 1D higher-order Dynamic Stiffness (DS) elements based on CUF are extended and applied to the free vibration analysis of train body. In the next section, CUF and PAT is introduced and higher-order models are formulated. The principle of virtual displacements is then used to derive the equations of motion and the natural boundary conditions, which are subsequently expressed in the frequency domain by assuming a harmonic solution. After the resulting system of ordinary differential equations of second order with constant coefficients is solved, the frequency dependent DS matrix of the system is derived. Finally, the algorithm of Wittrick and Williams is applied to extrapolate the free vibration characteristics of train body.

1D unified formulation
Preliminaries
Within the framework of the CUF, the displacement field $u(x, y, z; t)$ can be expressed as

$$u(x, y, z; t) = F_\tau(x, z)u_\tau(y; t), \quad \tau = 1, 2, \dots, M. \tag{1}$$

where F_τ are the functions of the coordinates x and z on the cross-section. u_τ is the vector of the generalized displacements, M stands for the number of the terms used in the expansion, and the repeated subscript, τ, indicates summation. The choice of F_τ determines the class of the 1D CUF model that is required and subsequently to be adopted. According to Eq. 1, TE (Taylor expansion) 1D CUF models consist of a MacLaurin series that uses the 2D polynomials $x^i z^j$ as F_τ functions, where i and j are positive integers. For instance, the displacement field of the second-order ($N = 2$) TE model can be expressed as

Aerospace Science and Engineering - III Aerospace PhD-Days   Materials Research Forum LLC
Materials Research Proceedings 33 (2023) 49-56   https://doi.org/10.21741/9781644902677-8

$$u_x = u_{x_1} + xu_{x_2} + zu_{x_3} + x^2u_{x_4} + xzu_{x_5} + z^2u_{x_6},$$
$$u_y = u_{y_1} + xu_{y_2} + zu_{y_3} + x^2u_{y_4} + xzu_{y_5} + z^2u_{y_6}, \qquad (2)$$
$$u_z = u_{z_1} + xu_{z_2} + zu_{z_3} + x^2u_{z_4} + xzu_{z_5} + z^2u_{z_6}.$$

The order N of the expansion is set as an input option of the analysis; the integer N is arbitrary and it defines the order the beam theory.

Governing equations of the N-order TE model and parallel axis theorem
The principle of virtual displacements is used to derive the equations of motion.

$$\delta L_{int} = \int_V \delta \epsilon^T \sigma dV = -\delta L_{ine}. \qquad (3)$$

where σ is stress, ϵ is strain, L_{int} stands for the strain energy and L_{ine} is the work done by the inertial ladings. δ stands for the usual virtual variation operator. After integrations by part, Eq. 3 becomes

$$\delta L_{int} = \int_L \delta \mathbf{u}_\tau^T \mathbf{K}^{\tau s} \mathbf{u}_s dy + [\delta \mathbf{u}_\tau^T \mathbf{\Pi}^{\tau s} \mathbf{u}_s]_{y=0}^{y=L}. \qquad (4)$$

where $\mathbf{K}^{\tau s}$ is the differential linear stiffness matrix and $\mathbf{\Pi}^{\tau s}$ is the matrix of the natural boundary conditions in the form of 3×3 fundamental nuclei. Due to space reasons, the $\mathbf{K}^{\tau s}$ matrix and $\mathbf{\Pi}^{\tau s}$ matrix are not expanded in detail which can be referred to [9]. However, the critical part of these matrices is the solution of the cross-sectional moment parameter $E_{\tau,\theta S,\zeta}^{\alpha\beta}$.

$$E_{\tau,\theta S,\zeta}^{\alpha\beta} = \int_\Omega \tilde{C}_{\alpha\beta} F_{\tau,\theta} F_{S,\zeta} d\Omega. \qquad (5)$$

where $\tilde{C}_{\alpha\beta}$ is the coefficient matrix related to the Young modulus, the Poisson ratio, and fiber orientation angle. For the integration of cross-section functions, in general, the Taylor expansion is only applicable to the whole cross-section. However, as shown in the Fig.1, the cross-section of the train body contains both the Cartesian coordinate system and the polar coordinate system, which need to be integrated separately. Therefore, we introduce the concept of the parallel axis theorem and the specific steps are as follows

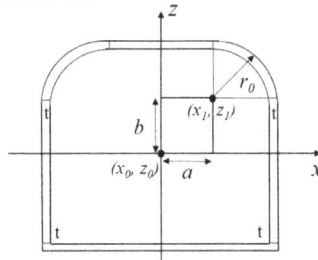

Fig.1 Cross-section of the train body

For the Taylor expanded body section integral, when the midpoint of the integral is (x_0, z_0), it can be expressed as

$$\int F_\tau F_S d\Omega = \sum_{m=0}^N \sum_{c=0}^N \int x^m z^n x^c z^d d\Omega. \qquad (6)$$

where $n=N-m$, $d=N-c$. Coordinates (x_0, z_0) have the following relationship with (x_1, z_1):

Aerospace Science and Engineering - III Aerospace PhD-Days   Materials Research Forum LLC
Materials Research Proceedings 33 (2023) 49-56   https://doi.org/10.21741/9781644902677-8

$$x_1 - x_0 = a, z_1 - z_0 = b. \tag{7}$$

Thus, to represent the circular area in the polar coordinates, each term of the summation in Eq. 6 can be rewritten as

$$\int_\Omega (x+a)^m (z+b)^n (x+a)^c (z+b)^d d\Omega$$
$$= \int \sum_{i=0}^M C_m^i x^i a^{m-i} \sum_{j=0}^n C_n^j z^j b^{n-j} \sum_{p=0}^t C_c^p x^p a^{c-p} \sum_{q=0}^d C_d^q z^q b^{d-q} d\Omega$$
$$= \sum_{i=0}^m \sum_{j=0}^n \sum_{p=0}^c \sum_{q=0}^d C_m^i C_n^j C_c^p C_d^q a^{m-i} b^{n-j} a^{c-p} b^{d-q} \int x^i z^j x^p z^q d\Omega. \tag{8}$$

where C is the binomial coefficient and the integral part in Eq. 8 can be further converted to solve in polar coordinates

$$\int x^i z^j x^p z^q d\Omega = \int r^{i+j+p+q} \cos^{i+p}\theta \sin^{j+q}\theta d\Omega$$
$$= \int_{r_1}^{r_0} \int_{\theta_1}^{\theta_2} r^{i+j+p+q} \cos^{i+p}\theta \sin^{j+q}\theta d\theta dr. \tag{9}$$

where r_0 is the radius of the outer side of the ring. Based on the above method, the cross-sectional moment parameter $E_{\tau,\theta S,\zeta}^{\alpha\beta}$ of the train body section can be derived.

Then, the virtual variation of the inertial work is given by

$$\delta L_{\text{ine}} = \int_L \delta \mathbf{u}_\tau \int_\Omega \rho F_\tau F_S d\Omega \ddot{\mathbf{u}}_s dy = \int_L \delta \mathbf{u}_\tau \mathbf{M}^{\tau s} \ddot{\mathbf{u}}_s dy. \tag{10}$$

where $\mathbf{M}^{\tau s}$ is the differential linear mass matrix. And the explicit form of the governing equations is

$$\delta u_{x\tau}: -E_{\tau S}^{66} u_{xs,yy} + \left(E_{\tau,xs}^{26} - E_{\tau S,x}^{26}\right) u_{xs,y} + \left(E_{\tau,x S_x}^{22} + E_{\tau,z S_z}^{44}\right) u_{xs}$$
$$-E_{\tau S}^{36} u_{ys,yy} + \left(E_{\tau,x}^{23} - E_{\tau S,x}^{66}\right) u_{ys,y} + \left(E_{\tau,x S_x}^{26} + E_{\tau,z S_z}^{45}\right) u_{ys}$$
$$+\left(E_{\tau,z S}^{45} - E_{\tau S,z}^{16}\right) u_{zs,y} + \left(E_{\tau,z S_x}^{44} + E_{\tau,x S_z}^{12}\right) u_{zs} = -E_{\tau S}^{\rho} \ddot{u}_{xs},$$

$$\delta u_{y\tau}: -E_{\tau S}^{36} u_{xs,yy} + \left(E_{\tau,xs}^{66} - E_{\tau S,x}^{23}\right) u_{xs,y} + \left(E_{\tau,x S_x}^{26} + E_{\tau,z S_z}^{45}\right) u_{xs}$$
$$-E_{\tau S}^{33} u_{ys,yy} + \left(E_{\tau,x}^{36} - E_{\tau S,x}^{36}\right) u_{ys,y} + \left(E_{\tau,x S_x}^{66} + E_{\tau,z S_z}^{55}\right) u_{ys}$$
$$+\left(E_{\tau,z}^{55} - E_{\tau S,z}^{13}\right) u_{zs,y} + \left(E_{\tau,x S_z}^{16} + E_{\tau,z}^{45}\right) = u_{zs} = -E_{\tau S}^{\rho} \ddot{u}_{ys}, \tag{11}$$

$$\delta u_{z\tau}: \left(E_{\tau,z}^{16} - E_{\tau S,z}^{45}\right) u_{xs,y} + \left(E_{\tau,x S,z}^{44} + E_{\tau,z S_x}^{12}\right) u_{xs}$$
$$+\left(E_{\tau,z}^{13} - E_{\tau S,z}^{55}\right) u_{ys,y} + \left(E_{\tau,x S_z}^{45} + E_{\tau,z S_x}^{16}\right) u_{ys} - E_{\tau S}^{55} u_{zs,yy}$$
$$+\left(E_{\tau,xs}^{45} - E_{\tau S,x}^{45}\right) u_{zs,y} + \left(E_{\tau,x S,x}^{44} + E_{\tau,z S_z}^{11}\right) u_{zs} = -E_{\tau S}^{\rho} \ddot{u}_{zs}.$$

where

$$E_{\tau S}^{\rho} = \int_\Omega \rho F_\tau F_S d\Omega. \tag{12}$$

Double over dots stand for the second derivative with respect to time (t). Letting $\mathbf{P}_\tau = \{P_{x\tau} \; P_{y\tau} \; P_{z\tau}\}^T$ to be the vector of the generalized forces, the natural boundary conditions are

$$\delta u_{x\tau}: P_{xs} = E_{\tau s}^{66} u_{xs,y} + E_{\tau s,x}^{26} u_{xs} + E_{\tau s}^{36} u_{ys,y} + E_{\tau s,x}^{66} u_{ys} + E_{\tau s,z}^{16} u_{zs},$$

$$\delta u_{y\tau}: P_{ys} = E_{\tau s}^{36} u_{xs,y} + E_{\tau s,x}^{23} u_{xs} + E_{\tau s}^{33} u_{ys,y} + E_{\tau s,x}^{36} u_{ys} + E_{\tau s,z}^{13} u_{zs}, \qquad (13)$$

$$\delta u_{z\tau}: P_{zs} = E_{\tau s,z}^{45} u_{xs} + E_{\tau s,z}^{55} u_{ys} + E_{\tau s}^{55} u_{zs,y} + E_{\tau s,x}^{45} u_{zs}.$$

For a fixed approximation order N, Eq. 11 and 13 have to be expanded using the indices τ and s in order to obtain the governing differential equations and the natural boundary conditions of the desired model.

In the case of harmonic motion, the solution of Eq. 11 is sought in the form

$$\mathbf{u}_s(y; t) = \mathbf{U}_s(y) e^{i\omega t}. \qquad (14)$$

where $\mathbf{U}_s(y)$ is the amplitude function of the motion, ω is an arbitrary circular or angular frequency, and i is $\sqrt{-1}$. The formulation of the equilibrium equations and the natural boundary conditions in the frequency domain can be obtained by substituting Eq. 14 into Eq. 11.

Dynamic stiffness formulation
In Section 2, the ordinary differential equations of the beam in free vibration have been derived and the procedure to obtain the Dynamic Stiffness (DS) matrix for a structural problem can be summarized as follows: (i) Seek a closed form analytical solution of the governing differential equations of the structural element; (ii) Apply a number of general boundary conditions equal to twice the number of integration constants in algebraic form, which are usually the nodal displacements and forces; (iii) Eliminate the integration constants by relating the amplitudes of the generalized nodal forces to the corresponding generalized displacements generating the DS matrix $\mathcal{K}$. For the sake of brevity, the expressions for the DS matrix $\mathcal{K}$ are not reported here, but can be found in standard texts, see for example Pagani [9]. It should be noted that the DS matrix consists of both the inertia and stiffness properties of the structure element unlike the FEM for which they are separately identified.

The DS matrix $\mathcal{K}$ is the basic building block to compute the exact natural frequencies of a higher-order beam. The DSM has also many of the general features of the FEM. In particular, it is possible to assemble elemental DS matrices to form the overall DS matrix of any complex structure consisting of beam elements. The global DS matrix can be written as

$$\bar{\mathbf{P}}_G = \mathcal{K}_G \bar{\mathbf{U}}_G. \qquad (15)$$

where $\mathcal{K}_G$ is the square global DS matrix of the final structure. For the sake of simplicity, the subscript "G" is omitted hereafter. The train body structure can be regarded as beams of different cross-sectional forms, and the whole train body structure can be obtained by simply assembling it like FEM. The boundary conditions can be applied by using the well-known penalty method (often used in FEM) or by simply removing rows and columns of the stiffness matrix corresponding to the degrees of freedom which are zeroes. Due to the presence of higher-order degrees of freedom at each interface, a multitude of boundary conditions can be applied at the required nodes.

The Wittrick–Williams algorithm is used to solve the transcendental (nonlinear) eigenvalue problem generated by the DSM. Once the natural frequencies are calculated and the associated global DS matrix is obtained, the complete displacement field can be generated as a function of x, y, z and the time t. Clearly, the plot of the required mode and required element can be visualized on a fictitious 3D mesh. By following this procedure it is possible to compute the exact mode shapes using just one element which is impossible in FEM.

Aerospace Science and Engineering - III Aerospace PhD-Days   Materials Research Forum LLC
Materials Research Proceedings 33 (2023) 49-56   https://doi.org/10.21741/9781644902677-8

Numerical Results

A train body with four types cross-sections such as the one shown in Fig. 2 is considered. The four types represent the cross-section of the body frame, window, door and end wall respectively. The material data are the Young modulus, $E = 75$GPa, the Poisson ratio, $v = 0.33$, material density, $\rho = 2700$kgm^{-3}. The cross-sectional data are L_1=1.6m, L_2=1.65m, L_3=1m, L_4=2m, W_1=3.3m, H_1=2.55m, H_2=0.425m, H_3=1.275m, H_4=0.4m, H_5=0.2m, H_6=1.7m, H_7=2m, H_8=0.8m, H_9=1.04m, R_1=0.85m, R_2=0.75m, t=0.02m. Distribution of cross-section types in the y-direction (lengthwise) is shown in Fig. 3. The bodies of 4 types cross-sections are combined into a complete train structure which the length is 18m.

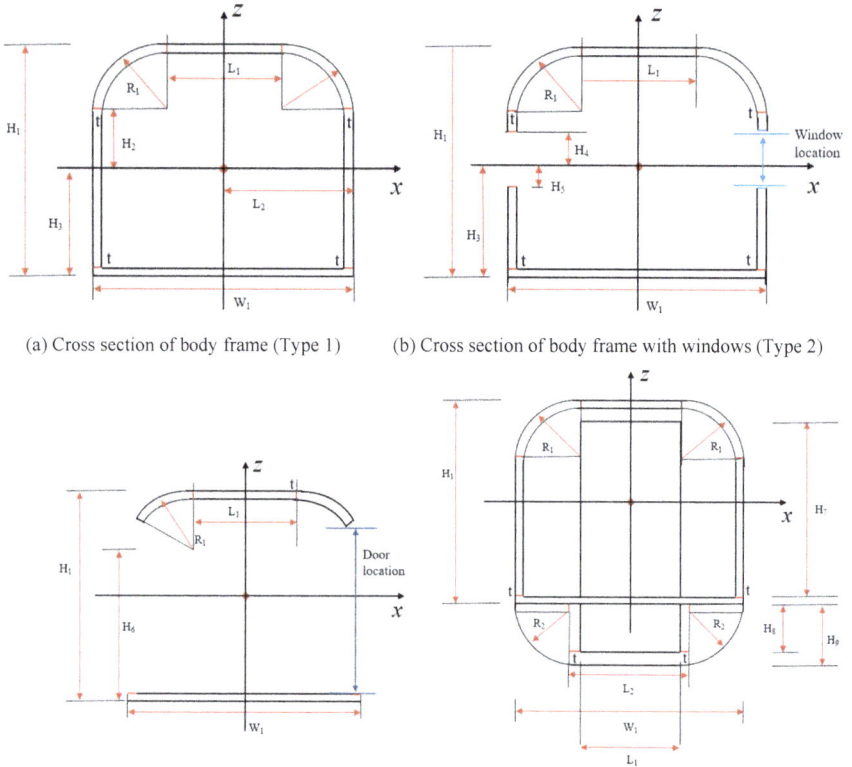

(a) Cross section of body frame (Type 1)   (b) Cross section of body frame with windows (Type 2)

(c) Cross section of body frame with doors (Type 3)   (d) Cross section of body frame with end wall (Type 4)
Fig. 2 Cross section of four typical train body frames

Aerospace Science and Engineering - III Aerospace PhD-Days
Materials Research Proceedings 33 (2023) 49-56

Materials Research Forum LLC
https://doi.org/10.21741/9781644902677-8

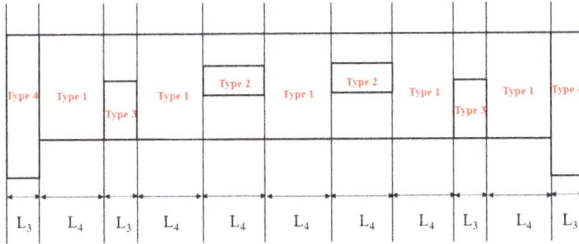

Fig.3 Distribution of cross-section types in the y-direction (lengthwise)

Table 1 shows the first 6 natural frequencies of the train body for free-free BCs. Classical Timoshenko beam method (TBM) as well as up to the fifth-order TE refined train body models by the present DSM approach are given in Table 2.

Table 1 First to sixth natural frequencies (Hz) for the FF train body.

	Mode 1[a]	Mode 2[b]	Mode 3[c]	Mode 4[d]	Mode 5[e]	Mode 6[f]
N=5	50.3449	59.7189	82.2272	109.2521	127.8564	145.3032
N=4	50.9998	60.0692	82.4901	111.4041	129.5863	145.3141
N=3	51.2261	60.1823	86.9639	113.2928	130.6047	145.3141
N=2	53.2889	61.870	87.8691	126.1918	140.4748	145.3489
N=1	53.2628	61.8535	89.5141	104.9372	126.0787	140.4051
TBM	-	61.8535	-	-	125.0255	140.4051

[a] First flexural mode on plane yz. [b] First flexural mode on plane xy. [c] First torsional mode.
[d] Second flexural mode on plane yz. [e] Second flexural mode on plane xy. [f] Second torsional mode.

Fig.4 shows the first six modes of the train body for free-free BCs. The Mode 1 is first flexural mode on plane yz, the Mode 2 is first flexural mode on plane xy, the Mode 3 is first torsional mode, the Mode 4 is second flexural mode on plane yz, the Mode 5 is second flexural mode on plane xy and the Mode 6 is second torsional mode. Moreover, it has been demonstrated that CUF TE higher-order models can deal with non-classical phenomena such as torsion, shear effects and couplings. Train body elasticity solutions can, in fact, be reproduced with CUF models if a sufficient number of terms are considered in the kinematic field of the beam theory.

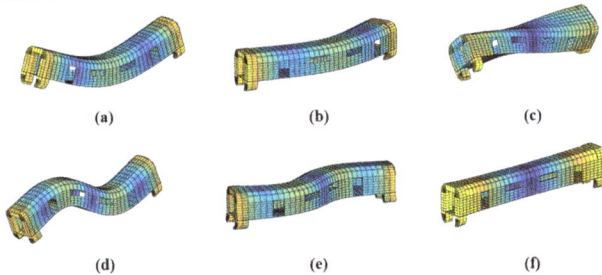

Fig.4 First flexural on plane yz (a), first flexural on plane xy (b), first torsional mode (c), second flexural on plane yz (d), second flexural mode on plane xy (e), second torsional mode (f) for the FF train body, N=4.

Aerospace Science and Engineering - III Aerospace PhD-Days   Materials Research Forum LLC
Materials Research Proceedings 33 (2023) 49-56   https://doi.org/10.21741/9781644902677-8

Conclusions

In the framework of CUF, the introduction of the parallel axis theorem in the Taylor expansion greatly improves its applicability: 1) geometrical boundary conditions can be applied in subdomains of the cross-section (and not only to the whole cross-section). 2) cross-sections can be divided into further beam sections and easily assembled. Combined with DSM, a high-order DS matrix is developed and the natural frequencies and mode shapes of the train body structure are calculated using the WW algorithm. Through further validation, the method can provide a powerful tool for the dynamic analysis and optimised design of laminated composite train body.

References

[1] Ribeiro D, Calçada R, Delgado R, Zabel MB, V. Finite-element model calibration of a railway

vehicle based on experimental modal parameters, Vehicle System Dynamics. 51 (2013) 821-856. https://doi.org/10.1080/00423114.2013.778416

[2] Zaouk AK, Marzougui D, Bedewi NE. Development of a Detailed Vehicle Finite Element Model Part I: Methodology, International Journal of Crashworthiness. 5 (2000) 25-36.

https://doi.org/10.1533/cras.2000.0121

[3] Yoon SC, Kim YS, Kim JG, Park SH, Lee HU. A Study on the Structural Fracture of Body

Structure in Railroad Car, Key Engineering Materials. 577 (2013) 301-304. https://doi.org/10.4028/www.scientific.net/KEM.577-578.301

[4] Kurtaran H, Buyuk M, Eskandarian A. Ballistic impact simulation of GT model vehicle door

using finite element method, Theoretical & Applied Fracture Mechanics. 40 (2003) 113-121. https://doi.org/10.1016/S0167-8442(03)00039-9

[5] Wang Wei, Xin Yong. Finite element modeling and modal analysis of vehicle frame, Mechanical Design and Manufacturing. 11 (2009) 53-54. https://doi.org/10.3969/j.issn.1001-3997.2009.11.024

[6] Shen Zhenhong, Zhao Honglun. Research on Finite Element Modeling Method of Rail Vehicle Sandwich Plate Structure, Electric Locomotive and Urban Rail Vehicle. 30 (2007) 42-45. https://doi.org/10.3969/j.issn.1672-1187.2007.01.013

[7] E. Carrera, A. Pagani, J.R. Banerjee. Linearized buckling analysis of isotropic and composite beam-columns by Carrera Unified Formulation and Dynamic Stiffness Method, Mechanics of Advanced Material and Structures. 9 (2016) 1092–1103. https://doi.org/10.1080/15376494.2015.1121524

[8] M. Dan, A. Pagani, E. Carrera. Free vibration analysis of simply supported beams with solid and thin-walled cross-sections using higher-order theories based on displacement variables, Thin-Walled Structures. 98 (2016) 478-495. https://doi.org/10.1016/j.tws.2015.10.012

[9] A. Pagani, E. Carrera, J.R. Banerjee, P.H. Cabral, G. Caprio, A. Prado. Free vibration analysis of composite plates by higher-order 1D dynamic stiffness elements and experiments, Composite Structures. 118.5 (2014) 654-663. https://doi.org/10.1016/j.compstruct.2014.08.020

Aerospace Science and Engineering - III Aerospace PhD-Days   Materials Research Forum LLC
Materials Research Proceedings 33 (2023) 57-60   https://doi.org/10.21741/9781644902677-9

Condition-based-maintenance for fleet management

Leonardo Baldo[1,a*], Andrea De Martin[1], Massimo Sorli[1], Mathieu Terner[2]

[1]Department of Mechanical and Aerospace Engineering, Politecnico di Torino, Corso Duca Degli Abruzzi, 24, Torino (TO)

[2]Leonardo S.p.A. - Aircraft Division, Commercial and Customer Services, Str. per Malanghero, 10072 Caselle Torinese (TO)

[a]leonardo.baldo@polito.it

Keywords: PHM, CBM, Maintenance, FCS, Fleet Management

Abstract. New "enlightened" and holistic maintenance strategies are shaping the industrial world from the inside, providing intelligent and focused solutions where high availability, reliability and safety are required. Maintenance planning and scheduling is an extremely daunting and multi-faceted task which involves competences from fairly different fields: customer support, quality, engineering, production, RAMS, cost estimation etc. In the aerospace sector, a significant percentage of Life Cycle Costs (LCCs) and, in particular, operating costs, are determined by Maintenance, Repair and Overhaul (MRO) activities and the relative asset unavailability due to down-time or turn-around-time [1,2]. This is the reason why currently there is an ongoing intense effort in the research community and in the industry towards new maintenance strategies which could overcome the limitations of preventive maintenance thus streamlining operations, without jeopardizing mission safety. This research project is hence spot on and focuses on the development of optimized maintenance strategies, built around the system health status.

Introduction

Preventive maintenance employs a very simple approach where the component is replaced after a number of cycles or hours. This approach is definitely safe (as significant safety margins and factors are applied) but it is beyond any doubt inefficient for both the Original Equipment Manufacturer (OEM) and the operator, which have to replace working parts. On top of that, the Integrated Logistic Support (ILS) chain must be somehow oversized to guarantee component readiness.

A plethora of maintenance solutions have hence engendered to pitch in these scenarios, bolstered by the parallel growth in sensors technologies and Industry 4.0: starting from Opportunistic Maintenance (OM), passing through Condition-Based Maintenance (CBM) and even approaching predictive maintenance. If OM proposes an intelligent regrouping of maintenance activities, CBM and predictive maintenance [3] tries to plan maintenance actions according to the real component/subsystem health status. If the idea behind the CBM concept is, at least, quite straightforward, the implementation is anything but simple. This requires the complex interaction of several entities which must be integrated in a single holistic framework to enable the vision of customized, pinpointing and tailored planning [1].

In fact, CBM involves different steps: data collection, data analysis, forecasting and acting. The selected component or subsystem must be monitored by appropriate sensors which transmit data that are logged and analyzed in a Prognostic and Health Management (PHM) perspective. Meaningful features that can successfully track down possible failures are extracted and examined through different algorithms. A central step is represented by Remaining Useful Life (RUL) predictions: leveraging on sophisticated algorithms, a probabilistic approach is often employed [4]. After that, a maintenance plan can be derived from RUL predictions, the relative confidence

Aerospace Science and Engineering - III Aerospace PhD-Days   Materials Research Forum LLC
Materials Research Proceedings 33 (2023) 57-60   https://doi.org/10.21741/9781644902677-9

levels and many other factors (e.g. hangar slots and availability, maintenance task flexible regrouping, criticality of the potential failure, regulations, etc.).

Before the algorithm could be employed, an extended off-line activity must be carried out with historical data to train the algorithms and verify the accuracy and confidence levels of the overall prognostic framework [4].

The goal of this Ph.D research project co-financed by Leonardo S.p.A. and Politecnico di Torino is to develop a CBM framework which could support Leonardo's customers and, at the same time, bringing the current state-of-the-art of maintenance strategies a step forward, employing cutting edge algorithms to real-life operational data. Furthermore, it has to be noted that interaction with an industrial partner is then pivotal to have a real-life feedback of the solution feasibility.

If in the civil aviation sector, the development of PBL (Performance Based Logistics) and CBM logics is difficult due to safety and reliability concerns, in the defence field all of this is compounded by additional complexities. Disruptive routines, alteration (or absence) of flight schedules and routes, 24-hour-a-day activities, excessive structural stresses, extended flight envelopes require an even more flexible maintenance planning, which should take into consideration specific mission profiles and requirements.

Material and Methods
Starting from a candidate aircraft for the analysis and the general overview of its subsystems architecture, the research is focused on algorithms and solutions which integrate PHM strategies with Remaining Useful Life (RUL) predictions in more complex frameworks for maintenance scheduling and planning [5].

Among different possible subsystems, the research will focus on primary flight control actuation systems and the relative non-linearities (e.g. friction, free-play [6] etc). To this end, a literature review is currently being carried out to highlight the potential PHM strategies applied to Electro-Hydraulic servo actuators for primary flight control [7], starting from diagnostic applications [8].

The first step of any PHM related activity starts with the analysis of the system architecture along with the RAMS-T industry department. A series of data ranging from technical publications to historical failure events as well as maintenance, operative and flight performances are currently being examined to identify possible opportunities and threats.

The central objective is to identify and assess the correlations that may exist between the operational data gathered from flight operations, maintenance reports, historical data, and the real-time data obtained from onboard sensors (downloaded on ground after landing or in the future potentially transferred live during flight). This research will encompass mapping out and defining these correlations with the aim of comprehending the frequency, types, severity, and occurrence of failures. Leveraging on engineering and logistics resources, algorithms will be designed and developed for CBM purposes of selected critical systems/items in view of optimizing fleet management. After appropriate simulations with developed models aimed at forecasting operation in nominal and degraded conditions, the final step will include the framework validation with real-life data, in order to assess the strategy performance in terms of prognostic capability and positive effects on the fleet availability.

Enabling technologies could range from Digital Twins (DT) [9,10,11] to high and low fidelity models which could predict the component behavior. Data-driven, model-based as well as hybrid solutions will be taken into consideration with specific literature reviews to select appropriate Machine Learning (ML) methodologies and computational approaches. Promising strategies include particle filtering, Long Short-Term Memory (LSTM) [12] or even Physics Informed Neural Networks (PNN) combined with classification algorithms like Support Vector Machines (SVMs) or random forests. Moreover, the approaches should take into consideration the "few-shot" phenomenon, which lead to the substantially unbalanced healthy-unhealthy datasets typical of PHM tasks [12,13]. During the development phase there will be an extensive use of modelling techniques based on physical laws and experimental data: thanks to high and low fidelity models, the expected component behavior can be outlined and compared with the actual trends. In this direction, even the initial steps towards the creation of components and subsystem DTs or reasoning models will be considered, exploiting CAD and physics-of-failure representation.

Since the selected aerospace application is extremely safety-critical, particular attention will be paid to the method traceability, explainability and interpretability selecting, if possible, Explainable Artificial Intelligence (ExAI) methods. In this way, operators and maintenance crews can interpret the results and understand why a particular decision has been taken, providing useful feedbacks and contributing to decision-making.

Results & Conclusion
The expected result is a tight integration between academia and industry, computational analyses and on-site experience, technical design and customer support, maintenance planning and prognostic results, enhanced by cutting edge data science methods. CBM and predictive maintenance are here to stay and this project is perfectly spot on with the research community, giving its own contribution to make the aviation world more efficient, safer and, at the same time, assuring top-notch asset performance.

References
[1] « Forestalling the Future with Predictive Maintenance » from https://www.oliverwyman.com/our-expertise/insights/2016/apr/oliver-wyman-transport---logistics-2016/operations/predictive-maintenance.html

[2] Mike Gerdes, Dieter Scholz, e Diego Galar, «Effects of condition-based maintenance on costs caused by unscheduled maintenance of aircraft», Journal of Quality in Maintenance Engineering 22, fasc. 4 (1 gennaio 2016): 394–417, https://doi.org/10.1108/JQME-12-2015-0062

[3] Scott, Michael J., Wim J. C. Verhagen, Marie T. Bieber, e Pier Marzocca. «A Systematic Literature Review of Predictive Maintenance for Defence Fixed-Wing Aircraft Sustainment and Operations». Sensors 22, fasc. 18 (gennaio 2022): 7070. https://doi.org/10.3390/s22187070

[4] Vachtsevanos, George, Frank Lewis, Michael Roemer, Andrew Hess, e Biqing Wu. Intelligent Fault Diagnosis and Prognosis for Engineering Systems: Vachtsevanos/Intelligent Fault Diagnosis. Hoboken, NJ, USA: John Wiley & Sons, Inc., 2006. https://doi.org/10.1002/9780470117842

[5] Jorben Pieter Sprong, Xiaoli Jiang, e Henk Polinder, «A Deployment of Prognostics to Optimize Aircraft Maintenance - A Literature Review: A Literature Review», Annual Conference of the PHM Society 11, fasc. 1 (22 settembre 2019), https://doi.org/10.36001/phmconf.2019.v11i1.776

[6] Michael Candon et al., «A Nonlinear Signal Processing Framework for Rapid Identification and Diagnosis of Structural Freeplay», Mechanical Systems and Signal Processing 163 (15 gennaio 2022): 107999, https://doi.org/10.1016/j.ymssp.2021.107999

[7] Andrea De Martin, Giovanni Jacazio, e Massimo Sorli, «Evaluation of Different PHM Strategies on the Performances of a Prognostic Framework for Electro-Hydraulic Actuators for Stability Control Augmentation Systems», in Annual Conference of the PHM Society, vol. 14, 2022, https://doi.org/10.36001/phmconf.2022.v14i1.3289

[8] Samuel David Iyaghigba, Fakhre Ali, e Ian K. Jennions, «A Review of Diagnostic Methods for Hydraulically Powered Flight Control Actuation Systems», Machines 11, fasc. 2 (25 gennaio 2023): 165, https://doi.org/10.3390/machines11020165

[9] Cordelia Mattuvarkuzhali Ezhilarasu, Zakwan Skaf, e Ian K Jennions, «Understanding the role of a Digital Twin in Integrated Vehicle Health Management (IVHM)», in 2019 IEEE International Conference on Systems, Man and Cybernetics (SMC), 2019, 1484–91, https://doi.org/10.1109/SMC.2019.8914244

[10] Sam Heim et al., «Predictive Maintenance on Aircraft and Applications with Digital Twin», in 2020 IEEE International Conference on Big Data (Big Data), 2020, 4122–27, https://doi.org/10.1109/BigData50022.2020.9378433

[11] Raymon van Dinter, Bedir Tekinerdogan, e Cagatay Catal, «Reference Architecture for Digital Twin-Based Predictive Maintenance Systems», Computers & Industrial Engineering 177 (1 marzo 2023): 109099, https://doi.org/10.1016/j.cie.2023.109099

[12] Ingeborg de Pater e Mihaela Mitici, «Developing Health Indicators and RUL Prognostics for Systems with Few Failure Instances and Varying Operating Conditions Using a LSTM Autoencoder», Engineering Applications of Artificial Intelligence 117 (1 gennaio 2023): 105582, https://doi.org/10.1016/j.engappai.2022.105582

[13] Maren David Dangut, Zakwan Skaf, e Ian K. Jennions, «Handling Imbalanced Data for Aircraft Predictive Maintenance Using the BACHE Algorithm», Applied Soft Computing 123 (1 luglio 2022): 108924, https://doi.org/10.1016/j.asoc.2022.108924

Aerospace Science and Engineering - III Aerospace PhD-Days   Materials Research Forum LLC
Materials Research Proceedings 33 (2023) 61-67   https://doi.org/10.21741/9781644902677-10

Investigation of the space debris environment for a sustainable evolution of the space around the earth

Andrea Muciaccia[1,a*], Mirko Trisolini[1], Lorenzo Giudici[1], Camilla Colombo[1]

[1] Department of Aerospace Science and Technology, Politecnico di Milano, Italy

[a]andrea.muciaccia@polimi.it

Keywords: Space Debris, Space Sustainability, Space Traffic Management, Space Surveillance and Tracking

Abstract. The sustainability of the space environment around the Earth is becoming an increasingly important issue in the space sector. Indeed, the space population is evolving over time. Therefore, careful mission design together with mitigation guidelines and policies are essential to regulate its evolution and to avoid the proliferation of derelict objects around the Earth. The main objective of this research is to connect different models that share the same goal: the sustainable evolution of the space environment around the Earth. In this view, the research focuses on the definition of metrics to assess the influence of missions (already occurred or planned) on the space environment and of a carrying capacity that the space can support, and on the characterization of in-orbit breakup events.

Introduction

The sustainability of the space environment around the Earth is becoming an increasingly important issue in the space sector. Indeed, the space population is evolving over time [1].

On one hand, there is the deployment of many satellites, including large constellations, that place many satellites in specific orbital regions. This requires new mitigation policies and careful mission design, with special attention to end-of-life strategies. To help this, several risk metrics are being developed to assess the impact of missions on the space environment, each of which seeks to capture the main elements influencing it. Bastida Virgili and Krag [2][3] proposed a criterion to select candidates for Active Debris Removal (ADR) missions, while Lewis [4] introduced a criterion which includes capacity and health scores to measure the efficacy of mitigation measures and the influence of spacecraft on the operational orbital region, respectively. Rossi et al. [5] proposed the criticality of spacecraft index to rank abandoned objects. Letizia et. [6] defined a risk indicator, ranking all space objects considering the effect of their fragmentation on other operative satellites.

On the other hand, new breakup events occur frequently increasing the background population of inactive objects. Examples are the recent CZ-6A breakup occurred on the 12th of November 2022 and the Cosmos 1408 breakup occurred on the 25th of November 2021. Indeed, some events are still difficult to predict (e.g., collision between objects) while others are unpredictable (e.g., explosion of a rocket body). These new uncontrolled objects, posing a threat to the population of objects orbiting the Earth, are to be tracked as soon as possible after the event to investigate their origin that is to determine the epoch and location of the event and the object(s) involved. In the past years, several tools have been developed to detect fragmentations. Romano et al. [7] and Andrisan et al. [8] developed tools which estimate the epoch and position of the breakup by studying the average distance between the objects in the debris cloud. Differently, Frey et al. [9] and Muciaccia et al. [10] focused their works on the long-term evolution of orbits (years) considering an averaged dynamic and determining the epoch of the breakup by detecting a convergence of objects in the space of inclination and right ascension of the ascending node. Dimare et al. [11] identify fragmentations by defining a similarity function of the orbital elements

Aerospace Science and Engineering - III Aerospace PhD-Days

Materials Research Proceedings 33 (2023) 61-67

Materials Research Forum LLC

https://doi.org/10.21741/9781644902677-10

of the observed objects. Once the characteristics of the event are known, risk analyses can be carried out by modelling the distribution of the fragments right after the event and its evolution over time, and by studying their interaction with other orbiting objects.

The main objective of this PhD research is to connect different models that share the same goal: the sustainable evolution of the space environment around the Earth. In this view, the research will focus first on the definition of metrics to assess the influence of missions (already occurred or planned) on the space environment and of a carrying capacity that the space can support. This is necessary to regulate the evolution of the population of active objects and to avoid overcrowding of specific regions of space, giving the possibility of use to future missions as well. Then, the research will investigate models to characterise breakup events. The latter is essential to limit the proliferation of space debris (i.e., uncontrolled objects) generated by collision between the fragments and the active satellites. Indeed, knowing the characteristics of fragmentation it is possible to define satellite at risk and thus to plan collision avoidance manoeuvers useful to decrease the effect of the fragmentation. By combining the models, we will then have monitoring of a large part of the population of objects orbiting the Earth.

A schema of the activities is shown in Figure 1, while a description of each activity is presented in the following sections.

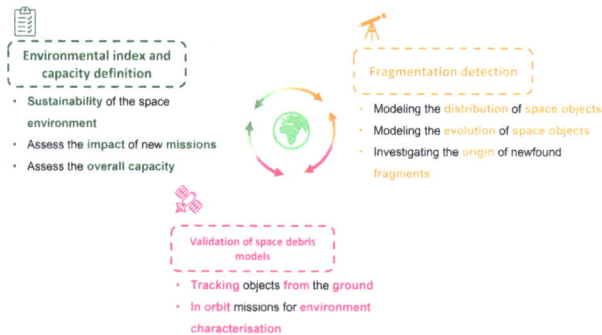

Figure 1. **Schema of the Ph.D. activities.**

Environmental index and capacity definition

The model evaluates the impact of a generic mission during its entire lifetime, taking into account several aspects of mission design.

First, the mission profile is divided into phases (e.g., operational, deorbiting, etc.) to investigate the weight of each on the total mission index, that is computed as

$$I_t = \int_{t_0}^{t_{EOL}} I \, dt + \alpha \cdot \int_{t_{EOL}}^{t_e} I \, dt + (1-\alpha) \cdot \int_{t_{EOL}}^{t_f} I \, dt \tag{1}$$

where I is the index evaluated at a single epoch, t_0 is the starting epoch, t_{EOL} is the epoch at which the operational phase ends, t_e is the epoch at which the disposal ends, t_f is the epoch at which the object would naturally decay from its initial orbit and α is a parameter associated to the reliability of the Post Mission Disposal (PMD) strategy and varies between 0 and 1. The first term is used to compute the index of all the phases before the PMD, while the latter is computed using the last two terms. The index at a single epoch is evaluated following the ECOB formulation [6]

Aerospace Science and Engineering - III Aerospace PhD-Days   Materials Research Forum LLC
Materials Research Proceedings 33 (2023) 61-67   https://doi.org/10.21741/9781644902677-10

$$I = p_c \cdot e_c + p_e \cdot e_e \qquad\qquad (2)$$

where p_c and p_e represent the collision and explosion probabilities, and e_c and e_e represent the collision and explosion effects, respectively. In case the satellite is active and can perform Collision Avoidance Manoeuvres (CAM), the evaluation of the index at a single epoch is computed as

$$I = \beta \cdot I_{CAM} + (1 - \beta) \cdot I_{no-CAM} \qquad\qquad (3)$$

where I_{cam} is the index at a single epoch when CAM capabilities are considered, I_{no-cam} is the index at a single epoch when No-CAM capabilities are considered, and β is the CAM efficacy (ranging from 0 to 1) and is considered fixed along the entire mission profile.

Grid definition
The parameters (i.e., the probability of collision and the collisions and explosions effects) previously introduced are computed on a grid based on Keplerian orbital elements. The set of orbital elements is not fixed but it varies depending on the orbital region under analysis. Indeed, each orbital region is characterised by a peculiar distribution of the objects. As an example, for the specific case of the LEO region, a two-dimensional grid in semi-major axis and inclination [6] is used. The grid is defined from 0 deg to 180 deg in terms of inclination and from 6771 km to 8371 km in terms of semi-major axis, and the selection of the bin size can be chosen arbitrarily (default cell size of 10 km in semi-major axis and 10 deg in inclination).

Probability of collision
The probability of collision is evaluated adopting a flux-based model of the space debris environment and exploiting the analogy with the kinetic gas theory [12]. The value of the average debris flux is extracted from ESA MASTER 8 [13], considering the debris population at a specified epoch. In addition, MASTER 8 is also exploited to evaluate the averaged impact velocity, used to filter out the flux of particles able to generate catastrophic collisions. Studies were carried out to investigate the influence of parameters on the value of the collision probability, such as the collision avoidance maneuver capabilities of satellites or the size of the trackable debris from the ground (see Fig 4).

Probability of explosion
The probability of explosion is derived from historical data from the ESA DISCOS database [14]. A preliminary investigation has been performed on the type of explosion events and the type of objects involved. From the information available in DISCOS, a list of event families along with a list of object classes have been defined. Then, two methods have been compared to compute the explosion probability: the Kaplan-Meyer estimator [6], commonly used in medical sciences to estimate the survival rate of patience, and the Nelson-Aalen estimator [15], used to directly evaluate the cumulative hazard rate function associated to the fragmentation events for the different classes of objects.

Fragmentation effect
The evaluation of the effects of a fragmentation is performed on a set of spacecraft targets that is representative of the entire population of active objects. The data of the operational satellites are extracted from ESA DISCOS, where information about the activity status, the orbital region and the orbital elements, and the physical properties (i.e., mass and area) can be retrieved. The targets are defined by looking at the distribution of the cross-sectional area of operational satellites on a grid in terms of Keplerian parameters (described before).

Then, the effect terms of both collisions (e_c) and explosions (e_e) depend on the characteristics of the fragmentation, and on the evolution of the cloud of debris (propagated using a continuum

Aerospace Science and Engineering - III Aerospace PhD-Days   Materials Research Forum LLC
Materials Research Proceedings 33 (2023) 61-67   https://doi.org/10.21741/9781644902677-10

approach [16]) and its interaction with the objects' population. Specifically, the resulting increase in the collision probability for operational satellites is used for the assessment of the consequences. The effects map is generated by evaluating the probability of collision with the representative targets. For each bin belonging to the grid, a fragmentation (collision or explosion) is generated and propagated for 15 years; over this time span, the cumulative probability of collision with the population of representative targets is estimated, and the effects e are computed as:

$$e = \frac{1}{A_{TOT}} \sum_{i=1}^{N_t} P_c(t = 15 \text{ ys}) \, A_i \qquad (4)$$

where A_{TOT} is the overall spacecrafts' cross-section, A_i is the cumulative cross-section of the objects belonging to the ith bin, and P_c is the collision probability.

Fragmentation detection

The model characterise the breakups that occurred in orbit by evaluating:

- Epoch and location of the event
- Involved object(s)
- Mass and energy associated tot he event (useful to model the distribution of the generated fragments)

Two methods are considered. A short-term investigation analysing the possibility of fragmentation in a window of days and making use of osculating orbital elements (SGP4 [17]) for the propagation of the orbital elements of the objects, and a long-term investigation analysing the possibility of fragmentation in a window of months or years and making use of mean orbital elements (PlanODyn [18]) for the propagation of the orbital elements of the objects.

The general workflow of the two methodologies is the same. First, a set of unknown objects is generated from public catalogues. All the objects in the set are then propagated backwards to study the evolution of their orbits, and thus to identify possible clusters in a specific phase space in terms of Keplerian orbital elements. Whenever a possible breakup is identified (i.e., a cluster is detected), the model examine the fragments selected to characterise them in terms of families by using a hierarchical clustering method [19]. In addition, a second set of objects including only satellites is scanned to identify the parent(s) of the fragmentation by comparing the location of the satellites at the estimated epoch of the event and the location of the event itself.

The difference between the two models lies in the way epoch and location of the fragmentation are estimated.

The short-term routine uses a triple-loop filter to identify a cluster of objects in terms of their proximity to each other. The filter, comparing two objects at a time, is composed by an apogee/perigee filter which checks that the relative geometry of the orbits can lead to close encounters. If this filter is passed, the model evaluates the minimum orbit intersection distance (MOID) [20]. If the MOID between the orbits of the two analysed objects is below a defined threshold, a last temporal filter is considered. The latter consists of generating angular windows around the MOID, then converting them into time windows using Kepler's equation, and finally checking the possibility of having both objects in the same window at the same time. This filter is coupled with the propagator to perform the investigation inside the window under analysis.

The long-term routine detects the fragmentation using the right ascension of the ascending node (RAAN) as study parameter. Indeed, near the event epoch, all the fragments generated will share this Keplerian orbital parameters, making it useful for the purpose of the analysis.

Aerospace Science and Engineering - III Aerospace PhD-Days   Materials Research Forum LLC
Materials Research Proceedings 33 (2023) 61-67   https://doi.org/10.21741/9781644902677-10

The information coming from the previous analysis (i.e., the epoch and the location of the fragmentation, and the parent(s)) are than used to characterise the fragmentation in terms of total mass and energy involved. The latter are used as initial condition to generate the cloud of fragments with the NASA standard breakup model [21].

Main results

The environmental index model can be used for several types of analysis. First, the index can be used to investigate the impact of a single mission on the space population. An example is shown in Figure 2, where the picture shows the evolution of the index over time.

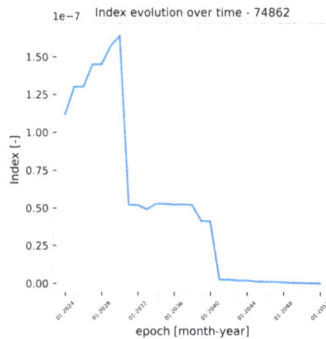

Figure 2. Index evolution over time of a payload.

Then, the same procedure can be applied to the entire population of orbiting objects to investigate the most critical regions and the share of the index associated to specific class of objects (e.g., rocket bodies). Figure 2 shows the distribution of the index (the marker size is proportional to the index value) on a semi-major axis and inclination grid.

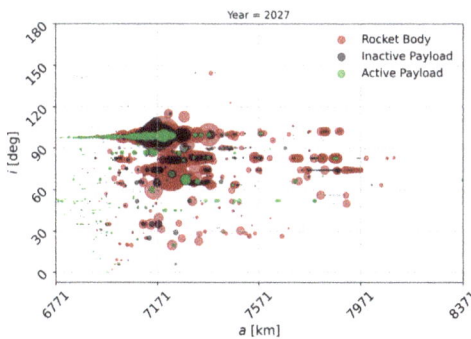

Figure 3. Index value for objects in LEO on a semi-major axis and inclination grid (marker size is proportional to the index value).

As visible from the picture, the most critical region is that at around 7171 km in terms of semi-major axis and 90° of inclinations.

Regarding the fragmentation event, several studies were conducted on past and recent fragmentation. An example of application is shown in Figure 2, considering the Cosmos 2251-Iridium 33 breakup. The initial set of objects included about 2000 objects (19 belonging to the collision event) available on SpaceTrack on 16th of February 2009 (i.e., 6 days after the event). The model was applied to the fragmentation and was able to properly detect the epoch (10th February 2009) and location of the fragmentation, along with the involved fragments and parent(s).

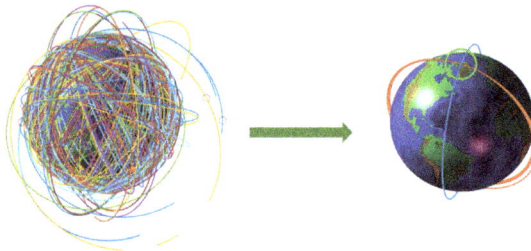

Figure 4. Cosmos 2251 - Iridium 33 breakup. Set of initial objects (left) and final set including only the involved families (right).

Acknowledgments

This research has received funding as part of the COMPASS project "Control for orbit manoeuvring by surfing through orbit perturbations" (Grant agreement No 679086), European Research Council (ERC) funded project under the European Union's Horizon 2020 research and from the European Space Agency contract 4000133981/21/D/KS.

References

[1] ESA Space Debris Office, "ESA's Annual Space Environment Report," 2022. [Online]. Available:
https://www.sdo.esoc.esa.int/environment_report/Space_Environment_Report_latest.pdf.

[2] Bastida Virgili B., Krag H., "Strategies for Active Removal in LEO", in Proceedings of the fifth European Conference on Space Debris, ESA Communications, ESA SP-672, CD-ROM, ESA Communication Production Office, Noordwijk, The Netherlands, 2009.

[3] Bastida Virgili B., Krag H., "Active Debris Removal for LEO Missions", in Proceedings of the sixth European Conference on Space Debris, ESA Communications, ESA SP-723, CD-ROM, ESA Communication Production Office, Noordwijk, The Netherlands, 2013.

[4] Lewis, H.G., "ACCORD: Alignment of Capability and Capacity for the Objective of Reducing Debris", FP7 Final Report, University of Southampton, 2014. URL:
http://cordis.europa.eu/docs/results/262/262824/final1-accordfinalreportsection4-1.pdf.

[5] Rossi A., Valsecchi G. B., Alessi E. M., "The Criticality of Spacecraft Index", Advances in Space Research 56(3):449-460, March 2015, doi: 10.1016/j.asr.2015.02.027. https://doi.org/10.1016/j.asr.2015.02.027

[6] Letizia, F., Lemmens, S., Virgili, B. B., Krag, H., "Application of a debris index for global evaluation of mitigation strategies," Acta Astronautica, Vol. 161, pp. 348-362, 2019. https://doi.org/10.1016/j.actaastro.2019.05.003

[7] M. Romano, A. Muciaccia, M. Trisolini, P. Di Lizia, C. Colombo, A. Di Cecco e L. Salotti, "PUZZLE software for the characterisation of in-orbit fragmentations", 8th European Conference on Space Debris, Darmstadt, Germany, Virtual Conference, Apr. 20-23, 2021.

[8] R. L. Andrisan, A. G. Ionita, R. D. Gonzalez, N. S. Ortiz, F. P. Caballero, and H. Krag, "Fragmentation event model and assessment tool (fremat) supporting on-orbit fragmentation analysis," in Proceedings of the 7th European Conference on Space Debris, 2017.

[9] S. Frey, C. Colombo, and S. Lemmens, "Advancement of a density-based debris fragment model and application to on-orbit break-up localisation," 36th IADC Plenary Meeting, 2018.

[10] A. Muciaccia, M. Romano, C. Colombo e M. Trisolini, "In-orbit fragmentations localisation: study and characterisation of the events", 16th International Conference on Space Operations, Cape Town, South Africa, 2021.

[11] L. Dimare, S. Cicalo, A. Rossi, E. M. Alessi, and G. Valsecchi, "In-orbit fragmentation characterization and parent bodies identification by means of orbital distances," in Proceedings of the 1st International Orbital Debris Conference (IOC), 2019.

[12] Su, S.-Y., and Kessler, D., "Contribution of explosion and future collision fragments to the orbital debris environment", Advances in Space Research, Vol. 5, pp. 25-34. https://doi.org/10.1016/0273-1177(85)90384-9

[13] Flegel, S., Gelhaus, J., Möckel, M., Wiedemann, C., and Kempf, D., "Maintenance of the ESA MASTER Model", Final Report, 2011

[14] ESA, Discosweb, https://discosweb.esoc.esa.int/objects.

[15] Wayne Nelson (1969) Hazard Plotting for Incomplete Failure Data, Journal of Quality Technology, 1:1, 27-52, DOI: 10.1080/00224065.1969.11980344 https://doi.org/10.1080/00224065.1969.11980344

[16] L. Giudici, M. Trisolini, Colombo, "Phase space description of the debris' cloud dynamics through continuum approach", in: Proc. 73rd International Astronautical Congress, Paris, France, 2022.

[17] T. Kelso, F. Hoots and R. Roehrich, ""Spacetrack report no. 3-models for propagation of norad element sets,"," NASA, Tech. Rep,, 1988.

[18] C. Colombo, "Planetary orbital dynamics (PlanODyn) suite for long term propagation in perturbed environment," in Proceedings of the 6th International Conference on Astrodynamics Tools and Techniques (ICATT), Darmstadt, Germany, Mar. 2016.

[19] V. Zappala, A. Cellino, P. Farinella and Z. Knezevic, "Asteroid families. I. Identification by hierarchical clustering and reliability assessment," The Astronomical Journal, vol. 100, p. 2030-2046, 1990. https://doi.org/10.1086/115658

[20] G. F. Gronchi, "An algebraic method to compute the critical points of the distance function between two keplerian orbits," Celest. Mech. Dyn. Astron, vol. 93, pp. 295-329, 2005. https://doi.org/10.1007/s10569-005-1623-5

[21] NASA, "Proper implementation of the 1998 NASA breakup model," Orbital Debris Quarterly News, vol. 15(4), p. 4-5, 2011.

Aerospace Science and Engineering - III Aerospace PhD-Days   Materials Research Forum LLC
Materials Research Proceedings 33 (2023) 68-75   https://doi.org/10.21741/9781644902677-11

Innovative design methodology with LTO noise prediction capabilities for future supersonic aircraft

Grazia Piccirillo[1,a] *

[1] Politecnico di Torino, Department of Mechanical and Aerospace Engineering, Corso Duca degli Abruzzi 24, 10129, Torino, Italy

[a]grazia.piccirillo@polito.it

Keywords: LTO Noise, Conceptual Design, Supersonic Aircraft, MORE&LESS Project

Abstract. To bring supersonic flight back to reality, it is crucial to demonstrate that future supersonic aircraft can reduce their environmental impact compared to the past. In line with this effort, the EU-funded MORE&LESS project is reviewing the environmental impact of supersonic aviation, by applying a multidisciplinary holistic framework to help check how enabling technologies of supersonic aircraft, trajectories, and operations comply with environmental requirements. The present study is part of this project and focuses on updating the traditional conceptual design process with methods for estimating the noise generated at airport level by supersonic aircraft. To achieve this, it is necessary to include some basic capability in the design flow, such as modelling for aircraft noise prediction and simulation of take-off and landing procedures. The approach used to consider these two elements is described in this paper. Preliminary results from Concorde-like case study are presented.

Introduction

It is well known that environmental impact was one of the most relevant factors leading to the retirement of Concorde, the first supersonic airliner. High fuel consumption, high noise levels, and the emission of nitrogen oxides (NOx) into the atmosphere raised concerns about public health and air pollution, causing Concorde to restrict its routes. Furthermore, the Concorde accident in 2000 reinforced concerns about the safety of the aircraft, bringing to the decision to permanently withdraw it from service in 2003.

Since then, significant progress has been made to better understand the environmental footprint associated with SuperSonic Transport (SST) operations [1, 2]. Consequently, research into the design of a new generation of greener supersonic aircraft have become more concrete in recent years. NASA is leading an experimental supersonic aircraft project called the "X-59 QueSST" to reduce noise pollution caused by the sonic boom [3]. Additionally, private companies, such as Boom Supersonic, are developing commercial supersonic aircraft projects for long-haul air travel [4]. Alongside these developments, regulatory efforts are being made to address the environmental impacts of supersonic aircraft. The International Civil Aviation Organization's Committee on Aviation Environmental Protection (CAEP) is collaborating with industry and research organizations to update existing environmental regulations and improve knowledge of the environmental consequences of introducing supersonic aircraft [5].

Aligned with this undertaking, the EU-funded MORE&LESS project aims at supporting Europe to shape global environmental regulations for future supersonic aviation [6]. The main goal is to review the environmental impact of supersonic aviation by applying a multidisciplinary holistic framework, called ESATTO, to help check how enabling technologies of supersonic aircraft, trajectories, and operations comply with environmental requirements. The routines and software tools used from aircraft design to trajectory simulation activities will be updated to extend their field of application to supersonic aircraft (considering the exploitation of both biofuel and liquid hydrogen as fuel) and then integrated into this unique framework. In this way, the ESATTO

Aerospace Science and Engineering - III Aerospace PhD-Days   Materials Research Forum LLC
Materials Research Proceedings 33 (2023) 68-75   https://doi.org/10.21741/9781644902677-11

framework will encompass different disciplines (aerodynamics, propulsion, aeroacoustics, pollutant emissions and environmental impact) and their mutual relationships, thus allowing to perform a multidisciplinary optimization of supersonic aircraft' trajectories and operations. Recommendations to suggest new guidelines to allow for the introduction of supersonic aircraft with the least possible environmental impact will follow consequently.

The initial stage of this multidisciplinary optimization approach is the design of the reference supersonic concept. To make the whole process more effective, an upgrade of the traditional conceptual design methodology could be needed, not only to improve the prediction capabilities of the aircraft performance but also to anticipate environmental impact evaluations and identify the effects of environmental constraints on aircraft design and determine the extent to which sustainability requirements are met or not.

This study addresses this need and contributes to the MORE&LESS project. Precisely, it aims at including new methods and models to support the introduction of noise analysis since the early stages of the design process for the next generation of supersonic aircraft. Therefore, it is focused on the estimation of Landing and Take-Off (LTO) noise at airport level. Indeed, one of the primary issues surrounding supersonic aircraft was the noise impact near airport areas, which far exceeds that of subsonic aircraft. High thrust and speed required by supersonic aircraft taking-off result in higher and unacceptable noise levels. Breakthrough technologies, improved performance and advanced flight procedures are acknowledged as potential measures to mitigate the noise footprint of such aircraft [7, 8]. To enable these assessments as early as possible in the design process, this paper suggests an innovative design approach that includes the capability to predict aircraft LTO noise in order to facilitate the identification of most promising concepts and procedures.

The document refers to ongoing activities and therefore a simplified approach has been adopted at this early stage, which will be refined in due course. Some key points have been addressed, including the implementation of an appropriate model for estimating the noise generated by supersonic aircraft and the evaluation of performance for defining LTO flight paths. The related noise model is based on semi-empirical relationships known in literature. Moreover, the application to the reference case-study will be used to demonstrate its capabilities in the estimation of aircraft emissions noise levels at the three certification measurement points and define possible flight procedures for noise certifications. The major outcome will be the estimation of the noise impact of future supersonic aircraft during the early stages of the design process. Finally, the results will serve to verify the technical feasibility of future supersonic concepts, identifying design guidelines and advanced flight procedures aimed at reducing noise at the airport level.

The method of analysis is disclosed in Section 2. The principles of conceptual design process are recalled and then the main steps of the proposed innovative workflow are described. Hence, the analyses needed to support the noise performance estimation will be described. After that, the method used to assess LTO noise in conceptual design is presented, specifying details about the aircraft noise model and its future updates. Then, further information about the flight procedure modelling is included. Lastly, Section 3 deals with the description of the current results and the future expected outcome. Conclusions are drawn in Section 4.

Method of analysis

To enable the environmental assessment of supersonic aircraft during the design process, it is necessary to review the traditional workflow, first verifying the ability to provide all necessary inputs related to vehicle geometry, operations, and performance. In doing this, the integrated multidisciplinary methodology for the design of high-speed aircraft adopted in ASTRID-H has been taken as a reference [9].

Once the overarching process has been described, the proposed method to assess aircraft noise levels at this stage of the project is disclosed. Supersonic aircraft noise model is the core of this method of analysis. During the initial phase of the activities, this model will be maintained as simple as possible, using semi-empirical methods found in literature with low accuracy for supersonic aircraft case-studies. Then, the model will be improved benefitting from the results of high-fidelity aeroacoustics simulations for the prediction of jet noise. Another relevant element is the introduction of operational procedures modelling capability for the selected case-study. This will open the possibility for the definition of advanced noise reduction procedures for novel aircraft concepts. The final objective of the described method is to evaluate noise levels at the three certification measurement points defined by ICAO. Details available at this stage of the study regarding the proposed approach, the models implemented, and the input/output data exchanged are reported in the next paragraphs.

Upgraded conceptual design process
Addressing the design of future sustainable supersonic aircraft requires the consideration of different and improved methods and models than those typically used for subsonic aircraft. However, from the earliest stages of design, several problems can be encountered, including the lack of reference data for preliminary aircraft sizing and engine characterization or adequate simplified models for evaluating some of the most critical phases of the mission (e.g., take-off, supersonic cruise). In the frame of this work, the methodology implemented in the conceptual design tool to support high-speed vehicles design developed at Politecnico di Torino, ASTRID-H, will be used as a reference, with the aim of integrating the developed models for environmental impact analysis in an updated version, helping to move towards ASTRID-H 2.0.

ASTRID (Aircraft on-board Systems sizing and TRade-off analysis in Initial Design) is a proprietary tool of the research group of Politecnico di Torino and it has been developed for almost a decade through research activities. This tool allows to carry out the aircraft conceptual and preliminary design, the sizing and integration of subsystems for a wide range of aircraft, from conventional to innovative configurations, mainly in the subsonic and low supersonic speed regime. Then, ASTRID-H is an extension dedicated to high-speed vehicle applications. Based on its structure, our focus is the conceptual design phase at vehicle level. To better characterize supersonic aircraft improvements in aerothermodynamic and propulsive modelling capabilities are ongoing, with the inclusion of medium to high fidelity routines. Specifically, these routines will contain surrogate models relying on more accurate databases for aircraft performance prediction. The surrogate models will be integrated in the conceptual design flow to refine the first guess data obtained during the first iteration loop for the conceptual design of the aircraft, ensuring reliable output data. The iterative procedure will serve to improve the thrust requirement estimation derived from the Matching Chart analysis and to provide a preliminary mission profile (Fig. 1). The new tool version will be tailored to civil supersonic concepts covering the entire range of supersonic aircraft speeds from Mach 2 to 5.

In this context, the objective is a further updating of the methodology with the introduction of environmental targets as high-level requirements. Of course, for this evaluation to be effective, it is necessary to include in the process relationships, models or methods that allow even an initial assessment of these aircraft characteristics. The surrogate models and the definition of a preliminary mission profile will support the fulfilment of this need. Environmental constraints for supersonic aviation may concern engine emissions, sonic boom and LTO noise. Regarding LTO noise, so that a complete analysis can be carried out towards a noise assessment at airport level, at least three elements must be introduced: an aircraft noise model (Aircraft noise model), the simulation of take-off and landing flight procedures (Departure/approach flight path), and the assessment of noise perceived on the ground at certification points (Noise at certification points), as indicated in the green box in Fig. 1.

Aerospace Science and Engineering - III Aerospace PhD-Days   Materials Research Forum LLC
Materials Research Proceedings 33 (2023) 68-75   https://doi.org/10.21741/9781644902677-11

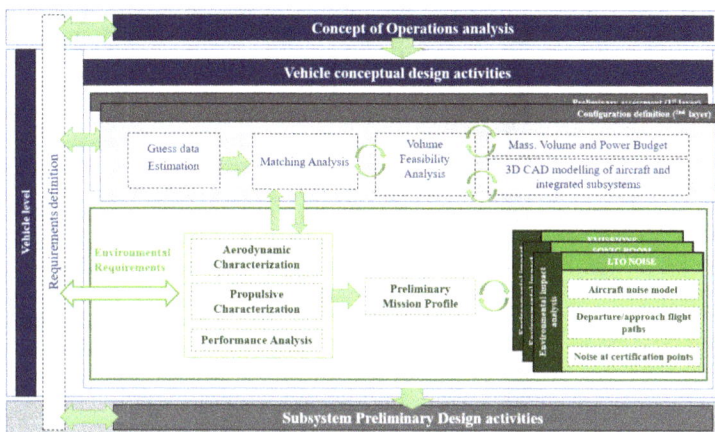

Fig. 1 – Upgrade of the rapid aircraft modelling methodology implemented in ASTRID-H for conceptual design of future supersonic aircraft.

LTO noise assessment

The simplified methodology for the assessment of LTO noise at conceptual design level to be applied is described in Fig. 2. The first step involves the characterisation of an acoustic model that is appropriate for supersonic aircraft. This aircraft noise model has to be based on a mathematical formalism that is flexible and applicable at a design stage when detailed information about the aircraft is not yet available. Such a feature has been met by the semi-empirical noise source models available in literature. At this early phase of study, the relationships present in the early versions of the Aircraft Noise Prediction Program (ANOPP) developed by NASA around the 1970s were applied [10]. Overall aircraft noise is predicted as an assembly of major noise sources, each modelled with an individual semi-empirical noise source model. Specifically, the considered contributions are related to airframe noise and engine noise; then, these major noise sources are further decomposed in wing, landing gear and vertical tail for airframe noise, while jet (considering both mixing and shock-associated noise) and fan for engine noise. The reference equations can be found in [11]. The implemented model will give as output the mean-square acoustic pressure (p^2) as a function of directivity angles and frequency for each of the noise components and for total acoustic pressure, obtained by summing each of the individual ones. The Sound Pressure Level (SPL) can be easily computed from it (Eq. 1).

$$SPL = 10 \; \log_{10} \left(\frac{p}{p_0} \right)^2 \tag{1}$$

With $p_0 = 0{,}000002$ Pa minimum audible sound pressure for the human ear. Although the implemented model is not highly accurate for supersonic case study applications, it proved to be reliable for general evaluations in the early stages of the project. To overcome this limitation, a study is underway to update the model by modifying the empirical parameters used to estimate jet noise, the dominant noise source, by the comparison with the results of simulations performed with more accurate models.

Aerospace Science and Engineering - III Aerospace PhD-Days   Materials Research Forum LLC

Materials Research Proceedings 33 (2023) 68-75   https://doi.org/10.21741/9781644902677-11

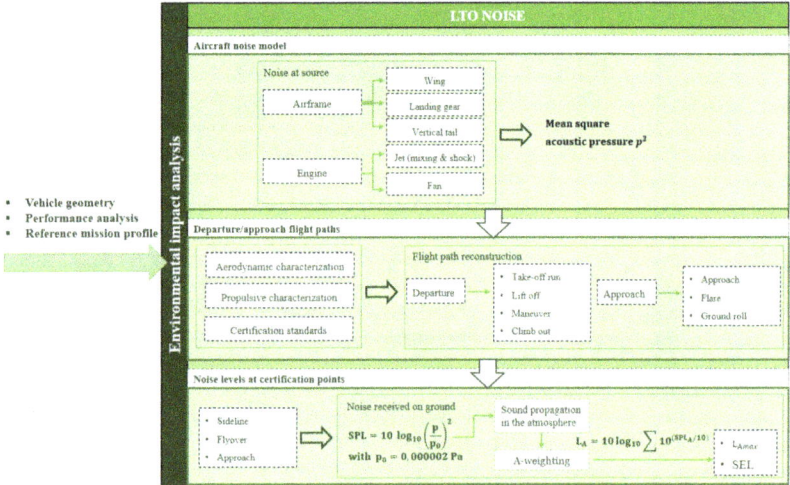

Fig. 2 – LTO noise assessment methodology

The calculated SPL refer to the noise at source. To predict the noise received on ground, flight paths for the LTO phases must be defined. Reliable input data are required to estimate the main operating parameters of the aircraft along defined trajectories (e.g. aerodynamic coefficients, thrust, speed). In this methodology, LTO trajectories are reconstructed from the segmentation of the flight path into sections with constant thrust and speed. This method allows a rapid evaluation of aircraft performance. The next step consists of selecting the noise measurement points defined for certification purposes, depicted in Fig. 3, and defined as follows:

1)   Sideline (full-power reference noise measurement point): the measurement point is along the line parallel to the axis of runaway centre line at 450 m, where the noise level is maximum during take-off.

2)   Flyover or Cutback (intermediate-power reference measurement point): the measurement point is along the extended runaway centre line at 6500 m from the start to roll.

3)   Approach (low-power condition): the measurement point on the ground it is along the extended runaway centre line at 2000 m from the threshold. This corresponds to a position 120 m vertically below the 3° descent path originating from a point 300 m beyond the threshold.

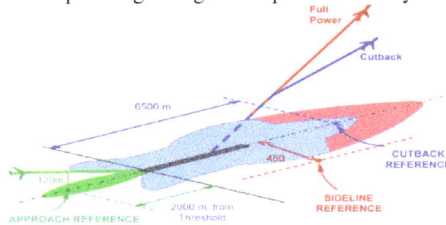

Fig. 3 - ICAO certification noise measurement points

Then, the received noise on ground is predicted considering the sound attenuation due to the propagation in the atmosphere (according to SAE ARP 866 B). Finally, the overall sound pressure level (L_A) is predicted from the A-weighted SPL (SPL_A) as indicated in Eq. 2:

Aerospace Science and Engineering - III Aerospace PhD-Days   Materials Research Forum LLC
Materials Research Proceedings 33 (2023) 68-75   https://doi.org/10.21741/9781644902677-11

$$L_A = 10\log_{10}\sum 10^{(SPL_A/10)} \tag{2}$$

Consequently, both related A-weighted single event L_{Amax} and exposure noise metrics SEL are derived from it and used as evaluation variables to verify the noise requirements.

Results

Preliminary results that can be obtained from the proposed methodology are shown in this section.

For the aircraft that will be the subject of this study, performance has been estimated for the definition of standard take-off and landing trajectories. Precisely, the considered aircraft is Case-Study 1a (CS1a), that is the reference concept for Mach 2 in the MORE&LESS project. This vehicle has a conventional supersonic aircraft with Concorde-like configuration and propelled with biofuel. High-level requirements are listed in Tab. 1, while aircraft configuration is in Fig. 4. Medium to high fidelity datasets for the aerodynamic and propulsive characterization are available for this vehicle. Therefore, this data has been used to verify the take-off and landing performance.

Tab. 1 – High level requirements for the selected case study

High-level requirements for Case Study 1a (CS1a)	
Mach @ cruise	2
Range	7200 km
Payload	120 pax
Fuel	Biofuel

Fig. 4 – CS1a configuration

Different flight paths based on a standard flight procedure have been reconstructed. The maximum thrust per engine is 90 kN, and a thrust rating of 65% has been considered for the climb-out phase. Some performance output values are listed in Tab 2. Otherwise, performance data for approach phase are in Tab. 3. The final trajectories are reported in Fig. 4.

Tab. 2 – Performance variables for take-off flight path

C_{Lmax}	V_{LOF}	L/D_{climb}	γ_{climb}	R/C	V_{climb}	Thrust rating
[-]	[m/s]	[-]	[deg]	[m/s]	[m/s]	[-]
0.959	114.08	8.6	$\cong 1$	2.08	128	100 % (TO) 65% (CL)
1.171	103.24	8.6	$\cong 1$	2.08	128	100 % (TO) 65% (CL)
1.309	97.65	8.6	$\cong 1$	2.08	128	100 % (TO) 65% (CL)
1.384	94.95	8.6	$\cong 1$	2.08	128	100 % (TO) 65% (CL)

Tab. 3 – Performance variables for approach flight path

M_{Lan}	V_{APP}	V_{TD}	$L/D_{descent}$	$\gamma_{descent}$	Landing distance	Thrust rating
[kg]	[m/s]	[m/s]	[-]	[deg]	[m]	[-]
65 % M_{TOW}	111.85	98.94	6.83	3	3282	30 % (APP)

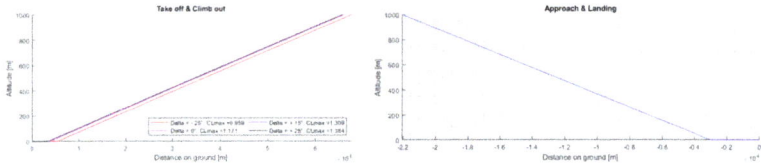

Fig. 5 – Take-off and Landing flight paths

To provide an indication of what is expected from the application of the proposed methodology, the results derived from Concorde case-study are shown in Fig. 6 from [11]. Noise levels in terms of L_{Amax} at the three certification points defined by ICAO will be derived from the simulation of take-off and landing trajectories. These will then be used to compare the values obtained with the current noise limits applied for supersonic aircraft.

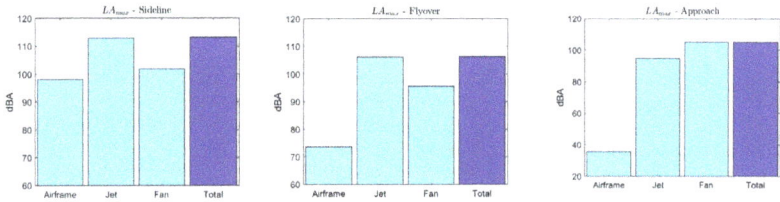

Fig. 6 – Noise levels at the three certification measurement points [10]

Conclusions

An approach for upgrading traditional design process for supersonic aircraft towards an innovative framework has been presented. Specifically, the paper focused on the introduction of LTO noise emissions estimations during conceptual studies. Simplified models and methods have been included for supersonic aircraft noise prediction and flight procedures modelling. Supersonic aircraft noise has been predicted using semi-empirical method well-established in literature, which consider the main aircraft noise sources as independent noise sources. The aircraft flight paths have been simulated dividing the aircraft trajectories in segments at constant thrust and speed. In this way, the aircraft noise levels at the three certification measurement points can be rapidly estimated. The case study to apply the approach is a supersonic aircraft with a Concorde-like configuration propelled with biofuel. Preliminary LTO trajectories have been calculated based on which noise levels will be estimated. To give an idea of the expected results, the estimations obtained considering the Concorde have been reported.

From the outcome of the current study, it is apparent that the conventional Concorde-like design is required by the incorporation of noise reduction measures from the earliest stage of the project, both from a design and operational point of view. Although these results pertain to ongoing

Aerospace Science and Engineering - III Aerospace PhD-Days   Materials Research Forum LLC
Materials Research Proceedings 33 (2023) 68-75   https://doi.org/10.21741/9781644902677-11

preliminary activities, they demonstrate the potential of introducing a noise estimation methodology in the early stages of the design process, paving the way towards a design methodology that includes environmental analysis of the aircraft. The most significant limitations identified involve the requirement for a certain level of accuracy in input data and the fidelity level of the models used with respect to the considered case study. Therefore, research on possible updates to the methodology is ongoing.

References

[1] S. Candel, «Concorde and the Future of Supersonic Transport,» Journal of Propulsion and Power, p. 10, 2023.

[2] Y. Sun e H. Smith, «Review and prospect of supersonic business jet design,» Progress in Aerospace Sciences, vol. 90, pp. 12-38, 2017. https://doi.org/10.1016/j.paerosci.2016.12.003

[3] L. R. Benson, Quieting the Boom: The Shaped Sonic Boom Demonstrator and the Quest for Quiet, NASA Aeronautics book series, 2013.

[4] Boom Supersonic, «BOOM,» [Online]. Available: https://boomsupersonic.com/overture. [March 2023].

[5] Internation Civil Aviation Organization (ICAO), «ICAO Committee on Aviation Environmental Protection,» in ICAO Environmental Report , 2022, p. 4.

[6] M. Project. [Online]. Available: https://www.h2020moreandless.eu/project/. [March 2023].

[7] J. J. Berton, D. L. Huff, K. Geiselhart e J. Seidel, «Supersonic Technology Concept Aeroplanes for Environmental Studies,» in AIAA, Special Session: Community Noise Impact from Supersonic Transports, Orlando, Florida, 2020. https://doi.org/10.2514/6.2020-0263

[8] J. J. Berton, S. M. Jones, J. A. Seidel e D. L. Huff, «Advanced Noise Abatement Procedures for a Supersonic Business Jet,» in International Symposium on Air Breathing Engines (ISABE), Manchester, United Kingdom, 2017.

[9] D. Ferretto, R. Fusaro e N. Viola , «A conceptual design tool to support high-speed vehicle design,» in AIAA, Aircraft Concept Design, Tools and Processes III, 2020. https://doi.org/10.2514/6.2020-2647

[10] B. J. Clark, «Computer Program To Predict Aircraft Noise Levels,» NASA Technical Paper 1913, 1981.

[11] G. Piccirillo, R. Fusaro, N. Viola e L. Federico, «Guidelines for the LTO Noise Assessment of Future Civil Supersonic Aircraft in Conceptual Design,» aerospace, 2022. https://doi.org/10.3390/aerospace9010027

Aerospace Science and Engineering - III Aerospace PhD-Days
Materials Research Proceedings 33 (2023) 76-82

Materials Research Forum LLC
https://doi.org/10.21741/9781644902677-12

Vibro-acoustic analysis and design optimization to improve comfort and sustainability of future passenger aircraft

Moruzzi Martino Carlo[1,a] *

[1]Università di Bologna, Italy*

[a]martinocarlo.moruzz2@unibo.it

Keywords: Aircraft Noise, Noise Reduction, Vibro-Acoustics, Carrera's Unified Formulation

Abstract. The challenges of recent years lead to the development of greener aircraft. However, the concept of environmental sustainability cannot be declined on its own; economic sustainability is also required by increasing the services provided to the passenger, particularly comfort. This work aims to propose a design process for a new sustainable aircraft that takes into account both environmental sustainability, by reducing the aircraft's emissions with a new configuration, and economic sustainability, by introducing a series of solutions to increase comfort during flight, with particular attention to acoustic comfort, i.e. internal noise.

Introduction

Recent social developments have brought new demands to the design of commercial passenger aircraft. The environmental aspect has become paramount, and the new aircraft of the future must be low emission. At present, the most promising technologies, in addition to aircraft optimization, see a complete overhaul of the current configuration: a cylindrical fuselage with lifting wings and a combustion-based propulsion system. Major research involves completely different configurations (blended wing body aircraft, etc.) and new propulsion systems (hydrogen, biofuel). However, these technologies, although certainly with very low emissions, lead to an increase in the aircraft's production and operating costs. An increase that has repercussions on airlines and thus on ticket prices for users. A further addition to the sustainability paradigm becomes necessary in order to make these new technologies available to all: which cannot only be environmental, but also economic. This aspect can be developed either by trying to reduce the cost of these technologies or by offering a better flight experience to the passenger, e.g., by increasing the comfort of the journey. In this paper we focus on the second aspect, which in addition to containing the seemingly inevitable increase in ticket costs due to the change in technology,[1,2] leads the aircraft to suffer less competition from other means of transport, such as high-speed trains or road transport, particularly for regional or short-haul routes.

The main location of the flight, and therefore what we will be dealing with, is the passenger cabin. Comfort during a flight depends on several factors and can be divided according to [1] into visual, interaction, postural, acoustic, or thermal comfort. Each of these depends on several factors, the former on ergonomics, interior design and lighting, acoustic comfort by the noise level and its spectrum, thermal comfort, often coupled with acoustic, is a function of cabin temperature and its variation. In this work, the focus has been on acoustic comfort, and thus on interior noise. In fact, several improvements can still be achieved in this field, for example by applying new generation absorbent materials. Furthermore, the probable innovation in the aviation sector will lead to a total

---

[1] https://www.iea.org/reports/the-future-of-hydrogen

[2] https://www.iea.org/data-and-statistics/charts/fossil-jet-kerosene-market-price-compared-with-hefa-aviation-biofuel-production-cost-2019-2020

Aerospace Science and Engineering - III Aerospace PhD-Days   Materials Research Forum LLC
Materials Research Proceedings 33 (2023) 76-82   https://doi.org/10.21741/9781644902677-12

change in the aircraft, and acoustic studies will be necessary to understand the new acoustic behavior of the aircraft.

The internal noise problem in aircraft can be broken down into three issues:

- how to evaluate this problem, therefore which parameters must be used to quantify the noise, understood not simply with sound, that is an acoustic pressure, but as a disturbance for the passengers;

- how to calculate numerically or experimentally, if possible, this noise inside a complex structure such as that of an aircraft;

- how to reduce, once quantified, this noise, without significantly changing the performance of the aircraft, i.e., by increasing its weight.

In design phases comfort evaluation can follow two different paths to measure it [2]. A human model-based measurement considers the human as a model, which generally has some maximum, minimum and average parameters. These can be obtained from physical and physiological laws (e.g., the size of a passenger for ergonomics or the maximum loud noise) or from statistical survey proposed to the users (the passengers, the crew, and the pilots). The advantage of this approach is the simplicity, and the fact that it can be applied at any step of the design process. However, these parameters do not consider subjective measurements and are unable to capture user's perception in terms of emotions. In order to improve comfort, it is important to have a user centered design process. This is possible through cabin or interior mock-up, where the user experience is evaluated with questionnaires surveys using psychometric scales. In the preliminary design phases, the main issue is the availability of large and detailed mock-ups. Nonetheless, new technologies as virtual reality and augmented reality can create virtual mock-ups in a more flexible and cheaper way, allowing to switch and submit to the user several design concepts and possible improvements [3]. In noise evaluation the human model-based measures are the standard practice, and they are used in this work. Usually, a noise level above 85 dB must be avoided for health reasons (the level is lowered for long exposure, as in residential area), we measure the noise in the possible positions of passengers during flight and we apply filters to acoustic pressure to take in account the human's ear sensibility (the A, B, C and D filters). A concept for human centered design process for noise evaluation is proposed in [4, 5].

The second open question can be simplified: studies of an experimental nature on complete or partial aircraft are almost impossible in the preliminary phases of the design, while in the more advanced phases they remain very expensive and complex, limiting the flexibility of the design process. Therefore, they are recommended only in the last stages. Numerical analyses are very useful; however, it is necessary to clearly define the methods used and their accuracy linked to the computational cost. Traditionally, the noise problem in the frequency domain is studied through Finite Element Method (FEM) and Boundary Element Method (BEM) when dealing with large acoustic cavities, for low frequency. High frequency problems are solved through Statistical Energy Analysis (SEA) methods. Middle frequencies are an open issue and hybrid models can be exploited. In this work we focus on low frequency noise, which is difficult to absorb or to stop with conventional material and acoustic solution. The method to study the problem and develop solutions is FEM. Nevertheless, in order to increase the accuracy of the problem and decrease the computational effort, the numerical analyses are carried inside the Carrera's Unified Formulation (CUF) framework, which exploits a class of powerful shell and beam theories [6].

The reduction of perceived noise can follow different paths: active and passive solutions have been proposed and studied for the noise aircraft problem [7, 8]. A very promising technology is Acoustic Metamaterials (AMMs). These materials provide optimal sound absorption properties in the chosen frequency ranges. They are composed of a host material, such as foam, and inclusions of another material. By playing with the size of these inclusions and their position, the acoustic and mechanical properties of the AMM are modified. This can be studied at a preliminary level in the CUF frame and then applied inside the passenger cabin and the aircraft itself, for example in the lining panel separating the fuselage and cabin.

In conclusion, alongside the more traditional elements in the design of a commercial aircraft, such as safety and performance, there is also sustainability, which to be truly sustainable must include both environmental sustainability, due to concerns about global warming and increased pollution in certain areas, and economic sustainability, in order to ensure the survival of the aviation market. If an increase in ticket prices might be inevitable by changing the propulsion system completely, i.e., switching to hydrogen or biofuel, it is possible to offer the user a better service, increasing comfort during the flight. The first step, proposed in this work, focuses on acoustic comfort and thus the reduction of noise in the passenger cabin.

CUF integration

In order to be used in the vibro-acoustic field, the CUF requires some integration, both in the formulation itself and in the procedure part within the software in which the formulation is developed, MUL2. In the CUF framework the fluid-structure coupling is integrated within the fluid matrices in terms of fundamental nuclei [6]. The major changes are summarized below:

- vibro-acoustic validation [9];

- acoustic boundary conditions and source [10];

- new adaptive finite elements [11], wh1ich allow to study complex geometries and variable thickness plates. Moreover, this concept lays the foundation for the development of non-homogeneous interfaces.

These integrations give the possibility to study vibro-acoustic problems in simple and complex structures in order to develop solutions at a preliminary level for noise reduction or to understand the spread of noise and vibration in advanced multi-layer structures. Other developments are being studied, regarding the concept of Adaptive Finite Element based on [12, 13].

Noise reduction solutions: AMM

Low frequencies, although attenuated by the human ear, are complex to absorb. This can be explained by the following simple relationship, valid for a plate, where the Transmission Loss TL is proportional to the thickness d and density ρ of the plate and the angular frequency $\omega = 2\pi f$ (or to the frequency f) of the acoustic signal:

$$TL \propto d \cdot \rho \cdot \omega \tag{1}$$

Therefore, for low-frequency noise, in order to have a high TL, it is necessary to use materials with high density or thick plates. Both cases result in an increase in system weight, which in the aeronautical field is to be avoided, as it is linked to an increase in fuel consumption.

Unconventional materials, such as AMMs, can achieve high TL values even in low frequency ranges due to their properties [14]. In this work, two types of AMMs are studied:

- a first one developed within the CUF and made by melamine foam with cylindrical inclusions [15], Fig. 1;

- a second produced using additive printing techniques (Fused Deposition Modeling FDM) and numerically studied using a Layer Wise (LW) approach [16], Fig. 2.

Aerospace Science and Engineering - III Aerospace PhD-Days   Materials Research Forum LLC
Materials Research Proceedings 33 (2023) 76-82   https://doi.org/10.21741/9781644902677-12

The AMMs studied exhibit very different behaviors, with the former reaching high TL values below 300 Hz, while the latter strongly attenuates noise around 1000 Hz, thus now in the mid-frequency range. In general, an attempt is made with these materials not to increase the weight of the system compared to conventional solutions applied in the cabin lining panel.

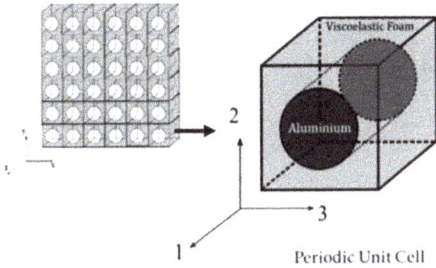

Figure 1: the first AMM used to reduce noise and developed in [15]

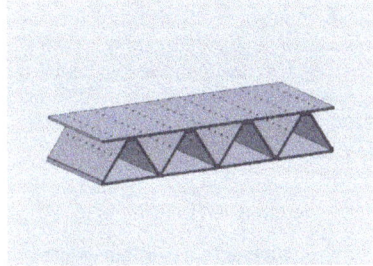

Figure 2: the second AMM produced with additively produced (FDM) in [16].

The aircraft passenger cabin

The passenger cabin is the center of the comfort problem, being the main, if not the only, place the user frequents. This is why, before moving on to its vibro-acoustic study, it is necessary to give a brief description from an acoustic point of view, within the fuselage, which holds the cabin. As described in Fig. 3 the cabin-fuselage system is composed by:

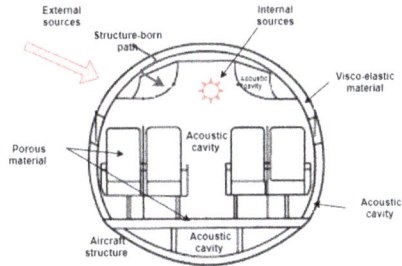

Figure 3: The sketch of a section of the fuselage and passenger cabin from an acoustic point of view. The different vibro-acoustic components and phenomena are underlined in order to show the complexity of the problem.

- the primary structure includes the aircraft skeleton, so the stringers, the frames and panels of the fuselage, the passenger and cargo decks supports and their panels on which the floor is placed, the bulkhead and wing box. Moreover windows frames, first glass pane and reinforcements are included in this subsystem;

- the secondary structure includes seats with their supports, lining, ceiling and dado panels of the cabin, overhead luggage compartment and luggage compartment in the cargo hold;

- cabin interiors, as the floor carpet, the galleys and the toilets;

- several subsystems, as the air conditioning and pressurization system, the electrical system, the thermal insulation system, etc;

- passengers, crew and luggage;

- acoustic cavities, in the fuselage there are several cavities filled by air, such as the passenger cabin, the cargo hold, the gap between the lining panel and the fuselage, the small air gap in the windows.

From an acoustic point of view, we must understand the contribution of each subsystem. Nevertheless, the high complexity of the system requires some degree of simplification, and some components or subsystems are neglected or simplified.

The passenger cabin FEM model generation

The above description shows the need to simplify the vibro-acoustics model of the fuselage and cabin, eliminating those subsystems that do not counteract the N&V point of view (at least apparently) and simplifying others, the description of which would be too complex, or because there is no detailed information about them. Finally, possible external and internal sources must also be considered. Again, in order to avoid over-complicating the model, it is necessary to include them as external sources, i.e., structural or acoustic loads. Therefore, the major limitations in generating the FEM model are:

- lack of adequate information, either because this is not available or because the systems in question have not yet been fully define in the design process;

- computational cost, in dynamic problems, increasing the maximum frequency of the problem is inversely proportional to the size of the elements, so increasing the maximum frequency increases the number of DoF;

- accuracy of analysis, some types of structures require advanced approaches and in general special attention must be paid to the size (1D, 2D or 3D) of the elements in the various components (for example, the lining panels require a 3D formulation for the core or a 2D-LW approach, the stringers, and frames shell elements in order to avoid a numerical increase in the stiffness).

Sensitivity study

Once the FEM model has been constructed, a sensitivity study can be carried out to assess how the internal noise varies by varying the configuration of the structure, inserting new acoustic solutions, such as the AMMs proposed here, and varying the acoustic sources.

The results partially reported in [17, 18] show the potential of AMMs with an important noise reduction. Furthermore, this reduction was demonstrated for different fuselage models. As an example, the sound pressure maps in the passenger cabin applying the AMM in [15] are shown in Fig. 4. The results are calculated at ear height of seated passengers. However, the computational cost remains a significant constraint in this type of analysis. Obviously, for the study of new configurations such as in [17] it is necessary to consider the extended concept of environmental and economic sustainability, as reported for a windowless configuration in [19]. Alongside a reduction in fuel consumption, the acoustic behavior did not change and therefore additional acoustic solutions were required, i.e., AMMs.

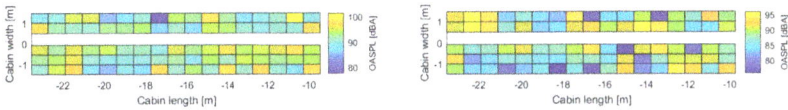

Figure 4: The Overall Pressure Level (OASPL) [dBA] maps on the positions of the seated passenger's heads in fuselage under an external complex pressure field [17]. The cabin of the model is composed by eighteen rows with five seats each one (from the bow on the right to the aft on the left). There is an increase in OASPL near the propeller position between row 2 and 5. (a) Lining panel core in Nomex in traditional fuselage. (b) Lining panel core in AMM in traditional fuselage.

Conclusions

In conclusion, the increase of comfort, including acoustic comfort, for the passenger during the flight can lead to an increased competitiveness of air transport with respect to other means of transport. Therefore, it is necessary to propose new solutions for noise reduction and tools to characterize noise development in the cabin. Therefore, it is necessary not only to study new solutions for noise reduction, such as AMMs, but to define a design process for the study of noise and vibration. This vibro-acoustic analysis must take into account the new innovative materials, both used in aeronautics and designed to reduce noise, and therefore use advanced approaches such as those available thanks to CUF (as Layer Wise approaches). In fact, this formulation makes it possible to save significant degrees of freedom and thus raise the maximum frequency of the problem or include new elements, without having a decrease in accuracy. In addition, the lower computational cost of analysis can lead to easier integration of aircraft acoustic optimization (also including vibro, aero and thermo-acoustics) in a multi-disciplinary optimization (MDO) process. Furthermore, in the future new tools for assessing the impact of noise on the comfort of passengers must be developed, placing the user at the center of the design process.

References

[1] D.P. Restuputri, K. Purnamasari, N. Afni, S. Legtria, E. Shoffiah, M. Septia, and I. Masudin. Evaluation of aircraft cabin comfort: Contributing factors, dissatisfaction indicators, and degrees of influence. AIP Conference Proceedings, 2453(1):020047, 2022. https://doi.org/10.1063/5.0094297

[2] S Bagassi, F Lucchi, F De Crescenzio, and S. Piastra. Design for comfort: aircraft interiors design assessment through a human centered response model approach. In 31st ICAS 2018 Proceedings, 2018.

[3] F. De Crescenzio, S. Bagassi, and F. Starita. Preliminary user centred evaluation of regional aircraft cabin interiors in virtual reality. Scientific Reports, 11, 05 2021. https://doi.org/10.1038/s41598-021-89098-3

[4] S. Santhosh, M.C Moruzzi, F. De Crescenzio, and S. Bagassi. Spatial sound system to aid interactivity in a human centred design evaluation of an aircraft cabin environment. In XXVI AIDAA Congress Proceedings, 2021.

[5] S. Santhosh, M.C. Moruzzi, F. De Crescenzio, and S. Bagassi. Auralization of noise in a virtual reality aircraft cabin for passenger well being using human centred approach. In 33rd ICAS 2022 Proceedings, 2022.

[6] E. Carrera, M. Cinefra, M. Petrolo, and E. Zappino. Finite element analysis of structures through unified formulation. Wiley, 2014. https://doi.org/10.1002/9781118536643

[7] A. Filippone. Aircraft Noise: Noise Sources, pp. 470-532. Cambridge Aerospace Series. Cambridge University Press, 2012. https://doi.org/10.1017/CBO9781139161893.019

[8] L.L. Beranek. The noisy dawn of the jet age. Sound and Vibration, 41:94-99, 2007.

[9] M. Cinefra, M.C. Moruzzi, S. Bagassi, E. Zappino, and E. Carrera. Vibro-acoustic analysis of composite plate-cavity systems via cuf finite elements. Composite Structures, 259:113428, 2021. https://doi.org/10.1016/j.compstruct.2020.113428

[10] M.C. Moruzzi, M. Cinefra, and S. Bagassi. Analysis of an acoustic monopole source in a closed cavity via cuf finite elements. Aerotecnica Missili & Spazio, Aug 2022. https://doi.org/10.1007/s42496-022-00129-2

[11] M.C. Moruzzi, M Cinefra, S Bagassi, and E. Zappino. Vibro-acoustic analysis of multi-layer cylindrical shell-cavity systems via cuf finite elements. In 33rd ICAS 2022 Proceedings, 2022. https://doi.org/10.1016/j.compstruct.2020.113428

[12] M. Cinefra. Non-conventional 1d and 2d finite elements based on cuf for the analysis of non-orthogonal geometries. European Journal of Mechanics/A Solids, 88:104273, 2021. https://doi.org/10.1016/j.euromechsol.2021.104273

[13] M. Cinefra. Formulation of 3d finite elements using curvilinear coordinates. Mechanics of Advanced Materials Structures, pages 1-10, 2020. https://doi.org/10.1080/15376494.2020.1799122

[14] S. Chen, Y. Fan, Q. Fu, H. Wu, Y. Jin, J. Zheng, and F. Zhang. A review of tunable acoustic metamaterials. Applied Sciences, 8(9), 2018. https://doi.org/10.3390/app8091480

[15] M. Cinefra, G. D'Amico, AG. De Miguel, M. Filippi, A. Pagani, and E. Carrera. Efficient numerical evaluation of transmission loss in homogenized acoustic metamaterials for aeronautical application. Applied Acoustics, 164:107253, 2020. https://doi.org/10.1016/j.apacoust.2020.107253

[16] M.C. Moruzzi, S. Bagassi, M. Cinefra, M. Corsi, and M. Rossi. Design of additively manufactured metamaterial for cabin noise and vibrations reduction. In XXVI AIDAA Congress Proceedings, 2021.

[17] M.C. Moruzzi, M. Cinefra, and S. Bagassi. Vibroacoustic analysis of an innovative windowless cabin with metamaterial trim panels in regional turboprops. Mechanics of Advanced Materials and Structures, 28:1-13, 2019. https://doi.org/10.1080/15376494.2019.1682729

[18] M.C. Moruzzi, M. Cinefra, S. Bagassi, and E. Carrera. Attenuation of noise in the cabin of a regional aircraft by metamaterial trim panels. In 32nd ICAS 20020 Proceedings, 2021.

[19] MC. Moruzzi and S. Bagassi. Preliminary design of a short-medium range windowless aircraft. International Journal on Interactive Design and Manufacturing (IJIDeM), 14(3):823-832, Sep 2020. https://doi.org/10.1007/s12008-020-00676-7

Aerospace Science and Engineering - III Aerospace PhD-Days   Materials Research Forum LLC
Materials Research Proceedings 33 (2023) 83-90   https://doi.org/10.21741/9781644902677-13

Sonic boom CFD near-field analysis of a Mach 5 configuration

S. Graziani

Politecnico di Torino, DIMEAS, corso Duca degli Abruzzi 24, Torino

samuele.graziani@polito.it

Keywords: Sonic Boom, Near-Field Evaluation, CFD

Abstract: The following paper compares experimental results obtained in free flight at Mach 4.7 within the MORE&LESS project of a configuration at Mach 5 with high-fidelity simulations based on CFD and propagation tools. The simulations replicate the flight and environmental conditions of the test days, and the CFD approach is based on dedicated workshops by NASA for accurate near-field study. Measurements are compared with CIRA acoustic microphones and contains four different stations positioned at a maximum of ten meters from the centreline of the trajectory.

Introduction and background

The world of civil aviation has changed dramatically over the past few decades. While traveling thousands of miles in a few hours has become easier and more affordable than a few decades ago, traveling faster than the speed of sound is in the air. The EU-funded MORE&LESS project is reviewing the environmental impact of supersonic aviation by applying a multidisciplinary holistic framework to help check how enabling technologies of supersonic aircraft, trajectories, and operations comply with environmental requirements.

As a result of recent technological advances for the second generation of supersonic civilian aircraft and given the future entry into service of the BOOM Overture aircraft, there is a need to set new standards regarding supersonic flight over the land for civilian aircraft purposes. Since 1973 commercial supersonic overland flight is prohibited in most countries, and the ability to break down this constraint is vital to the commercial success of the second generation of supersonic aircraft. Changing the current ban on supersonic flight overland with an international regulation with a noise emission ceiling is one of the goals of the industry itself. However, there is a need to accurately identify both the methods for calculating the noise emitted on the ground and to determine what the limit of acceptability might be to impose so as not to create excessive annoyance to the population. The following paper demonstrates the veracity of the methodology used to compute the grid by reproducing the experimental tests that were carried out at ISL, which is a partner of the MORE&LESS project in October 2022.

Sonic Boom Description and methodology in conceptual design phase

Any object traveling faster than the local speed of sound generates a disturbance in the atmosphere. Theoretically, for slender configurations, this phenomenon is governed by the linearized supersonic flow theory and computed from the supersonic area rule methods. On the other hand, for blunt bodies, such as a space shuttle, the aerodynamic flow is nonlinear, and the computation of the near field from the theoretical point of view is much more complex than in the previous case. [1,2,3,4]

The disturbance that is generated propagates through the atmosphere: in the region in the vicinity of the aircraft, there is the "near field" zone, in which the signature is a function only of the geometric characteristics and flight conditions of the vehicle and extends for a couple of lengths below the aircraft itself. In this small region, the atmospheric gradients do not have a significant role. In the "midfield" area the signature is a function of both geometric characteristics and disturbances related to atmospheric effects, in which there are significant nonlinear distortions of

Aerospace Science and Engineering - III Aerospace PhD-Days   Materials Research Forum LLC
Materials Research Proceedings 33 (2023) 83-90   https://doi.org/10.21741/9781644902677-13

the signature itself, finally, in the "far field" region the signature is a function of the propagation in the real atmosphere and has the typical N-wave shape. For a classical supersonic aircraft configuration, there is an initial compression at the nose of the aircraft in which the local pressure increases from p_0 by an amount Δp. Following this first compression, there is a slow expansion occurs until there is a pressure value slightly lower than atmospheric, and finally, at the tail, there is a new compression that re-establishes the local pressure value. For a ground observer, the acoustic response of the ear is composed of two different booms as the human ear can detect changes beyond a specific frequency, and it manages to identify sudden changes in pressure. If the interval between those two rapid compressions is below 0.10 seconds, the ear would not be able to distinguish between them, and they would seem like one single sound.

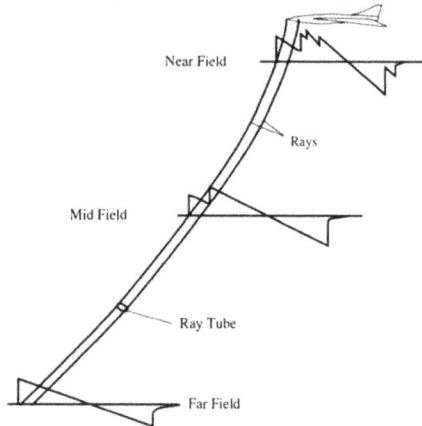

Figure 1: Sonic boom propagation through the atmosphere

In the slender body acoustic limit theory [5] , the near-field pressure can be calculated as:

$$\delta p(x - \beta r, r) = p_0 \cdot \frac{\gamma M^2 F(x - \beta r)}{\sqrt{2\beta r}}$$

(1)

In equation (1) δp is the overpressure with the wave, p_0 is the ambient pressure, x is the axial coordinate in body fixed, γ is the ratio of specific heats, M is the Mach number and $\beta = \sqrt{M^2 - 1}$. The quantity F is the acoustic source strength, it is based on linearized supersonic flow area rule theory and can be evaluated as:

$$F(x) = \frac{1}{2\pi} \int_0^x \frac{A''(\xi)}{\sqrt{x - \xi}} d\xi$$

(2)

In equation (2) A is the cross sectional area of the vehicle along cuts aligned with the Mach angle. Some early methods for studying sonic boom were based on Walkden's theory and involved a simplified study of the atmosphere. One formulation used for volume-induced sonic boom is:

$$\Delta p = K_r K_s \sqrt{p_v p_g} (M^2 - 1)^{\frac{1}{8}} \cdot \frac{D}{l^{\frac{3}{4}}} \cdot r^{-\frac{3}{4}}$$

(3)

Aerospace Science and Engineering - III Aerospace PhD-Days   Materials Research Forum LLC
Materials Research Proceedings 33 (2023) 83-90   https://doi.org/10.21741/9781644902677-13

In equation (3) K_r is the ground reflection factor and it is equal to 2.0, K_S is the aircraft shape factor, D is the aircraft equivalent diameter, p_v & p_g are the ambient pressure at the vehicle altitude and on the ground. The $\sqrt{p_v p_g}$ factor is the consideration to the fact that atmosphere is not uniform, while a complete adjustment for the atmosphere uses the theory of geometric acousticsAnother simplified model for the complete study of sonic boom is given by Carlson's method. It manage to study the sonic boom characteristics concerning both bow shock overpressure and time signature duration for different configurations for aircraft flying at an altitude of up to 76 km.[6]

The method contains many limitations and is easily applicable in the conceptual design phase to get an indication of the order of magnitude of the shock intensity. The methodology is valid for aircraft in level flight or moderate climb or descent flight phases, the effect of flight path curvature and acceleration is neglected, and it is just applicable for the classical N-wave in the far-field region. The formulation for the maximum bow shock overpressure is :

$$\Delta p_{max} = K_p K_r \sqrt{p_v p_g} (M^2 - 1)^{\frac{1}{8}} \cdot h_e^{-\frac{3}{4}} l^{\frac{3}{4}} K_s$$

(4)

Where K_p pressure amplification factor, K_r is the ground reflection factor, h_e is the effective altitude and K_S is the aircraft shape factor. The formulation for the time duration is equal to :

$$\Delta t = K_t \cdot \frac{3.42}{a_v} \cdot \frac{M}{(M^2 - 1)^{\frac{3}{8}}} \cdot h_e^{\frac{1}{4}} \cdot l^{\frac{3}{4}} \cdot K_s$$

(5)

Where K_t signature duration factor and a_v is the local speed of sound.

The core of the methodology is the calculation of the constant related to the shape factor: the first step is the calculation of the equivalent area due to volume, which can be defined with the cross-sectional area of the aircraft along the longitudinal axis. The second step involves the evaluation of the equivalent area due to lift, which can be calculated as the distribution of planform area along the longitudinal axis. The third step is the combination of these two measurements to obtain the total effective area of the aircraft, from which it is possible to go on to derive some parameters for deriving the shape factor, such as maximum effective area $A_{e\,max}$ and the effective length l_e.

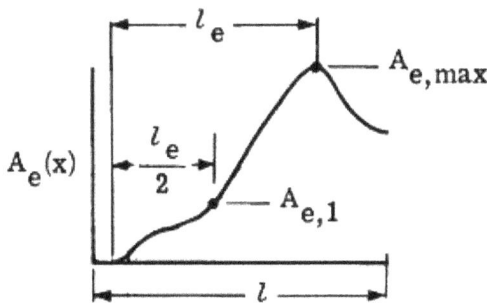

Figure 2 : Calculation of the effective area

In the final step, the aircraft shape factor may it is found by specific shape factor curve with the insertion of appropriate maximum effective area and effective length.

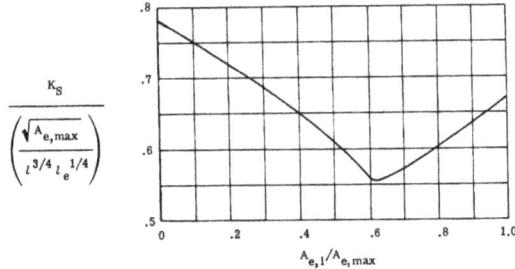

Figure 3 : Shape factor charts

Within the method, all the parameters of equations 4 and 5 can be calculated from dedicated graphs that are a function of Mach and altitude.

CFD mesh

As previously mentioned, supersonic flight over land is prohibited, and for there to be a chance for the economic success of the second generation of the supersonic vehicle there is a need to carefully define a standard in terms of regulations by imposing a maximum noise level. The study of the evolution of sound disturbance is usually divided into different regions to facilitate calculation. The area in the vicinity of the aircraft where shocks are formed and where there are numerous nonlinear phenomena such as shock-shock interactions, shock curvature, and crossflow is evaluated within Computational Fluid Dynamics (CFD). [7] However, it is impossible to study by CFD down to the ground because of the size of the domain and consequently, the large computing power required, so dedicated propagation models exist for detailed study. However, they need as input the results processed in the near-field region obtained by CFD.

In these sonic boom propagation methods, the details of the configuration geometry are less important than atmospheric variations and molecular relaxation phenomena. Particularly for the study of both the near field region and propagation methods, NASA, since 2014 has been conducting dedicated workshops every three years related to the study of these methods and verifying the goodness of the results by comparing the obtained data with wind tunnel values. For the case study, it was decided to follow the directions and suggestions of the last workshop for the creation of the grid: in particular, there is the creation of a hybrid one, with an unstructured core and a second region with a structured mesh. Concerning the structured mesh, there is an extrusion from the unstructured mesh of a series of layers to produce the grid elements that are aligned to the freestream Mach angle μ:

$$\mu = sin^{-1}\left(\frac{1}{M}\right)$$

$$(6)$$

Aerospace Science and Engineering - III Aerospace PhD-Days   Materials Research Forum LLC
Materials Research Proceedings 33 (2023) 83-90   https://doi.org/10.21741/9781644902677-13

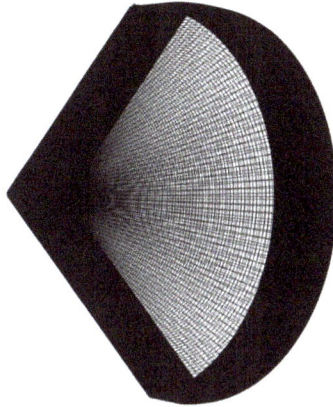

Figure 4 : Example of a CFD sonic boom grid

Case study

The near-field CFD simulations are based on one of the aircraft being studied within the MORE&LESS project and for which experimental data are available. Specifically, the aircraft is an appropriately scaled model of the MR5 aircraft, which consists of a re-design of the MR3 aircraft. For the MR5 aircraft, except for the length, all other dimensions are kept constant to the original configuration. In this way, the layout of the vehicle is modified, since its slenderness parameter is now different.

The final configuration of the MR5 aircraft has a length of 75 meters with a wingspan of 41 meters an MTOW of about 290 tons.

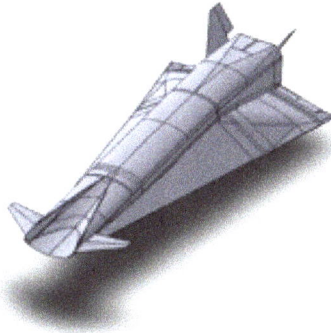

Figure 5 : MR5 aircraft configuration

As already mentioned above, the geometry that was studied by CFD is the same that was used in the experimental tests. This geometry, compared to the original configuration, is modified to avoid asymmetric lifting effects during the free-flight tests. First, the canards and the fins are removed. To maintain the bottom contour, which is responsible for the later investigated sonic-boom signature symmetry is obtained by mirroring the lower part to the top, which leads to a plane-symmetric model with no lift generation at zero angles of attack.

Aerospace Science and Engineering - III Aerospace PhD-Days   Materials Research Forum LLC
Materials Research Proceedings 33 (2023) 83-90   https://doi.org/10.21741/9781644902677-13

The model that is studied has the following characteristics:

I.   Mass equal to 501.7 g
II.   Length of 201.5 mm
III.   CG/Nose of 104.2 mm
IV.   Reference base surface 895.34 mm^2
V.   Equivalent base diameter 33.76 mm^2

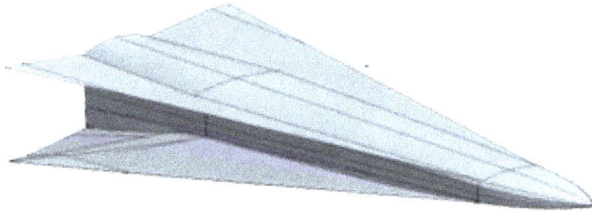

Figure 6 : Test case

The flight conditions that are studied by CFD are those of experimental tests, specifically the Mach number studied is 4.7 and the angle of attack is 0 deg. The geometry is modeled with the CAD program Solidworks and the meshes are generated with ICEMCFD 2020 R2. Regarding the mesh structure, as previously mentioned, the philosophy adopted during the NASA workshops devoted to the study of the near field for the sonic boom was followed. Thus, an inner cylinder formed by an unstructured mesh was constructed, in which the size of the elements is 10^{-3} m for the elements discretizing the surface of the aircraft and $4 \cdot 10^{-3}$ m for the elements of the cylinder itself: the total number of elements in the unstructured mesh is just over 8.5 million.

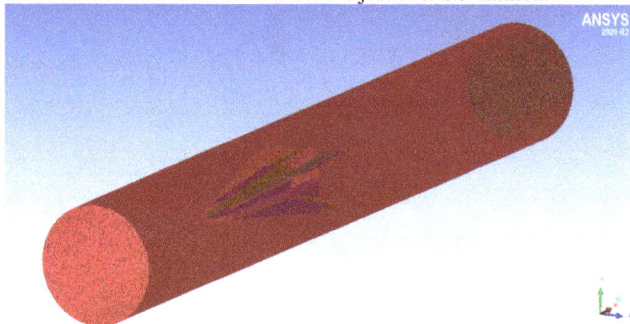

Figure 7 : Unstructured mesh with the particular of the aircraft

As for the generation of the structured mesh, 12 blocks were created for the correct description of the elements at a greater distance from the body.
Care was adopted for the generation of the structured part, to avoid the problem of the mesh interface.
The total number of elements in the structured mesh is just above 15 million.

Aerospace Science and Engineering - III Aerospace PhD-Days   Materials Research Forum LLC
Materials Research Proceedings 33 (2023) 83-90   https://doi.org/10.21741/9781644902677-13

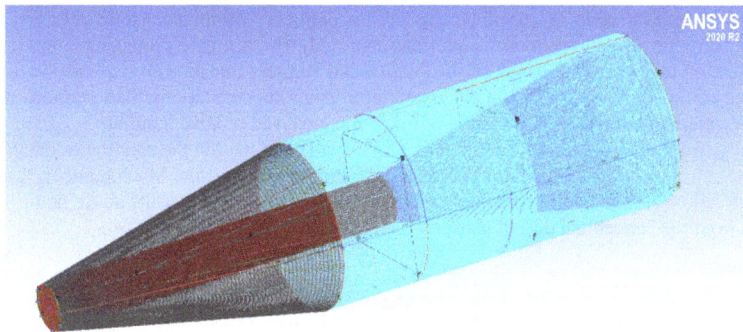

Figure 8 : Structured mesh aligned with Mach angle

Finally, the merge between the two meshes was performed going to create the interface surface, and after performing the necessary checks on the quality of the grid, the total number of elements is about 22.5 million.

Figure 9 : Final mesh

The numerical simulations will be performed with the finite volume code ANSYS FLUENT version 2021R1. As for the simulation, the implicit, density-based solver with double precision is used, and the fluid is considered an ideal gas. The first simulations to be studied are those related to the test carried out at ISL: specifically, the Mach number is kept equal to 4.7. The reference flight altitude was about 2 meters in the tests performed. It will also have the calculation of drag, lift, and moment coefficients to have a comparison with the results obtained through the use of an alternative CFD approach based on only unstructured elements performed by ISL.

Conclusion and future work

This paper aims to demonstrate the veracity of the methodology proposed within the workshops organized by NASA related to the study of sonic boom in the near field region through comparison with experimental data from ISL. Due to the low flight altitude at which the flight tests were conducted, there is no need to adopt a propagation code for the study of the evolution of sonic disturbance in the atmosphere. For future case studies, the following method through CFD will be adopted to study the near field region of other configurations within the MORE&LESS project, and using specific propagation codes, it will be possible to have the creation of databases regarding the sonic boom of different types of aircraft having very different mission profiles and cruise Mach numbers.

Aerospace Science and Engineering - III Aerospace PhD-Days   Materials Research Forum LLC
Materials Research Proceedings 33 (2023) 83-90   https://doi.org/10.21741/9781644902677-13

The propagation code is currently studied by partners within the MORE&LESS project (TUHH) through the "Propaboom" code that can study the acoustic propagation for loudness determination on the ground based on the Augmented Burgers Equation. It requires the nearfield signature obtained by CFD and can evaluate the propagation considering the variations through the atmosphere of temperature, pressure, horizontal winds, and relative humidity.

Finally, from the use of high-fidelity simulations and code, within the creation of databases of numerous aircraft differing in characteristics, configuration, range, and Mach number, there will be the creation of low-fidelity surrogate models suitable for estimating the sonic boom of new aircraft from the conceptual design phases.

References

[1] K. J. Plotkin, State of the art of sonic boom modeling, © 2002 Acoustical Society of America. https://doi.org/10.1121/1.1379075]

[2] D. J. Maglieri, K. J. Plotkin, Aeroacoustics of Flight Vehicles: Theory and Practice. Volume 1: Noise Sources , 1991

[3] Y. S. Pan, W. A. Sotomayer, Sonic Boom of Hypersonic Vehicles, AIAA Journal, 2012, https://doi.org/10.2514/3.50150

[4] R. Cowart, Developing noise standards for future supersonic civil aircraft, 2013 , https://doi.org/10.1121/1.4800343

[5] H.W. Carlson, D.J. Maglieri, Review of Sonic-Boom Generation Theory and Prediction Methods, The Journal of the Acoustical Society of America, 1972, https://doi.org/10.1121/1.1912901

[6] H.W. Carlson, Simplified Sonic Boom Prediction, TP1122, 1978

[7] M.A.Park, R.L.Campbell, A. Elmiligui, Specialized CFD Grid Generation Methods for Near-Field Sonic Boom Prediction, AIAA Journal, 2014

Aerospace Science and Engineering - III Aerospace PhD-Days   Materials Research Forum LLC
Materials Research Proceedings 33 (2023) 91-96   https://doi.org/10.21741/9781644902677-14

Acoustic metamaterial design for aeronautical purposes

Giuseppe Catapane[1,a] *

[1]PASTA-Lab (Laboratory for Promoting experiences in Aeronautical STructures and Acoustics), Università degli Studi di Napoli "Federico II, Via Claudio 21, Napoli, 80125, Italy

[a]giuseppe.catapane@unina.it

Keywords: Acoustic Metamaterials, Sound Absorption, Noise Suppression

Abstract. Labyrinth-shape quarter wavelength tubes are numerically studied under plane wave excitation, with analytical comparison. These labyrinth resonators (LRs) are tuned at 60, 90 and 120 Hz, and their sound absorption response exhibits maximum peak at those frequencies with high fidelity and performance. These objects can absorb tonal sources at very low frequencies, with an incredibly competitive thickness, resulting in the possibility of considering them for the design of acoustic liners for an aerospace engine, but also for the automotive and naval industries. They are put together to form an acoustic metamaterial which exhibits multiple tonal peaks, demonstrating that the performance of each resonator is not affected by their coupling.

Introduction

The importance of environmental noise control has increased with modern urbanization, transportation traffic and associated noise-induced health impairments, such as irritation, sleep disruption, or even ischemic heart disease [1]. In this context, the transportation industry has to cope with several not-trivial compromises without affecting performance of the present models

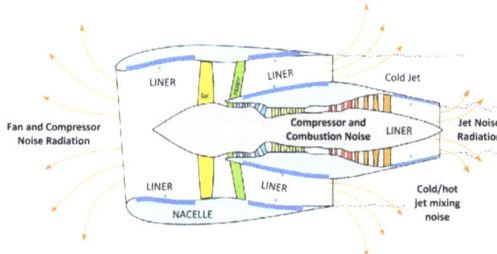

Figure 1: schematization of main noise source caused by an aircraft engine and actual disposition of conventional acoustic liners.

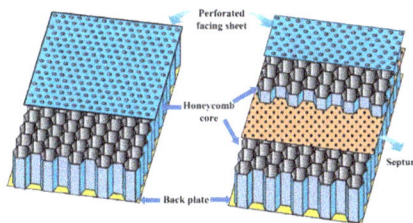

Figure 2: acoustic liner configurations.

Aerospace Science and Engineering - III Aerospace PhD-Days   Materials Research Forum LLC
Materials Research Proceedings 33 (2023) 91-96   https://doi.org/10.21741/9781644902677-14

and/or relative costs. For instance, an aircraft design procedure cannot neglect sound emission considerations and an aircraft will not be sold or be certified if too noisy. Nevertheless, sound reduction systems or sound packages should be as lightweight as possible and placed in available spaces that are usually limited. Acoustic liner represents the most used solution to suppress aircraft engine noise; they are generally placed in the internal sides of the nacelle of a turbofan engine where it is not passing hot air (Fig. 1). An acoustic liner is made by a sandwich structure with one or multiple layers of micro-perforated panels and honeycomb core for structural purpose (Fig. 2). The perforated plate is responsible of acoustic resistance, while cavities are accounted for acoustical reactance [2]. Since each cavity is divided by the others, liners in question are local reacting liners, with the resonance frequency tuned by Helmholtz Resonator formula, which depends on the holes and cavity geometry. The Helmholtz Resonator has just one resonance frequency, hence it cannot be considered as an effective solution for sound suppression, because its peak is narrow and can be affected by flow conditions. At the actual state-of-the-art, there are no available and feasible solutions which guarantee wide bandwidth low-frequency sound suppression with space accounted for acoustic purpose limited and weight increase that should be minimized.

The present study wants to present alternative unitary cells for the design of conventional liners. To this aim, quarter wavelength tubes (QWTs) represent interesting local resonators for low-frequency sound absorption. A quarter wavelength tube is an open-closed tube that has resonant frequencies when its length L is an odd-integer multiple of the quarter of the acoustic wavelength:

$$f_{res} = \frac{(2m - 1)c_0}{4L}. \qquad\qquad m = 1, 2, 3 \ldots \qquad\qquad (1)$$

In correspondence of these multiple resonance frequencies, high sound absorption occurs. To cope with their excessive length requirement for low-frequency application, their channel is stretched into labyrinth branches without affecting the resonance behavior (Fig. 3).

Theoretical background and numerical implementation
Labyrinth-type resonators are designed through analytical and numerical simulations for normal plane wave radiation excitation (PWR). The acoustic impedance of a labyrinth resonator (LR) is studied according to the analytical approach proposed by Magnani et Al. [3], where the labyrinth resonator is evaluated as a perforated plate followed by a QWT. The QWT of length L has an impedance equal to:

$$Z_{QWT} = -jZ_{eff}\cot(k_{eff}L_{eff}), \qquad\qquad (2)$$

modelled through Low Reduced Frequency model (LRF), introduced for the first time by Zwikker and Kosten [4], which takes into account viscous and thermal dissipation changing the sound wave Helmholtz equation. $Z_{eff} = \rho_{eff}c_{eff}$ and $k_{eff} = \omega/c_{eff}$ are respectively the effective impedance and the effective wavenumber retrieved by *lossy Helmholtz equation*, and L_{eff} is the effective

Aerospace Science and Engineering - III Aerospace PhD-Days   Materials Research Forum LLC
Materials Research Proceedings 33 (2023) 91-96   https://doi.org/10.21741/9781644902677-14

length of the labyrinth, which takes into account number of branches n and width of the channel d that are not considered in the QWT resonance frequency equation [5]:

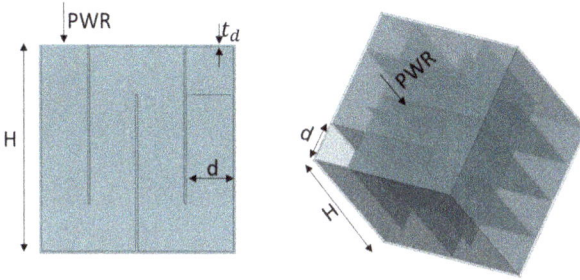

Figure 3: Labyrinth resonator excited by a plane wave radiation (PWR).

$$L_{eff} = L - (4 - \pi)\frac{d}{2}(n - 1). \tag{3}$$

The plate of thickness t_d with the square inlet hole of side-length d is studied through Johnson-Champoux-Allard (JCA) approach [6]. The impedance of the labyrinth resonator is:

$$Z_{LR} = \frac{1}{\phi_{inlet}}\left[Z_{dh_{JCA}}\frac{-jZ_{QWT}\cot\left(k_{dh_{JCA}}t_d\right) + Z_{dh_{JCA}}}{Z_{QWT} - jZ_{dh_{JCA}}\cot\left(k_{dh_{JCA}}t_d\right)}\right], \tag{4}$$

where $Z_{d_{JCA}} = \rho_{JCA}c_{JCA}$ and $k_{d_{JCA}} = \omega/c_{JCA}$ are respectively the impedance and the complex wavenumber of the perforated plate, with $c_{JCA} = \sqrt{K_{JCA}/\rho_{JCA}}$, ρ_{JCA} and K_{JCA} effective speed of sound, density and bulk modulus. $\phi_{inlet} = A_{hole}/A_{plate}$ is the perforatio ratio between the hole area and the plate area. The acoustic impedance is used for the evaluation of the sound absorption coefficient $\alpha = \Pi_{dissipated}/\Pi_{incident}$, which indicates the portion of the incident sound energy dissipated by the sample. The sound absorption of the labyrinth resonator will be calculated with:

$$\alpha_{LR} = \frac{4Re(Z_{LR}/Z_0)}{|Z_{LR}/Z_0|^2 + 2Re(Z_{LR}/Z_0) + 1}. \tag{5}$$

Several labyrinth resonators are modelled and analyzed through COMSOL Multiphysics, Pressure Acoustics Module, with first resonance peak at 60, 90, 120Hz (Figure 4a-c). Their height (thickness of the sample) is fixed at 100mm, to consider a competitive space for aeronautical application. The 60Hz labyrinth has lateral dimension of 97mm x 97mm, the 90Hz has lateral dimension of 145mm x 49mm and the 120Hz has lateral dimension of 73mm x 73mm. The numerical model is developed following the scheme of the equivalent experimental test with an impedance tube. A normal plane wave radiation excites the tube, and the sample is placed at the end of the tube backed by a rigid wall. Two probes with a relative distance s evaluate pressure in the tube, with P_2 at a distance x_2 respect to the sample, and P_1 at a distance $x_1 = x_2 + s$. The

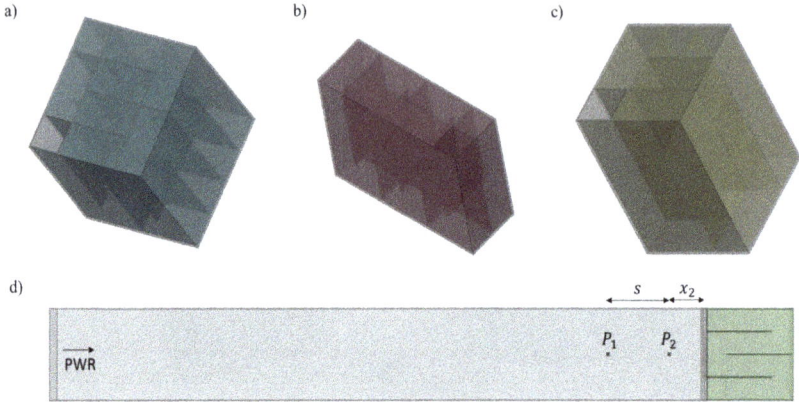

Figure 4: a) labyrinth model with first resonance peak at 60Hz; b) labyrinth model with first resonance peak at 90Hz; c) labyrinth model with first resonance peak at 120Hz; d) COMSOL model of the impedance tube for sound absorption measurements.

sound absorption coefficient is estimated according to the ISO 10534-2 1998 [7], with R reflection coefficient:

$$R = \frac{P_1/P_2 - e^{-jk_0 s}}{e^{jk_0 s} - P_1/P_2} e^{2jk_0 x_1}, \qquad \alpha = 1 - |R|^2. \qquad (6)$$

Results and discussion

Numerical simulations with analytical check are plotted for each labyrinth resonator (Fig. 5). The sound absorption estimated through numerical analyses consistently matched the analytical method. Each resonator has harmonic peaks, and their sound suppression is extremely interesting for low-frequency application on the engine nacelles, or to suppress tonal noise of diesel engines at low frequency for industrial application like naval and automotive. The amplitude of each peak depends on several parameter, but the most effective is the perforation ratio (the portion of the area of the inlet respect to the total area excited by the plane wave). The 120Hz labyrinth shows the worst first-peak among the three cases, but this can be easily solved by increasing its perforation ratio. A labyrinth-type metamaterial made with several samples of labyrinth resonator is studied to cover each resonance and take advantage of its harmonic behavior. The designed metamaterial is based on previous resonators, in particular with one 60Hz sample, two 90Hz sample and three 120Hz resonators (Fig. 6a). The multiple labyrinth resonator metamaterial keeps the absorption peak of each QWT in correspondence of its resonances, with an intuitive change of amplitude due to perforation ratio, affected by the number of repetitions per each labyrinth (Fig. 6b). For instance, the 60Hz labyrinth is just one, and its absorption is less effective than the case with just the 60Hz resonator. In the same way, the 120Hz effect is consistently enhanced. This result opens to several hybrid solutions which can embed multiple labyrinth resonators tuned at different frequencies, with the object to optimize their perforation ratio, finding a compromise between number of resonances and tonal behavior of the system. Indeed, a higher number of samples with different tuning frequency can lead to broadband sound absorption at low frequency, but the amplitude will

Aerospace Science and Engineering - III Aerospace PhD-Days
Materials Research Proceedings 33 (2023) 91-96

Materials Research Forum LLC
https://doi.org/10.21741/9781644902677-14

be reasonably reduced; on the other hand, a combination of few labyrinths implies tonal sound suppression.

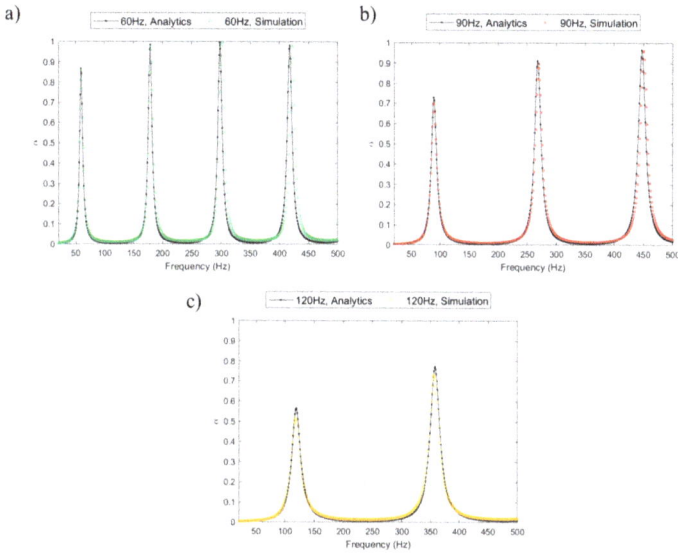

Figure 5: a) Numerical results with analytical comparison for a labyrinth resonator designed with first resonance at 60Hz; b) numerical and analytical results for 90Hz labyrinth resonator; c) numerical and analytical results for 120Hz resonator.

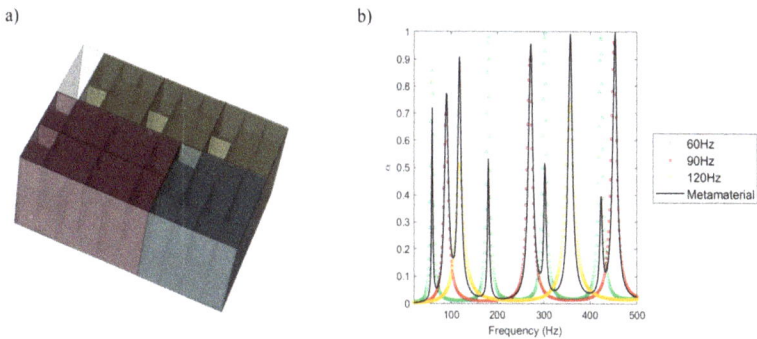

Figure 6: a) multiple labyrinth acoustic metamaterial; b) sound absorption plot of the proposed acoustic metamaterial.

Aerospace Science and Engineering - III Aerospace PhD-Days   Materials Research Forum LLC
Materials Research Proceedings 33 (2023) 91-96   https://doi.org/10.21741/9781644902677-14

Conclusions

Labyrinth resonators are presented as an innovative solution for the design of acoustic liners. Their competitive performance, flexibility and space requirements can be very attractive for an aeronautical application aiming to reduce engine noise. Numerical simulation represents a preliminary step; the sound absorption performance of the single labyrinth resonator and the complete acoustic metamaterial will be experimental tested inside impedance tube and under diffuse acoustic field to evaluate their performance under random excitation. The model will be 3D printed with several materials like PLA and resin. The final scope is to merge a comprehensive knowledge of models like this to design innovative acoustic liners that can improve low-frequency performance of pre-existing solution without affecting weight and space.

References

[1] L. Cao, Q. Fu, Y. Si, B. Ding, and J. Yu, "Porous materials for sound absorption," *Composites Communications*, vol. 10. Elsevier Ltd, pp. 25–35, Dec. 01, 2018. https://doi.org/10.1016/j.coco.2018.05.001.

[2] X. Q. Ma and Z. T. Su, "Development of acoustic liner in aero engine: a review," *Science China Technological Sciences*, vol. 63, no. 12. Springer Verlag, pp. 2491–2504, Dec. 01, 2020. https://doi.org/10.1007/s11431-019-1501-3.

[3] A. Magnani, C. Marescotti, and F. Pompoli, "Acoustic absorption modeling of single and multiple coiled-up resonators," *Applied Acoustics*, vol. 186, Jan. 2022. https://doi.org/10.1016/j.apacoust.2021.108504.

[4] C. Zwikker and C. W. Kosten, "Sound Absorbing Materials," *Elsevier Publishing company*, 1949.

[5] Giuseppe Catapane, Dario Magliacano, Giuseppe Petrone, Alessandro Casaburo, Francesco Franco, and Sergio De Rosa, "Labyrinth Resonator Design for Low-Frequency Acoustic Meta-Structures," in *Recent Trends in Wave Mechanics and Vibrations*, 2022, pp. 681–694. https://doi.org/10.1007/978-3-031-15758-5_70.

[6] N. Atalla and F. Sgard, "Modeling of perforated plates and screens using rigid frame porous models," *J Sound Vib*, vol. 303, no. 1–2, pp. 195–208, Jun. 2007. https://doi.org/10.1016/j.jsv.2007.01.012.

[7] "IS0 10534-2 1998: Acoustics-Determination of sound absorption coefficient and impedance in impedance tubes-Part 2: Transfer-function method," 1998.

Aerospace Science and Engineering - III Aerospace PhD-Days   Materials Research Forum LLC
Materials Research Proceedings 33 (2023) 97-103   https://doi.org/10.21741/9781644902677-15

Developing a methodology for Co-creation in aviation industry using extended reality technologies

Sandhya Santhosh[1,a] *, Francesca De Crescenzio[1]

[1]University of Bologna, Italy

[a]sandhya.santhosh2@unibo.it

Keywords: Co-creation, Extended Reality, Aviation Industry, Digital Methods

Abstract: In the recent times the concept of Co-creation, involving stakeholders, customers, and partners in the design, development, and delivery of products and services, has gained importance in various industries, including aviation. This PhD research paper presents the concept of co-creation in the aviation sector, focusing on its key elements and the potential of extended reality (XR) technologies as enablers. The paper highlights collaboration, interaction, and user experience as the essential elements of co-creation and discusses existing co-creation platforms and tools. Traditional co-creation platforms have limitations, prompting the proposal to leverage XR technologies such as Virtual Reality (VR), Augmented Reality (AR), and Mixed Reality (MR) as future co-creative platforms. These immersive technologies offer the potential to enhance collaboration, visualization, and engagement in a virtual environment. The study explores how XR tools can aid digital co-creation methods and whether XR systems can induce effective co-creation sessions. In this context, the research develops a multi-user real-time XR environment that fosters collaborative and interactive co-creation aiming to enhance engagement, communication, and idea generation. This paper includes use-case scenarios for adapting such environments in view of investigating the potential of XR tools to improve the quality and novelty of ideas, reduce communication challenges, and decrease time and costs in the aviation industry's co-creation process.

Introduction

The aviation sector is a highly competitive and rapidly evolving environment thus facing various challenges ranging from long life cycles, high service quality, needs to customer value. To be competitive in this sector, manufacturers need to continuously accommodate new strategies to ensure the final products meet the needs of the customers and users [1]. As technology has emerged at an expeditious pace, many airlines have issues to think outside the box in order to develop innovative ideas for new products and services. By engaging with customers and other stakeholders, aviation companies can gain a deeper understanding of their needs and preferences, identify opportunities for improvement, and create more effective solutions. The concept of involving stakeholders, customers and partners in the design, development, and delivery of the products and services is called "Co-creation". This collaborative approach seeks to create better products that meet their expectations and requirements. Developing a customer-centric culture is essential for any organization that wants to improve its service quality. This involves creating a culture where the needs and expectations of customers are always taken into consideration [2]. This research commenced towards studying insights to emerging design practices with co-creation in view of delivering innovation to aviation industry and improving the process by adapting cutting edge technologies in building co-creation environments/platforms.

Co-Creation at a glance

Co-creation has been a business topic till today, but, however, is emerging to numerous domains including the aviation industry. It can be defined as *collaborative development of new value*, the

Aerospace Science and Engineering - III Aerospace PhD-Days   Materials Research Forum LLC
Materials Research Proceedings 33 (2023) 97-103   https://doi.org/10.21741/9781644902677-15

value being products, solutions, risks and services. The key elements of co-creation include Collaboration, Interaction and User experience. How and Where to collaborate, How to interact and What will be the experience are the three main research questions this PhD research is oriented. Studies have anticipated the development of tools and methods that would extend the definition of co-creation to various domains [3]. Co-creation has become an important aspect of product design in other industries such as automotive, fashion, technology and aviation. For example, companies such as BMW and Audi used this concept to involve customers in the design process of their cars in order to collect feedback on various aspects of design. LEGO adapted to customer co-creation allowing customers to contribute ideas for a rocket model through an online platform [4][5].

Co-creation is foreseen to involve diverse participants ranging from researchers, engineers, designers, non-designers to end users. During this PhD research, it has been established that due to the diversified notion of user participation during a product development process, the platform where they co-create and the tools used play an important role in facilitating active and creative participation of the users.

The power of co-creation lies in how well the users are collaborating and interacting in the environment/ platform. Development of such co-creative platform or environment which is capable of accomplishing the three key elements of co-creation has been unprecedented and has not been given much of importance. In this context, it becomes essential to identify the technological tools that allow non-designers to go beyond the validation of already designed products and to play an active role in generating ideas and directly building solutions [6]. Through the survey of the literature, it has been observed that visual representations of the product during its development helps to foresee any issues and contributes to leading edge solutions and the aspect of simultaneous work and real-time sharing induces co-creation by reducing time and costs. Bringing both these aspects to a platform could overall improve co-creation [7].

Co-Creation Platforms

There are various co-creation platforms and tools available today to enable business and individuals to collaborate in creating, developing and testing the products and services. Some of the existing popular co-creation platforms are Idea Scale (an idea management platform that helps business to collect and evaluate ideas from employees, customers and partners), InnoCentive (a crowdsourcing platforms to solve complex problems), Canva (a design collaboration platform enabling teams to create and edit designs collaboratively) etc. All these platforms or tools are based on the concepts of Netnography & Crowdsourcing [8]. It can also be observed that these are also web and network based and limits user experience towards the aspect of co-creation .

Alongside of advanced science and technology, digital or virtual worlds for co-creation has gained momentum in offering a wide range of co-creation aspects by enhancing collaboration and improving engagement between the participants [9]. Digital co-creation platforms are web-based tools and software solutions that enable the collaborative creation and developments of products and services online. These platforms facilitate the exchange of ideas, data and feedback between different stakeholders allowing them to work together in real-time regardless of their location. Virtual worlds for co-creation systems have been studied in the past and featured the importance of user experience in these environments which induce quality of interactions thus positive relationships between companies and customers [10].

With immense growth in technology, digital technologies have emerged to immersive technologies such as Virtual Reality (VR), Augmented Reality (AR) and Mixed Reality (MR). Immersive technologies create and extend reality by immersing a user in a completely digital environment (VR), or adds and augments users surrounding by adding digital elements (AR) or both (MR). Extended Reality (XR) serves as an umbrella term for all the existing and upcoming reality-virtuality technologies[11]. These technologies offer high potential to virtually collaborate, visualize and co-create in an imaginary unknown world. Through this PhD, I propose XR

Aerospace Science and Engineering - III Aerospace PhD-Days   Materials Research Forum LLC
Materials Research Proceedings 33 (2023) 97-103   https://doi.org/10.21741/9781644902677-15

technologies as potential co-creation enabler [7]. The research leverages as a collaborative and visual approach to service and product design creating many synergies as ideas, insights, expertise and vision. The research explores the below points:

    a. How XR tools can aid digital co-creation methods?
    b. Can XR system induce co-creation session?

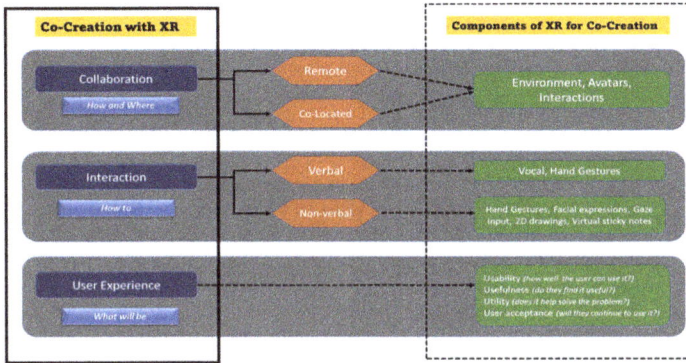

Fig. 1 Outline of components of Co-creation mapped to XR components

There are multiple reasons for enabling XR system as co-creation platform such as real-time sharing, product involvement, reduced communication challenges, media rich information, reduced time and costs etc. However, there are more pressing investigation to be performed in understanding if XR enabled co-creation session:

- Improve quality of ideas?
- Improve novelty of ideas?
- Reduce communication challenges?
- Provide better understanding of design?
- Reduce time and costs?

To back up the above questions, this PhD study has outset the number of aspects XR tools can offer and map them towards the key elements of co-creation as in Fig. 1. Taking this in to account, I have developed a multi-user real-time extended reality environment for collaborative and interactive co-creation platform. This platform is expected to serve as an effective and efficient solution for co-creation sessions when compared to conventional co-creation sessions by inducing co-creation in several ways such as providing immersive environments allowing participants to experience a shared virtual space fostering collaboration, enhancing communication by providing shared visual language and tools for collaboration, increasing engagement and motivation by providing visually rich and interactive environment, increasing accessibility by allowing remote participants to join and participate in real-time and real-time feedback allowing participants to quickly see the results and adjust their ideas accordingly.

This tool can be used in different types of activities in an innovation project. It's potential and role will also vary according to the various stages of the product development process and the roles of stakeholders during the process.

Aerospace Science and Engineering - III Aerospace PhD-Days   Materials Research Forum LLC
Materials Research Proceedings 33 (2023) 97-103   https://doi.org/10.21741/9781644902677-15

XR-Co-creation Platform Development and Use Cases

The XR process is an ideation model to design thinking and, following this process a system and workflow has been developed for multi-user Mixed Reality device, Microsoft HoloLens2. The application developed can be expanded to remote collaboration and necessary components of interaction to build a successful co-creation platform. The system is developed on UNITY software and is built within the framework of a networking and anchor solution for local position of virtual content and absolute position of user respectively in the 3D shared space. It ideally consists of a table which is common for every HoloLens2 and any content placed on this table is shared between all users. Multiple users can share this unique system and the scene can be customized for any use case of a co-creation sessions. In this regard, some examples of the scenes where this XR co-creation platform can be implemented in order to improve user creativity are as below.

Use case Scenario1:

The first application is in the aviation domain and regards the validation of the design of a cabin interior environment. It is built upon a scenario of design review and task analysis setting and is drawn between two potential users, a flight attendant and a human factors expert. The scene is designed to happen in front of a full scale immersive and interactive Galley of a Regional Aircraft. The task of flight attendant picking up a water jar and placing it on the platform of the galley while a second user acting the role of human factor expert human factors is observing over the shoulder and collecting feedback in real-time is organized. Till now, human factors experts are known to be present outside the loop while any task is on-going unless the physical prototype of the product is available. Through this mixed reality space, both users can simultaneously visualize, collaborate and interact with the full-scale product which is in design phase. Due to the nature of the scene, a co-located version of the platform has been considered. Both users can anticipate any issues related to task, design, reachability, comfort, wellbeing etc. Overall, this scene is to be evaluated for created value and if the highlighted co-creation elements are satisfactory[12].

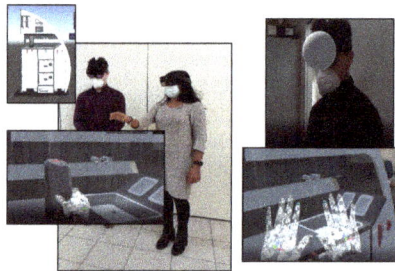

Fig. 2 Multi user co-located collaborative and interactive platform for Aircraft Galley use case

Use- Case Scenario2:

The Multi-user XR environment developed was applied to a use-case in aviation domain in regard to the validation of the design of a aircraft cabin interior. It is built up for the scene of a product tester role of the user [13] in order to test the compliance of the new seat design configuration for cabin crew and passenger activities of a concept test bed, Flying-V aircraft, developed in collaboration between Delft University of Technology (TUDELFT), Airbus and KLM Royal Dutch Airlines [14].

Aerospace Science and Engineering - III Aerospace PhD-Days   Materials Research Forum LLC
Materials Research Proceedings 33 (2023) 97-103   https://doi.org/10.21741/9781644902677-15

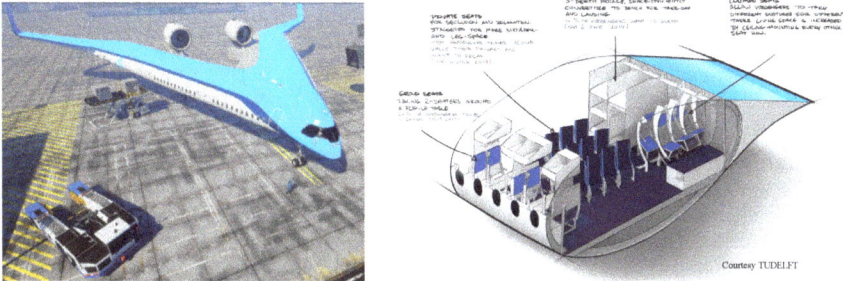

Fig. 3 Flying-V configuration

Flying-V is a new design of an energy-efficient long-distance aircraft with its passenger cabin, cargo hold and fuel tanks integrated in its wing structure creating a spectacular V-shape. Its improved aerodynamic shape and reduced weight contributes for 20% less fuel than the Airbus A350, today's most advanced aircraft [15].With its V-shaped configuration as in Fig. 3, the wings of the airplane are at an angle of 26 degrees with respect to the direction of the flight. In addition, the aerodynamic shape of the wing cross section lays out the passenger cabin configuration and is different when compared to the traditionally configured circular fuselage.

The new oval configuration has boosted insights to designers on seat design and innovations inside the Flying-V fuselage. This led to the idea of a new design called "staggered seats" which are designed for more leg room space and individual arm rest making more shoulder-to-shoulder space between passengers. Staggered seats are also placed at an angle with respect to the direction of flight which makes the seats not to fall in a single line but in an angled row [16].

In this context, it was observed that XR platform can be adapted in order to enact and verify certain tasks of the cabin crew and passenger and get the opinions of the users before the manufacturing of the prototype.

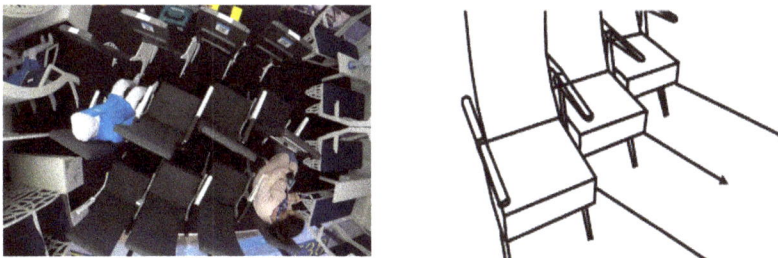

Fig. 4 Flying-V Staggered Seats

The aim of co-creation being shared value and creativity, the XR system is planned to be analyzed for the same user experience by adding the key elements of co-creation. By involving real users in this scenario, qualitative or quantitative data are planned to be collected in the following aspects to strengthen my proposal of XR technology as future innovative co-creation platform:

Aerospace Science and Engineering - III Aerospace PhD-Days   Materials Research Forum LLC
Materials Research Proceedings 33 (2023) 97-103   https://doi.org/10.21741/9781644902677-15

a. Does the platform improve the efficacy and efficiency of co-creation and support in creating value?

b. The impact of the key elements (collaboration, interaction and user experience) of co-creation to achieve a.

c. By satisfying a. & b., does it help accelerate product development process and product innovation?

Conclusion and Future Work

The PhD study highlights the importance of innovation and the ongoing digital transformation in the aviation industry along with proposing cutting edge technologies for consolidating concepts like co-creation. It emphasizes on Extended Reality (XR) technologies and proposes XR tools as future co-creation environments or platforms. In order to substantiate the proposal, I have developed a multi-user XR system for collaborative and interactive environment in order to improve user experience thus, co-creation experience. This system has been adapted to different scenarios in order to analyze for its effectiveness through appropriate methods and collecting data. It is expected to shift collaboration from a flat one-dimensional experience into a multi-dimensional one accelerating the aviation industry.

References

[1] POLAT, İnci., Value Co-Creation and Passenger Loyalty in the Context of the DART Model: The Mediating Role of Perceived Service Newness. Journal of Aviation 5, no. 2 (2021): 219-229. https://doi.org/10.30518/jav.1001127

[2] A.F. Payne, K. Storbacka, and P. Frow, Managing the co-creation of value, J. of the Acad. Mark. Sci., 36, 83-96, 2008. https://doi.org/10.1007/s11747-007-0070-0

[3] Sanders, Elizabeth B-N., and Pieter Jan Stappers. "Co-creation and the new landscapes of design." Co-design 4, no. 1 (2008): 5-18. https://doi.org/10.1080/15710880701875068

[4] "LEGO to launch production of fan-created Saturn V moon rocket model | collectSPACE," collectSPACE.com. http://www.collectspace.com/news/news-060916a-saturn-v-lego-ideas.html (accessed Feb. 02, 2022).

[5] "LAUNCH of the BMW Group Co-Creation Lab | The Making-of Innovation." https://michaelbartl.com/2010/09/11/launch-of-the-bmw-group-co-creation-lab-2/ (accessed Feb. 14, 2022).

[6] S. Alimamy, K. R. Deans, and J. Gnoth, The Role of Augmented Reality in the Interactivity of Co-Creation: A Critical Review, Int. J. Technol. Hum. Interact., vol. 14, no. 3, pp. 88-104, Jul. 2018. https://doi.org/10.4018/IJTHI.2018070106

[7] El-Jarn, H., & Southern, G.: Can co-creation in extended reality technologies facilitate the design process?. Journal of Work-Applied Management (2020). https://doi.org/10.1108/JWAM-04-2020-0022

[8] Bartl, M., Jawecki, G., & Wiegandt, P.: Co-creation in new product development: conceptual framework and application in the automotive industry. In: Conference Proceedings R&D Management Conference-Information, Imagination and Intelligence, Manchester Vol. 9, pp. 1-9, 2010, June.

[9] Prahalad, C. K., & Ramaswamy, V.: Co-creation experiences: The next practice in value creation. Journal of interactive marketing, 18(3), 5-14, 2004. https://doi.org/10.1002/dir.20015

[10] Kohler, T., Fueller, J., Matzler, K., Stieger, D., & Füller, J.: Co-creation in virtual worlds: The design of the user experience. MIS quarterly, 773-788, 2011. https://doi.org/10.2307/23042808

[11] Santhosh, S., De Crescenzio, F. and Vitolo, B., 2022. Defining the potential of extended reality tools for implementing co-creation of user oriented products and systems. In Design Tools and Methods in Industrial Engineering II: Proceedings of the Second International Conference on Design Tools and Methods in Industrial Engineering, ADM 2021, September 9-10, 2021, Rome, Italy (pp. 165-174). Springer International Publishing. https://doi.org/10.1007/978-3-030-91234-5_17

[12] Santhosh, S. and De Crescenzio, F., 2022, September. A Mixed Reality Application for Collaborative and Interactive Design Review and Usability Studies. In Advances on Mechanics, Design Engineering and Manufacturing IV: Proceedings of the International Joint Conference on Mechanics, Design Engineering & Advanced Manufacturing, JCM 2022, June 1-3, 2022, Ischia, Italy (pp. 1505-1515). Cham: Springer International Publishing. https://doi.org/10.1007/978-3-031-15928-2_131

[13] Nambisan, S. and Nambisan, P., 2008. How to profit from a better virtual customer environment'. MIT Sloan management review, 49(3), p.53.

[14] "FLYING-V" https://www.tudelft.nl/en/ae/flying-v (accessed Mar 01. 2023)

[15] Vink, P., Rotte, T., Anjani, S., Percuoco, C., Vos, R.: Towards a hybrid comfortable passenger cabin interior for the flying V aircraft. Int. J. Aviat. Aeronaut. Aerosp. 7(1), 1 2020. https://doi.org/10.15394/ijaaa.2020.1431

[16] Vink, P., Anjani, S., Percuoco, C., Vos, R. and Vanacore, A., 2021, May. A Staggered Seat is Beneficial for the Flying V Aircraft. In Proceedings of the 21st Congress of the International Ergonomics Association (IEA 2021) Volume III: Sector Based Ergonomics (pp. 184-190). Cham: Springer International Publishing. https://doi.org/10.1007/978-3-030-74608-7_24

[17] Cherry, E. and Latulipe, C., 2014. Quantifying the creativity support of digital tools through the creativity support index. ACM Transactions on Computer-Human Interaction (TOCHI), 21(4), pp.1-25. https://doi.org/10.1145/2617588

[18] Laugwitz, B., Held, T. and Schrepp, M., 2008. Construction and evaluation of a user experience questionnaire. In HCI and Usability for Education and Work: 4th Symposium of the Workgroup Human-Computer Interaction and Usability Engineering of the Austrian Computer Society, USAB 2008, Graz, Austria, November 20-21, 2008. Proceedings 4 (pp. 63-76). Springer Berlin Heidelberg.

[19] Verleye, K., 2015. The co-creation experience from the customer perspective: its measurement and determinants. Journal of Service Management. https://doi.org/10.1108/JOSM-09-2014-0254

Aerospace Science and Engineering - III Aerospace PhD-Days   Materials Research Forum LLC
Materials Research Proceedings 33 (2023) 104-109   https://doi.org/10.21741/9781644902677-16

MATLAB code for highly energetic materials

A. Cucuzzella[1*], Y. Caridi[1*], S. Berrone[1], L. Rondoni[1], U. Barbieri[2], L. Bancallari[2]

[1]Politecnico di Torino, Dipartimento di Scienze Matematiche "G. L. Lagrange"

[2]MBDA Italia S.p.A

*andrea.cucuzzella@polito.it, yuri.caridi@polito.it

Keywords: Detonation, Chemical Equilibrium, Energetic Material, Rayleigh-Hugoniot Relations

Abstract. Detonations represent high-speed chemical reactions characterized by rapid propagation, accompanied by a release of high-pressure energy. This transformative process converts unreacted explosive materials into stable product molecules, reaching a steady state known as the Chapman-Jouguet (CJ) state. This study aims to effectively describe the detonation phenomenon in energetic materials through the application of the CJ theory. Using a computational approach, we developed a MATLAB code to calculate the minimum detonation velocity (DCJ) of the explosive and analyze product expansion under constant entropy conditions.

Introduction

Detonations are high-speed chemical reactions that are characterized by their rapid propagation and high pressure accompanied by high-speed energy release. After the passage of the detonation shock wave, the reactive medium is transformed into high temperature and pressure products. Hence, detonation can be thought of as a path that transoforms the unreacted explosive into stable product molecules at the Chapman– Jouguet (CJ) state, which is a steady, chemical equilibrium shock state that conserves mass, momentum, and energy. Such characteristics make detonation a phenomenon with several potential fields of application. Nowadays, detonation seems to represent a turning point for the field of hypersonic propulsion. Several propulsion systems have been designed to exploit the detonation of energetic material to generate thrust. Among the most important ones are scramjets and pulse detonation engines (PDE), which, despite not yet being available on the international market, offer significant advantages over current architectures. The research presented aims, among its objectives, to lay the foundations for a new theory capable of reducing the uncertainty of these systems and promoting the development of such technologies in the next decade. The conservation's laws represent a fundamental tool for analyzing the properties of energetic materials. By substituting the conservation equation of mass into the conservation of momentum, removing the velocity of the products, it is possible to obtain a relationship between pressure and specific volume called the Rayleigh line :

$$p = p_0 + \rho_0^2 D^2 (v_0 - v)$$

$$(1)$$

where D represents the velocity of the shock wave through the explosive.

Methods

The aim of this study is to effectively describe the detonation phenomenon of energetic materials through the CJ theory. We developed a MATLAB code to compute the minimum detonation velocity (D_{CJ}) of the explosive and the expansion at constant entropy of the products once the equation of state (EOS) of the mixture is provided. The products under consideration are nine gaseous chemical species (H2, N2, O2, CO, NO, H2O, CO2, NH3, CH4) and one solid (C(s)). For

Aerospace Science and Engineering - III Aerospace PhD-Days   Materials Research Forum LLC
Materials Research Proceedings 33 (2023) 104-109   https://doi.org/10.21741/9781644902677-16

each of these species, it is necessary to provide values for their enthalpy and entropy under standard conditions, as well as their variations as a function of temperature at standard pressure. To accomplish this, the Shomate Equation has been selected, which takes the following form for both enthalpy and entropy :

$$s^0(T) = A * ln(t) + B * t + C * \frac{t^2}{2} + D * \frac{t^3}{3} - \frac{E}{2t^2} + G$$

(2)

$$h^0(T) = A * t + B * \frac{t^2}{2} + C * \frac{t^3}{3} + D * \frac{t^4}{4} - \frac{E}{t} + F$$

(3)

where t represents temperature expressed in Kelvin divided by one thousand, and the coefficients (A, B, C, D, E, F) were obtained from the NIST database.

Another fundamental step is represented by the choice of the equation of state. For the gaseous phase, simulations were carried out using both the equation of state for ideal gases and the Becker-Kistiakowsky-Wilson equation of state (BKW) in order to evaluate the differences that the two solutions bring in terms of explosive properties. The BKW EOS has the following form :

$$p = \frac{\rho RT}{W_{mix}} \cdot Z(\rho, T, x_{i,eq})$$

(4)

where the variable Z is called compressibility factor and it represents how much the mixture under consideration deviates from an ideal one :

$$Z = 1 + \chi e^{\beta \chi} \qquad \chi = \frac{\rho \kappa \sum k_i x_{i,eq}}{W_{mix}(T + \theta)^\alpha}$$

(5)

The parameters α, β, θ, and κ are constants computed by interpolating experimental results. Lastly, concerning the term k_i, it denotes the molar covolume of the chemical species that are being considered. For the solid phase, there are two possible paths. The first one is the simplest and less computationally demanding, it consists of introducing the hypothesis that the solid phase is incompressible. The second path, instead, takes into account compressibility effects through an equation of state. In particular, in the present case, Cowan's solid equation of state was chosen to analyze the compressibility effects of graphite at high pressures and temperatures :

$$p = P(\eta) + a(\eta)T + b(\eta)T^2$$

(6)

where η represents the ratio between the density of the solid phase, for a generic temperature and pressure, and the density of the solid phase under reference conditions. $P(\eta)$, $a(\eta)$, and $b(\eta)$ represent polynomials as a function of the density of the solid under examination.

The code is structured into three main modules :

- Explosion State
- Detonation State
- Isoentropic Expansion State

Aerospace Science and Engineering - III Aerospace PhD-Days   Materials Research Forum LLC
Materials Research Proceedings 33 (2023) 104-109   https://doi.org/10.21741/9781644902677-16

The first module, called Explosion State, is the most computationally demanding and it is responsible for representing the Hugoniot of the products in the Clapeyron plane :

$$e(n_i, T) - e_0 = \frac{1}{2}(p(n_i, T) + p_0)(v_0 - v)$$

(7)

where the internal energy of the system is computed for a solid-gas double-phase mixture, and n_i represents the chemical composition in moles of the equilibrium products. The module includes an internal loop that computes the chemical equilibrium through a constrained minimization algorithm of the Helmholtz free energy. The constraints introduced in this phase are two :

- The population constraints, which require that the number of particles of each chemical element must be conserved during detonation
- The limit to the number of moles that each chemical species can have, as negative values would not be physically meaningful and therefore cannot be considered.

Concerning the second module, called Detonation State, it is responsible for interpolating the points computed to obtain a functional expression of the Hugoniot of the products. Subsequently, through the minimization algorithm, the tangent point between the Hugoniot of the products and the Rayleigh line is computed as a function of the detonation velocity (D). According to the CJ theory, the value of D that brings the two curves to be tangent is called the CJ detonation velocity and it represents a very important characteristic parameter of explosives, namely the minimum shock wave velocity that generates a detonation of the energetic material. Similarly, from the found CJ point, it is possible to compute the thermodynamic properties of the mixture, including pressure and temperature. There is another important parameter that is determined in this phase : the entropy at the CJ point, which serves as an input for the next module. Indeed, by knowing this parameter, it is possible to graphically represent the isoentropic expansion of the products starting from the CJ point in the p-v plane, assuming that the mixture is in chemical equilibrium at every point. The third module, called Isoentropic Expansion State takes care of this last operation through a zero-finding algorithm applied to the following equation :

$$S(v, T) = S_{CJ}$$

(8)

Results
One of the first results obtained from the code concerns to the possibility of highlighting the differences between an ideal gas mixture and a real one. The figure in the next page (Figure 1) shows partial results of the CJ theory obtained for the RDX explosive :

Aerospace Science and Engineering - III Aerospace PhD-Days   Materials Research Forum LLC
Materials Research Proceedings 33 (2023) 104-109   https://doi.org/10.21741/9781644902677-16

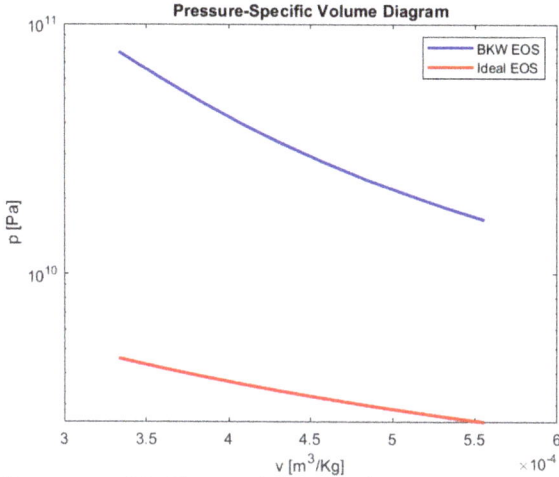

Figure 1: Comparison of the Hugoniot behaviors of RDX products using ideal and BKW equations of state

For a given specific volume, the two solutions diverge by approximately one order of magnitude, with the discrepancy increasing as the system pressure rises. The result obtained is extremely important as it suggests that the ideal gas assumption is not suitable for describing the detonation process. This outcome is in line with expectations, as detonation involves extremely high pressures and the ideal gas assumption is not an accurate representation of a mixture under such conditions.

Now, numerical results obtained for various types of energetic materials are presented and compared with experimental data taken from Mader [2] :

Table 1: Comparison between the results obtained from the numerical model and experimental data for two different types of explosives.

	Densità [Kg/m³]	CJ Point	Experimental Results	Model Results
RDX	1,8	D [m/s]	8754	9069
		T [K]	2587	2181
		γ	2,98	3,21
	1	D [m/s]	5981	6068
		T [K]	3600	3344
		γ	2,48	2,62
HMX	1,9	D [m/s]	9100	8878
		T [K]	2364	2423
		γ	3,03	3,37

Aerospace Science and Engineering - III Aerospace PhD-Days   Materials Research Forum LLC
Materials Research Proceedings 33 (2023) 104-109   https://doi.org/10.21741/9781644902677-16

The detonation velocity at the CJ point is the parameter that is less affected by imperfections related to the model. Moreover, it is possible to see how the detonation velocity strongly depends on the initial density of the explosive under examination, from the results obtained for RDX. In particular, for not excessively large values of density, this dependence is linear. Thus, it is possible to obtain a functional expression of the detonation velocity at the CJ point as the initial density of the energetic material changes.

Finally, it is possible to observe (figure 2) that the isoentropic expansion curve starts from the CJ point :

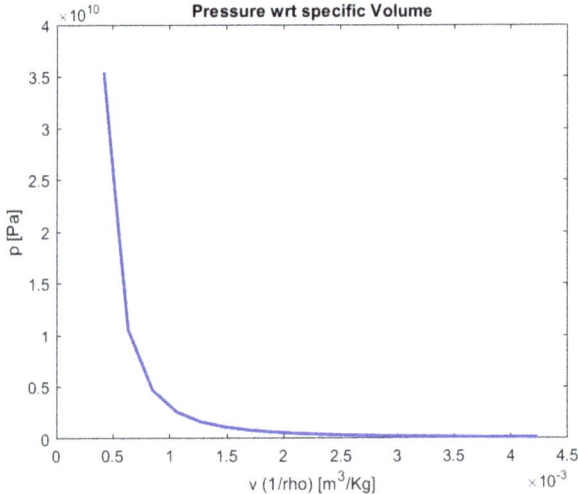

Figure 2: Expansion behavior of RDX for an isoentropic process.

It is possible to show that the three curves obtained (Rayleigh line, Hugoniot of products, and isoentropic expansion) must be tangent at the CJ point, this behavior is clearly visible in the following picture :

Aerospace Science and Engineering - III Aerospace PhD-Days   Materials Research Forum LLC
Materials Research Proceedings 33 (2023) 104-109   https://doi.org/10.21741/9781644902677-16

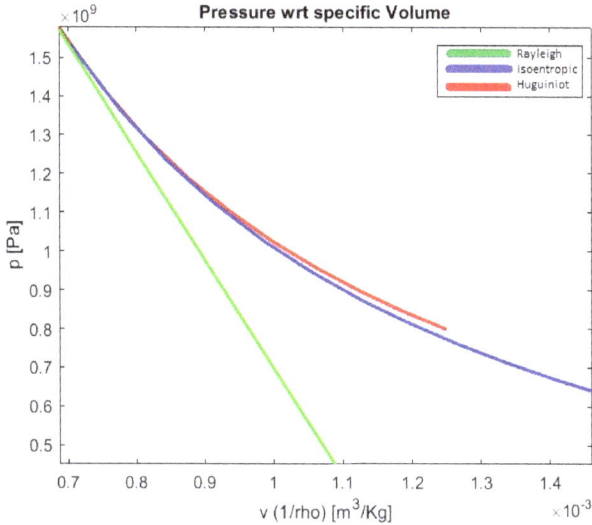

Figure 3 : Qualitative analysis between Rayleigh line, products Hugoniot and isentropic expansion in the neighborhood of the CJ point

References

[1] Shock Wave Science and Technology Reference Library, Vol. 6 Detonation Dynamic, F.Zhang, (2012)

[2] Numerical Modeling of Explosives and Propellants, C.Mader, (2008)

[3] Modular software for modelling the ideal detonation of explosives, Mathematical Engineering in Industry, T.L. Freeman, I.Gladwell M.J.Braithwaite, W.B. Brown,P.M.Lynch and I.B.Barker (1991)

Aerospace Science and Engineering - III Aerospace PhD-Days

Materials Research Proceedings 33 (2023) 110-117

Materials Research Forum LLC

https://doi.org/10.21741/9781644902677-17

Probing the surface of Ganymede by means of bistatic radar with the JUICE mission

Brighi Giancorrado[1,a] *

[1]Department of Industrial Engineering, Alma Mater Studiorum – University of Bologna, 47121 – Forlì (FC), Italy

[a]giancorrado.brighi@unibo.it

Keywords: JUICE, Ganymede, Bistatic Radar, RMS Surface Slope, Dielectric Constant

Abstract. Scheduled to get to the Jovian system in 2031, the ESA mission JUICE will explore Jupiter and its icy moons with unprecedented detail. Among other radar systems, the JUICE radio science experiment (3GM) will have the technical capability to probe the uppermost few meters of Ganymede's surface carrying out bistatic observations of the moon. In this work, we present a preliminary assessment of X-band downlink bistatic observations of Ganymede by means of JUICE High Gain Antenna during part of its final orbital phase around the moon. For the chosen time window, terrains between 60° N and 60° S appear to produce detectable echoes under the assumption of Ganymede reflecting as a water-ice rich perfectly conducting sphere. For the future activity of the 3GM experiment, more detailed surveys will be extended to the whole mission, comparing different available antennas and frequency bands, and accounting for the priority of different instruments and scientific objectives.

Introduction

In the wide framework of radio science experiments, employed for decades to remotely probe celestial bodies' interiors [1] [2] [3], ephemerides [4], rings and atmospheres [5] [6] [7], bistatic radar observations were successfully employed since the end of the sixties to study the surface of planets, satellites and asteroids [8]. When a radio signal is sent to a target body and reflected from its surface, terrain's features with scales proportional to radar wavelengths interact with it. A proper processing of the specular reflection from the planet can provide information about surface roughness in terms of root-mean-square (RMS) slope, near-surface relative dielectric constant and porosity of the target [8].

On its way to launch from the Guiana Space Center in April 2023, the ESA mission JUICE will reach the Jovian system in 2031 to explore the gas giant and its icy moons with unprecedented detail, with a particular emphasis on Ganymede. Among other instruments, the spacecraft will be equipped with the 3GM radio science experiment to address scientific objectives pertaining to gravity, geophysics and atmospheric science through radio tracking and radio occultations [9]. Given the capabilities of 3GM, downlink bistatic radar observations of Ganymede, in both X ($\lambda \approx 3.6$ cm) and Ka-band ($\lambda \approx 0.9$ mm), could help constraining the surface composition and near-surface structure of the two major geologic terrain types of the icy moon, i.e. the bright and dark regions [9].

In this work a preliminary assessment of downlink bistatic radar observations of Ganymede by means of JUICE is proposed. The investigation of a short time window is brought as an example, but the same type of analysis can be carried out for any other phase of the mission. Alongside the potential coverage, the main constraints to both feasibility of observation and expected quality of the scientific results are described in details, and supported by some useful, yet simple, analytical formulas.

Aerospace Science and Engineering - III Aerospace PhD-Days   Materials Research Forum LLC
Materials Research Proceedings 33 (2023) 110-117   https://doi.org/10.21741/9781644902677-17

After a brief introduction to the JUICE mission and its 3GM experiment, an overview of bistatic radar experiment is provided. The opportunity investigation is shown in the final section, followed by some conclusions.

The JUICE Mission

The JUICE (JUpiter ICy moons Explorer) mission is a large-class interplanetary mission of European Space Agency (ESA) Cosmic Vision 2015-2025 program that will provide the most comprehensive exploration of the Jovian system to address the theme of the emergence of habitable worlds around gas giants [9]. From 2031 to 2035, JUICE will orbit around Jupiter to enforce and deepen our knowledge of the giant gaseous planet and its three icy moons: Europa, Callisto and the prime target of the mission, Ganymede. A mindfully planned synergy between 10 state-of-the-art instruments and the 3GM radio science experiment will undertake a thorough investigation of the chemical composition, structure and evolution of the satellites' surfaces, inner layers and cores, trying to shed some light on the subsurface oceans believed to hide beneath the icy shell of the moons [10].

Scheduled for launch in April 2023, the JUICE orbiter will reach the Jovian system in July 2031 after an 8-year long cruise. Once in orbit around Jupiter, the spacecraft will perform a Jupiter Tour made of 2 flybys of Europa, 9 flybys of Ganymede and 21 flybys of Callisto. From December 2034, the JUICE mission will enter its final 9-month long phase orbiting Ganymede along elliptical (GEO) and circular orbits (GCO): the Ganymede orbital phase [11]. Among other scientific objectives, this orbital phase will give to JUICE the chance of sounding the icy crust of Ganymede to explore for the presence of water and reveal information about the moon's surface structure, origin and evolution. A powerful Ice Penetrating Radar (RIME) will work alongside infrared and ultraviolet spectrometers (MAJIS,UVS), a visible camera (JANUS), a laser altimeter (GALA) and the 3GM experiment to constrain Ganymede's surface properties and geology [9].

The 3GM radio science experiment will rely on the radio link between the on-board radio communication system and Earth stations to retrieve meaningful physical information about space bodies from observables of transmitted radio signals. The nominal investigations that the JUICE orbiter is scheduled to perform by means of 3GM are accurate measurements of the Galilean moons' gravity and radio occultations of neutral atmospheres and ionospheres of Jupiter, Callisto, Europa and Ganymede. Observations of Jupiter's rings physical properties can be carried out as experiments of opportunity, alongside occultations of the Io plasma torus and bistatic radar probing of the surface of the icy moons [9] [12].

Bistatic Radar Experiments

The theory behind quasi-specular bistatic radar (BSR) experiments was developed by Fjeldbo in [12], where he modeled reflections from a rough planet's surface with gaussian statistics using a simplified physical optics model, the Kirchhoff Approximation (KA). Then, he retrieved surface information from the modeled echoes. The planet's surface is assumed to feature i) isotropic and homogeneous behavior, ii) correlation length larger than λ and iii) negligible subsurface scattering [8]. The model was proved to work fairly well for various planetary bodies like the Moon, Mars and Venus [13] [14] [15].

Figure 1 shows a downlink specular near-forward bistatic geometry, where the spacecraft acts as the transmitter and the ground stations on Earth receive echoes from the target planet. In this traditional geometry, chosen for this work, the JUICE orbiter would transmit an unmodulated circularly polarized signal to the target body, Ganymede.

Aerospace Science and Engineering - III Aerospace PhD-Days   Materials Research Forum LLC
Materials Research Proceedings 33 (2023) 110-117   https://doi.org/10.21741/9781644902677-17

If the planet's surface satisfies the aforementioned assumptions, the majority of the echoes from

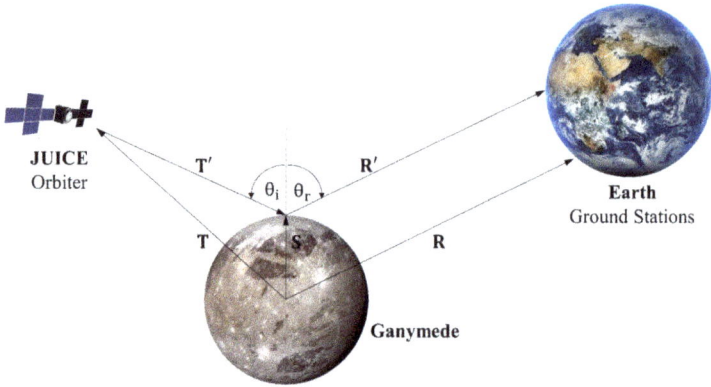

Figure 1 JUICE bistatic observation of Ganymede. Example geometry of a downlink forward specular bistatic radar experiment

the moon are expected to travel in the specular direction ($\theta_i = \theta_r$ in Figure 1) [8]. If JUICE is pointed towards the moving specular point on Ganymede (**S** in Figure 1), instantaneously constrained by the relative geometry between JUICE, Ganymede and Earth, echoes will be successfully received on ground. After the reflection, the polarized signal will be broadened in frequency because of surface roughness and will be polarized in both the original (SC) and an orthogonal circular sense (OC) compared to the incoming wave [8].

The differential Doppler between different regions of the reflecting area over the target planet induces the frequency broadening of the echoes. While the distance between transmitter (**T**) and receiver (**R**) from the specular point (**S**) over the target planet are changing, the signal travelling via the specular point undergoes a Doppler shift:

$$f_s = -\frac{1}{\lambda}\frac{d}{dt}\{|T'| + |R'|\}. \tag{1}$$

Differently from Figure 1, in the real world signals transmitted by an antenna travel through solid angles in space. A finite area of Ganymede will be illuminated, and reflections from that can be described, using a physical optics perspective, as a summation of discrete specular contributions arising from variously tilted facets with an effective length-scale of $\approx 100\lambda$ [8]. The average adirectional tilting of these facets is a measure of surface roughness, and the first output of BSR experiments: the RMS slope.

For a perfectly smooth sphere, the facet where the specular point lies will be the only one contributing to the specular echo. As rougher spheres are considered, facets away from the specular point will be properly tilted to contribute to the overall echo headed to Earth, and their contribution will be Doppler shifted differently from the specular point's one. As the spatial distribution of these properly oriented reflectors gets broader with roughness, the power spectrum of echoes gets broader in frequency. The spectral broadening can be related to the RMS slope (ζ) with the following formula:

Aerospace Science and Engineering - III Aerospace PhD-Days   Materials Research Forum LLC
Materials Research Proceedings 33 (2023) 110-117   https://doi.org/10.21741/9781644902677-17

$$B = 4\sqrt{\ln(2)}\left(\frac{V_s\zeta}{\lambda}\right)\cos\theta. \tag{2}$$

Where V_S is the specular point velocity over the planet's surface, λ is the radar wavelength and θ is the incidence angle of observation. With some knowledge of the problem geometry, the RMS slope can be hence obtained from the half-power bandwidth (B) of a received echo [12].

One last remark about the inference of the RMS slope is that for a rougher surface reflectors actively contributing to the overall echo are distributed over a spatially broader area around the specular point. If a fraction of these glints is not illuminated their contributions will not appear in the echo and the roughness will be underestimated. The finiteness of the transmitting antenna FOV introduces an upper bound to the observable RMS slope. If the RMS slope is larger than that, it is said that the observation was carried out in beam-limited conditions [8].

The near-surface relative permittivity can obtained from the power ratio of the orthogonally polarized components of the received signal. Referring again to Figure 1, if transmitter (**T**) and receiver (**R**) performances and positions are known, the reflected power at the receiver can be expressed by the bistatic radar equation:

$$P_R = \frac{P_T G_T}{4\pi|T|^2}\sigma\frac{G_R\lambda^2}{(4\pi|R|)^2}. \tag{3}$$

Where P_T is the transmitted power, G_T and G_R are respectively transmitter and receiver gains, and σ is the planet radar cross section [8]. The latter depends on many unpredictable properties of the reflecting surface, such as the different scattering mechanisms effectively contributing to planet's echoes, the observation's geometry and the surface relative permittivity. The dielectric constant contributes to the radar cross section as a reflectivity, which describes how much power, of the original circularly polarized P_T, is reflected in the same sense (P_{SC}) and in the orthogonal sense (P_{OC}) of circular polarization [16]. The relation between relative permittivity and circular polarization reflectivity is described by the Fresnel reflection coefficients [16].

A simple model for the radar cross section of a rough planet is derived in [12] under the KA assumptions. For a generic sense of circular polarization with reflectivity Γ, the radar cross section is taken as follows [13].

$$\sigma = \frac{4\pi|T|^2 R_P\cos\theta}{(R_P\cos\theta + 2|T'|)(R_P + 2|T'|\cos\theta)}*\Gamma. \tag{4}$$

With R_p radius of the target planet. Combining together equations 3 and 4 it can be observed that the ratio between orthogonally polarized powers equals the ratio between circular polarization reflectivities. If we plug this equality into the expression of the Fresnel reflection coefficients the final expression that links relative permittivity (ε) and circular power ratio (CPR) can be written as follows:

$$\varepsilon = \sin^2\theta\left(1 + \frac{\tan^2\theta}{CPR}\right). \tag{5}$$

Hence, the near-surface permittivity of a reflecting surface can be retrieved from the ratio between the power of the two orthogonally polarized components of the surface's echoes [8].

Aerospace Science and Engineering - III Aerospace PhD-Days   Materials Research Forum LLC
Materials Research Proceedings 33 (2023) 110-117   https://doi.org/10.21741/9781644902677-17

It is not straightforward to associate a specific surface composition to a relative dielectric constant, especially for a body like Ganymede. The icy moon is expected to have a water-ice dominant near-surface composition with various fractions of non-ice contaminants [17] [18]. The relative permittivity of pure water-ice is 3.1, but this value varies with ice porosity, purity and temperature. This adds some complexity to the analysis of BSR observations, and increases the importance of a synergy between radar instruments on board JUICE. If a robust knowledge of the local surface composition is available, various models can be employed to describe the dependence of relative permittivity on porosity or ice purity [8] [19]. The surface porosity, for example, could be retrieved from the difference between the observed relative permittivity and the nominal value of 3.1 for water-ice.

JUICE Bistatic Radar Opportunities

The JUICE orbiter has the technical capabilities to perform downlink specular bistatic observations of Ganymede by means of its 3GM experiment. The spacecraft is equipped with a fixed High Gain Antenna (HGA) and a steerable Medium Gain Antenna (MGA). With the help of an on-board Ultra Stable Oscillator (USO), both the antennas are able to transmit and receive in X- ($\lambda \approx 3.6$ cm) and Ka-band ($\lambda \approx 0.9$ mm) [9]. The ground segment for the experiment consists of the 35m diameter stations of the ESTRACK network.

In the following, a preliminary assessment of bistatic observations of Ganymede during the GEO phase using the HGA transmitting in X-band is proposed. Both the feasibility and an expected quality of the scientific outputs are addressed.

Since the employment of JUICE instruments for other experiments is not taken into account, the only constraint to the feasibility of the observation is that the transmitting HGA antenna is capable of tracking the specular point. The specular point exists whenever the spacecraft is not behind Ganymede with respect to Earth, and its velocity over the planet's surface depends on the relative motion of JUICE, Ganymede and Earth. Since the HGA is fixed with the spacecraft, its capability to track the specular point depends on the maximum angular velocity the spacecraft can undergo during the mission. The upper bounds chosen for the analysis were the maximum angular velocity and acceleration that the spacecraft will encounter during the Jupiter Tour [11].

The location of the specular point at a given time instant is uniquely determined by the positions of JUICE, Ganymede and Earth, and was computed solving a nonlinear system of three equations with a strong geometrical meaning [20]. Going back to the labels of Figure 1, the following statements can be made about the specular point **S**:

    1) It lies over the surface of Ganymede's ellipsoid;
    2) **T'** and **R'** lie on the same plane;
    3) $\theta_I = \theta_R$.

Once the specular point is located, its velocity and the incidence angle observation can be computed.

In terms of quality of the scientific outputs of BSR, different conclusions can be drawn for different surface properties. The RMS slope can be satisfactorily retrieved as long as echoes are detectable and no beam-limitation occurs. To account for the latter, a maximum observable RMS slope was computed for the transmitting antenna FOV throughout the phase, using the expression of an effective reflecting area increasing with the RMS slope [8].

Aerospace Science and Engineering - III Aerospace PhD-Days   Materials Research Forum LLC
Materials Research Proceedings 33 (2023) 110-117   https://doi.org/10.21741/9781644902677-17

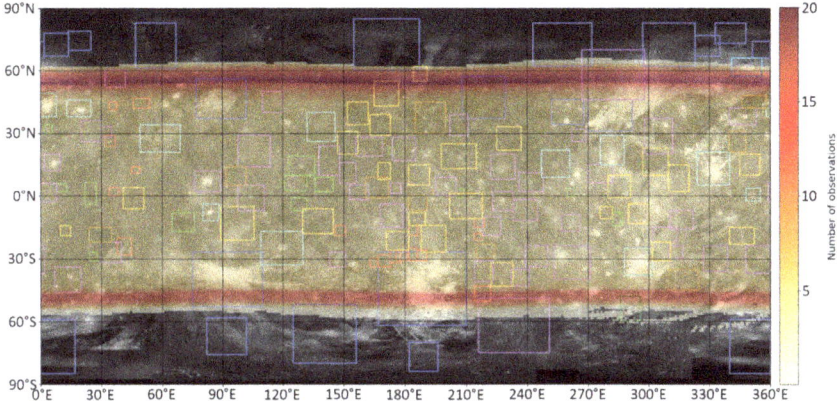

Figure 2 BSR observations coverage during the GEO phase (22/12/34 - 13/05/35). Areas shaded from yellow to red are illuminated by JUICE HGA assuming it to perfectly track the specular point. Different colors relate to a different number of passes over the same region. Colored squares highlight regions of interest over Ganymede according to [10]. Blu: polar deposits; Red: patarae, Green: dark ray craters, Purple: impact craters; Cyan: bright ray craters; Gold: bright terrains; Brown: ancient dark terrain

The dielectric constant can be retrieved from the reflected signals if both the orthogonally polarized components are detectable. For a terrain of known relative permittivity ε, the Brewster angle is the incidence angle that brings the circular power ratio to unity. If the observation is performed far from the Brewster angle, one of the polarized components will get weak, and its power computation will be more uncertain. For an accurate retrieval of relative permittivity the bistatic observation should be carried out close to the Brewster angle [16]. For a surface relative permittivity to be 3.1, the Brewster angle is roughly 60°.

Figure 2 shows the bistatic radar coverage during the elliptical part of the Ganymede Orbital phase when spacecraft angular velocity and acceleration are below 0.16°/s and 0.27e-3°/s², and when the incidence angle of observation is between 50° (CPR≈0.3) and 70° (CPR≈3). With a time resolution of 60s, the antenna beam pattern of JUICE pointing towards the specular point is plotted.

Assuming a relative permittivity of 3.1 to compute the surface reflectivity and combining equations 3 and 4, the amount of power received on Earth in both polarizations was estimated using the technical parameters shown in Tab. 1. The system noise temperature was assumed to be 28.5 K, which is a nominal value for Malargüe station at elevation of 50°, X-band and clear sky condition [11]. The signal-to-noise ratio (SNR) throughout the phase is always above 22.5 dBHz, which is a satisfying value to see echoes. This result should be considered an upper bound, and does not guarantee echoes will be actually visible, since some of the incoming power may not follow the model of specular reflection proposed by Fjeldbo. Earth-based radar observations of the icy moons hint at the presence of strong subsurface scattering phenomena which would escape Fjeldbo's assumptions, but specular echoes can't be ruled out [17] [22]. As it was the case for other planetary bodies, the validity of the model can be addressed only with new actual observations.

Aerospace Science and Engineering - III Aerospace PhD-Days   Materials Research Forum LLC
Materials Research Proceedings 33 (2023) 110-117   https://doi.org/10.21741/9781644902677-17

Table 1 Technical parameters of the bistatic radar radio link between JUICE and 35m Malargüe antenna [11]. *The power assumed for transmission in X-band is the full power of the Deep Space Transponder (DST)

	Transmit Power, P_T [W]	Transmit Gain, G_T [dBi]	Receive Gain, G_R [dBi]	Noise Temperature, T_{sys} [K]	Wavelength, λ [cm]	Half-power beamwidth, θ_{BW} [°]
JUICE HGA X-band	52*	43.85	68.2	28.5	3.6	1

Conclusions

The JUICE orbiter could be the first spacecraft to carry out BSR observations of one of the icy moons of Jupiter. This experiment would allow for an independent retrieval of surface roughness and relative permittivity which is a very precious return to characterize the surface of the moon, especially looking forward to a synergy with the other on-board radar instruments of the mission.

The purpose of this work was to show how a preliminary assessment of bistatic radar opportunities could be developed for a given mission and target planet. JUICE is taken as an example here, but the same analysis described in this abstract may be applied on other missions. With few simple formulas it was possible to make a case study that provides inputs to discuss when and how it could be better to observe Ganymede by means of bistatic radar.

For the future activity of the 3GM experiment, analogous considerations will be made for the whole mission, comparing the performances of the two antennas at the two possible frequency bands. More advanced considerations should be made about effective maneuverability and availability of the antennas before scheduling these observations. An extended and more detailed version of this analysis could be a starting point to find interesting time windows where BSR observations could be performed with a good expected scientific return.

As a final remark, these remain experiments of opportunity. The final scheduling of these observations will not be solely based on link-budget considerations, but a good compromise should be found between the expected scientific return and the many other scientific objectives of the JUICE mission as a whole: one of the most ambitious space missions of this century.

References

[1] M. Zannoni, D. Hemingway, P. Tortora and L. G. Casajus, "The gravity field and interior structure of Dione," Icarus, vol. 345, p. 113713, 2020. https://doi.org/10.1016/j.icarus.2020.113713

[2] L. Gomez Casajus et al., "Gravity Field of Ganymede After the Juno Extended Mission," Geophysical Research Letters, vol. 49, no. 24, pp. 1-10, 2022. https://doi.org/10.1029/2022GL099475

[3] L. Gomez Casajus et al., "Updated Europa gravity field and interior structure from a reanalysis of Galileo tracking data," Icarus, vol. 358, p. 114187, 2021. https://doi.org/10.1016/j.icarus.2020.114187

[4] V. Lainey et al., "Resonance locking in giant planets indicated by the rapid orbital expansion of Titan," Nature Astronomy, vol. 4, pp. 1053-1058, 2020. https://doi.org/10.1038/s41550-020-1120-5

[5] D. Buccino et al., "Ganymede's Ionosphere Observed by a Dual-Frequency Radio Occultation With Juno," Geophysical Research Letters, vol. 49, no. 23, 2022. https://doi.org/10.1029/2022GL098420

[6] A. Moirano et al.,"Morphology of the Io Plasma Torus From Juno Radio Occultations," Journal of Geophysical Research: Space Physics, vol. 126, no. 10, pp. 1-35, 2021. https://doi.org/10.1029/2021JA029190

[7] E. Gramigna et al.,"Analysis of NASA's DSN Venus Express radio occultation data for year 2014," Advances in Space Research, vol. 71, no. 1, pp. 1198-1215, 2023. https://doi.org/10.1016/j.asr.2022.10.070

[8] R. A. Simpson, "Spacecraft Studies of Planetary Surfaces Using Bistatic Radar," IEEE Transactions of Geosceince and Remote Sensing, pp. 465-482, 1993. https://doi.org/10.1109/36.214923

[9] European Space Agency, "ESA JUICE definition study report /Red Book. JUICE JUpiter ICy moons Explorer Exploring the emergence of habitable worlds around gas giants," ESA, 2014.

[10] K. Stephan et al., "Regions of interest on Ganymede 's and Callisto's surfaces as potential targets for ESA's JUICE mission," Planetary and Space Science, vol. 208, 15 November 2021.

[11] A. Boutonnet, A. Rocchi and W. Martens, "JUICE-Jupiter Icy moons Explorer Consolidated Report on Mission Analysis (CReMA)," ESA, 2022.

[12] G. Fjeldbo, "Bistatic-Radar Methods for Studying Planetary Ionospheres and Surfaces," 1964.

[13] R. A. Simpson and G. L. Tyler, "Viking Bistatic Radar Experiment: Summary of First-Order Results Emphasizing Noth Polar Data," Icarus, 1981. https://doi.org/10.1016/0019-1035(81)90139-1

[14] L. G. Tyler and H. T. Howard, "Dual-Frequency Bistatic-Radar Investigations of the Moon with Apollos 14 and 15," Journal of Geophysical Research, vol. 78, no. 23, pp. 4852-4874, 1973. https://doi.org/10.1029/JB078i023p04852

[15] R. A. Simpson et al., "Venus Express bistatic radar: High-Elevation anomalous reflectivity," Journal of Geophysical Research, vol. 114, 2009. https://doi.org/10.1029/2008JE003156

[16] R. A. Simpson et al. "Polarization of Bistatic Radar Probing of Planetary Surfaces: Application to Mars Express Data," Proceedings of the IEEE, pp. 858-874, 2011. https://doi.org/10.1109/JPROC.2011.2106190

[17] B. Hapke, "Coherent Backscatter and the Radar Characteristics of Outer Planet Satellites," Icarus, vol. 88, no. 2, pp. 407-417, 1990. https://doi.org/10.1016/0019-1035(90)90091-M

[18] G. J. Black, D. B. Campbell and P. D. Nicholson, "Icy Galilean Satellites: Modeling Radar Reflectivites as a Coherent Backscatter Effect," Icarus, vol. 151, no. 2, pp. 167-180, 2001. https://doi.org/10.1006/icar.2001.6616

[19] E. Heggy, G. Scabbia, L. Bruzzone and R. T. Pappalardo, "Radar probing of Jovian icy moons: Understanding subsurface water and structure detectability in the JUICE and Europa missions," Icarus, vol. 285, pp. 237-251, 2017. https://doi.org/10.1016/j.icarus.2016.11.039

[20] G. Brighi, "Cassini Bistatic Radar Experiments: Preliminary Results on Titan's Polar Regions," Aerotecnica Missili e Spazio, vol. 102, pp. 59-76, 2022. https://doi.org/10.1007/s42496-022-00135-4

[21] R. A. Simpson, L. G. Tyler and G. G. Schaber, "Viking Bistatic Radar Experiments: Summary of Results in Near-Equatorial Regions," Journal of Geophysical Research, vol. 89, no. B12, pp. 10385-10404, November 1984. https://doi.org/10.1029/JB089iB12p10385

[22] R. A. Simpson and G. L. Tyler, "Surface properties of Galilean satellites from bistatic radar experiments," NASA, Washington, Reports of Planetary Geology and Geophysics Program, 1990, 1991.

Aerospace Science and Engineering - III Aerospace PhD-Days   Materials Research Forum LLC
Materials Research Proceedings 33 (2023) 118-125   https://doi.org/10.21741/9781644902677-18

Helicopter tracking error and input aggression for point tracking tasks under boundary avoidance situations

Qiuyang Xia[1,2,a,b]

[1]Politecnico di Milano, Via La Masa 34, Milano, 20156, Italy

[2]Beihang University, 37 Xueyuan Rd., Haidian Dist., Beijing, 100191, China

[a]qiuyang.xia@polimi.it, [b]qiuyang.xia@buaa.edu.cn

Keywords: Rotorcraft-Pilot Coupling, Boundary Avoidance Tracking, Point Tracking, Human-Machine Interaction

Abstract. Helicopters are broadly applied in complex and harsh task environment such as rescue mission and firefighting. These tasks require helicopters to operate in ground proximity, keep tracking the target while avoid obstacles to avoid trashing. The combination of point tracking and boundary avoidance tracking can be utilized to describe this task condition. This study implemented a simulation task on MATLAB and Simulink and utilized a simplified helicopter dynamic model to investigate point tracking and boundary avoidance tracking tasks. The analysis of variance (ANOVA) and regression analysis were used to analyze the effects of task conditions on participants' tracking error and input aggression. Results demonstrated that the overall tracking error had a negative correlation with input aggression, and that participants tended to have higher input aggression and lower tracking error near the boundary.

Introduction

During the process of creating and operating modern high performance rotorcraft, engineers and pilots must anticipate and manage unfavorable occurrences known as "Rotorcraft-Pilot Coupling" (RPCs)[1]. These phenomena emerge from the undesired and atypical coupling between the pilot and the rotorcraft and can lead to instabilities that are both oscillatory and non-oscillatory, reducing handling quality, increase structural strength requirements, and sometimes resulting in disastrous accidents.

Boundary-avoidance Tracking is a pilot-task model proposed by Gary[2], [3], which indicates that in the process of performing flight tasks, pilots not only need to complete the task of "maintaining specific parameters" (point tracking) but also typically need to "avoid certain parameters" (boundary avoidance. Researchers believe that boundary avoidance behavior has a strong correlation with the critical phenomenon of PIO, which previous point tracking models cannot correctly.

In the process of task execution, the situation awareness of the human-machine system has a significant impact on task performance[4]. The task design of this study includes explicit tracking tasks and direct data acquisition. Therefore, point tracking error and input aggression were used to evaluate the performance of participants, reflecting their situation awareness.

This study is based on a simple hardware flight simulator. The core of the task design of this study lies in the randomness of the task. In several previous studies about pilot boundary avoidance model[5], [6], periodic tasks were used, which would cause participants to learn the regularity of the task goals and predict the action trajectory of the task goals, thus affecting the objectivity of the experiment and model fitting.

This research investigated pilot's response to point tracking and boundary avoidance in a simulated flight task, and the main aim of this study lies in the tracking performance and input

Aerospace Science and Engineering - III Aerospace PhD-Days   Materials Research Forum LLC
Materials Research Proceedings 33 (2023) 118-125   https://doi.org/10.21741/9781644902677-18

aggression under different task conditions regarding both point tracking and boundary avoidance tracking.

Method

Figure 1. The joystick used for simulation tasks.

Fourteen participants volunteered and took part in the experiment tests. The joystick utilized in this study had two main sticks. The left stick only moved in vertical direction, simulating the "collective" inceptor for helicopters, and the right stick moved in both vertical and horizontal directions, simulating the "cyclic" inceptor for helicopters. The simulated task was developed and operated on a laptop. The joystick was connected to the laptop using a USB cable from the joystick.

The tasks were designed based on a helicopter tracking task and the concepts of "point tracking" and "boundary avoidance tracking". During the task two types of information were displayed on the monitor, meaning "point" and "boundary".

The GUI was designed in svg format. The participants would see an interface as Figure 2.

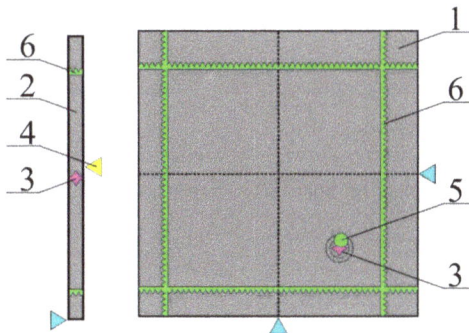

Figure 2. GUI Interface.

Elements displayed in the interface is explained as below:
1. Square scale, indicates the cyclic stick.
2. Vertical scale, indicates the collective stick.
3. Purple diamonds, point tracking target indicators.
4. Triangle indicator, displays the "response" of the participants controlling the collective stick.
5. Dot indicator, displays the "response" of the participants controlling the cyclic stick.
6. Sawtooth boundaries, for boundary avoidance tracking tasks.

Aerospace Science and Engineering - III Aerospace PhD-Days   Materials Research Forum LLC
Materials Research Proceedings 33 (2023) 118-125   https://doi.org/10.21741/9781644902677-18

Among them, 4 and 5 would change color according to the distance between target and response, as an indicator for the participants to adjust their controlling strategies. 6 would also change color if the response was close to the boundaries.

This research used Simulink module integrated in MATLAB 2022a to generate and output target and boundary movement signals, at the same time transfer input signals. The Simulink terminates the task when participants hit the boundary.

Target movement parameters were set random in a certain limit, consequently, target movements were random. An example of one-axis target movement was shown in Figure 3. During the whole experiment, participants individually experienced a set of unpredictable random tasks, while tasks were consistent among all the participants.

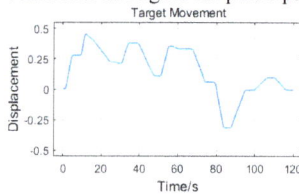

Figure 3. Target Movement.

Boundary movement patterns could be configured as "discrete" or "continuous". An example of one-axis boundary movement was shown in Figure 4. In the figure, the difference between "discrete" and "continuous" was illustrated.

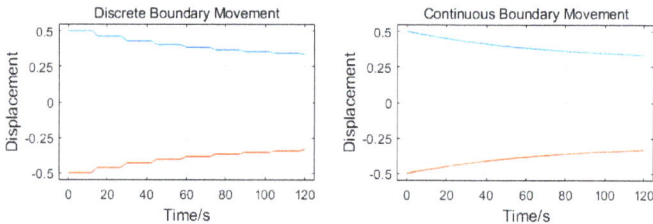

Figure 4. "Discrete" and "continuous" boundary movement.

Shift movement was applied on both target and boundary. Figure 5 shows the boundary movement with shifting.

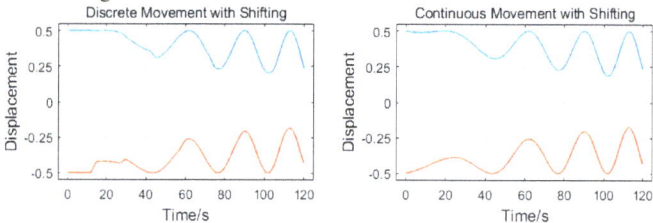

Figure 5. Boundary movement with shifting.

The participants were instructed to operate only the cyclic stick. The participants performed a sequence of 5 different types of tasks, each type 3 runs (except for Task 0 which the participants can repeat as many times as they wished). The types of tasks were described below:

a) Task 0: point tracking task only with no terminate condition other than task duration.

Aerospace Science and Engineering - III Aerospace PhD-Days   Materials Research Forum LLC
Materials Research Proceedings 33 (2023) 118-125   https://doi.org/10.21741/9781644902677-18

b) Task 1: boundary avoidance task, boundary movement is "discrete", no shifting.
c) Task 2: boundary avoidance task, boundary movement is "continuous", no shifting.
d) Task 3: boundary avoidance task, boundary movement is "discrete" with shifting.
e) Task 4: boundary avoidance task, boundary movement is "continuous" with shifting.

In this study, tracking error and input aggression was utilized to evaluate the performance and control strategy of the participants. The calculation of these indicators is as below:

$$error = target - response \tag{1}$$

$$aggression = \sqrt{\frac{1}{t_1 - t_2} \int_{t_1}^{t_2} \left| \dot{\delta}(t) \right|^2 dt} \tag{2}$$

where:

target – sequence of target movement signals.

response – sequence of response signals.

$\delta(t)$ – sequence of input signals, and $\dot{\delta}(t)$ is the time derivative.

t_1, t_2 – starting and ending time of a time interval for analyzing aggression.

To evaluate the performance and aggression in certain period (whole task run, or specific condition, for example), the root mean square value of error and aggression is calculated:

$$errorRMS = \sqrt{\frac{1}{n} \sum_i error_i^2} \tag{3}$$

$$aggressionRMS = \sqrt{\frac{1}{n} \sum_i aggression_i^2} \tag{4}$$

Different "conditions" were defined to distinguish different groups of situations the participants encountered during the tasks:

a) Group 1

"*Approach*": The dot moves towards one of the boundaries.

"*Leave*": The dot moves in a reversed direction of one of the boundaries.

b) Group 2

"*Near*": The dot locates between one of the boundaries and "boundary thresholds".

"*Away*": The dot locates outside boundary thresholds.

c) Group 3

"*Approach and Near*": Conditions that meet both "*Approach*" and "*Leave*" at the same time.

"*Leave or Away*": Conditions that meet "*Leave*" or "*Away*".

Figure 6 demonstrates the above conditions in an intuitive way.

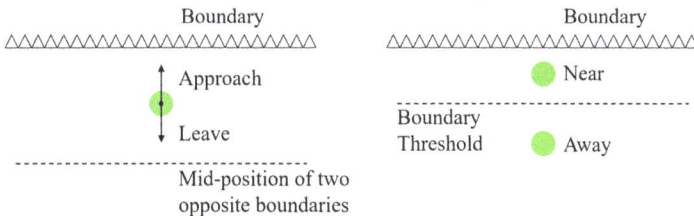

Figure 6. Demonstrations of "conditions".

The analysis of this paper is based on a boundary threshold of 0.1. which is low capture difficult tasks, and large enough to sample enough amount of data to be analyzed.

Statistical analyses were done with task runs that didn't fail, to extract data that fully represented participants performances under pressure.

Aerospace Science and Engineering - III Aerospace PhD-Days   Materials Research Forum LLC
Materials Research Proceedings 33 (2023) 118-125   https://doi.org/10.21741/9781644902677-18

Results
a) Group 1

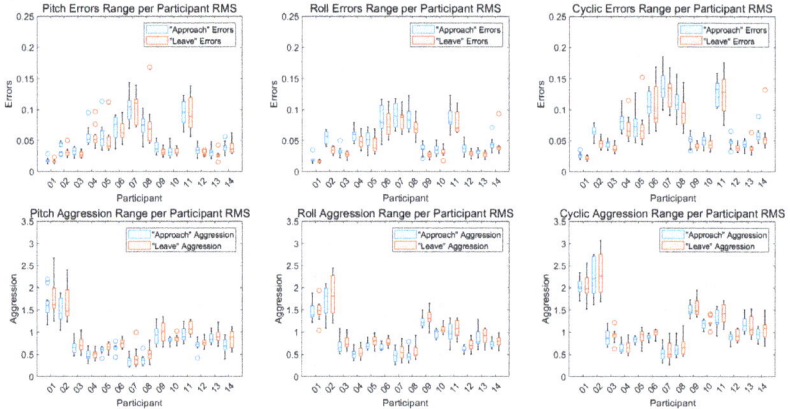

Figure 7. Error and Aggression under "Approach" and "Leave" conditions.

Table 1. ANOVA analysis between "Approach" and "Leave" Conditions

Difference Significance: Approach and Leave Conditions

Tracking Error

Pitch			Roll			Cyclic		
Source	F-value	P-value	Source	F-value	P-value	Source	F-value	P-value
Condition	4.15	0.042	Condition	33.78	<0.001	Condition	13.80	<0.001
Participant	58.58	<0.001	Participant	70.93	<0.001	Participant	76.19	<0.001
Condition*Participant	0.29	0.993	Condition*Participant	1.03	0.421	Condition*Participant	0.45	0.951

Input Aggression

Pitch			Roll			Cyclic		
Source	F-value	P-value	Source	F-value	P-value	Source	F-value	P-value
Condition	18.32	<0.001	Condition	14.38	<0.001	Condition	4.97	0.027
Participant	89.52	<0.001	Participant	98.80	<0.001	Participant	132.63	<0.001
Condition*Participant	0.42	0.963	Condition*Participant	0.17	1.000	Condition*Participant	0.24	0.997

The tracking errors under "*Approach*" condition were slightly larger than that under "*Leave*" condition for most participants. But the results showed no statistical significance (p>0.05). The input aggressions under "*Approach*" condition were slightly larger than that under "*Leave*" condition, though the significance of the difference is not observed (p>0.05).

The result showed that participants were more likely to control the stick less aggressively when they tried to follow the target that was getting close to the boundary to avoid hitting the boundary.

b) Group 2

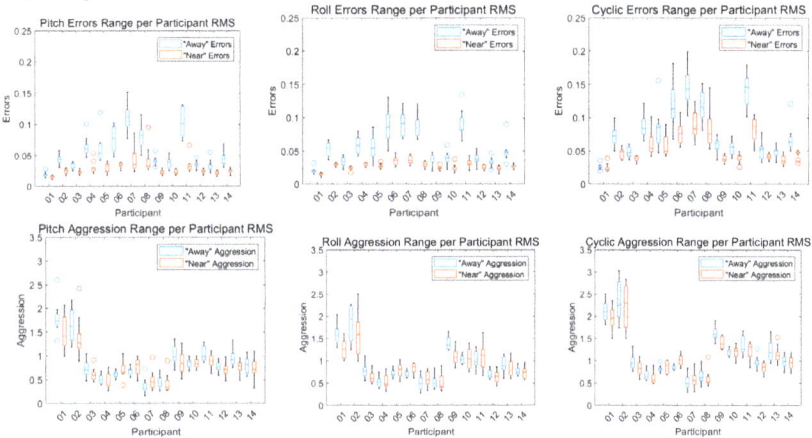

Figure 8. Error and Aggression under "Near" and "Away" conditions.

Table 2. ANOVA analysis between "Near" and "Away" Conditions

Difference Significance: Near and Away Conditions

Tracking Error

Pitch			Roll			Cyclic		
Source	F-value	P-value	Source	F-value	P-value	Source	F-value	P-value
Condition	335.44	<0.001	Condition	519.03	<0.001	Condition	162.14	<0.001
Participant	43.93	<0.001	Participant	44.27	<0.001	Participant	63.15	<0.001
Condition*Participant	13.48	<0.001	Condition*Participant	19.71	<0.001	Condition*Participant	5.24	<0.001

Input Aggression

Pitch			Roll			Cyclic		
Source	F-value	P-value	Source	F-value	P-value	Source	F-value	P-value
Condition	12.34	<0.001	Condition	10.77	0.001	Condition	9.26	0.003
Participant	70.31	<0.001	Participant	72.41	<0.001	Participant	142.73	<0.001
Condition*Participant	2.10	0.014	Condition*Participant	3.04	<0.001	Condition*Participant	1.10	0.355

The tracking errors under "*Near*" condition are significantly lower than that under "*Away*" condition ($p<0.001$) for all participants. Participants' behavior during the test runs showed that they exerted greater effort to control the stick, maintaining point tracking task while prevent hitting the boundary.

The input aggressions presented inconsistent results. For some participants, the input aggressions were larger under "*Near*" condition, others were lower. The difference of the input aggressions was significant ($p<0.05$).

c) Group 3

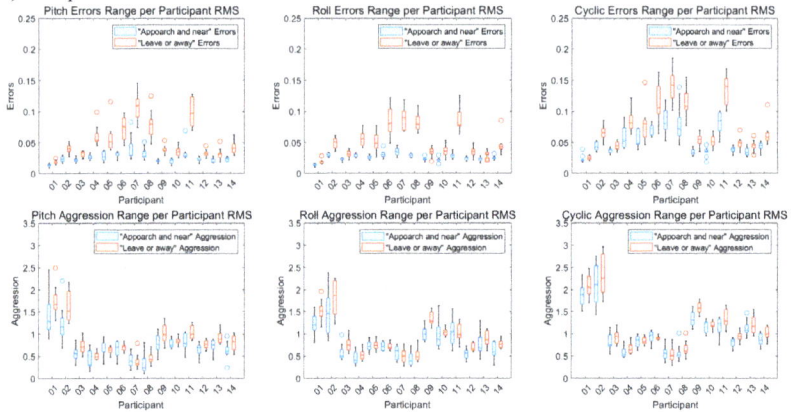

Figure 9. Error and Aggression under "Approach and Near" and "Leave or Away" conditions.

Table 3. ANOVA analysis between "Approach and Leave" and "Near or Away" Conditions

Difference Significance: Approach and Near and Leave or Away Conditions

Tracking Error

Pitch			Roll			Cyclic		
Source	F-value	P-value	Source	F-value	P-value	Source	F-value	P-value
Condition	392.54	<0.001	Condition	537.74	<0.001	Condition	167.23	<0.001
Participant	45.36	<0.001	Participant	50.57	<0.001	Participant	68.64	<0.001
Condition*Participant	15.91	<0.001	Condition*Participant	23.29	<0.001	Condition*Participant	5.79	<0.001

Input Aggression

Pitch			Roll			Cyclic		
Source	F-value	P-value	Source	F-value	P-value	Source	F-value	P-value
Condition	42.93	<0.001	Condition	35.88	<0.001	Condition	25.72	<0.001
Participant	62.66	<0.001	Participant	66.40	<0.001	Participant	140.93	<0.001
Condition*Participant	1.95	0.025	Condition*Participant	2.11	0.014	Condition*Participant	1.01	0.444

The tracking errors under "*Approach and Near*" and "*Leave or Away*" condition showed a similar trend as under "*Near*" and "*Away*" condition mainly because tracking errors under "*Approach*" condition and "*Leave*" condition showed no significant difference. The difference of tracking error here also showed statistical difference ($p < 0.001$). Inconsistent results were also observed for aggression under this group. The difference between conditions and participants are significant for pitch and roll axis($p < 0.05$), but not the composed cyclic($p > 0.05$). Different participants applied different input strategies under severe task conditions, resulted in different point tracking performance.

Conclusion

This research featured a simulation task design based on the concepts of point tracking and boundary avoidance tracking, and data analysis method to investigate pilots' point tracking performance and input aggression. Several results could be drawn from this study. Since the boundary avoidance tracking was introduced to the task, participants presented different input strategy indicated as aggression, and resulted in different tracking error. When the target was near the boundary, the participants presented significantly lower tracking errors ($p < 0.001$), and the input aggressions are also lower for the pitch and roll axis respectively ($p < 0.05$). In summary, this research demonstrated the relationship among task condition, input strategy, and task performance.

Task design, data analyzing method, and results could inspire related research in pilot's biodynamic feedthrough, human-machine interaction, and rotorcraft-pilot coupling.

References

[1] M. D. Pavel *et al.*, "Adverse rotorcraft pilot couplings—Past, present and future challenges," *Progress in Aerospace Sciences*, vol. 62, pp. 1–51, Oct. 2013. https://doi.org/10.1016/j.paerosci.2013.04.003

[2] W. Gray, "Boundary-Escape Tracking: A New Conception of Hazardous PIO:," Defense Technical Information Center, Fort Belvoir, VA, Sep. 2004. Accessed: Dec. 10, 2021. [Online]. Available: http://www.dtic.mil/docs/citations/ADA427054

[3] W. Gray, "Boundary Avoidance Tracking: A New Pilot Tracking Model," in *AIAA Atmospheric Flight Mechanics Conference and Exhibit*, San Francisco, California, Aug. 2005, p. 5810. https://doi.org/10.2514/6.2005-5810

[4] M. R. Endsley, "A Systematic Review and Meta-Analysis of Direct Objective Measures of Situation Awareness: A Comparison of SAGAT and SPAM," *Hum Factors*, vol. 63, no. 1, pp. 124–150, Feb. 2021. https://doi.org/10.1177/0018720819875376

[5] L. Lu and M. Jump, "Pilot modelling for boundary hazard perception and reaction study," in *43rd European Rotorcraft Forum, ERF 2017*, Politecnico di Milano, Bovisa, Milano, Italy, Jan. 2017, vol. 1, pp. 640–652. Accessed: Mar. 06, 2023. [Online]. Available: https://livrepository.liverpool.ac.uk/3009474

[6] H. Ji, L. Lu, M. D. White, and R. Chen, "Advanced pilot modeling for prediction of rotorcraft handling qualities in turbulent wind," *Aerospace Science and Technology*, vol. 123, p. 107501, Apr. 2022. https://doi.org/10.1016/j.ast.2022.107501

Aerospace Science and Engineering - III Aerospace PhD-Days   Materials Research Forum LLC
Materials Research Proceedings 33 (2023) 126-133   https://doi.org/10.21741/9781644902677-19

Exploring the use of a technology scouting methodology to integrate innovative solutions from startups into an aerospace industry

Giovanna Carrera[1,a] *, Luca Mastrogiacomo[1,b]

[1] Department of Production and Management, Politecnico di Torino, Torino

[a]giovanna.carrera@polito.it, [b]luca.mastrogiacomo@polito.it

Keywords: Open Innovation, Startups, Scouting Process, Roadmap, Technology Evaluation

Abstract. Technological innovation is extremely important in the industrial world, as it allows companies to remain competitive and improve the efficiency, productivity, and sustainability of their activities. Companies that invest in technological innovation can obtain numerous advantages, including improved product quality, cost reduction, greater flexibility, and the ability to quickly respond to customer needs. An approach to open innovation that has become widespread in recent years is to conduct technology scouting through the vast array of solutions provided by innovative startups. With a large number of startups proposing new technologies in aerospace sector, could be challenging for companies to identify the most promising solutions. Therefore, a structured methodology for evaluating technology proposed by startups is essential to ensure the identification and implementation of the best solutions and the effective allocation of resources. This paper presents a case study that describes the process of technology scouting proposed by startups within an aerospace industry company, based on a company-defined roadmap.

1. Introduction

The aerospace industry is constantly pushing the boundaries of technology, and startups have a significant role to play in this process. That is why it is important for aerospace companies to conduct scouting of technology proposed by startups, in order to stay at the forefront of innovation. During each "innovation call," regardless of the proposed technology, there are many candidate startups, and the need to evaluate and choose the one that best aligns with the company's trend is very important. The purpose of this paper is to present a case study in which a technological scouting methodology has been applied.

A structured methodology for evaluating technology proposed by startups allows for the continuous improvement of the evaluation process and builds trust and confidence among all stakeholders.

The entire scouting process is based on a company-defined roadmap. The steps begin when the company defines the problem in which it is interested in scouting, and continue until the selection of the best alternative among those presented.

2. Scouting Process: Roadmap

The industry scouting process for the selection of a technology proposed by a startup is a multi-step process that involves several phases. It involves a combination of factors, such as the alignment of the technology with the company's business needs, the potential impact of the technology, and the startup's ability to execute on their plans.

The entire scouting process can be imagined as a set of activities to be followed in succession.[1] The roadmap on which it develops is shown in Fig 1. It is characterized by two nested processes: the scouting process and the technology evaluation methodology process. The two processes are complementary:

- The first process (colored in red) concerns all activities carried out internally by the company or potentially by startup acceleration companies and relates to the selection, screening, and finally management of collaboration contracts with the chosen alternatives.
- The second process (colored in light blue) is a more complex and articulated process. It includes step-by-step evaluation activities of the technology proposed by the startup.
- The following paragraphs, describe detail the different steps that make up the roadmap. Each of these phases has been implemented according to the pattern shown in the Fig 1.

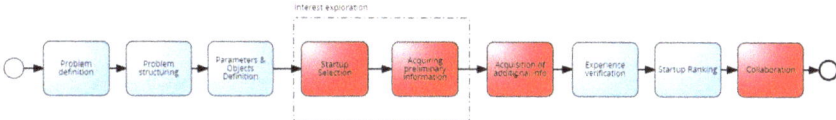

Figure 1 Roadmap

2.1   Problem Definition

Problem definition is the process of clearly identifying and understanding the problem or need that technology is intended to address, Fig 2.

2.1.1 Understanding and defining business needs & Priority processing

Structuring a collection of business needs is important for innovation scouting because it helps to identify the specific areas where new technology is needed and where it can have the most impact. By having a clear understanding of business needs, companies can better focus their innovation scouting efforts on areas that are greatest relevant to their operations and where they can add the most value. This allows for a more efficient use of resources and a higher likelihood of successfully identifying and implementing new technology that can help the company achieve its goals.

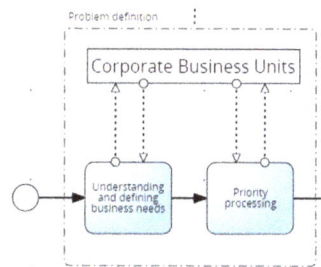

Figure 2 Problem Definition

At the corporate level, the relevant business unit has prepared a "needs sheet" which is a collection of all the requirements. It highlights the reference problem, the possible directions of solutions, and the main elements that should characterize the technological solution proposed by the startup.

Following the receipt of this document, the open innovation pilot team takes care of elaborating the data, defining reference keywords of the chosen topic, organizing the scouting mode, and structuring various support tools for the entire process.

2.2   Problem Structuring

Problem structuring is the second phase of the process, in which the problem is organized and categorized in a way that allows a comprehensive analysis (see Fig 3).

During this phase, with the support of the Business Unit related to the topic selected during the Problem Structuring phase, a panel of experts has been structured, to participate in the evaluation of the startup candidates.

The selection of the panel was done carefully, as the final selections of the scouting process will depend on it. In this case, the expert panel for the evaluation of startup candidates has been purposely formed in a multidisciplinary way in order to obtain a broad evaluation of the candidate companies.

In this project the panel is made up of an expert in the topic from the relevant corporate Business Unit, a full professor from the Politecnico di Torino, part of the department of management and production engineering, a Ph.D student with sector expertise, and two professionals from the Open Innovation Team who manage the scouting project.

2.3 Parameters & Objects Definition

This phase of the roadmap is the process of defining key parameters and objectives that will be used to evaluate the proposed technology, (see Fig 4).

Before going into the details of this phase, it is important to specify its structure. At the company level, regardless of the scouting process, which we will refer to as Dealflow from now on, it was decided to evaluate the candidate startups based on four macro areas which we will define as "parameters". Specifically, the evaluation parameters chosen by the management to assess the suitability objective are Team, Finance, Community and Technology.

Regarding the first three parameters, the company decided to keep the defining criteria for each group unchanged for every Dealflow. The only parameter subject to change for each Dealflow is that related to technology. This parameter has different needs based on the type, field of application, difficulty of implementation, and structure.

In this phase, the panel of experts is tasked with deciding the weight to give to each parameter and its associated criteria (Team, Finance, Community, and Technology). The panel of experts will also define the criteria and objectives that will be used to evaluate the proposed technology. This step is critical in determining the technology's value proposition and potential for success.

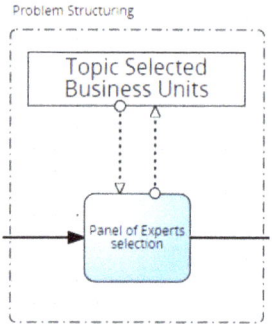

Figure 3 Problem

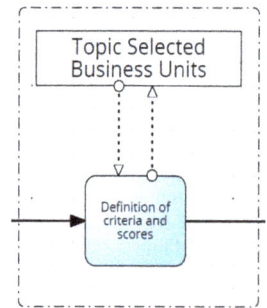

Figure 4 Parameters & Objects Definition

2.3.1 Definition of Criteria and Scores

This stage of the process is structured into the following subphases:

- *Definition of parameters weight s and consistency verification*

The first step is to assign a weight to each parameter in the evaluation areas. Through a brainstorming technique, the panel of experts expressed the importance ratios of each parameter relative to the others.

Subsequently, the consistency of the weights was verified through pairwise comparison using the Analytic Hierarchy Process (AHP) technique, followed by the eigenvector method[2,3,4]. The result, shown in Figure 5, indicates that the evaluators expressed a strong preference for the parameter related to Technology, which was assigned a weight of 62%, while the parameters of Team, Finance, and Community were assigned final levels of importance of 18%, 10%, and 10%, respectively, Fig 5.

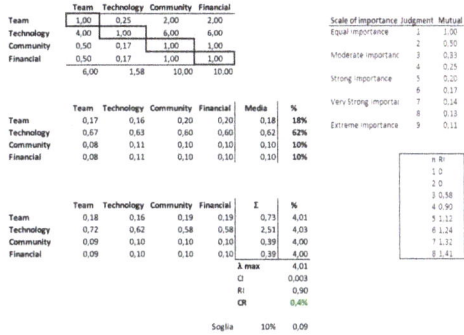

	Team	Technology	Community	Financial
Team	1,00	0,25	2,00	2,00
Technology	4,00	1,00	6,00	6,00
Community	0,50	0,17	1,00	1,00
Financial	0,50	0,17	1,00	1,00
	6,00	1,58	10,00	10,00

Scale of importance	Judgment	Mutual
Equal importance	1	1.00
	2	0.50
Moderate importance	3	0.33
	4	0.25
Strong importance	5	0.20
	6	0.17
Very Strong importance	7	0.14
	8	0.13
Extreme importance	9	0.11

	Team	Technology	Community	Financial	Media	%
Team	0,17	0,16	0,20	0,20	0,18	18%
Technology	0,67	0,63	0,60	0,60	0,62	62%
Community	0,08	0,11	0,10	0,10	0,10	10%
Financial	0,08	0,11	0,10	0,10	0,10	10%

n	RI
1	0
2	0
3	0.58
4	0.90
5	1.12
6	1.24
7	1.32
8	1.41

	Team	Technology	Community	Financial	Σ	%
Team	0,18	0,16	0,19	0,19	0,73	4,01
Technology	0,72	0,62	0,58	0,58	2,51	4,03
Community	0,09	0,10	0,10	0,10	0,39	4,00
Financial	0,09	0,10	0,10	0,10	0,39	4,00
				λ max	4,01	
				CI	0,003	
				RI	0,90	
				CR	0,4%	
				Soglia	10%	0,09

Figure 5 Definition of parameters weights

- *Criteria associated with the parameters Team, Finance, and Community*

As previously mentioned, the criteria associated with the three parameters related to the areas of Team, Finance, and Community are unchanged for each Dealflow, Fig 6.

Figure 6 Criteria & Parameters

- *Definition of Technology Parameter Criteria:*

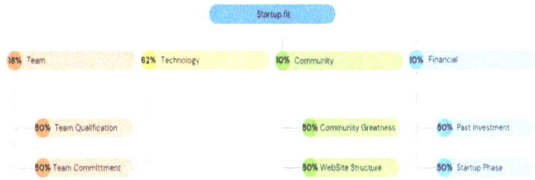

The crucial part of this phase of the process is to define the criteria for evaluating the different technologies proposed by the startups.

This phase of the process has been structured in a mixed *Delphi-NTG-Brainstorming* approach. All these techniques were leveraged to achieve the methodology with the best outcome[5].

- *Delphi: Sharing the topic with the panel of experts*

The Delphi technique is a method used to gather and process information from a group of experts in order to reach a consensus on a specific issue[6].

An invitation to a meeting has been sent to the panel of experts, attaching the topic of the meeting and highlighting the main objective of the scouting process in question.

At this point, the experts become more aware of the topic and begin to have an idea of what the truly important aspects are for the company and what the main objectives are to achieve.

- *Nominal technique group NTG - Presentation of solutions*

The nominal group technique is a method used to collect and prioritize information and ideas from a group of people. In the case study under consideration, the respondents are the 5 people who make up the previously defined panel of experts. This technique can be used in the selection of criteria for evaluating technology proposed by startups.

The previously defined panel of experts generates a list of suitable criteria for evaluating the performance of startups. The group worked together to prioritize the criteria in order to assess the technology proposed by startups effectively.

- *Brainstorming - Final Selection of Suitable Criteria*

In this phase of the process, once N criteria have been defined, the panel of experts is called to come together to refine the choices made in order to obtain a smaller number of criteria, which will make it easier to start the evaluation process in a more streamlined way. The experts of the scouting process chose to use a brainstorming technique.

CRITERIA EVALUATION	Technology Expert	Full Professor	Ph.D. Student, Team Digital Factory	Ph.D Student, Team Open Innovation	Member of Open Innovation Team	FINAL SCORE
Accuracy:	2	4	3	4	5	18
Industry Fit:						0
Cost-Effectiveness:	1			1		2
Security:						0
Technological advancements:		3	5	5		13
Real-time Capability:	4	5	4	3	4	20
Differentiation:						0
Data Management:	5				3	8
Go-to-market strategy:					1	1
Customer traction:			1			1
Forecast optimization:						0
Customizability:						0
Certifiability:						0
Flexibility:						0
Intellectual Property:		2				2
Risk Evaluation:	3	1	2	2	2	10
Competitive landscape:						0

Table 1 Final Selection of Criteria

Once all the ideas have been generated, the group then goes through each one, grouping similar ideas together and eliminating duplicates. From there, the group may use a voting process or other criteria to prioritize the ideas and select the final set for the decision-making process.

In this case, the criteria deemed most suitable for the scouting case in question for the evaluation of technology in the digital twin field were the following in Table 1.

Each evaluator was asked to rank the criteria they deemed most appropriate by assigning a score from 1 to 5 (with 1 being less important and 5 being more important). The sum of votes for each criterion was then calculated horizontally. As shown in Table 1, three macro-groups of criteria can be identified. The first is characterized by all those criteria that did not receive any scores. This was a result of the ideation process that led many evaluators to

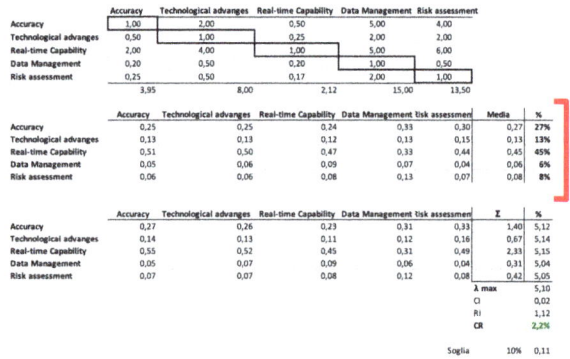

	Accuracy	Technological advanges	Real-time Capability	Data Management	Risk assessment
Accuracy	1,00	2,00	0,50	5,00	4,00
Technological advanges	0,50	1,00	0,25	2,00	2,00
Real-time Capability	2,00	4,00	1,00	5,00	6,00
Data Management	0,20	0,50	0,20	1,00	0,50
Risk assessment	0,25	0,50	0,17	2,00	1,00
	3,95	8,00	2,12	15,00	13,50

	Accuracy	Technological advanges	Real-time Capability	Data Management	Risk assessment	Media	%
Accuracy	0,25	0,25	0,24	0,33	0,30	0,27	27%
Technological advanges	0,13	0,13	0,12	0,13	0,15	0,13	13%
Real-time Capability	0,51	0,50	0,47	0,33	0,44	0,45	45%
Data Management	0,05	0,06	0,09	0,07	0,04	0,06	6%
Risk assessment	0,06	0,06	0,08	0,13	0,07	0,08	8%

	Accuracy	Technological advanges	Real-time Capability	Data Management	Risk assessment	Σ	%
Accuracy	0,27	0,26	0,23	0,31	0,33	1,40	5,12
Technological advanges	0,14	0,13	0,11	0,12	0,16	0,67	5,14
Real-time Capability	0,55	0,52	0,45	0,31	0,49	2,33	5,15
Data Management	0,05	0,07	0,09	0,06	0,04	0,31	5,04
Risk assessment	0,07	0,07	0,08	0,12	0,08	0,42	5,05

λ max	5,10
CI	0,02
RI	1,12
CR	2,2%
Soglia	10%	0,11

Figure 7 AHP Process

reconsider their ideas and prefer others. The second macro-group consists of criteria that received low scores such as cost efficiency, go-to-market strategy, customer traction, and intellectual property. Finally, there are criteria characterized by a higher score that have emerged as the object of greater interest by unanimity. The latter, highlighted in gray in the table, are Accuracy, Technological Advancements, Real-Time Capability, Data Management, and Risk Evaluation. The final choice therefore fell on 5 out of the total 17 criteria previously proposed.

- *AHP process*

After the selection, the weight was calculated again through pairwise comparison using the AHP technique [2,3,4]. With the help of the panel of experts, the levels of importance necessary to complete the pairwise comparison matrix were unanimously decided by the evaluators through brainstorming. In order to verify the consistency of the weights assigned to these criteria, the Consistency Index was calculated using the eigenvector technique. This procedure resulted in an acceptable consistency among the parameters, with the CI value of 0.022 being lower than 10%, Fig 7.

2.4   Interest Exploration

The phase of the Interest exploration process corresponds to the moment when the company launches the Business Scouting call. This is nested with different steps as shown in the Fig 8.

Through various search engines, Startups with technologies deemed of interest for the examined business problem were selected. In the case at hand, Dealflow, led to the identification of 20 startups with technology that has good potential for the company.

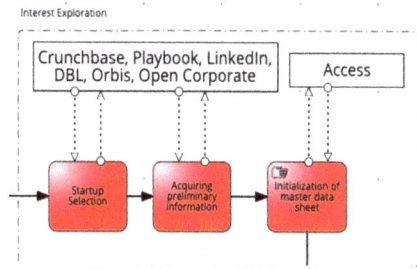

Figure 8 Interest Exploration

2.5   Acquisition of additional Data

In this phase of the process, the previously selected companies must be evaluated based on their complementarity with the company's needs, Fig 9. After storing the list of startups related to the scouting flow and their preliminary information in a database using a management tool, a report containing the profile sheets of each startup is generated. This reference report is then sent to a panel of experts for a preliminary evaluation. The aim of this initial assessment is to exclude those companies that are definitely not suitable for the scouting call and that, according to the experts, may not have the required skills. For the startups that are deemed valid at first

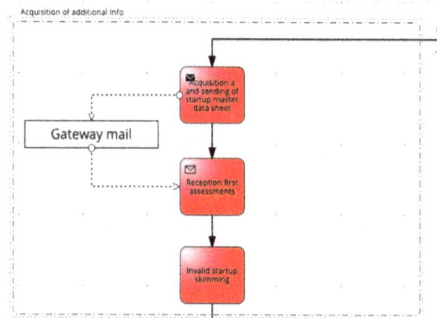

Figure 9 Acquisition of additional Data

glance, further information will be acquired to deepen the scouting process. This type of evaluation is mainly carried out by the direct stakeholders of the relevant business unit who evaluate the technological aspect of the startup and its adaptability to the company's core business. The first screening by the expert panel led to a screening of approximately 70% of the candidates. The scouting process continued with 7 candidates.

2.6   Experience Verification

The "Experience Verification" phase of the roadmap is the process of verifying the startup's experience and track record in the field, Fig.10.

Company's open innovation team prepared several questionnaires to be submitted to selected startups with questions aimed at deepening the topics of interest.

If the preliminary evaluation, supported by the initial information, is positive, after having further examined the characteristics of the companies, they can proceed with the first individual meeting with the panel of experts. Otherwise, they will be excluded from the final selection but will still remain within the company database for future reconsideration. The experience verification phase aims to understand the actual technology of the company, attend presentations by the candidates, delve into the topic of technology, the startup's mission, and the organizational and work method. In the case study, a Pitch Day was set up, during which each startup presented their activities, objectives, and proposed technology. The set of meetings was held entirely within a single day in which the company attended, in succession, the 7 individual startup pitches, each lasting 30 minutes, followed by 5 minutes in which the panel of experts expressed their evaluation. To do this, the individuals managing the scouting process prepared a questionnaire to send to the panel

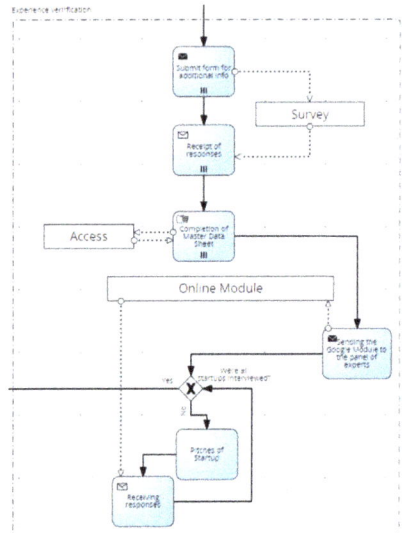

Figure 10 Experience Verification

of experts so that at the end of each pitch, they could give their evaluation of the startup in a simple and immediate way. This questionnaire was produced using an Online Module to facilitate the creation, sending, and collection of results and is divided into four modules. It is divided into five sections and aims to collect scores on a Likert scale ranging from 1 to 7 for each previously defined criterion [7]. The Likert scale range captures the intensity of individuals' perception for the different proposed startups. In the present case, and following a careful literature analysis, it was decided to use a 7-point Likert scale of preference [8]. The Online Module organization has provided for the presence of 5 sections related to the selection of evaluation parameters: Startup header, Technology, Team, Community, Financial.

2.7 Startup Ranking
The "Startup Ranking" phase of the roadmap is the process of ranking the startups based on the criteria/parameters established in the previous steps. This step is important as it is complex at the same time. Many approaches are possible, one of the possible ones has been chosen here to determine which technologies and startups warrant further

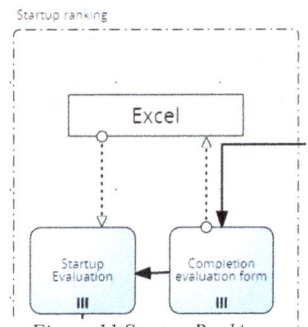

Figure 11 Startup Ranking

investment and support (see Fig 11). Automatically, the previous Form, presented a final result for each criteria, given by the average of the scores obtained by the 7 components of the expert panel. The results obtained from the Likert scale were analyzed as a whole and weighted again according to the relative importance of the parameters in question. The final calculation shown in the table resulted in a ranking of suitability for each startup, taking into account all previously analyzed evaluation parameters and criteria. In order to select only the most suitable startups, an acceptance threshold of 4,2 (arbitrary chosen) was set, which represents the mathematical sufficiency (highlighted in green in Fig 12).

Conclusions

The use of a technology scouting methodology to integrate innovative solutions from startups into the aerospace industry is crucial to remain competitive and improve the efficiency, productivity, and sustainability of the company's activities. Through the case study presented, we discussed a structured methodology for

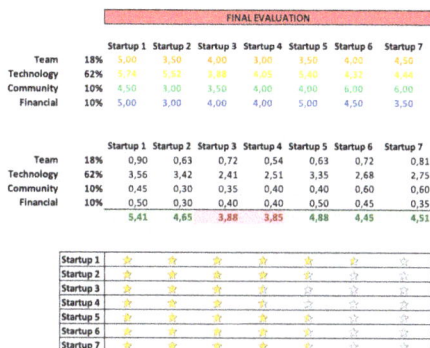

		Startup 1	Startup 2	Startup 3	Startup 4	Startup 5	Startup 6	Startup 7
Team	18%	5,00	3,50	4,00	3,00	3,50	4,00	4,50
Technology	62%	5,74	5,52	3,88	4,05	5,40	4,32	4,44
Community	10%	4,50	3,00	3,50	4,00	4,00	6,00	6,00
Financial	10%	5,00	3,00	4,00	4,00	5,00	4,50	3,50

		Startup 1	Startup 2	Startup 3	Startup 4	Startup 5	Startup 6	Startup 7
Team	18%	0,90	0,63	0,72	0,54	0,63	0,72	0,81
Technology	62%	3,56	3,42	2,41	2,51	3,35	2,68	2,75
Community	10%	0,45	0,30	0,35	0,40	0,40	0,60	0,60
Financial	10%	0,50	0,30	0,40	0,40	0,50	0,45	0,35
		5,41	4,65	3,88	3,85	4,88	4,45	4,51

Figure 12 Final Evaluation

evaluating technology proposed by startups. By continuously improving the evaluation processes, companies can build trust and confidence among all stakeholders, and ultimately achieve success in their innovation efforts.

To date, while the proposed technology scouting methodology has shown promising results, it is important to acknowledge its limitations. These limitations are mainly due to the assumptions made during the development of the methodology, such as the use of specific aggregation methods and the Analytic Hierarchy Process (AHP) for evaluation. Future work should focus on deepening the understanding of these limitations and exploring alternatives approaches.

References

[1] Kerr, C., Phaal, R.,. Technology roadmapping: Industrial roots, forgotten history and unknown origins. Technol. Forecast. Soc. Change 155, 1–16, 2020. https://doi.org/10.1016/j.techfore.2020.119967

[2] T. L. Saaty, "The Analytic Hierarchy Process", McGraw-Hill, 1980. https://doi.org/10.21236/ADA214804

[3] Kim, H. S., & Kim, S. J. An integrated approach for the selection of maintenance criteria using the analytic hierarchy process and Delphi method. Reliability Engineering & System Safety, 130, 29-37, 2015.

[4] T. L. Saaty, "Theory and Applications of the Analytic Network Process", RWS Publications, 2005.

[5] E. Ordoobadi and F.G.E. Talavera "Using NTG for Requirements Elicitation in Health Information Systems: A Case Study" ,2009.

[6] Xiong, R., Wang, J., & Chen, Y. Selection of renewable energy projects based on a combination of the Delphi method and the analytic hierarchy process, 2017.

[7] Likert, R. A Technique for the Measurement of Attitudes. Archives of Psychology, 22(140), 1– 55.], 1932.

[8] Lewis, A., & Smith, D. Defining Higher Order Thinking. Theory into Practice, 32, 131-137,1993. https://doi.org/10.1080/00405849309543588

High-fidelity simulation of a supersonic parachute for Mars descent

L. Placco[1*], F. Dalla Barba[2] and F. Picano[1,2]

[1] Centro di Ateneo di Studi e Attività Spaziali 'Giuseppe Colombo' (CISAS), Università degli Studi di Padova, via Venezia 15, 35131 Padua, Italy

[2] Department of Industrial Engineering, Università degli Studi di Padova, via Venezia 1, 35131 Padua, Italy

*luca.placco@unipd.it

Keywords: Supersonic Parachute, Supersonic Flows, Large Eddy Simulation, Fluid-Structure Interaction

Abstract. The project aims to characterize the unsteady dynamics of the parachute-capsule in a supersonic flow during the descent phase on planetary entry. Presently, Large-Eddy Simulation in combination with an Immersed-Boundary Method is employed to analyze the time-evolving flow of a rigid supersonic parachute trailing behind a reentry capsule during the descent phase through Mars atmosphere. The flow is simulated at Ma = 2 and Re = 10^6. A massive GPU parallelization is employed to allow a very high fidelity solution of the multiscale turbulent structures present in the flow that characterize its dynamics. We show how the interaction of wake turbulent structures with the bow shock produced by the supersonic decelerator induces strong unsteady dynamics. This unsteady phenomenon called 'breathing instability' is strictly related to the ingestion of turbulence by the parachute's canopy and is responsible of drag variations and structure oscillations observed during previous missions and experimental campaigns. The next steps will take into account the flexibility of the parachute.

Introduction

The recent unsuccessful European missions (i.e. ExoMars 2016) proved how the prediction and the understanding of the dynamics of the descent capsule under the effect of a supersonic decelerator is still an open question in the active research scene that revolves around space exploration. The failure of Schiaparelli EDM landing indeed was ultimately caused by an improper evaluation of the coupled oscillatory motions existing between the descent module and the deployed parachute. The models and the experimental evaluations that were employed to predict the general behaviour of the capsule under the effect of a supersonic decelerator proved to be insufficient, triggering the premature end of the mission [1]. In this context, the main aim proposed by this research activity is to develop a novel technique to study effectively how compressible and turbulent flows interact with non-rigid structures, to properly evaluate and predict their non-steady behaviour. The interaction of a flexible body surrounded by a flow is considered as a constant presence in many different disciplines, despite being a very specific condition. It is acknowledged as fluid-structure interaction and plays also a major role in the study of flows that involve both space and suborbital applications. This is most true when considering planetary atmospheric reentry or the use of decelerators and active flying devices to extend the observation time of a probe. The non-linearity that characterises both the fluid and the solid behaviour proves to be still a challenge when approaching the subject and this becomes further complicated when considering unsteady compressible flows.

Aerospace Science and Engineering - III Aerospace PhD-Days   Materials Research Forum LLC
Materials Research Proceedings 33 (2023) 134-139   https://doi.org/10.21741/9781644902677-20

Case study and intended approach

The research community that involves fluid-structure interaction matters have always been present in the computational scene; the rising interest also on its implementation on advanced conditions is attracting further attention, especially for being applied on state-of-art cases and problems that have an immediate impact on the active missions and studies.

This is especially true for space programs, given the recent technological advancements concerning both exploration and robotics (i.e. the latest NASA's Mars Perseverance and the Chinese probe Tianwen-1). The typical case study of a supersonic decelerator (parachutes and inflated aeroshells) is representative for both the complexity of the study and for its scientific importance; ExoMars 2016 mission, which contained the Schiaparelli descent module, failed to demonstrate the entry, descent and landing procedure as, according to the Schiaparelli Anomaly Inquiry Report ([1]), the major event that brought to the saturation of the Inertial Measurement Unit was the highly dynamic oscillatory motion caused by an improper evaluation of the disk-gap-band type parachute that deployed at Mach 2.

The description of this phenomenon is very elaborate, being affected by several uncertainties such as atmosphere fluctuations, unsteady flow dynamics and structure oscillations [2],[3]. At the current time experimental procedures involving the design of these devices struggle to observe the properties of these problems since scale effects and the artificial environment recreated by wind tunnels are not able to replicate the effective operative conditions [4]. Flight tests instead are considered impractical, so they can only be employed at the end of the design process to obtain the validation that cannot be obtained in a simulated environment [5]. Even standard computational approaches appear to be limited in the provided results: Reynolds Averaged Navier Stokes technique (RANS), which is the engineering standard for design and validation campaigns, is unable to correctly reproduce the non-stationary nature of such turbulent flow conditions [4].

To overcome such limitations, the proposed approach involves more accurate methodologies which are Large-Eddy Simulations (LES) as they can properly represent the time-evolving multi-scale dynamics of the flow. They were always considered unfeasible due to the prohibitive computational cost required to run the analysis; however, the advent of GPU technology for parallel computing is enabling the usage of Large-Eddy Simulations [6], making them the most groundbreaking approach now available and increasing dramatically both the spatial and temporal resolution of the obtained results thus improving the quality and the quantity of the captured details. Furthermore, Immersed Boundary Methods, which are fundamental to deal effectively with moving solid boundaries, have received more attention recently [7], enabling the application of computational strategies for compressible flows that are able to solve the fluid-solid interface of porous thin structures and shells. Still, all of the analysis that have been performed on the matter up to now were not able to draw together all of these novel aspects, limiting the findings. For this matter, the ultimate aim of this ongoing research activity is to perform high-fidelity simulations of the full deployment and inflation sequence of a parachute for planetary descent in a supersonic regime as it interacts with the turbulent wake generated by the forebody.

Special attention will be given to the modeling of the typical 'breathing' behaviour of the supersonic decelerator during the descent phase of a reentry spacecraft. To achieve these results, a novel technique to deal with the fluid-structure interaction of compressible flows and thin membranes is in the process of development, starting from the existing Immersed Boundary Methods' strategies.

As a starting point for the implementation of the final configuration, a large-eddy simulation of a rigid mock-up parachute trailing behind a reentry capsule has been performed, showing both the potential of the LES approach and the primary dependence of the breathing phenomenon to the interaction of the turbulent wake of the descent module with the front bow shock produced by the inflated decelerator. We present the preliminary results obtained from the simulation.

Aerospace Science and Engineering - III Aerospace PhD-Days   Materials Research Forum LLC
Materials Research Proceedings 33 (2023) 134-139   https://doi.org/10.21741/9781644902677-20

Computational approach and simulation setup

Compressible Navier-Stokes equations are solved with the high-order finite difference solver STREAmS [6]. Turbulent structures are ultimately identified using the implicit large eddy simulation (ILES) approach; in this way, conventional LES turbulence modeling has been omitted, using instead the numerical dissipation given by the numerical discretization as artificial viscosity acting at small scales. Thus, the three-dimensional compressible Navier-Stokes equations solved are the following:

$$\frac{\partial \rho}{\partial t} + \frac{\partial (\rho u_j)}{\partial x_j} = 0.$$

$$\frac{\partial (\rho u_i)}{\partial t} + \frac{\partial (\rho u_i u_j)}{\partial x_j} + \frac{\partial p}{\partial x_i} - \frac{\partial}{\partial x_j}\left(\mu \left(\frac{\partial u_i}{\partial x_j} + \frac{\partial u_j}{\partial x_i} - \frac{2}{3}\frac{\partial u_k}{\partial x_k}\delta_{ij} \right) \right) = 0. \quad (1)$$

$$\frac{\partial (\rho E)}{\partial t} + \frac{\partial (\rho E u_j + p u_j)}{\partial x_j} + \frac{\partial}{\partial x_j}\left(\lambda \frac{\partial T}{\partial x_j} \right) + \frac{\partial}{\partial x_j}\left(\mu \left(\frac{\partial u_i}{\partial x_j} + \frac{\partial u_j}{\partial x_i} - \frac{2}{3}\frac{\partial u_k}{\partial x_k}\delta_{ij} \right)u_i \right) = 0.$$

where ρ is the density, u_i denotes the velocity component in the i Cartesian direction ($i = 1,2,3$) and p is the thermodynamic pressure. With the intent of reproducing the effect of Mars' atmosphere, the fluid is considered as an ideal gas of CO_2; the ratio between the specific heat at constant pressure C_p and the specific heat at constant volume C_v is set to 1.3 while Prandtl number is 0.72. $E = C_v T + u_i^2/2$ represents the total energy per unit mass and the dynamic viscosity μ is assumed to follow the generalized fluid power-law. The thermal conductivity λ is related to μ via the Prandtl number with the following expression: $\lambda = C_p\mu/Pr$. Convective and viscous terms are discretized using a sixth-order finite difference central scheme while flow discontinuities are accounted through a fifth-order WENO scheme. Time advancement of the ODE system is given by a third-order explicit Runge-Kutta/Wray algorithm. No-slip and no-penetration wall boundary conditions on the body are enforced through an Immersed-Boundary Method (IBM) algorithm. The simulation was performed at $Ma = 2$ and $Re = 10^6$ to simulate the condition at which the parachute deploys. The reference fluid properties associated to the free-stream condition correspond to an altitude of about 9 km from the planet surface and have been obtained using a simulated entry and descent trajectory through the Mars atmosphere of a generic reentry probe [2].

The flow domain (fig. 1) selected to perform this first simulation has a size of $L_x = 20D$, $L_y = 5D$, $L_z = 5D$, where D is the maximum diameter of the descent module; parachute diameter is set to $2.57D$. the mesh is a rectilinear structured grid that consists of $N_x \cdot N_y \cdot N_z = 2560 \cdot 840 \cdot 840$ nodes. The grid density changes in both axial and transverse directions, gaining resolution in the central portion of the domain; the position of the capsule nose is set at $[1D,0,0]$ while the parachute center lies at $[10D,0,0]$. Computations have been carried out on CINECA Marconi100 cluster, allowing the domain parallel computing on a total of 64 GPUs.

Aerospace Science and Engineering - III Aerospace PhD-Days   Materials Research Forum LLC
Materials Research Proceedings 33 (2023) 134-139   https://doi.org/10.21741/9781644902677-20

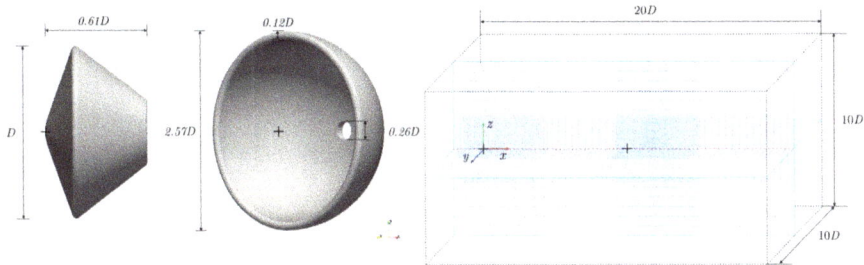

Figure 1: Setup of the simulation domain and dimensions of the descent module and parachute.

Results

In figure 2 we observe the two-dimensional instantaneous flow field obtained by isolating the $y = 0$ slice from the full 3D domain; Mach number contours are shown. Subsonic flow regions (in red), sonic regions (in white) and supersonic areas (in blue) can be identified. We observe the generation of two bow shocks ahead of the capsule and the canopy. Subsonic regions are concentrated around the axis of symmetry, behind the two bow shocks. The highest velocity is reached by the flow exiting the vent, with a Mach number of $Ma \cong 3.5$. The flow at the vent section is sonic. Pushed by the high pressure within the canopy and finding a larger passage section, it rapidly accelerates to the highest Mach number of the flow field. The high value associated to the Reynolds' number results in the generation of an extended wake behind the capsule. Near wake regions are characterized by subsonic recirculations while supersonic regime supersedes as the flow progresses towards the outflow. After interacting with the canopy bow shock, the flow develops into vortical structures all around the canopy and at its back, producing large subsonic regions.

The capsule bow shock is steady, as well as the subsonic region between the shock and the capsule. On the contrary, the flow appears more unstable in the canopy region: the source of this instability is the turbulence shock interaction occurring due to the passage of the turbulent wake of the descent module through the bow shock of the parachute. The intensity of turbulence carried by the wake is amplified as the flow travel across the canopy bow shock (turbulence ingestion by the shock), leading to large fluctuations of momentum and pressure. In addition, being disrupted by the irregularities of the wake, the canopy bow shock does not reach a steady state and shows an oscillatory motion. This motion is related to the parachute breathing cycle, which is present regardless the parachute rigidity. The breathing motion involves inhomogeneous pressure/density fluctuations, leading to large drag variations, despite the canopy area remaining constant. Main cause of the breathing cycle seems to be the aerodynamic interaction between capsule wake and canopy bow shock. In real applications of flexible parachutes, canopy deformations even couple with this interaction, amplifying drag variability. The deformability effect of thin porous membranes will be taken into account in the following steps of the research project. Figure 3 shows the different phases of the cycle that surrounds the periodic motion of the front bow shock along the flow direction: an increasing density inside the canopy pushes the shockwave away, allowing a larger flux to escape from the canopy (from [1] to [2]). Thus, this creates a decrease in the density that in turn draws back in the shockwave ([2] to [3]) and restarts the cycle ([3] to [4]). Further details on the flow dynamics will be given during the conference.

Aerospace Science and Engineering - III Aerospace PhD-Days
Materials Research Proceedings 33 (2023) 134-139

Materials Research Forum LLC
https://doi.org/10.21741/9781644902677-20

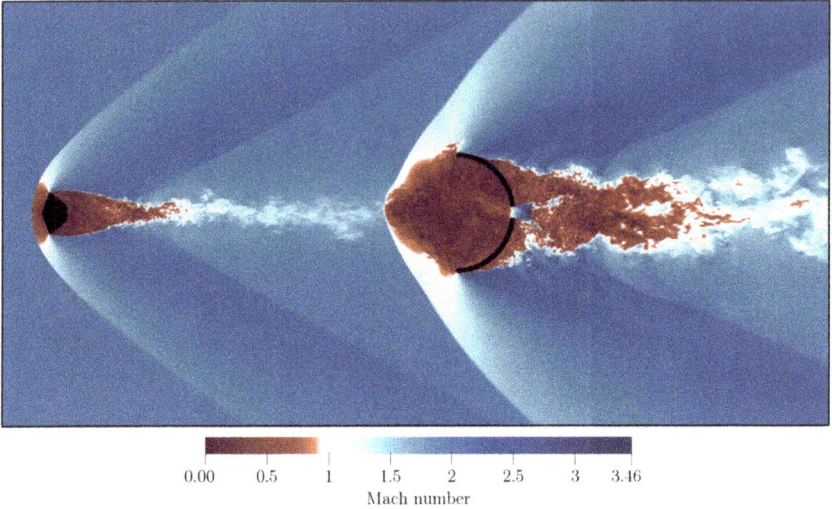

Figure 2: *Instantaneous Mach contours (y = 0 cross section) of the simulated flow domain.*

Conclusions

The present work proposes an high-fidelity time-evolving simulation of the interaction between the turbulent wake of a supersonic descent module and a generic rigid artificially thick decelerator. We show how the critical 'breathing' instability associated to supersonic parachutes is intrinsically connected to the interaction of the turbulent wake flow of the descent module and the front bow shock produced by the decelerator. To overcome the limitation of the current setup and further extended the representation of its dynamics, the implementation of a novel immersed boundary method technique is in progress. This will require the solution of fluid-structure interaction of compressible supersonic flows and flexible thin membranes. The new framework will involve an extension of the current IBM module and a finite element method model to deal with flexible moving boundaries (zero-thickness), representing the very thin structure of the simulated decelerator. In this way, the approach in development will allow to represent properly both the entire deployment sequence and the system unsteadiness in all its components, thus providing the full representation of the 'breathing' phenomenon.

Aerospace Science and Engineering - III Aerospace PhD-Days   Materials Research Forum LLC
Materials Research Proceedings 33 (2023) 134-139   https://doi.org/10.21741/9781644902677-20

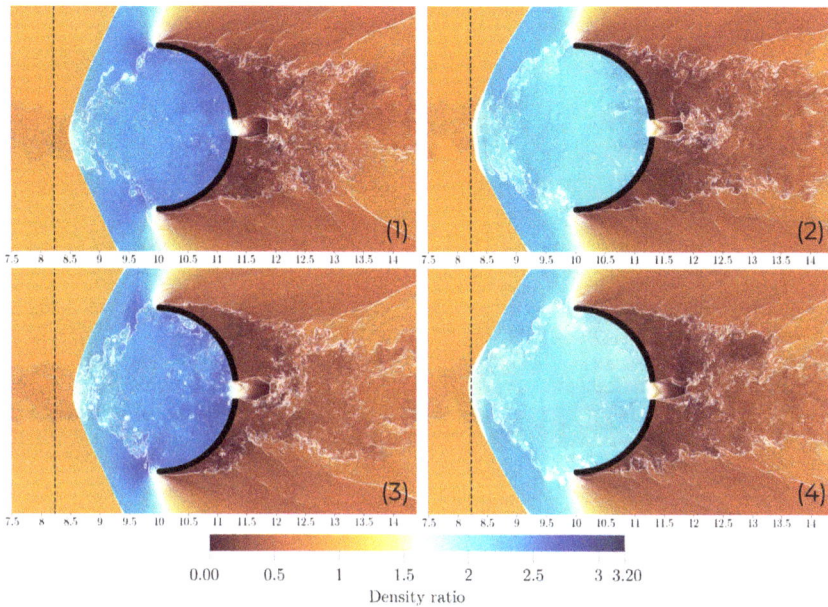

Figure 3: Instantaneous density ratio contours (y = 0 cross section) at different progressive timestep around the parachute canopy.

References

[1] T. Tolker-Nielsen. EXOMARS 2016 - Schiaparelli Anomaly Inquiry, 2017.

[2] A. Aboudan, G. Colombatti, C. Bettanini, F. Ferri, S. Lewis, B. Van Hove, O. Karatekin, and Stefano Debei. Exomars 2016 schiaparelli module trajectory and atmospheric profiles reconstruction. Space Science Reviews, 214: 97, 08 2018. https://doi.org/10.1007/s11214-018-0532-3

[3] X. Xue and Chih-Yung Wen. Review of unsteady aerodynamics of supersonic parachutes. Progress in Aerospace Sciences, 125:100728, 2021. ISSN 0376-0421. https://doi.org/10.1016/j.paerosci.2021.100728

[4] Nimesh, Dahal. Study of pressure oscillations in supersonic parachute. International Journal of Aeronautical Space Sciences, (19):24-31, 2018. https://doi.org/10.1007/s42405-018-0025-3

[5] B. S. Sonneveldt, I. G. Clark, and C. O'Farrell. Summary of the Advanced Supersonic Parachute Inflation Research Experiments (ASPIRE) Sounding Rocket Tests with a Disk-Gap-Band Parachute, AIAA 2019-3482. https://doi.org/10.2514/6.2019-3482

[6] M. Bernardini, D. Modesti, F. Salvadore, and S. Pirozzoli. Streams: a high-fidelity accelerated solver for direct numerical simulation of compressible turbulent flows. Computer Physics Communications, 263:107906, 2021. https://doi.org/10.1016/j.cpc.2021.107906

[7] H. Yu and C. Pantano. An immersed boundary method with implicit body force for compressible viscous flow. Journal of Computational Physics, 459:111125, 2022. https://doi.org/10.1016/j.jcp.2022.111125

Aerospace Science and Engineering - III Aerospace PhD-Days   Materials Research Forum LLC
Materials Research Proceedings 33 (2023) 140-147   https://doi.org/10.21741/9781644902677-21

Recent advances in dynamical modeling and attitude control of flexible spacecraft

David Paolo Madonna[1,a] *

[1]Department of Mechanical and Aerospace Engineering, Sapienza University of Rome, Via Eudossiana 18, 00184 Rome, Italy

[a]davidpaolo.madonna@uniroma1.it

Keywords: Flexible Spacecraft, Stress Stiffening, Kane's Method, Nonlinear Feedback Attitude Control, Control Momentum Gyroscopes

Abstract. This work is focused on some recent advances on spacecraft dynamical modeling and attitude control. First, the problem of correctly linearizing the elastic behavior of spinning flexible spacecraft is discussed, and an example is presented that addresses this topic from both the analytical and the numerical point of view. Then, attitude control is considered and a new feedback law for single axis alignment is presented, together with the inclusion of accurate modeling of the actuation dynamics. Moreover, numerical simulations for the reorientation of a flexible multibody spacecraft are provided.

Introduction

Current space missions require predicting the spacecraft dynamics with considerable reliability. Among the various components of a spacecraft, subsystems like payload, structures, and power depend heavily on the dynamic behavior of the satellite during its operational life. Therefore, to ensure that the results obtained through numerical simulations correspond to the actual behavior, an accurate dynamical model must be developed. Reducing the margins of uncertainty implies the possibility of carrying out proper sizing of various features of the system, while on the other side it allows enhancing the mission performance.

Although in several cases it is acceptable to model the spacecraft as a single rigid body, in specific applications the satellite must be modeled as a multibody system composed of both rigid and flexible elements, to get higher accuracy. This occurs because the dynamic behavior of a spacecraft can be affected by many factors, including structural deformations, vibrations, and disturbances that arise from its interaction with the space environment.

In this perspective, when dealing with flexible spacecraft that undergo a high-speed rotation, it may become necessary to include the stress stiffening in the model, which is the increase in stiffness due to internal stresses produced by the inertial loads [1]. This phenomenon is included in the dynamical model only if the nonlinear elastic behavior of the flexible bodies is considered. In Section II, the procedure to correctly linearize the dynamical equations of a flexible spacecraft to preserve the stress stiffening effect is discussed and the error produced by a premature linearization is investigated.

Once a detailed model has been obtained, a suitable control architecture must be designed and tested on the complete model to ensure proper control of the spacecraft for both orbit and attitude control. In Section III, nonlinear feedback control strategies are investigated, which have the advantage of ensuring convergence for large-angle maneuvers even in attitude tracking scenarios [2]. In addition to control strategies, actuation is also studied, with a focus on momentum exchange devices, specifically single-gimbal control momentum gyroscopes. Suitable steering laws are analyzed to achieve accurate tracking and prevent singularities [3].

Aerospace Science and Engineering - III Aerospace PhD-Days   Materials Research Forum LLC
Materials Research Proceedings 33 (2023) 140-147   https://doi.org/10.21741/9781644902677-21

Stress stiffening

The stress stiffening phenomenon, commonly observed in high-speed flexible rotating satellites, has a significant impact on the dynamical behavior and stability of the spacecraft [4]. Specifically, a flexible spacecraft subject to fast rotational motion shows a dynamic stiffening effect induced by the stresses generated by the inertia loads. However, this is frequently overlooked in dynamical modeling when attitude control is designed, and the dynamic equations are linearized around the equilibrium position of the satellite. Of course, neglecting this phenomenon in case of spinning satellites can lead to a wrong representation of the real behavior of the structure and consequently errors in the elastic displacements as shown in some examples reported in Ref [5].

Kane's formulation of multibody spacecraft dynamics introduces partial velocities, to relate the Newton/Euler dynamical quantities to generalized velocities [6]. In this framework, stress stiffening is taken into account by linearizing the dynamical equations after the extraction of partial velocities. Unfortunately, this procedure is not easy to implement for complex spacecraft where the flexible elements can be described from the static and dynamic point of view using discrete formulations such as Finite Element Modeling (FEM). To overcome this challenge, it is recommended to first perform a linear static analysis of the flexible elements using FEM to determine the stress configuration under given inertial loading conditions. Then, this stress configuration can be used to compute the increment in strain energy due to nonlinear elastic deformations, to obtain a "stress stiffening matrix", which can be added a posteriori to the linear stiffness, with the final aim of deriving the complete equations of motion for the flexible elements [7].

A preliminary analysis is conducted to evaluate the effectiveness of the procedure described above, using a rotating cantilever beam as a case study (see Fig.1). The beam is assumed to undergo planar motion, while flexibility is modeled using a single elastic bending mode. This simple example allows for a clear identification of the contribution of stress stiffening in the dynamical equations.

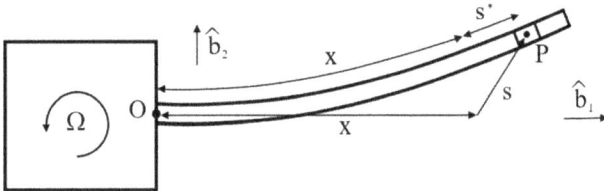

Figure 1: sketch of a rotating cantilever beam

Similarly to the more general case, in this example nonlinear velocities are obtained by including the nonlinear elastic behavior of the structure in the analysis. Referring to Fig. 1, if x is the distance OP when the beam is undeformed and neglecting the axial elastic displacement (as it has been assumed in this analysis), i.e. letting $s^* = 0$, one can write that

$$x = \int_0^\xi \sqrt{1 + \left(\frac{\partial s_2(\sigma,t)}{\partial \sigma}\right)^2}\, d\sigma \tag{1}$$

where $\mathbf{s} = \begin{bmatrix} s_1 & s_2 & 0 \end{bmatrix}^T$ is the elastic displacement, $\xi = x + s_1$ and σ is a dummy variable. J is introduced as

$$J(\sigma, t) \equiv 1 + \left(\frac{\partial s_2(\sigma, t)}{\partial \sigma} \right)^2 . \tag{2}$$

The bending elastic displacement is decomposed through the well-established modal decomposition approach [8] as

$$s_2(x, t) = \sum_{i=1}^{\nu} \varphi_{2i}(x) q_i(t) , \tag{3}$$

where ν denotes the number of flexible degrees of freedom of the beam, $q_i(t)$ is the i-th modal amplitude and $\varphi_{2i}(x)$ is the i-th eigenfunction associated with the bending motion. The following time derivative is obtained from Eq. 1:

$$\dot{s}_1 = \frac{1}{\sqrt{J(\xi, t)}} \sum_{i=1}^{\nu} \left\{ -\sum_{j=1}^{\nu} q_j(t) \int_0^{\xi} \frac{1}{\sqrt{J(\sigma, t)}} \left(\frac{\partial \varphi_{2i}(\sigma)}{\partial \sigma} \right)^2 d\sigma \right\} q_i(t) . \tag{4}$$

Hence, although axial elasticity is not included in the model by assumption, the component of the elastic displacement along $\hat{b}_1$ is nonzero because of the correlation with the bending displacement, which would not have appeared in the equations if only linear strain-displacement relations had been considered. Using Eq. 4, the following nonlinear velocities are obtained:

$$\mathbf{v}_P = \tilde{\mathbf{\Omega}} \mathbf{r}_{OP} + \dot{\mathbf{s}} \tag{5}$$

where $\mathbf{\Omega} = \begin{bmatrix} 0 & 0 & \Omega \end{bmatrix}^T$ and $\mathbf{r}_{OP}$ is the distance from O to P in the deformed configurations. It is possible to extract nonlinear partial velocities, which now can be correctly linearized with respect to small elastic displacements assuming the following form:

$$\mathbf{V} = \begin{bmatrix} -\beta_{11}(x) q_1(t) \\ \varphi_{21} \\ 0 \end{bmatrix} \tag{6}$$

where

$$\beta_{11}(x) = \int_0^x \left(\frac{\partial \varphi_{21}(\sigma)}{\partial \sigma} \right)^2 d\sigma . \tag{7}$$

Instead, extracting the partial velocities from linearized expressions of the velocity (i.e. Eq. 5 with $\dot{s}_1 = 0$) implies obtaining the following incomplete vector:

Aerospace Science and Engineering - III Aerospace PhD-Days   Materials Research Forum LLC
Materials Research Proceedings 33 (2023) 140-147   https://doi.org/10.21741/9781644902677-21

$$\overline{\mathbf{V}} = \begin{bmatrix} 0 \\ \varphi_{21} \\ 0 \end{bmatrix} , \tag{8}$$

which leads to a lack of terms in the dynamical equations. Once the correct partial velocities have been extracted, the linearized velocity can be derived to obtain the acceleration of the points of the beam

$$\mathbf{a}_{\mathrm{p}} = \overline{\mathbf{V}} \ddot{\mathbf{q}}(t) + \mathbf{a}_{\mathrm{p}}^{(R)} \tag{9}$$

where $\mathbf{a}_{\mathrm{p}}^{(R)}$ collects the terms of $\mathbf{a}_{\mathrm{p}}$ that do not depend on the time derivatives of generalized velocities. Hence, the following form of Kane's dynamical equations is used [9]:

$$\int_0^L \rho(x) \mathbf{V}^{\mathbf{T}} \overline{\mathbf{V}} \, dx \, \ddot{\mathbf{q}}(t) + \int_0^L \rho(x) \mathbf{V}^{\mathbf{T}} \mathbf{a}_{\mathrm{p}}^{(R)} \, dx + \omega_b^2 \mathbf{q}(t) = 0 , \tag{10}$$

where L is the beam length, $\rho(x)$ is the beam mass per unit of length and ω_b is the frequency of the bending mode. Then, one obtains the final governing equation,

$$\ddot{q}(t) + \left\{ \omega_b^2 - \Omega^2 + \lambda \Omega^2 \right\} q(t) = -\dot{\Omega} \int_0^L \rho(x) x \varphi_{21}(x) \, dx . \tag{11}$$

The term

$$\lambda = \int_0^L \rho(x) \beta_{11}(x) x \, dx > 1 \tag{12}$$

is responsible for the stress stiffening. It is worth noting that coupling between the rigid rotation and the elastic displacement of the beam, here represented by the amplitude of the modal shape q(t), has two opposite effects. The first effect, associated with the linearized flexible dynamics and represented by the term $\left\{ \omega_b^2 - \Omega^2 \right\} q(t)$ is responsible of the "reduction" of the internal stiffness of the beam. The second effect, corresponding to $\left\{ \lambda \Omega^2 \right\} q(t)$ is related to the centrifugal action that stretches and stiffens the beam. It is important to observe that if $\Omega^2 = \omega_b^2$ the "effective stiffness" of the beam would vanish if the centrifugal term were omitted leading to a completely misrepresentation of the real dynamic behavior of the rotating beam.

Table 1: beam's physical properties

Density (ρ)	$1.2 \left[kg / m \right]$
Length (L)	$10 \left[m \right]$
Young's modulus (E)	$7 \cdot 10^{10} \left[N / m^2 \right]$
Area moment of inertia (I)	$2 \cdot 10^{-7} \left[m^4 \right]$
First bending frequency (ω_b)	$0.6044 \left[Hz \right]$

In this study, numerical simulations are conducted based on the data presented in Table 1, to investigate the elastic behavior of a cantilever beam subject to a rotation motion following a cubic law. The maximum tip displacements (with and without stiffening effects) are plotted against the ratio of the beam angular velocity to its first bending natural frequency, as shown in Fig. 2. As stated above the results indicate that linearizing the partial velocities prematurely leads to a significant increase in error even when $\Omega / \omega_b < 1$. In Fig. 3 the time histories of tip displacement for $\Omega = 6 \text{ rad} / \text{s}$ are reported.

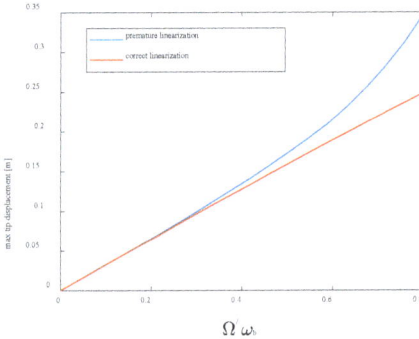

Figure 2: max tip displacement vs angular velocity

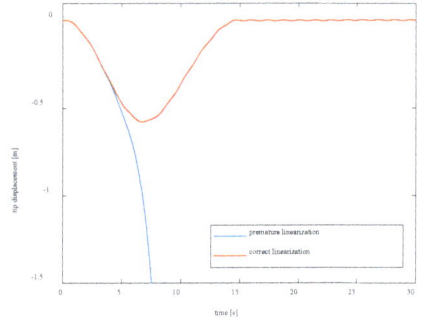

Figure 3: tip displacement for $\Omega = 6 \text{ rad} / \text{s}$

With a premature linearization one obtains wrong results, while the beam shows a completely different behavior. Hence, considering the stress stiffening effects allows reducing the margin of safety while designing space systems that undergo fast motion conditions.

Nonlinear attitude control applied to spacecraft dynamics.

It is well known that accurate dynamical modeling of multibody spacecraft is essential in designing their attitude control system. The motion of the spacecraft components, including robotic arms and steerable solar panels, as well as elastic oscillations of the structures, profoundly impact the attitude dynamics. Therefore, these factors must be considered during the synthesis of the attitude control, which is the topic addressed in this section.

In the following, a significant attention is paid on nonlinear feedback control laws, which ensure convergence in terms of attitude and angular velocity for large-angle maneuvers even when tracking is required [10]. In particular, the triaxial feedback control torque can be represented in its nonlinear form as

$$\mathbf{T}_C = \tilde{\omega}\mathbf{J}_C\omega - \mathbf{M}_C + \mathbf{J}_C\dot{\omega}_C - \mathbf{J}_C\mathbf{A}^{-1}\mathbf{B}\omega_D - \text{sgn}\{q_{e_0}(t_0)\}\mathbf{J}_C\mathbf{A}^{-1}\mathbf{q}_e \ , \tag{13}$$

where ω is the angular velocity of the spacecraft, $\mathbf{J}_C$ is the moment of inertia computed with respect of the center of mass, $\mathbf{M}_C$ is the vector of known disturbance torques, ω_C is the commanded angular velocity vector, $\omega_D = \omega - \omega_C$ is the simplified error angular velocity, $\{q_{e_0}, \mathbf{q}_e\}$ is the error quaternion associated with the misalignment between the commanded frame and the body frame and $\mathbf{A}$ and $\mathbf{B}$ are gain matrices that must be positive definite ($\mathbf{A}$ must also be

Aerospace Science and Engineering - III Aerospace PhD-Days   Materials Research Forum LLC
Materials Research Proceedings 33 (2023) 140-147   https://doi.org/10.21741/9781644902677-21

symmetric); finally, the sign of $q_{e_0}(t_0)$ is introduced to choose the shortest path to reach the desired attitude.

When it is sufficient to drive only a single axis toward a desired direction, a reduced-attitude control law can be applied to further improve the results, as shown in Refs. [9,11]. A new reduced-attitude control law is derived and presented by the author in Ref. [12]. The expression of this control law is the following:

$$T_C = \tilde{\omega} J_C \omega - M_C + J_C \left[{}_{B \leftarrow C} \mathbf{R} \, \dot{\omega}_C - \tilde{\omega}_e \, {}_{B \leftarrow C} \mathbf{R} \, \omega_C \right] - J_C A^{-1} B \omega_e - J_C A^{-1} \mathbf{g}\left(q_{e_0}^M, \mathbf{q}_e^M \right) \tag{14}$$

$$\mathbf{g}\left(q_{e_0}^M, \mathbf{q}_e^M \right) = 2 \begin{bmatrix} 0 \\ q_{e_0}^M q_{e_2}^M + q_{e_1}^M q_{e_3}^M \\ q_{e_0}^M q_{e_3}^M - q_{e_1}^M q_{e_2}^M \end{bmatrix} \tag{15}$$

where ${}_{B \leftarrow C} \mathbf{R}$ is the rotation matrix from the commanded frame to the body frame, $\omega_e = \omega - {}_{B \leftarrow C} \mathbf{R} \, \omega_C$ is the error anglar velocity and $\left\{ q_{e_0}^M, \mathbf{q}_e^M \right\}$ is a modified error quaternion. In the case it is required to drive the body axis $\hat{b}_1$ toward the commanded axis $\hat{b}_{C,1}$, the modified error quaternion is associated with the misalignment between the commanded frame and a frame obtained from the body frame through an eigenaxis rotation, for which

- the eigenangle ϕ is defined as $\phi = \cos^{-1}\left(\hat{b}_1 \cdot \hat{b}_{C,1} \right)/2$, and
- the eigenaxis corresponds to $\hat{e} = \dfrac{\hat{b}_1 \times \hat{b}_{C,1}}{\|\hat{b}_1 \times \hat{b}_{C,1}\|}$.

The global asymptotic stability of the control law reported in Eqs. 16-17 is proven by the author in Ref. [12].

Moreover, the actuators' dynamics of the momentum exchange devices are also considered by the author in [9,12], to implement a high-fidelity control law. In particular, single gimbal control momentum gyroscopes (SG-CMGs) are used because they can provide the desired torque requiring a reduced amount of power compared to other actuators. However, their main drawback is that they suffer from specific singular configurations, which prevent from providing the desired torque to the spacecraft.

In Refs. [9,12], accurate steering laws are used to reduce the error in the control torque introduced by the actuator dynamics. In particular, further terms that take the detailed model of the actuators into account are added to the steering law commonly used in the literature [3], to significantly improve the accuracy of the actuators' torque. Furthermore, singularity avoidance algorithms are studied and applied to this enhanced steering law to escape from singular configurations. Specifically, a small error in the commanded torque is introduced to make the gyros move away from singularity. The magnitude of this error is opportunely reduced through a singularity direction avoidance (SDA) pseudoinverse law. Moreover, sizing of the pyramidal arrays of SG-CMGs is also investigated, to guarantee storage of angular momentum in any direction so that the desired angular velocities can be approached and achieved [13].

Aerospace Science and Engineering - III Aerospace PhD-Days   Materials Research Forum LLC
Materials Research Proceedings 33 (2023) 140-147   https://doi.org/10.21741/9781644902677-21

A Monte Carlo campaign is carried out to simulate attitude maneuvers for a large flexible spacecraft that mounts a pyramidal array of SG-CMGs, steered using the techniques described before. In particular, the performance of the new reduced-attitude control law is compared the one obtained using triaxial control law of Eq. 15, and results are reported in Figs. 4 and 5.

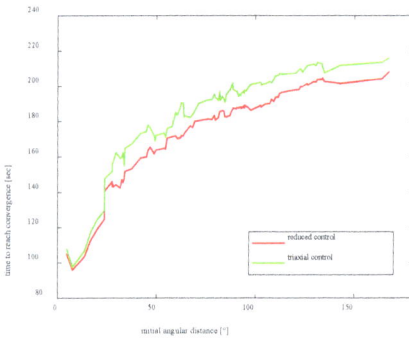

Figure 4: convergence time vs initial angular distance

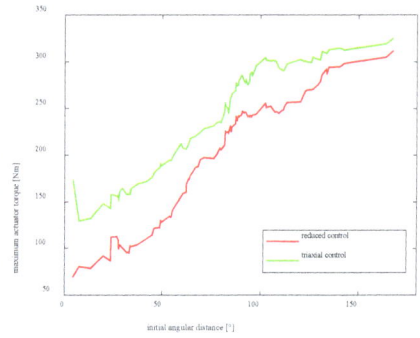

Figure 5: max torque vs initial angular distance

From inspection of Figs. 4-5, it is apparent that the reduced-attitude control law is always preferable (assuming that single-axis control is required) because it ensures faster convergence while requiring lower torques.

Conclusions

In the first part, a discussion about the correct linearization of dynamical equations of a rotating flexible body highlights how stress stiffening affects the equations of motion, and the effect of neglecting this phenomenon is numerically evaluated. Then, techniques to design the attitude control system are discussed, also dealing with sizing and steering of arrays of single gimbal control momentum gyroscopes. The performance of a new nonlinear feedback control law is compared to that of an existing law, to point out its advantages in terms of convergence time and torque requirement.

References

[1] T. R. Kane, R. Ryan, A. K. Banerjee, Dynamics of a Cantilever Beam Attached to a Moving Base, Journal of Guidance, Control, and Dynamics, Volume 10, Issue 2, 1987. https://doi.org/10.2514/3.20195

[2] A. H. de Ruitter, C. Damaren, J. Forbes, Spacecraft Dynamics and Control: An Introduction, Wiley, 2013. https://doi.org/10.5860/choice.51-0301

[3] F. A. Leve, B. Hamilton, M. Peck, Spacecraft Momentum Control Systems, Springer, 2015. https://doi.org/10.1007/978-3-319-22563-0

[4] P. Santini, Stability of flexible spacecrafts, Acta Astronautica, Volume 3, Issues 9–10, Pages 685-713, 1976. https://doi.org/10.1016/0094-5765(76)90106-5

[5] A. K. Banerjee, Flexible Multibody Dynamics: Efficient Formulations and Applications, Wiley, 2016. https://doi.org/10.1002/9781119015635

[6] T. R. Kane, P. W. Likins, D. A. Levinson, Spacecraft Dynamics, McGraw-Hill, 1983.

Aerospace Science and Engineering - III Aerospace PhD-Days   Materials Research Forum LLC
Materials Research Proceedings 33 (2023) 140-147   https://doi.org/10.21741/9781644902677-21

[7] R. D, Cook, Concepts and Applications of Finite Element Analysis, McGraw-Hill, 1985.

[8] S. Rao, Vibration of Continuous Systems, Wiley, 2007.
https://doi.org/10.1002/9780470117866

[9] D. P. Madonna, M. Pontani, P. Gasbarri, Attitude Maneuvers of a Flexible Spacecraft for Space Debris Detection and Collision Avoidance, Proceedings of the 73rd International Astronautical Congress, Paper IAC-22,C1,2,1,x68387, 2022.

[10] M. Pontani, Booklets and lecture notes of Space and Astronautical Engineering course "Advanced Spacecraft Dyanmics"

[11] M. Pontani, I. Napoli, A New Guidance and Control Architecture for Orbit Docking via Feedback Linearization, Proceedings of the 73rd International Astronautical Congress, Paper IAC-21-C1.1.14, 2021.

[12] D. P. Madonna, M. Pontani, P. Gasbarri, Nonlinear Attitude Maneuvering of a Flexible Spacecraft for Space Debris Tracking and Collision Avoidance, submitted to Acta Astroanutica.

[13] A. Mony, H. Hablani, G. Sharma, Control Momentum Gyro (CMG) Sizing and Cluster Configuration Selectionfor Agile Spacecraft10th National Symposium and Exhibition on Aerospace and Related Mechanisms, 2016.

Aerospace Science and Engineering - III Aerospace PhD-Days   Materials Research Forum LLC
Materials Research Proceedings 33 (2023) 148-155   https://doi.org/10.21741/9781644902677-22

Analysis of composite beams, plates, and shells using Jacobi polynomials and NDK models

Daniele Scano[1, a *]

[1]Mul2 Lab, Department of Mechanical and Aerospace Engineering, Politecnico di Torino, Corso Duca degli Abruzzi 24, 10129 Torino, Italy

[a]daniele.scano@polito.it

Keywords: Finite Element Method, Beam Models, Plate Models, Shell Models, Jacobi Polynomials, Node-Dependent Kinematics, Carrera Unified Formulation

Abstract. In this work, hierarchical Jacobi-based expansions are explored for the static analysis of multilayered beams, plates, and shells as structural theories as well as shape functions. Jacobi polynomials, denoted as $P_p^{(\gamma,\theta)}$, belong to the family of classical orthogonal polynomials and depend on two scalars parameters γ and θ, with p being the polynomial order. Regarding the structural theories, layer wise and equivalent single-layer approaches can be used. It is demonstrated that the parameters γ and θ of the Jacobi polynomials are not influential for the calculations. These polynomials are employed in the framework of the Carrera Unified Formulation (CUF), which allows to generate of finite element stiffness matrices straightforwardly. Furthermore, Node-dependent Kinematics is used in the CUF framework to build global-local models to save computational costs and obtain reliable results simultaneously.

Introduction

As modern engineering requires complicated and computationally expensive structural static analyses, appropriate 1D and 2D structural theories and Finite Element (FE) shape functions can be adopted to diminish the computational costs. The Carrera Unified Formulation (CUF) [1] is a versatile method to build 1D and 2D models. The governing equations can be derived and expressed in a compact way and are invariant from the adopted structure theory.

Considering the beam theories, Euler-Bernoulli Beam Model (EBBM) [2] and Timoshenko Beam Model (TBM) [3] are the classical formulations. For both, the cross-section is considered to be rigid in its plane. For EBBM, the shear deformation is neglected, while it is considered constant along the cross-section in the case of TBM. In the domain of CUF theories, Carrera and Giunta [4] used Higher Order Theories (HOT) derived from the Taylor polynomials. Furthermore, Carrera et al. [5] used Lagrange-like expansions over the cross-section. Concerning the FE models, Carrera et al. [1] used two-, three- and four-node Lagrange-like shape functions in the CUF framework.

As far as 2D plate and shell FEs are considered, Thin Plate Theory (TPT) and Thin Shell Theory (TST) are the classical models, see Kirchhoff [6]. The line remains orthogonal to the plate/shell reference surface in these models. When the transverse shear deformation is added, the Reissner–Mindlin [7,8] (also known as First-Order Shear Deformation Theory, FSDT) theory can be built. Carrera [9] proposed general HOT from the Taylor polynomials for the analysis of plates and shells for the CUF framework. The classical theories can be derived through penalization techniques from first-order Taylor. Furthermore, Carrera et al. [5] used Lagrange-like expansions along the thickness direction. Finally, Carrera et al. [1] used four-, eight- and nine-node FEs to study composite plates.

Two approaches can be used when dealing with laminates: Equivalent Single Layer (ESL) and Layer-Wise (LW) models. In the first one, the number of unknowns is unaffected by the number of layers, while in the second one, they depend on the layers, see Carrera [10].

Aerospace Science and Engineering - III Aerospace PhD-Days   Materials Research Forum LLC
Materials Research Proceedings 33 (2023) 148-155   https://doi.org/10.21741/9781644902677-22

Jacobi polynomials are utilized for building shape functions and structural theories for the analysis of beams, plates, and shells in CUF. Carrera et al. [11] first used these polynomials to build structural theories in the framework of CUF. They originate several polynomials changing the two parameters γ and θ, i.e., Legendre, Chebyshev, see the book of Abramowitz and Stegun [12]. Szabo et al. [13] proposed a hp-version of FE derived from Legendre (i.e., γ and θ equal to zero) polynomials for beam, plate, and solid. Zappino et al. [14] compared Legendre and Lagrange shape functions for 2D plate elements. Concerning the expansion functions in CUF, Pagani et al. [15] used Legendre for 2D cross-section in beam formulation, while Carrera et al. [16] studied plates with 1D expansions from Chebyshev polynomials.

Using enhanced models improves solutions' accuracy, but it increases costs. It is possible to use refined models in specific parts and low-fidelity models for the other part of the structure without using any mathematical artifices. Carrera and Zappino [17] first presented a global-local analysis for beams called Node-Dependent Kinematics (NDK). This was extended to laminated composite plates and shells by Zappino et al. [18] and Li et al. [19], respectively.

Hierarchical Jacobi polynomials for beams, plates, and shells

In the framework of CUF, Hierarchical Jacobi (HJ) polynomials have been used to build shape functions and structural theories for beams, plates, and shells. These elements have the interesting capability to use hierarchical features.

Jacobi polynomials are formulated using recurrence relations, see [12]. The Jacobi polynomials are described by the following expression:

$$P_p^{(\gamma,\theta)}(\zeta) = \left(A_p + B_p\right)P_{p-1}^{(\gamma,\theta)}(\zeta) - C_p P_{p-2}^{(\gamma,\theta)}(\zeta) \tag{1}$$

where γ and θ are two scalar parameters, and p stands for the polynomial order. The formula is evaluated in the natural plane $\zeta = [-1, +1]$. The first values are $P_0^{(\gamma,\theta)}(\zeta) = 1$ and $P_1^{(\gamma,\theta)}(\zeta) = A_0\zeta + B_0$. The explicit expressions of the scalars A_p, B_p, and C_p can be found in [12].

One-dimensional functions. It is possible to use HJ polynomials for building theories of structure along the thickness (z-axis) for plate and shell formulations, see Fig. 1 (a). Similarly, Jacobi-like shape functions can be adopted along the y-axis for the beam formulation, see Fig. 1 (b). For both cases, the building procedure is the same. However, for the sake of simplicity, one-dimensional shape functions are first considered. In this case, two kinds of polynomials are used along the y axis: vertex (or node) and edge. Basically, there are two vertexes and a number of edge modes that depends on the polynomial order of the chosen elements.

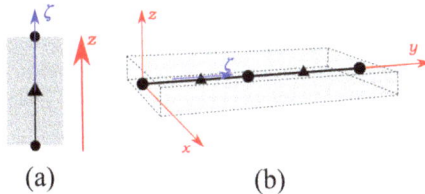

(a)   (b)

Figure 1: circle represents a vertex expansion, whereas triangle is an edge expansion. Theory of structure for plate and shell (a), shape functions for beam (b).

Given that formulas are formally identical for both cases $L_m(\zeta)$ is used to indicate the expansions. When the HJ polynomials are used as structural theories, $P_\tau(\zeta)$ is used, while $N_i(\zeta)$ is adopted for the shape functions. The hierarchic functions are defined as follows:

Aerospace Science and Engineering - III Aerospace PhD-Days   Materials Research Forum LLC
Materials Research Proceedings 33 (2023) 148-155   https://doi.org/10.21741/9781644902677-22

$$L_1(\zeta) = \frac{1}{2}(1 - \zeta)$$
$$L_2(\zeta) = \frac{1}{2}(1 + \zeta)$$
$$L_m(\zeta) = \phi_{m-1}(\zeta), m = 3,4, \dots, p + 1 \qquad (2)$$

with

$$\phi_j(\zeta) = (1 - \zeta)(+\zeta)P_{j-2}^{(\gamma,\theta)}, j = 2,3, \dots, p \qquad (3)$$

where p indicates the polynomial order. Given the following property

$$L_m(-1) = L_m(+1) = 0, m \geq 3 \qquad (4)$$

The function $L_m(\zeta)$, m= 3, 4, … are named bubble functions or edge expansions.

Two-dimensional functions. It is possible to use HJ polynomials for building theories of structure in the cross-section (*x-z* plane) for beam formulation, see Fig. 2 (a). Similarly, Jacobi-like shape functions can be adopted over the *x-y* plane for the plate and shell formulations, see Fig. 2 (b). For both cases, the building procedure is the same. For the sake of simplicity, two-dimensional shape functions are first considered. In this shape functions, three kinds of polynomials are used: vertex, edge, and internal. There are four vertex modes that vanish at all nodes but one. Contrarily, the number of edge modes changes according to the polynomial order of the FE, and they vanish for all sides of the domain but one. Finally, the internal modes are included from the fourth-order polynomial. They vanish at all sides. See [16] for more information.

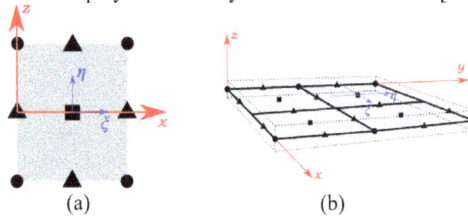

Figure 2: circle represents a vertex expansion, whereas triangle is an edge expansion and square indicates an internal expansion. Theory of structure for beam (a), shape functions for plate and shell (b).

Given that formulas are formally identical for both cases $L_m(\zeta)$ is used to indicate the expansions. When the HJ polynomials are used as structural theories, $P_\tau(\zeta)$ is used, while $N_i(\zeta)$ is adopted for the shape functions. The vertex modes are written as follows:

$$L_m(\xi, \eta) = \frac{1}{4}(1 - \xi_m\xi)(1 - \eta_m\eta), m = 1,2,3,4 \qquad (5)$$

where ξ and η are calculated in the natural plane between -1 and +1, and ξ_m and ξ_m and η_m are the vertexes. From $p \geq 2$, the edge modes arise in the natural plane as follows

Aerospace Science and Engineering - III Aerospace PhD-Days   Materials Research Forum LLC
Materials Research Proceedings 33 (2023) 148-155   https://doi.org/10.21741/9781644902677-22

$$L_m = \frac{1}{2}(1-\eta)\phi_p(\xi), m = 5,9,13,18,\dots \tag{6}$$
$$L_m = \frac{1}{2}(1+\xi)\phi_p(\eta), m = 6,10,14,19,\dots$$
$$L_m = \frac{1}{2}(1+\eta)\phi_p(\xi), m = 7,11,15,20,\dots$$
$$L_m = \frac{1}{2}(1-\xi)\phi_p(\eta), m = 8,12,16,21,\dots$$

where p represents the polynomial degree of the bubble function ϕ_j. Internal expansions are inserted for $p \geq 4$, they vanish at all the edges of the quadrilateral domain. There are $(p-2)(p-3)/2$ internal polynomials. By multiplying 1D edge modes, L_m internal expansions are built. For instance, considering the fifth-order polynomials, three internal expansions are found, which are

$$L_{17} = \phi_2(\xi)\phi_2(\eta), \ 2+2 = 4$$
$$L_{22} = \phi_3(\xi)\phi_2(\eta), \ 3+2 = 5 \tag{7}$$
$$L_{23} = \phi_2(\xi)\phi_3(\eta), \ 2+3 = 5$$

Refined element based on CUF
In this section, the Carrera Unified Formulation (CUF) is presented for beams, plates, and shells. Multilayered beam, plate and shell structures are shown in Fig. 3.

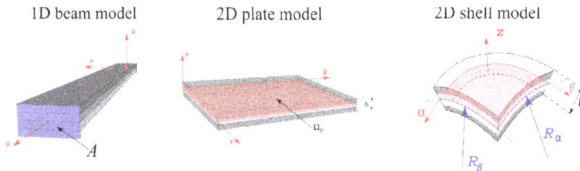

Figure 3: Generic multilayered beam, plate, and shell structures.

Concerning the beam model, the cross-section A lays on the $x - z$ plane of a Cartesian reference frame, whereas the beam axis is placed along the y direction. Contrarily, the plate model uses the z coordinate along the thickness direction, and the coordinates x and y indicate the in-plane mid-surface Ω_0. The 2D shell uses a curvilinear reference system (α, β, z) to account for the curvatures R_α and R_β. The displacement vector for the models is introduced in the following

$$\mathbf{u}^k(x,y,z) = \{u_x^k, u_y^k, u_z^k\}^T, \ \mathbf{u}^k(\alpha,\beta,z) = \{u_\alpha^k, u_\beta^k, u_z^k\}^T \tag{8}$$

where k indicates the layer. The stress, $\boldsymbol{\sigma}^k$, and strain, $\boldsymbol{\epsilon}^k$, vectors are defined as

$$\boldsymbol{\sigma}^k = \{\sigma_{xx}^k, \sigma_{yy}^k, \sigma_{zz}^k, \sigma_{xz}^k, \sigma_{yz}^k, \sigma_{xy}^k\}^T \quad \boldsymbol{\epsilon}^k = \{\epsilon_{xx}^k, \epsilon_{yy}^k, \epsilon_{zz}^k, \epsilon_{xz}^k, \epsilon_{yz}^k, \epsilon_{xy}^k\}^T$$
$$\boldsymbol{\sigma}^k = \{\sigma_{\alpha}^k, \sigma_{\beta\beta}^k, \sigma_{zz}^k, \sigma_{\alpha z}^k, \sigma_{\beta z}^k, \sigma_{\alpha\beta}^k\}^T \quad \boldsymbol{\epsilon}^k = \{\epsilon_{\alpha\alpha}^k, \epsilon_{\beta\beta}^k, \epsilon_{zz}^k, \epsilon_{\alpha z}^k, \epsilon_{\beta z}^k, \epsilon_{\alpha\beta}^k\}^T \tag{9}$$

The displacement–strain relations are expressed as

$$\boldsymbol{\epsilon}^k = \mathbf{b}\mathbf{u}^k \tag{10}$$

where $\mathbf{b}$ is the matrix of differential operators, see [1] for more information. The constitutive relation for linear elastic orthotropic materials reads as:

$$\boldsymbol{\sigma}^k = \mathbf{C}^k \boldsymbol{\epsilon}^k \tag{11}$$

Aerospace Science and Engineering - III Aerospace PhD-Days   Materials Research Forum LLC
Materials Research Proceedings 33 (2023) 148-155   https://doi.org/10.21741/9781644902677-22

where $\mathbf{C}^k$ is the material elastic matrix, see Bathe [20] for the explicit form.

The 3D displacement field $\mathbf{u}^k(x, y, z)$ of the 1D beam and 2D plate and $\mathbf{u}^k(\alpha, \beta, z)$ shell models can be expressed as a general expansion of the primary unknowns in Table 1.

Table 1: CUF Formulation. τ=1, ..., M.

Formulation	3D Field	CUF Expansion	
1D beam	$\mathbf{u}^k(x, y, z)$	$F_\tau^k(x, z)$	$\mathbf{u}_\tau^k(y)$
2D plate	$\mathbf{u}^k(x, y, z)$	$F_\tau^k(z)$	$\mathbf{u}_\tau^k(x, y)$
2D shell	$\mathbf{u}^k(\alpha, \beta, z)$	$F_\tau^k(z)$	$\mathbf{u}_\tau^k(\alpha, \beta)$

F_τ are the expansion functions of the generalized displacements $\mathbf{u}_\tau^k$ the summing convention with the repeated indexes τ is assumed and M denotes the order of expansion. Thanks to this formalism, it is possible to choose a generic structural theory freely. As explained in the previous sections, Taylor, Lagrange, and Jacobi polynomials can be used. Furthermore, the last two polynomials can be adopted in both ESL and LW approaches.

The Finite Element Method (FEM) is adopted to discretize the generalized displacements $\mathbf{u}_\tau^k$. Thus, recalling equations described in Table 1, they are approximated as displayed in Table 2.

Table 2: Finite element method. i=1, ..., N_n

Formulation	3D Field	FEM + CUF Expansion		
1D beam	$\mathbf{u}^k(x, y, z)$	$N_i(y)$	$F_\tau^k(x, z)$	$\mathbf{u}_{\tau i}^k$
2D plate	$\mathbf{u}^k(x, y, z)$	$N_i(x, y)$	$F_\tau^k(z)$	$\mathbf{u}_{\tau i}^k$
2D shell	$\mathbf{u}^k(\alpha, \beta, z)$	$N_i(\alpha, \beta)$	$F_\tau^k(z)$	$\mathbf{u}_{\tau i}^k$

N_i stand for the shape functions, the repeated subscript i indicates summation, N_n is the number of the FE nodes per element and $\mathbf{u}_{\tau i}^k$ are the following vectors of the FE nodal parameters:

$$\mathbf{u}_{\tau i}^k = \left\{ u_{x_{\tau i}}^k, u_{y_{\tau i}}^k, u_{z_{\tau i}}^k \right\}^T, \quad \mathbf{u}_{\tau i}^k = \left\{ u_{\alpha_{\tau i}}^k, u_{\beta_{\tau i}}^k, u_{z_{\tau i}} \right\}^T \tag{12}$$

A further step can be made if the cross-sections functions are anchored to the nodes of beam elements. In this way, the so-called Node-dependent kinematics (NDK) method can be performed. Substantially, each FE node has its own structural theories. Thence, the 3D field is modified as

$$\mathbf{u}^k = N_i F^{ki} \mathbf{u}_{\tau i}^k \tag{13}$$

FE governing equations

The Principle of Virtual Displacements is used for a static analysis, and it reads:

$$\int_{V_k} \delta \boldsymbol{\epsilon}^T \boldsymbol{\sigma} dV_k = \int_{V_k} \delta \mathbf{u}^{kT} \mathbf{p}^k dV_k \tag{14}$$

where $\mathbf{p}$ is the external load. When a cartesian frame is used $dV_k = dxdydz$, where for a curvilinear reference system $dV_k = H_\alpha H_\beta d\alpha d\beta dz$. The left-hand side is the variation of the internal work, while the right-hand side is the virtual variation of the external work. The real and the virtual systems are used, and displacements and virtual displacement are written as

$$\mathbf{u}^k(x, y, z) = N_i F_\tau^{ki} \mathbf{u}_{\tau i}^k, \quad \delta \mathbf{u}^k(x, y, z) = N_j F_s^{kj} \delta \mathbf{u}_{sj}^k \tag{15}$$

By using the CUF-type displacement functions in Eq. (13), the geometric relations in Eq. (10), and constitutive equations Eq. (11), the following expression can be obtained:

$$\mathbf{K}_{ij\tau s}^{k}\mathbf{u}_{\tau i}^{k} = \mathbf{P}_{sj}^{k} \tag{16}$$

where $\mathbf{K}_{ij\tau s}^{k}$, a 3X3 matrix, is the fundamental nucleus (FN) of stiffness matrix, and $\mathbf{P}_{sj}^{k}$, a 3X1 vector, represents the FN of the load vector. See [1] for the explicit form of the components of the stiffness matrix for each formulation and the assembly procedure.

Results

Concerning the structural theories, TP indicates Taylor with order P, and LHJP indicates layer-wise Jacobi of Pth polynomial order, whereas EHJP stands for equivalent single-layer Jacobi. For the shape functions only the acronym, JP is used.

A three-layered composite plate subjected to a sinusoidal pressure (see Fig. 4) with b/h=4 is studied with Jacobi-like polynomials along the thickness, see Pagano [21]. Nine-node Lagrangian shape functions are used for the FE mesh. The CUF based results were presented in [11].

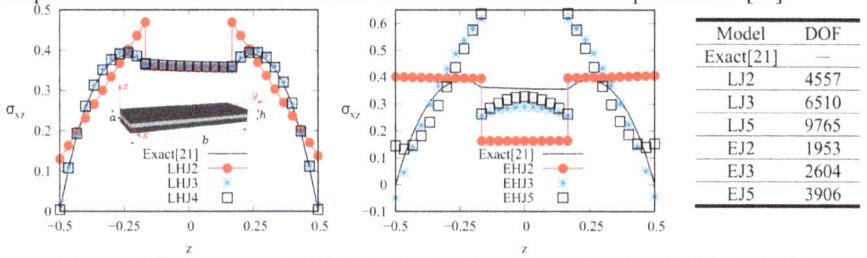

Model	DOF
Exact[21]	—
LJ2	4557
LJ3	6510
LJ5	9765
EJ2	1953
EJ3	2604
EJ5	3906

Figure 4: Shear stresses in [a/2, 0, z] of three-layer composite plate for LW and ESL.

As a second example, a thin-walled cylinder is analysed by using beam and shell formulations, see Fig. 5. This case is taken form Carrera et al. [22]. In this case, Jacobi-like shape functions are adopted, while Taylor polynomials are used as structural theories.

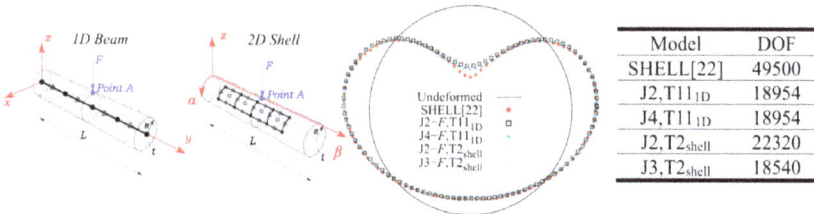

Model	DOF
SHELL[22]	49500
J2,T11$_{1D}$	18954
J4,T11$_{1D}$	18954
J2,T2$_{shell}$	22320
J3,T2$_{shell}$	18540

Figure 5: Comparison for beam and shell formulations for thin-walled cylinder. Deformed cross-section at the midspan of the hollow cylinder.

It is shown that the parameters γ and θ of the Jacobi polynomials are not influential for the calculations. Thence, Legendre-like polynomials can be adopted without loss of generality.

Finally, a cantilever beam is considered, see [23, 24]. Fig. 6 shows the axial stresses near the clamped section. NDK models with Legendre-Legendre combination are compared with uniform models. In this case, the following notation is HLE5$^{\times a}$-HLE1$^{\times b}$, where a and b represent the number of nodes of the beam elements adopting the corresponding kinematics. Forty cubic Lagrange-like finite elements are used along y axis.

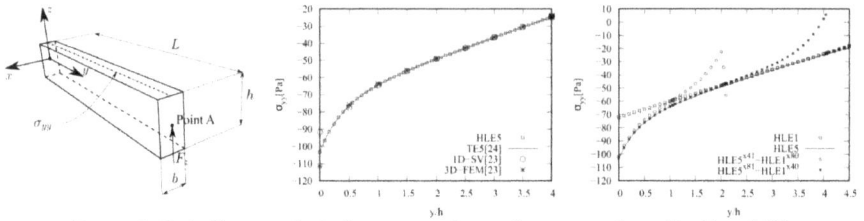

Figure 6: End-effects analysis for compact beam. Stresses evaluated in [0, y, h/2].

In proximity of the clamped section, the results calculated with NDK models are near to those referred to uniform kinematic HLE5 model.

Conclusions

The Carrera Unified Formulation (CUF) permits to build a huge number of models, by adopting different shape functions and structural theories in a hierarchical and coherent manner. In the present work, Jacobi polynomials have been included as shape functions and structural theories in analysis of beam, plates, and shells. It has demonstrated, however, that γ and θ of the Jacobi polynomials are not influential for the calculations. Concerning the structural theories, using the equivalent single layer approach for the Lagrange and Jacobi-based expansions is useful to reduce the computational time. Furthermore, it is possible to use an advanced global-local analysis, that is Node-Dependent Kinematics, which can link different structural theories in the same finite element.

References

[1] E. Carrera, M. Cinefra, M. Petrolo, and E. Zappino. Finite element analysis of structures through unified formulation. John Wiley & Sons, 2014.https://doi.org/10.1002/9781118536643

[2] L. Euler. Methodus inveniendi lineas curvas maximi minimive proprietate gaudentes sive solutio problematis isoperimetrici latissimo sensu accepti, volume 1. Springer Science & Business Media, Berlin, Germany, 1952.

[3] S.P. Timoshenko. On the transverse vibrations of bars of uniform cross section. Philosophical Magazine, 43:125-131, 1922.https://doi.org/10.1080/14786442208633855

[4] E. Carrera and G. Giunta. Refined beam theories based on a unified formulation. International Journal of Applied Mechanics, 02(01):117-143, 2010.https://doi.org/10.1142/S1758825110000500

[5] A. Pagani, E. Carrera, R. Augello, and D. Scano. Use of Lagrange polynomials to build refined theories for laminated beams, plates and shells. Composite Structures 2021;276.https://doi.org/10.1016/j.compstruct.2021.114505

[6] G. Kirchhoff. ˝Über das Gleichgewicht und die Bewegung einer elastischen Scheibe. Journal für die reine und angewandte Mathematik, 40:51-88, 1850.https://doi.org/10.1515/crll.1850.40.51

[7] E. Reissner. The effect of transverse shear deformation on the bending of elastic plates. Journal of Applied Mechanics, 12:69-77, 1945.https://doi.org/10.1115/1.4009435

[8] R.D. Mindlin. Influence of rotary inertia and shear on flexural motions of isotropic, elastic plates. Journal of Applied Mechanics-transactions of the ASME, 18:31-38, 1951.https://doi.org/10.1115/1.4010217

[9] E. Carrera. Developments, ideas, and evaluations based upon Reissner's Mixed Variational Theorem in the modeling of multilayered plates and shells. Appl. Mech. Rev., 54(4):301-329, 2001.https://doi.org/10.1115/1.1385512

[10] E. Carrera. Evaluation of layerwise mixed theories for laminated plates analysis. AIAA J 1998;36(5):830-9.https://doi.org/10.2514/2.444

[11] E. Carrera, R. Augello, A. Pagani, and D. Scano. Refined multilayered beam, plate and shell elements based on Jacobi polynomials. Composite Structures, 304:116275, 2023.https://doi.org/10.1016/j.compstruct.2022.116275

[12] M. Abramowitz and I.A. Stegun. Handbook of Mathematical Functions with Formulas, Graphs, and Mathematical Tables. Dover Publications, 1964.

[13] B. Szabo, A. Duester, and E. Rank. The p-Version of the Finite Element Method, Chapter 5. John Wiley & Sons, 2011.

[14] E. Zappino, G. Li, A. Pagani, E. Carrera, and A.G. de Miguel. Use of higher-order Legendre polynomials for multilayered plate elements with node-dependent kinematics. Composite Structures, 202:222-232, 2018. Special issue dedicated to Ian Marshall.https://doi.org/10.1016/j.compstruct.2018.01.068

[15] A. Pagani, A.G. de Miguel, M. Petrolo, and E. Carrera. Analysis of laminated beams via unified formulation and Legendre polynomial expansions. Composite Structures, 156:78- 92, 2016. 70th Anniversary of Professor J. N. Reddy.https://doi.org/10.1016/j.compstruct.2016.01.095

[16] E. Carrera, M. Cinefra, and G. Li. Refined finite element solutions for anisotropic laminated plates. Composite Structures, 183:63-76, 2018. In honor of Prof. Y. Narita.https://doi.org/10.1016/j.compstruct.2017.01.014

[17] E. Carrera and E. Zappino. Analysis of complex structures coupling variable kinematics one-dimensional models. In: ASME 2014 International Mechanical Engineering Congress and Exposition. American Society of Mechanical Engineers; 2014.https://doi.org/10.1115/IMECE2014-37961

[18] E. Zappino, G. Li, A. Pagani, and E. Carrera. Global-local analysis of laminated plates by node-dependent kinematic finite elements with variable ESL/LW capabilities. Composite Structures 2017; 172:1-14.https://doi.org/10.1016/j.compstruct.2017.03.057

[19] G. Li, E. Carrera, M. Cinefra, A. G. de Miguel, A. Pagani, and E. Zappino. An Adaptable Refinement Approach for Shell Finite Element Models Based on Node-Dependent Kinematics, Composite Structures, 2018.https://doi.org/10.1016/j.compstruct.2018.10.111

[20] K.J. Bathe. Finite Element Procedure. Prentice hall, Upper Saddle River, New Jersey, USA, 1996.

[21] N.J. Pagano. Exact solutions for composite laminates in cylindrical bending. Journal of Composite Materials, 3(3):398-411, 1969.https://doi.org/10.1177/002199836900300304

[22] E. Carrera, G. Giunta, and M. Petrolo. A modern and compact way to formulate classical and advanced beam theories. Developments in Computational Structures Technology, 75-112, 2010.https://doi.org/10.4203/csets.25.4

[23] N. Ghazouani and R. El Fatmi. Higher order composite beam theory built on Saint-Venant's solution. Part II-Built-in effects influence on the behavior of end-loaded cantilever beams. Composite Structures, 93(2): 567-581.https://doi.org/10.1016/j.compstruct.2010.08.023

[24] E. Carrera, M. Petrolo, and E. Zappino. Performance of CUF approach to analyze the structural behavior of slender bodies. Journal of Structural Engineering, 138(2):285-297, 2012.https://doi.org/10.1061/(ASCE)ST.1943-541X.0000402

Aerospace Science and Engineering - III Aerospace PhD-Days   Materials Research Forum LLC
Materials Research Proceedings 33 (2023) 156-162   https://doi.org/10.21741/9781644902677-23

A metamodel based on basis spline hyper-surfaces for thermal simulation of the wire arc additive manufacturing process

Mathilde Zani[1,a], Enrico Panettieri[1,b], Marco Montemurro[1,c*]

[1]Arts et Métiers Institute of Technology, Université de Bordeaux, CNRS, INRA, Bordeaux INP, HESAM Université, I2M UMR, F-33405 Talence

[a]mathilde.zani@ensam.eu, [b]enrico.panettieri@endam.eu, [c]marco.montemurro@ensam.eu

Keywords: Additive Manufacturing, Metamodel, Industry 4.0, Numerical Simulation

Abstract. Wire arc additive manufacturing an additive manufacturing process that allows producing large metal parts in a layer-by-layer fashion assuring a high deposition rate. However, the set of parameters that control the process leads to phenomena that are difficult to understand and predict. The work of this paper is to propose a metamodel based on basis spline entities to approximate the thermal response of the WAAM process for different combinations of deposition parameters. The aim is to reduce the computational cost, and obtain results without actually solve a complete FE model for any combination of parameters in the design space.

Introduction

In the aerospace industry, the amount of wasted material generated during manufacturing is called the buy-to-fly (BTF) ratio, which is defined as the ratio of the mass of raw material to the mass of the finished part [1]. In order to reduce material waste and thereby reduce the BTF ratio, Additive Manufacturing (AM) technology is an environmentally friendly manufacturing alternative to conventional manufacturing processes. Among metal AM processes, wire arc additive manufacturing (WAAM) is one of the most promising technologies in terms of deposition rate [2] allowing the production of large near net-shape metal parts with complex geometry by depositing weld beads in a layer-by-layer strategy [3].

Despite these advantages, the quality of parts manufactured by WAAM is highly affected by the thermal and mechanical phenomena occurring during the process, which are influenced by its main parameters. Furthermore, the understanding of the relationships between the physical phenomena and the parameters governing the process (together with the interaction between these parameters) represents a challenging task [2,4]. Accordingly, process simulation is a powerful tool to address such issues, allowing the simulation of the effect of different deposition parameters and, thus, optimising the process.

From a simulation perspective, WAAM technology is typically simulated using a transient thermomechanical Finite Element (FE) analysis with progressive material addition. However, the computational time associated with this analysis can become prohibitive, especially when the influence of process parameters on the thermomechanical properties of the material must be integrated into the design process. As discussed by Ding et al. [5] this usually results in a reduction in the effectiveness gains of WAAM process numerical modelling. In addition, due to the prohibitive computational costs related to FE non-linear thermomechanical analyses, such a modelling strategy cannot assess the sensitivity of the temperature field and residual strain/stress fields within manufactured parts to the main process parameters. Therefore, appropriate abaci should be used at the preliminary design phase to predict the behaviour of the resulting material in terms of stiffness, thermal conductivity, thermal expansion coefficients, etc., since calculated values of these parameters cannot be obtained in a reasonable time using this type of modelling strategy. Accordingly, Ding et al. [5], Montevecchi et al., [6], Michaleris [7] proposed different methodologies to reduce computational costs, while keeping a reasonable level of accuracy.

Aerospace Science and Engineering - III Aerospace PhD-Days   Materials Research Forum LLC
Materials Research Proceedings 33 (2023) 156-162   https://doi.org/10.21741/9781644902677-23

Therefore, a trade-off must be found between the computational costs and the precision required for a given application [8].

In this context, metamodels are efficiently employed to capture the influence of the main parameters of the process on the manufactured parts and to obtain results having a level of accuracy as good as the one related to non-linear thermomechanical FE models. Generally, a metamodel consists of the definition of a parametric hyper-surface that is capable of approximating (or interpolating) some data [9,10] without knowing the explicit physical equations of the problem at hand. In comparison with other metamodeling approaches [10], Non-Uniform Rational Basis Splines (NURBS) entities offer many unique advantages [9].

This research proposes a metamodel based on Basis spline (B-spline) entities (a sub-class of NURBS hyper-surfaces) applied to thermal analyses of the WAAM process. The main goal is to analyse the thermal response of the process as a function of different deposition parameters. The temperature histories are monitored during the simulations at different points on the substrate and approximated through a B-spline hyper-surface. The hyper-surface is built as a result of an optimisation procedure generalising the one proposed in previous works [9].

The paper is organised as follows. Section 2 briefly introduces the main features of WAAM process and gives the description of the non-linear thermal problem that will be used to build the metamodel. In Section 3 the fundamentals of B-spline entities are recalled, and the algorithm employing for generating the metamodel is presented. The results are presented in Section 4. Lastly, Section 5 ends the paper with conclusions and prospects.

Finite element numerical model

WAAM is a Direct Energy Deposition process in which the material is directly deposed on a substrate and locally heated by a heat source. To better understand the process behaviour and improve the final quality of the part, FE models have been adopted to analyse and simulate the non-linear phenomena occurring in the WAAM process.

In this work, only the numerical model of the thermal problem is developed to analyse the thermal history as a function of different deposition parameters, notably the torch speed also referred to as travel speed (TS); the power of the welding Q; and the deposition rate expressed as the Wire feed Speed (WFS). Moreover, the material is deposed maintaining a constant volume for all the analyses.

The numerical model of the WAAM process deals with the progressive heating of the deposited material, the progressive material addition, and the thermal dependencies of the chosen material.

Firstly, a 3D transient non-linear heat transfer analysis is performed using the commercial software ABAQUS®. Moreover, the heat transfer from the arc to the molten pool is described employing an equivalent heat source model. Generally, for 3D AM simulations, the volumetric heat source proposed by Goldak et al., [11] is used as it capable of modelling the three-dimensional phenomena occurring in the molten pool. The Goldak heat source reads:

$$q(x,y,z) = \frac{Q6\sqrt{3}f_\zeta}{a_\zeta bc\pi^{3/2}} e^{\left(-3\left(\frac{x^2}{a_\zeta^2} + \frac{y^2}{b^2} + \frac{z^2}{c^2}\right)\right)}, \tag{1}$$

with $\zeta = f$ *(front) for* $x \geq 0$ *and* $\zeta = r$ *(rear) for* $x < 0$*, and* $f_f + f_r = 2$ to ensure continuity at the source origin x= 0. A schematic representation of the Goldak double ellipsoid heat source is represented in Fig.1.

Aerospace Science and Engineering - III Aerospace PhD-Days   Materials Research Forum LLC
Materials Research Proceedings 33 (2023) 156-162   https://doi.org/10.21741/9781644902677-23

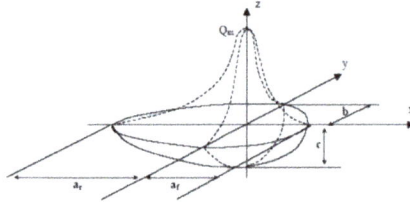

Figure 1 Double ellipsoid Goldak heat source) [11].

A second issue in the WAMM process simulation is the material deposition modelling. In this work, the progressive elements activation technique is employed [12]. This method, available in ABAQUS® software, allows activating the elements as a function of time and space and relating them to the movement of the heat source [13].

The FE model used to generate the database for the B-spline metamodel is built following the guidelines available in the work of [5]. It consists of a four-layer wall deposited along the centre line of the base plate, as shown in Fig. 2.

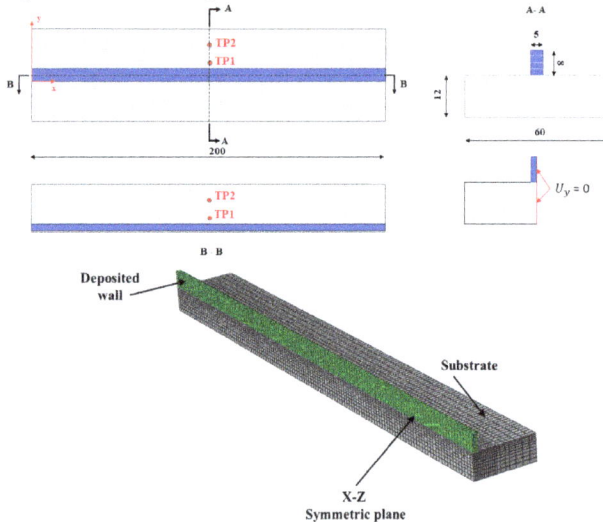

Figure 2 FE numerical model from [5]

The material used for both the substrate and the wall is a mild steel with material properties dependent of the temperature taken from literature [14].

To reduce the computational costs of the analysis, only half of the model in the X-Z plane is considered. Lastly, linear brick elements with eight nodes (DC3D8) are used for the thermal simulation with meshes of size 2 mm× 0.833 mm× 0.667 mm for the bead and the area near the welding line, and a coarsened mesh far from the wall to reduce the total number of elements. The complete details of the FE model, together with the initial thermal and boundaries conditions can be found in [5].

The goal of the metamodel is to approximate the temperature value at given locations, i.e., on the nodes where thermocouples TP1 and TP2 are placed as illustrated in Figure 2, as function of

Aerospace Science and Engineering - III Aerospace PhD-Days   Materials Research Forum LLC
Materials Research Proceedings 33 (2023) 156-162   https://doi.org/10.21741/9781644902677-23

process parameters. Accordingly, a series of WAAM simulations is run considering suitable intervals of the main process parameters, i.e., the input variables of the metamodel. The selected values are listed in Table 1, for a total of 1000 simulations.

Table 1 Values of deposition parameters

TS [mm/s]	Q [W]	t_{cool} [s]	time [s]
[4.4, 6.2, 8.4, 10,10.68, 11, 11.2, 12, 13.2, 16.4]	[364, 447.6, 571.1, 700.3, 849.4, 890.1, 1282.3, 1537.4, 1826.3, 2302.2]	[2,5,10,15,20,30, 40,50,60,100]	[0,496]

As the process parameters vary, there is a modification in the bead behaviour with a consequent change in its temperature history [2]. Inasmuch as the interest is creating a database for the B-spline metamodel, the heat source dimension and parameters are kept constants for all the simulations and equal to the one of [5]. Moreover, the heat step time increment is considered equal to 1 sec.

Finally, the geometry of the model, which should vary as the deposition parameters change is kept constant. This strong assumption is justified by the fact that the material is deposited at a constant volume and with the progressive element technique. For more details on this matter, the interested reader is addressed to [15].

The results of the temperature profile at the two thermocouples TP1, TP2 are presented in section 4.

B-spline metamodel of WAAM thermal problem

The B-spline entities are used as a tool for creating a metamodel that generates a response surface capable of fitting a given set of target points (TPs) X_i. This section briefly presents the theoretical background of B-spline entities, and the basics of the implemented algorithm to evaluate the metamodel.

Theoretical background

A B-spline hyper-surface is a polynomial-based function defined over a domain of dimension N domain to a codomain of dimension M, , $H : \mathbb{R}^N \to \mathbb{R}^M$ [9]. The mathematical formula of a generic B-spline hyper-surface reads:

$$H(\zeta_1, \dots \zeta_N) := \sum_{i_1=0}^{n_1} \cdots \sum_{i_N=0}^{n_N} N_{i_1,p_1}(\zeta_1) \times \dots \times N_{i_N,p_N}(\zeta_N) P_{i1\dots iN} \quad , \tag{2}$$

where $\zeta_k \in [0,1]$ is the kth dimensionless coordinate, whereas $P_{i_1,\dots,i_N} := \left\{ P_{i_1,\dots,i_N}^{(1)}, \dots, P_{i_1,\dots,i_N}^{(M)} \right\}$ are the control points (CPs) that constitute the control hyper net.

For each parametric direction $k = 1, \dots, N$, $N_{i_k,p_k}(\zeta_k)$ represents the Bernstein's polynomial of order p_k and defined recursively as discussed in [16].

The choice of using B-spline entities as a metamodel strategy stems from its ability to well approximate problem non-linearities, and to be applicable to MIMO systems. The first property derives from the local support characteristic of the blending functions. Indeed, thanks to this property, each control point affects only a restricted portion of the domain wherein the B-spline is defined. For more information about this topic, the reader is addressed to [16].

Aerospace Science and Engineering - III Aerospace PhD-Days   Materials Research Forum LLC
Materials Research Proceedings 33 (2023) 156-162   https://doi.org/10.21741/9781644902677-23

Implemented algorithm.

The algorithm, originally implemented by [9] has been generalised to any combination of dimensions N and M of the B-spline hyper-surface domain and codomain, respectively.

The algorithm main steps are schematically represented in Fig. 3.

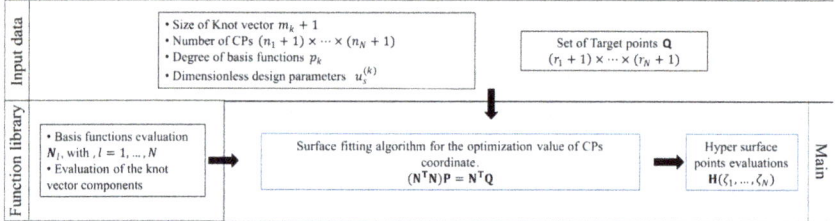

Figure 3 Algorithm flowchart. The hyper-surface surface fitting problem for the optimization of the CPs coordinates is taken from [9].

More precisely, once all the necessary input data have been provided, several routines are called to evaluate the basis functions, the optimal values of the CPs (constituting the control hyper-net), and then, the hyper-surface. The database of TPs is determined via numerical analyses conducted through the FE model presented in Section 2, and the metamodel will approximate the thermal history $T(t,TS,Q, t_{cool})$ at the two thermocouples for different combinations of input variables.

Results and discussion

This section presents the result of the B-spline metamodel in terms of thermal history $T(t,TS,Q, t_{cool})$ at the thermocouple TP1, TP2. For brevity, the results are shown only for TP1.

Fig. 4 represents the approximated temperature contour plots as a function of the inputs. The coefficient $R^2 = 1 - \frac{RSS}{TSS}$, which represents a measure of the quality of the approximation of the metamodel, is equal to 0.977. RSS is the residual sum of squares and TSS is the total sum of squares. This means that the fitting capability of the B-spline metamodel is accurate (the closer R^2 to the unit the more accurate is the approximation).

Furthermore, the computational cost to obtain this approximation is well in excess of 100% lower than the computational time of the FE simulations.

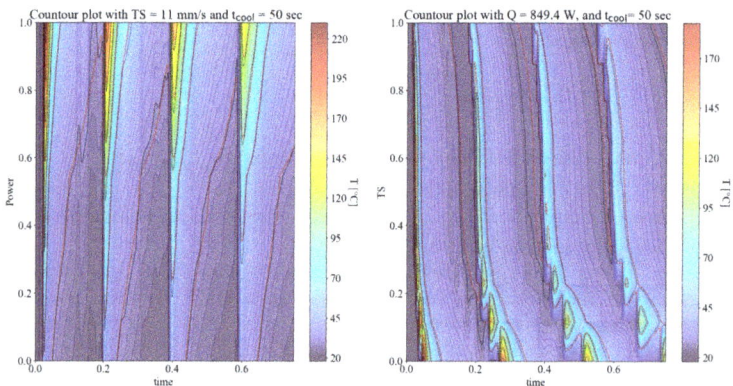

Figure 4 Contour plot of the nodal temperature for thermocouple TP1 resulting from the B-spline hyper-surface for a constant TS (on the left), a constant power (on the right)

Aerospace Science and Engineering - III Aerospace PhD-Days   Materials Research Forum LLC

Materials Research Proceedings 33 (2023) 156-162   https://doi.org/10.21741/9781644902677-23

However, the main purpose of the metamodel is to obtain the results for input values not included in domain described by the 1000 simulations, without solving the entire numerical model. For this reason, four sets of TS, Q, t_{cool} not included in the initial set have been chosen to test the accuracy of the approximated method. The results, together with the four parameter sets are shown in Fig. 5.

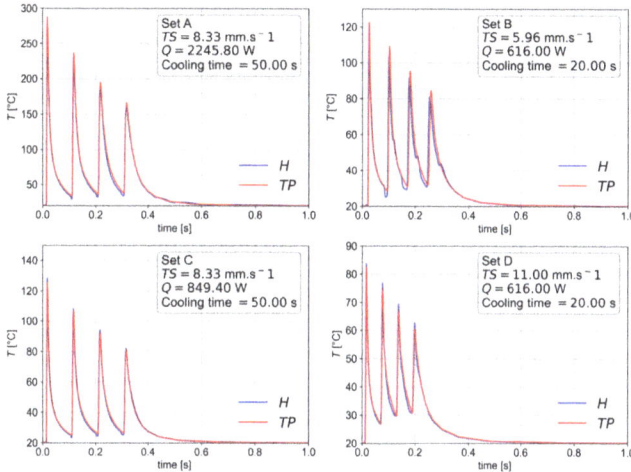

Figure 5 Temperature history at thermocouple TP1 for the four sets of deposition parameters constituting the validation set.

Overall, the approximated temperature history match quite well the simulated results.

Conclusions and Prospect

In this paper, a metamodel based on B-spline entities has been applied to the thermal simulation of the WAAM process to approximate the temperature history at different locations of a 3D model. The proposed metamodeling strategy allows approximating with a good accuracy the numerical results while considerably reducing the computational cost. Moreover, it has been possible to obtain temperature profiles for sets of values not included in the initial database. It must be highlighted that the generated metamodel is able to predict the results of the non-linear finite element model even if some discrepancies are present due to the chosen approximation method. Indeed, B-spline entities are not able to correctly approximate distributions of data characterised by strong non-linearities and/or discontinuities on the local tangent vector direction. To overcome this issue, the presented metamodeling strategy should be extended to the most general case of NURBS entities by including the weights related to each control point and the inner components of the knot vectors among the design variables. Research is ongoing on this aspect.

References

[1] Lockett, Helen; Ding, Jialuo; Williams, Stewart and Martina, Filomeno (2017). Design for Wire + Arc Additive Manufacture: design rules and build orientation selection. Journal of Engineering Design, 28(7-9) pp. 568–598, https://doi.org/10.1080/09544828.2017.1365826

[2] Ding, D., Pan, Z., Cuiuri, D. and Li, H. (2015), "Wire-feed additive manufacturing of metal components: technologies, developments and future interests", The International Journal of

Advanced Manufacturing Technology, Vol. 81, pp. 465–481, https://doi.org/10.1007/s00170-015-7077-3.

[3] Chergui, M.A. (2021), Simulation Based deposition Strategies Evaluation and Optimization in Wire Arc Additive Manufacturing, Theses, Université Grenoble Alpes. URL: https://tel.archives-ouvertes.fr/tel-03273221

[4] Wu, B., Pan, Z., Ding, D., Cuiuri, D., Li, H., Xu, J. and Norrish, J. (2018), "A review of the wire arc additive manufacturing of metals: properties, defects and quality improvement", Journal of Manufacturing Processes, Vol. 35, pp. 127–139, https://doi.org/10.1016/j.jmapro.2018.08.001

[5] Ding, J., Colegrove, P., Mehnen, J., Ganguly, S., Sequeira Almeida, P., Wang, F. and Williams, S. (2011), "Thermo-mechanical analysis of wire and arc additive layer manufacturing process on large multi-layer parts", Computational Materials Science, Vol. 50 No. 12, pp. 3315–3322, https://doi.org/10.1016/j.commatsci.2011.06.023

[6] Montevecchi, F., Venturini, G., Grossi, N., Scippa, A. and Campatelli, G. (2017), "Finite element mesh coarsening for effective distortion prediction in wire arc additive manufacturing", Additive Manufacturing, Vol. 18, pp. 145–155, https://doi.org/10.1016/j.addma.2017.10.010.

[7] Montevecchi, F., Venturini, G., Grossi, N., Scippa, A. and Campatelli, G. (2017), "Finite element mesh coarsening for effective distortion prediction in wire arc additive manufacturing", Additive Manufacturing, Vol. 18, pp. 145–155, https://doi.org/10.1016/j.addma.2017.10.010.

[8] Wang, L., Chen, X., Kang, S., Deng, X. and Jin, R. (2020), "Meta-modeling of high-fidelity fea simulation for efficient product and process design in additive manufacturing", Additive Manufacturing, Vol. 35, p. 101211, https://doi.org/10.1016/j.addma.2020.101211.

[9] Audoux, Y., Montemurro, M. and Pailhès, J. (2020b), "Non-uniform rational basis spline hyper-surfaces for metamodelling", Computer Methods in Applied Mechanics and Engineering, Vol. 364, p. 112918, https://doi.org/10.1016/j.cma.2020.112918.

[10] Baillargeon, S. (2005), "Le krigeage: revue de la théorie et application à l'interpolation spatiale de données de précipitations", http://hdl.handle.net/20.500.11794/18036

[11] Goldak, J., Chakravarti, A. and Bibby, M. (1984), "A new finite element model for welding heat sources", Metallurgical transactions B, Vol. 15 No. 2, pp. 299–305, https://doi.org/10.1007/BF02667333.

[12] Smith, M. (2020), "Abaqus/standard user's manual, version 6.20"

[13] Song, X., Feih, S., Zhai,W., Sun, C.N., Li, F., Maiti, R.,Wei, J., Yang, Y., Oancea, V., Brandt, L.R. et al. (2020), "Advances in additive manufacturing process simulation: Residual stresses and distortion predictions in complex metallic components", Materials & Design, Vol. 193, p. 108779, https://doi.org/10.1016/j.matdes.2020.108779

[14] Michaleris, P., DeBiccari, A. et al. (1997), "Prediction of welding distortion", Welding Journal- Including Welding Research Supplement, Vol. 76 No. 4, p. 172s, https://www.osti.gov/biblio/477340

[15] Cambon, C. (2021), Étude thermomécanique du procédé de fabrication métallique arc-fil : approche numérique et expérimentale, Theses, Université Montpellier.URL: https://tel.archives-ouvertes.fr/tel-03396711

[16] Piegl, L. and Tiller, W. (1996), The NURBS book, Springer Science & Business Media, https://doi.org/10.1007/978-3-642-97385-7

Aerospace Science and Engineering - III Aerospace PhD-Days
Materials Research Proceedings 33 (2023) 163-170

Materials Research Forum LLC
https://doi.org/10.21741/9781644902677-24

Comparison of predictive techniques for spacecraft shock environment

Ada Ranieri[1,a] *

[1]Department of Mechanics, Mathematics and Management (DMMM), Polytechnic University of Bari, Bari, Italy

[a]ada.ranieri@poliba.it

Keywords: Shock Prediction, Model-Based, Data-Driven, Statistical Energy Analysis, Artificial Neural Network

Abstract. Shock loads are very high amplitude and short duration transient loads. They are produced in space structures by pyrotechnic devices placed in the launchers initiating the stage or fairing separations. While the spacecraft structure is not susceptible to this range of frequencies, electrical units could be seriously damaged during the launch phase. Shock tests are performed on the electrical units to verify if they withstand the transient loads, thus their compliance to the requirements. To understand how the input force evolves from the launcher-spacecraft interface to the equipment of interest, a model of the dynamical behaviour of the spacecraft at high frequencies has to be developed. An initial approach constitutes the implementation of a mathematical model through the use of Statistical Energy Analysis (SEA). The results of a 4 degrees of freedom model using SEA will be shown. The model can be further developed by the integration of data-driven techniques. In this work a description of two different approaches is presented, that include model-based and data-driven methods. Finally, a cross-cutting potential solution is briefly introduced; it will combine experimental data with a mathematical model as to convey them in the training database of an Artificial Neural Network algorithm. The hybrid solution will possibly turn out as a reliable and efficient way to break down time and costs of the shock test campaign.

Introduction

During launch, the spacecraft experiences full-frequency dynamic loads, which go from 10 up to 10k Hz. The full-frequency band bring about great discrepancy of structure responses at different frequencies. This phenomenon is more evident at high frequencies, where shock occurs. The structure no longer shows a deterministic behaviour, and a statistical approach is needed. A way to boost the spacecraft development process could be to identify a technique that predict the structure response of any spacecraft when it is subjected to shock loads for the entire frequency range that the spacecraft encounters during launch. The integration of the technique in the design phase of the spacecraft structure could facilitate, shorten, or even avoid, the mechanical shock test.

The case study is a multi-Launcher satellite, meaning that it is designed to be compatible with more than one launcher. A satellite test campaign is usually customised for every launch because the requirements are given by the Launcher Authority, which differ from case to case. Having a multi-Launcher satellite means that it is qualified on an envelope of multiple launchers' requirements and there is no need to retest it for every mission.

The multi-launcher Satellite is composed by:

- Bus Module (BM), that supports mechanically the overall Payload Module.
- Payload Module (PM). The structure is mated onto the BM and includes the specific supporting structure required to link (thermo-)mechanically the Payload System to Platform Module.

Aerospace Science and Engineering - III Aerospace PhD-Days   Materials Research Forum LLC
Materials Research Proceedings 33 (2023) 163-170   https://doi.org/10.21741/9781644902677-24

The study case is the Structural Model (SM) of the Multi-Launcher satellite. The SM is designed to be mechanically representative of the spacecraft and it is used to perform qualification tests. A drawing of the SM is shown in Fig. 1. The SM is composed by two parts:

- Structural part
- Dummy masses, that simulate the mechanical properties of the equipment and the harness.

The peculiarity of this item is its recurrent platform, meaning that the platform can be adapted to multiple launchers and payloads.

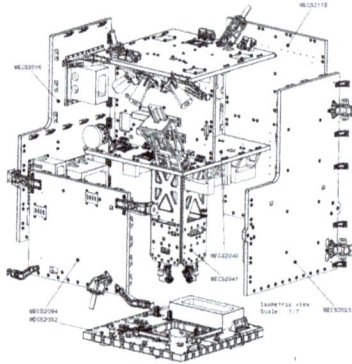

Fig. 1 – Structural Model of the spacecraft

The spacecraft undergoes the shock input represented in Fig. 2, resulting from an envelope of multiple launchers.

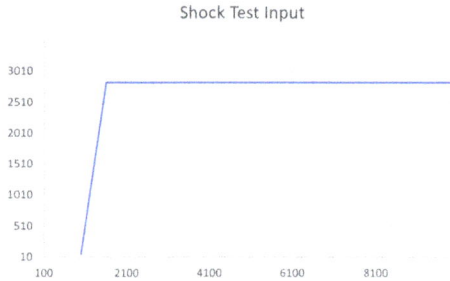

Fig. 2 - Shock input

To compute the response of the spacecraft to any shock load, it is necessary to identify the model that describes the real system behaviour and captures only the essential features of interest, leaving out everything else. If the essential dynamic of the system is known, it is possible to take a white box approach to create a model, using first principles and physics or multi-body technique. In this case, the use of a simplified description of a system is called model-based design. On the opposite side of the spectrum, there is the black box method that is mainly used to make prediction of the behaviour of a system when the dynamics are unknown, but a certain amount of data is available. However, there is a cross between the two approaches. Some rough knowledge of the

Aerospace Science and Engineering - III Aerospace PhD-Days   Materials Research Forum LLC
Materials Research Proceedings 33 (2023) 163-170   https://doi.org/10.21741/9781644902677-24

system can help in the decision of the right structure. This is called a grey box method because it is possible to use approximate knowledge of the physics of the system, through the use of numerical investigation techniques, to set up the initial problem and then using data to learn the remaining portion of the model structure or parameters set.

This work explores the white and black approaches that can be implemented for the study case. Finally, a hybrid solution that combines SEA and Artificial Neural Network (ANN) will be introduced. The overall problem is summed up in Fig. 3.

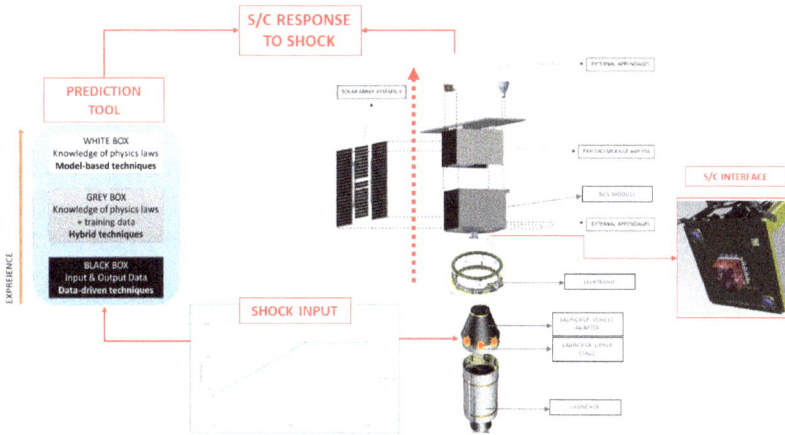

Fig. 3 - Illustration of the problem

State of the art

The major issue of shock prediction is due to its wide frequency range. The main difference is encountered between low and high frequency. At low frequency identical structures produce the same response, and it is possible to use a deterministic approach. For low frequency range, Finite Element Method (FEM) is one of the most mature and accepted numerical tool. On the other hand, at high frequencies, the modes have very small effective mass and there is high modal density and modal overlap. Thus, a minor difference between two identical structures could results in totally different response and the structure does not longer show a deterministic and predictable behaviour. FEM starts to be too sensitive to details and less accurate since the mesh must decrease in size resulting in to match the smallest characteristic deformations that occur a such frequencies [1], resulting in high computation costs [2]. A statistical approach needs to be applied. The most common technique is the Statistical Energy Analysis (SEA): the main idea in SEA is that a complex structure (e.g., a spacecraft) is described as a network of subsystems where the stored and exchanged energies are analysed [3].When dealing with mid frequency there is no universally accepted method as the structure does show neither a deterministic nor a chaotic behaviour. Accordingly, [4] proposed an improved methodology based on the Hybrid Finite Element-Statistical Energy Analysis (FE-SEA) method. Since SEA is able to deal with high frequency problems, combining FEM and SEA it is possible to cover the entire frequency range with rationality and sufficient accuracy of the prediction results. The hybrid method can predict the middle and high frequency shock response more effectively and reasonably, and the computational

Aerospace Science and Engineering - III Aerospace PhD-Days   Materials Research Forum LLC
Materials Research Proceedings 33 (2023) 163-170   https://doi.org/10.21741/9781644902677-24

efficiency is greatly improved, compared with the traditional FEM. [5] proposed the Virtual Mode Synthesis Simulation (VMSS) method, where the dynamical system is numerically convoluted with a measurement or simulated excitation force to obtain the dynamic response in the time domain. This numerical method resulted to be suitable to solve the problem of transient and high frequency environment prediction. [6] combined SEA in conjunction with VMSS to predict the dynamic response of a low altitude earth observation satellite during launch vehicle separation. In industry, The *Unified Approach And Practical Implementation Of Attenuation Rules For Typical Spacecraft Shock Generated Environments* [7] is a common technique that uses experimental data and allows to determine the acceleration at the mounting points of most critical components. This method traces the path of the shock load that propagates from its source (i.e., the spacecraft interface), where the acceleration is known, to the locations of the critical units. The attenuation factor is computed for each section of the path. It depends mainly on distance, angle, type of structure (e.g., honeycomb, skin-frame, monocoque, etc.) and presence of joints. These data have been collected mainly during experimental activities performed over the years in the European Space Agency (ESA).

The last several years have seen the academic field increasingly focused on data analysis subjects [8]. The effort arose from the fact that physics-based models have a relatively high computational demand, and are unsuitable for probabilistic, risk-based analyses. Quite the opposite, the data-driven methodology is simple in principle and easy to implement. Hence, a new data-driven approach makes use of Artificial Intelligence (AI) in building models that would replace the model-based techniques describing physical systems. The main advantage is that data-driven techniques work well with black box approaches, when the physical behaviour is unknown. Nevertheless, this activity is close in its objectives to traditional approaches to modelling and follows the traditionally accepted modelling steps.

Data-driven modelling comprehends a wide range of techniques, which include Machine learning (ML) and Artificial Neural Network (ANN). ANN turns out to be a powerful means because is able to store large amounts of experimental information to solve poorly defined problems that have eluded solutions by conventional computing techniques. This technology takes inspiration from a simplified biological neural network: an artificial neuron receives a signal then processes it through mathematical functions and sends it to the connected neurons, which are typically aggregated in layers. The output of each layer is the weighted sum of the outputs from the previous one. During the learning or training process, the weighting factors are modified so that the calculated output match the actual output, trying to minimise the error. An ANN is thus a dynamic system. This feature would result in a highly robust system where, changing the information stored in one element will have little effect on the final output. [9] proposed an ANN model for complex contacting bodies that, compared with the conventional model-based methods, is simple in principle and easy to implement. The method demonstrates great advantages of ANN when the internal mechanisms are unknown or too complicated to be explored so far. Contrary to conventional model-based techniques, neural networks can learn from example and generalize solutions to new representations of a problem, can adapt to small changes in a problem, are tolerant to errors in the input data, can process information rapidly, and are modular between computing systems. Neural networks cannot, however, guarantee success in finding an acceptable solution, and a limited ability to rationalise the solutions provided[10]. Moreover, ANN has been applied successfully to solve many difficult and diverse problems by training them with the feedforward and back-propagation algorithm. [11] presented a method for solving both ordinary differential equations (ODE's) and partial differential equations (PDE's) that relies on the function approximation capabilities of feedforward neural networks and results in the construction of a solution written in a differentiable, closed analytic form. The method could be extended in the abovementioned study case, for example to solve eigenvalue problems for differential operators.

Aerospace Science and Engineering - III Aerospace PhD-Days   Materials Research Forum LLC
Materials Research Proceedings 33 (2023) 163-170   https://doi.org/10.21741/9781644902677-24

[12] investigated the use of black-box ODE solvers as a model component, allowing explicit control of the trade-off between computation speed and accuracy.

Statistical Energy Analysis

In the first part of the work a traditional model-based method has been investigated. Statistical Energy Analysis (SEA) is a well-known method notably used for acoustics problems in the sixties. Later, it has been introduced in shock prediction to overcome FEM limitations at high frequencies. SEA describes behavior of complex systems by a set of energy-balanced equations between the various domains (subsystems) of the analyzed system. SEA assumes both perfect diffusion and weak coupling of subsystem vibrations. It makes use of wave propagation theory rather than modal approach, in order to decrease the computational cost at high frequencies. Coefficients of energy exchange driving the equations are predicted classically from analytical wave theory by decomposing modes into uncorrelated waves (diffusion of energy).

A single subsystem is considered as a separated part of the structure to be analysed. Any excitation acting on the subsystem can be characterised by the resulting power input P_i into the subsystem. If power is injected, the subsystem stores vibrational energy W_i. In practice, there will be also a power loss P_{ii} due to dissipation. This power loss may be related to the stored energy by the Damping Loss Factor (DLF) η_i by

$$P_i = \omega * \eta_i * W_i \quad . \tag{1}$$

If we consider a coupled subsystem, they share vibrational energy, in addition to the previous formula. So, the first subsystem will have two types of dissipation, one towards the external and one flowing to the other subsystem. The same phenomena occur in the reverse direction, so the Exchanged Power P_{ij} is the same for directly coupled subsystems.

$$P_{ij} = P_{ji} = \omega * \eta_{ij} * W_i \quad , \tag{2}$$

with η_{ij} known as the Coupling Loss Factor (CLF), indicating how well the subsystems are connected with each other. The global energy balance of the whole system can be written as

$$P_i = \omega * \eta_{ij} * E_i + \sum_{j=\neq}^{n} \omega * \eta_{ij} * E_i - \eta_{ji} * E_j \quad , \tag{3}$$

where ω represents the angular center frequency and Ei and Ej are the frequency energy levels of the subsystems. It can also be written in compact matrix form:

$$P = \omega[\eta]E, \tag{4}$$

where $[\eta]$ is known as Damping Loss Matrix. Finally, we have the reciprocity relation:

$$n_i * \eta_{ij} = n_j * \eta_{ji} \quad , \tag{5}$$

with n_i and n_j being the modal densities. Eq. 4 and Eq. 5 constitute the basic SEA equations.

As soon as matrix $[\eta]$ is known, it can be used repeatedly to predict the response of the subsystems for any given vector of injected powers at a negligible computational cost. It constitutes a reduced model, which describes the global system in terms of the energy content of its subsystems. This is what renders it well suited for high frequency simulations, where local indicator results are inefficient.

As an example, the method is applied on a 4 Degrees-of-Freedom (DoFs) system. The system is composed of two aluminium beams connected on their extremities with an angle of 90°. The

Aerospace Science and Engineering - III Aerospace PhD-Days   Materials Research Forum LLC
Materials Research Proceedings 33 (2023) 163-170   https://doi.org/10.21741/9781644902677-24

system presents four subsystems, each one related to a different exchange of energy in terms of longitudinal and flexural waves, as shown in Fig. 4. It has been chosen to neglect the energy transmission of longitudinal waves between the two beams.

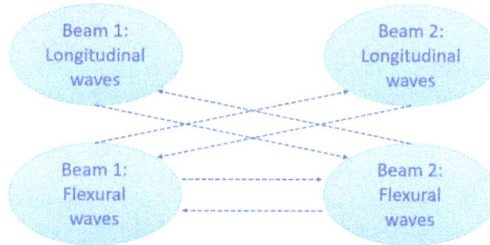

Fig. 4 - SEA subsystems

Basics SEA equations have been computed and the has been determined for each subsystem, as shown in Figure. The major difference between deterministic methods and SEA is that the former uses displacements, velocities and accelerations computed in determined locations and frequencies as parameters. The latter express averaged global variables. The resulting average square velocity, as shown in Fig. 5, has been averaged over points of observation, points of excitation and frequency.

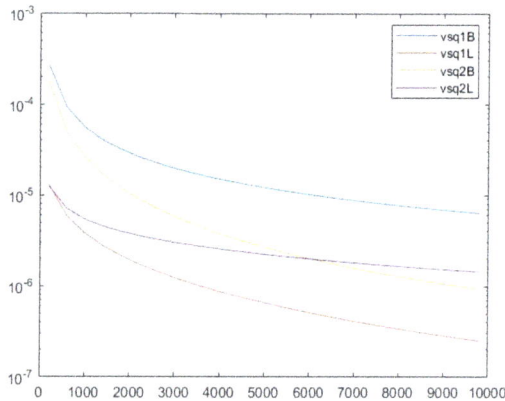

Fig. 5 - Average square velocities for bending and flexural waves of systems 1,2

The following step will be to adapt the SEA model using the previously presented shock input and converting the resulting average square velocity in terms of averaged acceleration. The model can be adapted for CFRP honeycomb sandwich panels, simulating the Structural Model mechanical properties. In this case a 6 DoF system should be modelled, as shear forces will appear. The results will be compared with the experimental method which is currently used in the industry [7]. Both the approaches computer the attenuation path of the shock load from its source to the location of interest.

Aerospace Science and Engineering - III Aerospace PhD-Days   Materials Research Forum LLC
Materials Research Proceedings 33 (2023) 163-170   https://doi.org/10.21741/9781644902677-24

Afterwards, the model forms the basis to implement an hybrid technique, that will involve a ANN algorithm. It will be trained with shock test data history that has been collected over the years and combined with the SEA model.

Conclusion

To avoid damage of the sensitive units during launch, the spacecraft must be tested thoroughly before it is sent into space. During the various phases of its development the spacecraft and its component parts undergo extensive testing. In the development process of a spacecraft, the shock test phase is time-consuming and expensive. It could be replaced by a mathematical model that predict, with certain assumptions, the structure response to dynamic loads. However, accurate prediction of the shock environment is critical to system and structure design, due to its broad frequency range and high acceleration levels. The combination of model-based and data-driven techniques to predict the structure response to shock loads could result in a reliable and efficient way to break down time and costs of the shock test campaign. This work presented SEA and ANN as potential solutions. A SEA model has been represented as a starting point for further development.

References

[1] M. Gherlone, D. Lomario, M. Mattone, and R. Ruotolo, "Application of wave propagation to pyroshock analysis," IOS Press, 2004.

[2] H. Zhao *et al.*, "The shock environment prediction of satellite in the process of satellite-rocket separation," *Acta Astronaut*, vol. 159, pp. 112–122, Jun. 2019. https://doi.org/10.1016/j.actaastro.2019.03.017

[3] E. Sarradj, "Energy-based vibroacoustics: SEA and beyond."

[4] X. Wang, W. Liu, X. Li, and Y. Sun, "The shock response prediction of spacecraft structure based on hybrid fe-sea method," *Applied Sciences (Switzerland)*, vol. 11, no. 18, Sep. 2021. https://doi.org/10.3390/app11188490

[5] E. C. Dalton and B. S. Chambers, "Analysis and validation testing of impulsive load response in complex, multi-compartmented structures," in *Collection of Technical Papers - AIAA/ASME/ASCE/AHS/ASC Structures, Structural Dynamics and Materials Conference*, 1995, vol. 2, pp. 759–767. https://doi.org/10.2514/6.1995-1243

[6] D.-O. Lee, J.-H. Han, H.-W. Jang, S.-H. Woo, and K.-W. Kim, "Shock Response Prediction of a Low Altitude Earth Observation Satellite During Launch Vehicle Separation," *International Journal of Aeronautical and Space Sciences*, vol. 11, no. 1, pp. 49–57, Mar. 2010. https://doi.org/10.5139/ijass.2010.11.1.049

[7] "Space engineering Mechanical shock design and verification handbook ECSS Secretariat ESA-ESTEC Requirements & Standards Division Noordwijk, The Netherlands," 2015.

[8] R. S. Kenett, A. Zonnenshain, and G. Fortuna, "A road map for applied data sciences supporting sustainability in advanced manufacturing: The information quality dimensions," *Procedia Manuf*, vol. 21, pp. 141–148, 2018. https://doi.org/10.1016/J.PROMFG.2018.02.104

[9] J. Ma, S. Dong, G. Chen, P. Peng, and L. Qian, "A data-driven normal contact force model based on artificial neural network for complex contacting surfaces," *Mech Syst Signal Process*, vol. 156, Jul. 2021, doi: 10.1016/J.YMSSP.2021.107612.

[10] I. Flood and N. Kartam, "Neural Networks in Civil Engineering. II: Systems and Application," *Journal of Computing in Civil Engineering*, vol. 8, no. 2, pp. 149–162, Apr. 1994 https://doi.org/10.1061/(asce)0887-3801(1994)8:2(149)

[11] I. E. Lagaris, A. Likas, and D. I. Fotiadis, "Artificial neural networks for solving ordinary and partial differential equations," *IEEE Trans Neural Netw*, vol. 9, no. 5, pp. 987–1000, 1998. https://doi.org/10.1109/72.712178

[12] R. T. Q. Chen, Y. Rubanova, J. Bettencourt, and D. Duvenaud, "Neural Ordinary Differential Equations," 2018.

Aerospace Science and Engineering - III Aerospace PhD-Days   Materials Research Forum LLC
Materials Research Proceedings 33 (2023) 171-176   https://doi.org/10.21741/9781644902677-25

Enabling strategies for safe proximity operations to uncooperative and non-collaborative objects in Low Earth Orbit

Giacomo Borelli[1,a] *, Camilla Colombo[2], Gabriella Gaias[3]

[1] Ph.D. student, Politecnico di Milano, Via La Masa 52, Milano, Italy

[2] Ph.D.Supervisor, Associate Professor, Politecnico di Milano, Via La Masa 52, Milano, Italy

[3] Ph.D.Co-Supervisor, Assistant Professor, Politecnico di Milano, Via La Masa 52, Milano, Italy

[a]giacomo.borelli@polimi.it

Keywords: Proximity Operations, Spacecraft Trajectory Optimisation, Flight Safety, On-Orbit Servicing

Abstract. In-orbit servicing, transportation and removal activities are on the way to revolutionize the space economy and space exploitation. Particularly for the near-Earth environment these activities are considered important in the near- and long-term future to ensure the sustainability of space activities. In this paper one of the many challenges facing this new expanding field is addressed, namely the safe and robust design of proximity operations with non-collaborative and uncooperative objects. Guidance and control methods are developed to improve the safety of various proximity operations phases, starting from the far-range approach at tens of kilometres to closer approach distances of few tens of meters. Furthermore, a guidance and control method of a servicer platform to cope with the uncontrolled tumbling motion of the target object is proposed. Here a contactless control approach exploiting safe relative trajectories and the thruster plume impingement is used to reduce the angular motion of the uncontrolled target.

Introduction

Proximity operations play an important role in future mission architectures in the On-Orbit Servicing, Assembly and Manufacturing (OSAM) and Active Debris Removal (ADR) domain. A paradigm shift between monolithic one-use assets towards OSAM activities in space is recognised as both profitable and efficient for the future space economy by the global space community. Despite a rich heritage of Rendezvous and Proximity Operations (RPOs) to cooperative targets, advances in the design of operations to uncooperative and non-collaborative targets are instrumental for a systematic implementation of autonomous RPOs within OSAM activities in the future. An uncooperative target is defined as an object in space that is not capable of aiding the knowledge of its state to another active object, for example with an inter-satellite link. A non-collaborative target is defined as a space object which cannot change its state to aid the OSAM activities, i.e. control its attitude or orbit. One of the key enablers for autonomous proximity operations to uncooperative and non-collaborative targets is flight safety. In fact, any anomaly with respect to the nominal profile or any contingency at spacecraft level will cause the triggering of safety measures, ultimately leading to chaser s/c in safe mode, thus potentially endangering the platforms and/or the completion of the mission. Such situations are not unknown to past missions. In the JAXA robotic demonstration mission ETS-VII [1], anomalies during an experiment caused the spacecraft to abort operations and position itself at 2.5 km distance from the target while investigating the issue. In 2005 during DART mission, the chaser unexpectedly used all the on-board propellant and during the retirement manoeuvres a collision with the target was detected [2]. More recently in early 2022 ELSA-d demonstration failures in the thrusters' assembly caused the chaser to move away at a safe distance from the target and a consequent re-assessment and re-planning of rendezvous and docking demonstration operations [3].

Aerospace Science and Engineering - III Aerospace PhD-Days   Materials Research Forum LLC
Materials Research Proceedings 33 (2023) 171-176   https://doi.org/10.21741/9781644902677-25

This Ph.D. research stems from the challenges encountered in the proximity operations domain to develop novel strategies to enable the safe and systematic implementation of OSAM activities in the future. The main research question is:

"How can we improve safety and robustness of proximity operations design for systematic application in future mission applications?"

In the research activities, the following key drivers and requirements are considered:
- **Safety**: The strategies shall ensure the safety of the whole mission operations.
- **Autonomy**: The service shall be able to perform the operations in diminishing the dependency from ground support as much as possible.
- **Efficiency**: The strategies shall be cost effective, both from a mission architecture point of view and from a spacecraft in orbit resources (propellant) point of view.
- **Reliability**: The strategies shall be robust to orbit conditions.

The research focuses on the relative mission design and Guidance Navigation and Control (GNC) aspects of proximity operations, specifically in the conditions where the target is uncooperative and non-collaborative targets. The research is organized in three main blocks, shown in Figure 1, which are deemed as instrumental to a safe approach to an uncooperative and non-collaborative target:

   I.   Approach GNC design to uncooperative and non-collaborative objects
   II.   Management of target tumbling motion
   III. Safe inspection planning

In the next sections the research performed or planned for each block is described .

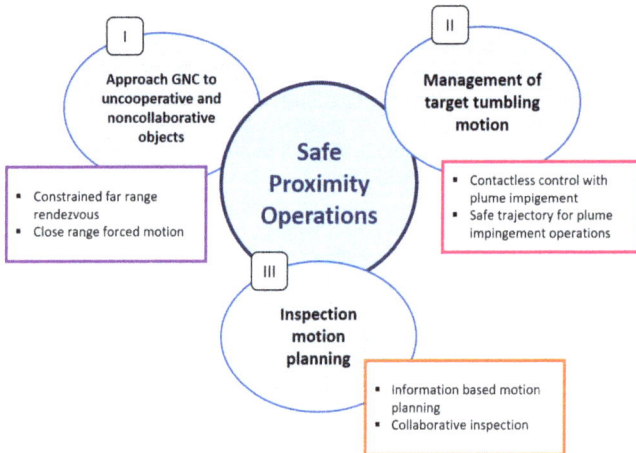

Figure 1. Schematic block diagrams of the Ph.D. activities.

Aerospace Science and Engineering - III Aerospace PhD-Days   Materials Research Forum LLC
Materials Research Proceedings 33 (2023) 171-176   https://doi.org/10.21741/9781644902677-25

Approach to Guidance Navigation and Control

The activities in the approach GNC design focus mainly on the guidance and control aspects of the approach trajectory. Specifically, two phases of the approach are identified as:

• **Far-range approach**: from the first detection of the target with the onboard sensors up, approximately around 30-50 km.

• **Close-range approach**: starting at around few hundred meters when the chaser is required to final approach the target condition in the target body.

In the far range a guidance and control strategy were developed to enhance the performance of the Angles Only (AO) navigation filter, and published in [4]. In fact, from very far distances the chaser often has to rely only on Line Of Sight (LOS) measurements for its navigation solution, which result in a system with very poor observability. A guidance scheme optimising the propellant consumption and the observability enhancement feature is developed which improved the navigation solution and thus the control errors of the approach actuated in a Model Predictive Control (MPC) fashion.

In the close-range approach, novel formulation of safety constraints are developed extending the concept of Eccentricity and Inclination (E/I) vector separation to ensure passive abort safety ensuring more challenging scenarios. The work was presented in [5]. In proximity flight in the range of few meters, more complex trajectory final condition requirements, such as synchronization or complex reconfiguration, prevent the trajectory to be designed with a stringent geometry requirement such as the spiral approach. The conditions are formulated in function of ROEs, which are able to guarantee both a implementation advantage in the optimisation procedure used and a increased level of safety of flight. The safety measure is expressed for conditions of safety at time a given time ti as follows:

• **Point-Wise Safety (PWS):** Chaser's trajectory at time ti is said to be PWS safe if it is outside a geometrical KOZ defined around the target only at the time instant ti.

• **Passive Abort Safety (PAS):** Chaser's trajectory at ti is said to be PAS safe for a time interval ΔT if it is outside a geometrical KOZ around the target at time ti, and it will remain outside such KOZ also after a ΔT time interval of uncontrolled flight starting at ti.

• **Active Collision Safety (ACS) :** Chaser's trajectory at ti is said to be ACS safe if at time ti is outside a geometrical KOZ around the target, and it will remain outside such KOZ even for a $\Delta TACS + \Delta T$ time interval after ti . The intervals $\Delta TACS$ and ΔT are respectively the controlled collision avoidance portion and the uncontrolled portion of the trajectory after ti.

The safety constraints are expressed explicitly in function of the ROE state and included in a guidance scheme based on a Sequential Convex Programming (SCP) algorithm. The latter methods have been developed thanks to its efficiency in solving nonlinear programming problem with limited amount of computational resources, useful for an onboard implementation of the algorithm. In Figure 3 are shown the trajectories designed considering the novel ROE based safety constraints with respect to a purely fuel optimal trajectory for a test case of synchronization to a rotating target. The target hold point angular velocity to synchronise the chaser with was considered in this example as 0.5 deg/s. In Figure 4 it is shown the projection in the RN plane of motion of the failure trajectories stemming from the nominal trajectory. This demonstrate the efficacy of the algorithm to grant PAS in terms of RN separation.

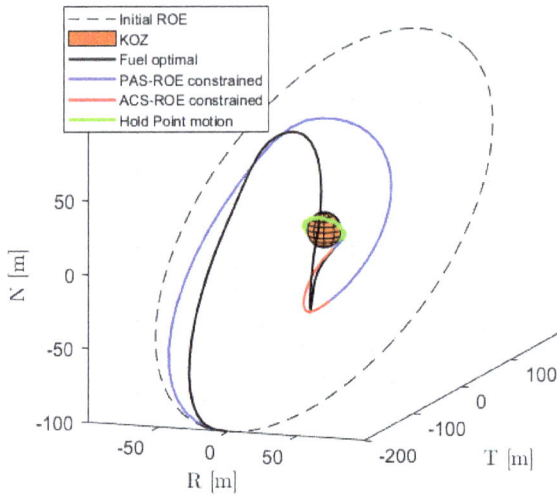

Figure 2. Trajectories in RTN of the fuel optimal and safety constrained case of a synchronisation test case. In blue and red are respectively the PAS and ACS section of the safety constrained solution, while the fuel optimal solution is displayed in black.

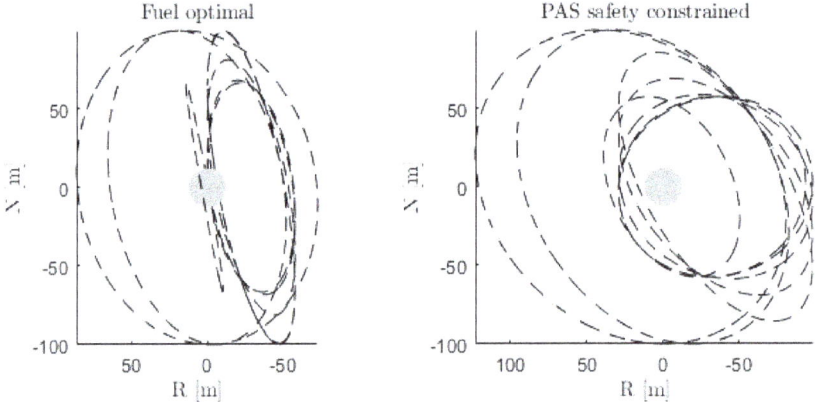

Figure 3. RN projection of the future uncontrolled trajectories correspondent to the ROE of PAS constrained nodes for synchronisation test case. The fuel optimal solution (left) and the safety constrained solution (right) are reported.

Management of the tumbling motion

When planning close-proximity operations the tumbling motion of the target is very influential in granting the possibility of performing the servicing/capture task. However, for targets with fast tumbling rates the energetic level of the synchronisation trajectory required, and the collision risks due to appendages quickly rotating pose great risk to the feasibility of operations. In this block of the Ph.D. research a strategy to detumble a target spacecraft using the plume impingement of the

Aerospace Science and Engineering - III Aerospace PhD-Days   Materials Research Forum LLC
Materials Research Proceedings 33 (2023) 171-176   https://doi.org/10.21741/9781644902677-25

chaser's thruster [5]. A tool for simulating the impingement actions of a monopropellant plume was developed and a guidance and control algorithm developed for the damping of the residual rotational angular momentum by controlling the thruster pointing and firing. In the example shown in Figure X, the study proved the feasibility of damping the tumbling rate below 0.1 deg/s for tumbling rates as fast as 11 deg/s for a constellation satellite with a 1 N hydrazine thruster. This will enable the management of the dangerous tumbling motion for further operations, by simply using the thrusters' already onboard the platform.

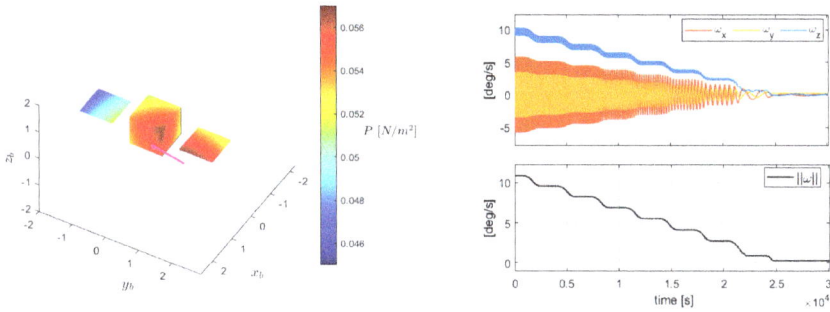

Figure 4. Pressure field on the target satellite due to a 1 N hydrazine thruster. Angular rate history of the target subject to the plume impingement control.

Safe inspection planning
In this block of the research the design of the trajectory guidance for an optimized inspection phase is developed. An inspection phase is the phase of a OSAM or ADR mission where the chaser performs inspection of the target satellite to characterize fully its state, orbital and rotational, and its physical status, i.e. damages or features. This block will be part of the 3rd PhD year and will focus on the design of trajectories for inspection to optimise the information gained during the observations. The ROE framework will be used to design a sequence of fly-around trajectory that fulfil the safety requirements described in the previous sections, but at the same time maximise the observation output of the onboard sensors.

Conclusions
The output of this Ph.D. project is the advancement in the safe and systematic design of proximity operations for future in-orbit servicing and removal mission in Low Earth Orbit (LEO). The results of the first two year provided advanced strategy to cope with trajectory safety using novel formulations in relative orbital elements, and strategy to manage the tumbling motion of a non-collaborative target. As a future step, the inspection phase will be designed with focus on the safety of operations.

Acknowledgments
The research of this Ph.D. has received funding from the European Research Council (ERC) under the European Union's Horizon2020 research and innovation programme as part of project COMPASS (Grant agreement No 679086).

Aerospace Science and Engineering - III Aerospace PhD-Days   Materials Research Forum LLC
Materials Research Proceedings 33 (2023) 171-176   https://doi.org/10.21741/9781644902677-25

References

[1] Kawano, I., Mokuno, M., Kasai, T., and Suzuki, T., "Result of Autonomous Rendezvous Docking Experiment of Engineering Test Satellite-VII," Journal of Spacecraft and Rockets, Vol. 38, No. 1, 2001, pp. 105–111. https://doi.org/10.2514/2.3661

[2] "Overview of the DART Mishap Investigation Results For Public Release, URL : http://www.nasa.gov/pdf/148072main_DART_mishap_overview.pdf

[3] https://astroscale.com/astroscales-elsa-d-mission-successfully-completes-complex-rendezvousoperation/

[4] Borelli, G., Gaias, G., and Colombo, C., "Rendezvous and proximity operations design of an active debris removal service to a large constellation fleet,"Acta Astronautica,Vol. 205, 2023, pp. 33–46. https://doi.org/10.1016/j.actaastro.2023.01.021

[5] Borelli, G., Gaias, G., and Colombo, C., "SAFETY IN FORCED MOTION GUIDANCE FOR PROXIMITY OPERATIONS BASED ON RELATIVE ORBITAL ELEMENTS", AAS/AIAA 33rd Space Flight Mechanics Meeting, Austin, TX, US.

[6] Borelli, G., Gaias, G., and Colombo, C., "ROTATIONAL CONTROL WITH PLUME IMPINGEMENT TO AID THE RIGID CAPTURE OF AN UNCOOPERATIVE FAILED SATELLITE", 2020 AAS/AIAA Astrodynamics Specialist Conference, Lake Tahoe, CA, US.

Aerospace Science and Engineering - III Aerospace PhD-Days   Materials Research Forum LLC
Materials Research Proceedings 33 (2023) 177-184   https://doi.org/10.21741/9781644902677-26

Mechanical properties of additively manufactured lattice structures through numerical BCC cell characterisation

Giuseppe Mantegna[1,a] *

[1]Università degli Studi di Enna Kore, Facoltà di Ingegneria e Architettura, Cittadella universitaria, 94100 Enna – Italy

[a]giuseppe.mantegna@unikorestudent.it

Keywords: Additive Manufacturing, Periodicity, 3D Lattice, FEM

Abstract. Additive Manufacturing (AM) technologies have acquired significant attention in the modern industry due to their versatility in creating custom-designed components with complex shapes utilising multiple materials with minimal material waste. Moreover, the full potential of AM technologies relies on creating meta-materials, structured materials with distinctive mechanical properties designed for specific purposes and optimised throughout various regions of the structure. For instance, it is possible to design topology-optimised structures through periodic lattice cells by varying the cell types, dimensions or relative volume fractions [1]. In this way, structures with graded or separate regions can be manufactured in one single manufacturing process, significantly reducing the design-to-production time and allowing a rapid iteration and design optimisation [2]. As previously mentioned, lattice structures are formed by a unit cell properly repeated with an ordered topology to achieve desired mechanical properties. Amongst the different cell types, thanks to its simple configuration and ease of print with AM technologies, the Body-Centered Cubic (BCC) structure is frequently used [3]. Several studies are present in the literature to predict and evaluate the behaviour of this cell type analytically, numerically and by experiments showing its excellent specific mechanical properties [4]–[6]. This paper investigates the mechanical properties of additively manufactured structures, comparing experimental results with numerical simulations conducted using various modelling approaches, including a full 3D model, a simplified 1D and 2D approach, and homogenized models. At first, the mechanical properties of a lattice BCC structure are considered, followed by the investigation of a sandwich structure featuring a lattice core. At this stage, the skins and the core are 3D printed in Polylactic acid PLA; the lattice core structure is composed of BCC unit cells. Finally, the mechanical performances of the AM structures are compared to assess the accuracy and reliability of the different modelling approaches.

Finite Element Analyses

To determine the mechanical properties of the BCC cells and lattice structures, numerical simulations were carried out in the commercial software Ansys. In this study, various modelling approaches are presented to assess the robustness of the models and minimize computational expenses for future simulations on macro-scale structures. More specifically, the study compares full 3D analyses with simplified 1D and 2D models, as well as homogenized models based on the previous analyses' outcomes. Three major numerical configurations are considered. At first the mechanical properties of the single BCC unit cell are evaluated. As a periodic element repeated in the structure, proper periodic and boundary conditions must be ensured for the single BCC cell analyses. For this reason, a specific routine has been developed in Ansys Parametric Design Language (APDL) to ensure a double periodicity on the 3D FEM simulations, while ad-hoc boundary conditions were imposed for the simplified beam 1D models. Subsequently compression tests on a cubic BCC structure are considered followed by bending tests on a sandwich structure consisting of a BCC lattice core sandwiched between two skins.

Aerospace Science and Engineering - III Aerospace PhD-Days   Materials Research Forum LLC
Materials Research Proceedings 33 (2023) 177-184   https://doi.org/10.21741/9781644902677-26

Single BCC Cell – Compression and Shear tests

Figure 1 displays the single Body-Centered Cubic (BCC) unit cell which serves as the fundamental periodic element for the cubic and sandwich structures. The geometric properties of this BCC unit cell are listed in Table 1. As in other studies [4], [5], due to the bending-dominated behaviour of the BCC cell in compression, the 1D beam model struts are divided into multiple sections to take into account the rigidity of the nodes. According to [4], for a BCC cell, it is possible to retrieve the stiff beam length h_{comp} as a function of the diameter-cell size ratio. However, when the cell is subjected to a shear loading, each strut is both bending and stretched loaded. In this condition, the overmentioned approach has been corrected by numerically calibrating the h-d ratio. For this study, a stiff beam length $h_{shear} = 0.35$ is selected. Further studies are currently being conducted to optimise the length according to the local stress distribution.

Table 1 – BCC Cell nominal size.

Cell size, L [mm]	10.0
Strut diameter, d [mm]	2.0
Stiff beam length, h_{comp} [mm][4]	0.93
Stiff beam length, h_{shear} [mm]	0.35

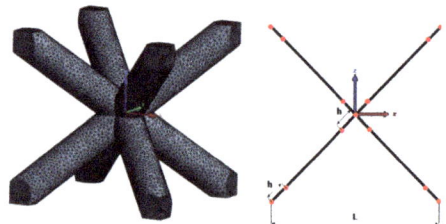

Figure 1 – 3D and 1D BCC cells.

As mentioned, FDM PLA are considered for the numerical analyses whose properties, retrieved on experimental tests performed on dog-bone specimens, are $E = 3132\ MPa$, $\sigma_{yield} = 20\ MPa$ with a density $\rho = 1275\ kg/m^3$.

The 3D models use double periodic conditions along the x and y directions modelled through constraints equations and implemented through a specific APDL code, as outlined in previous work [7] and schematically shown in Figure 2.a.

Since the 1D model does not have multiple nodes on the lateral faces, periodic conditions are enforced through remote points coupling for the compression test as shown in Figure 3. The boundary conditions used for the compression and shear analyses in both 3D and 1D models are presented in Figure 2.b and Figure 3, respectively.

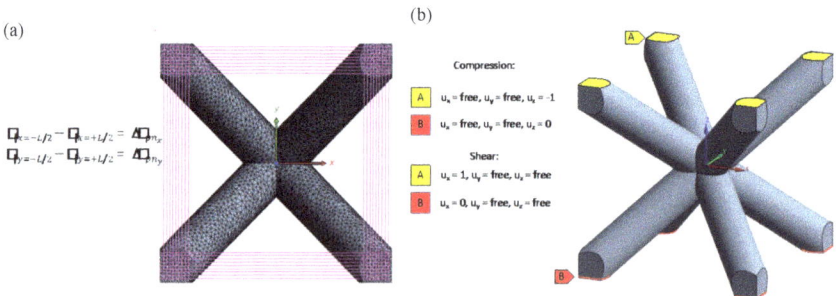

Figure 2 – 3D model Compression/Shear (a) Periodic conditions; (b) Boundary conditions.

Compression	Shear
$h_{comp} = 0.93$ mm	$h_{shear} = 0.35$ mm

Nodes$_{z\sim-L/2}$: u_x = free, u_y = free, u_z = 0

Nodes$_{z=-L/2}$: u_x = free, u_y = free, u_z = -1

Fixed Rotation on nodes

Nodes$_{x\sim-L/2}$: u_x Coupled with RP1
Nodes$_{x\sim-L/2}$: u_x Coupled with RP2
Nodes$_{y\sim-L/2}$: u_y Coupled with RP1
Nodes$_{y\sim-L/2}$: u_x Coupled with RP2

Nodes$_{z\sim-L/2}$: u_x = 0, u_y = 0, u_z = 0
Nodes$_{z=-L/2}$: u_x = 1, u_y = 0, u_z = 0

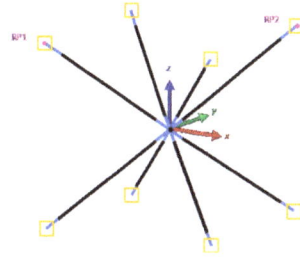

Figure 3 – 1D Compression/Shear Boundary and Periodic conditions

From the 3D, model an equivalent elastic modulus $E_{3D}^* = 23.5\ MPa$ is retrieved with a Poisson's ratio $v_{3D}^* = 0.47$ and a shear modulus $G_{3D}^* = 82.4\ MPa$. The simplified 1D model gives an equivalent elastic modulus $E_{1D}^* = 22.6\ MPa$, a Poisson's ratio $v_{1D}^* = 0.46$ and a shear modulus $G_{1D}^* = 82.7\ MPa$. The 1D numerical model accurately replicate the behaviour of the 3D numerical, resulting in a significant reduction in computational costs. This reduction is due to the decreased number of model elements, namely 192 elements as opposed to 47192 elements employed in the 3D model. Moreover, the absence of constraints equations that pair each node on the lateral faces of the 3D model contributes significantly to the reduction of computational effort. Figure 4 shows the directional deformation contour maps for the 3D and 1D compression and shear tests.

(a)

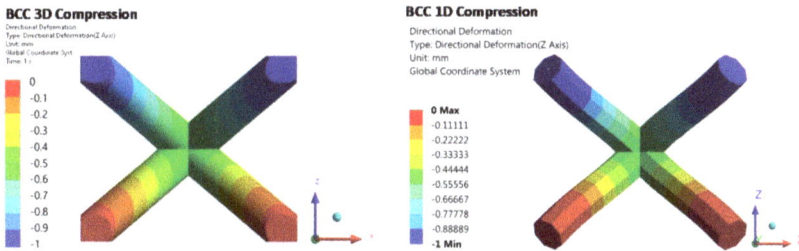

(b)

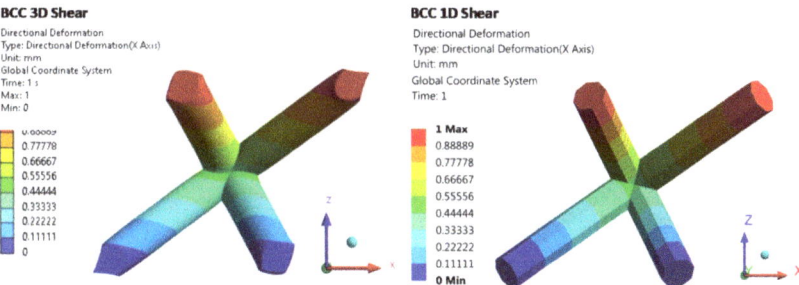

Figure 4 – Directional deformation contour maps of 3D and 1D Compression (a) and Shear (b).

Aerospace Science and Engineering - III Aerospace PhD-Days   Materials Research Forum LLC
Materials Research Proceedings 33 (2023) 177-184   https://doi.org/10.21741/9781644902677-26

BCC Structure Compression tests

Following the numerical analyses on the single BCC cell, the BCC structure showed in Figure 5 was considered and investigated under a compression loading.

Table 2 – BCC Cell and Coupon nominal size.

Cell size, l [mm]	10.0
Strutt diameter, d [mm]	2.0
Stiff beam length, h_{comp} [mm]	0.93
Coupon dimensions, $L \times W \times H$ [mm]	40x40x40
Coupon Volume fraction, V^*/V	0.1784

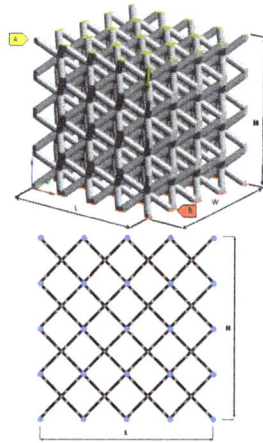

Figure 5 – 3D and 1D Compression coupons.

For this analysis, no periodicity conditions were given. A displacement along the *z*-direction is given at the top surface (Figure 5, Surface A) while the bottom surface (Figure 5, Surface B) has constrained displacements along the z-direction. Both the 3D and 1D numerical analyses give consistent results with the single BCC cell results: an apparent modulus $E_{3D} = 23.6\ MPa$ and $E_{1D} = 22.7\ MPa$ is retrieved, with a negligible error of 0.4% in relation to the corresponding modulus values obtained from the single BCC cell analyses.

The periodicity and boundary conditions imposed on the 3D and 1D models for the compression test analysis thus give coherent results and can be used to characterise the macro-behaviour of the structure from the unit cell.

The equivalent stress contour map for the 3D model is presented in Figure 6.a while the combined stress for the 1D model is shown in Figure 6.b showing the most stressed areas of the structure despite being different mechanical entities. Similar to the single cell configuration, the compression loading configuration stresses the structure mainly in bending. As expected, the node joints situated at the faces of the coupon are the main locations of stress concentration on both the 1D e 3D model since they are not constrained by other adjacent unit cells outwards of the coupon faces. Notably, in the 3D model, the inner joint area experiences a more elevated stress level in comparison to the outer area, with an increase in stress ranging between 15% to 30%. However, this stress distribution pattern cannot be accurately predicted by the 1D model, which only accounts for a single node located at the strut end and, therefore, does not differentiate between an inner and outer joining area.

(a)   (b)

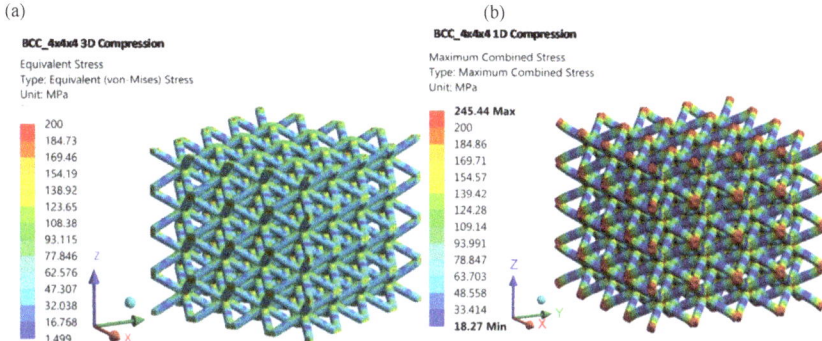

Figure 6 – (a) Equivalent stress 3D model and (b) Maximum combined Stress 1D model contour maps.

Sandwich structure flexure test

A Three Point Bending Test (3PBT) on a sandwich structure is considered as final numerical analysis. More specifically, the sandwich consists of two skins with a nominal thickness of $t_s = 1\ mm$ enclosing a BCC lattice core formed by $20 \times 3 \times 1$ cells with dimension as reported in Table 1.

Three numerical models are considered to study this test, namely 3D model, 1D/2D model and Homogenised model. The homogenised 3D model, Figure 7.a, is composed of three solid plates. The FDM PLA properties previously used are imposed to the two external skins, while the homogenized core is characterized by equivalent mechanical properties obtained from the single cell 3D analysis, namely $E^* = 23.5\ MPa$, $v^* = 0.47$ and $G^* = 82.4\ MPa$. As illustrated in Figure 7.b, only half specimen is considered for all of them. A symmetry plane, by constraining the displacements along the x-direction is thus imposed.

(a)   (b)

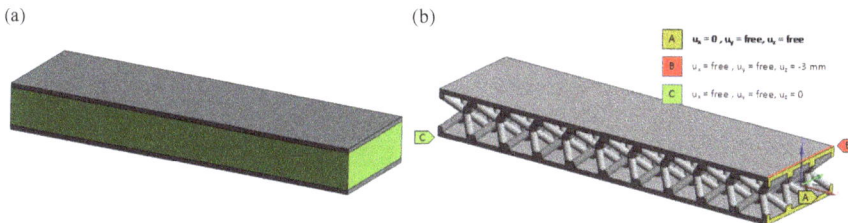

Figure 7 –3PBT Boundary Conditions. (a) 3D homogenised model, (b) 3D model and boundary conditions.

The load-displacement diagram in Figure 8 compares the mechanical response of the three models. The numerical bending rigidities, calculated as $D = Load/Deflection$, for each model are: $D_{3D} = 38.66\ N/mm$, $D_{hom} = 37.97\ N/mm$ and $D_{1D/2D} = 37.36\ N/mm$.

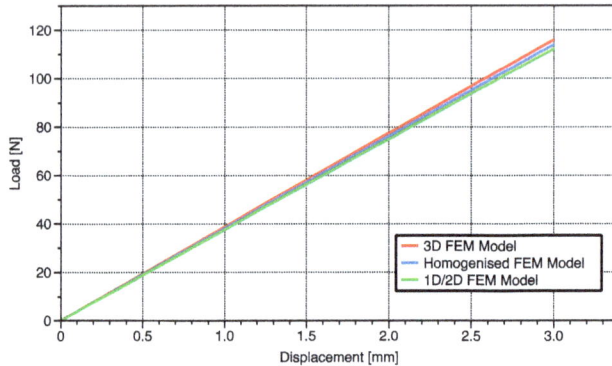

Figure 8 – 3PBT load-displacement diagram, numerical results.

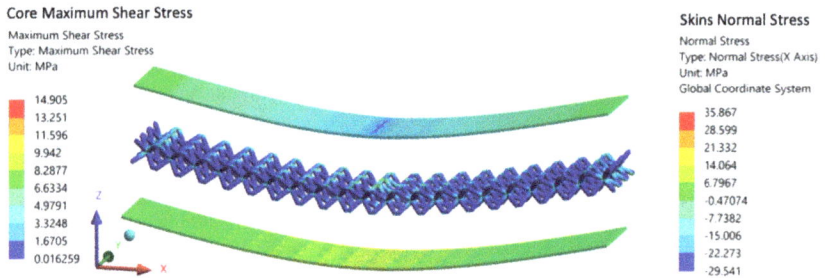

Figure 9 – 3D Stress distribution contour map exploded-view: Core shear stress and Skins normal stress.

The exploded-view in Figure 9 depicts the stress distribution of the sandwich structure, wherein the lower skin experiences compression longitudinal stress, while the upper skin is subjected to compression. Notably, the core bears the highest shear stresses underneath the supports located at the end of the specimen.

Experimental Tests and comparison with numerical results

Compression tests

A series of experimental tests on the 3D coupons was performed to validate the numerical results. The aforementioned coupons were manufactured in Polylactic acid PLA through fused deposition modelling (FDM) with a base print speed of $50\ mm/s$ and a temperature of the extruder of $200\ °C$. For the sake of reproducibility, three coupons were tested.

The compression response of the BCC coupons was compared with the numerical results in Figure 10. The graph depicts three distinct phases of lattice structures under compression, which are consistent with previous research findings [5], [8]. The linear elastic regime is visible in the initial section of the graph, followed by an elastic-plastic collapse at the centre. The right-hand side of the diagram, characterized by a steep increase in stress, corresponds to the densification phase. Table 3 summarise the mechanical properties retrieved from the experimental tests. A high level of correspondence between the numerical and experimental results is evident, with a 4.5% error for the 3D model and a negligible 0.5% error for the 1D model.

Aerospace Science and Engineering - III Aerospace PhD-Days   Materials Research Forum LLC
Materials Research Proceedings 33 (2023) 177-184   https://doi.org/10.21741/9781644902677-26

Table 3 – Experimental compression tests results.

Weight [g]	Relative density	Equivalent Young Modulus [MPa]	Equivalent Yield Modulus [MPa]	Maximum Strength [MPa]
14.617	0.1791	21.06	0.76	1.01
14.67	0.1797	23.56	0.83	1.17
14.65	0.1795	23.11	0.80	1.20

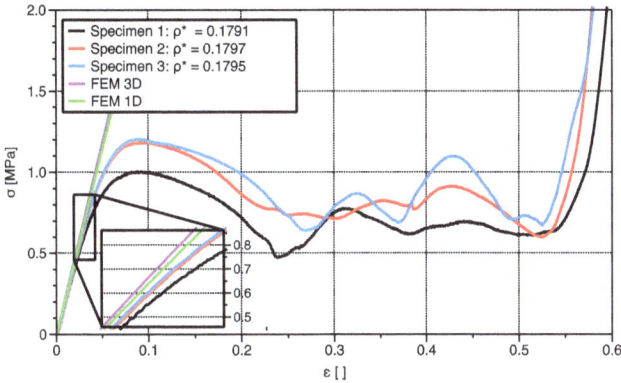

Figure 10 - Compression tests: Numerical-experimental stress-strain curves comparison.

During the experimental tests, the first and second specimens showed a shear deformation band, as depicted in Figure 11.a and Figure 11.b, as a result of compressive buckling that provoked the diagonal shear deformation. The third sample, on the other hand, showed a uniform layer-by-layer failure from the uppermost layers and subsequently propagating to the lower layers, with the central region remaining intact until the final stages of the deformation process.

Figure 11 – Deformation modes of BCC lattice structure. Shear bands details.

Three-Point bending test

An experimental test of the 3PBT was also performed to check the validity of the numerical results. The test arrangement, shown in Figure 12.b , considers a span between the supports is $S = 200 \, mm$ while the load is applied in the centre of the specimen.

Figure 12.a compares the experimental load-displacement curve with the numerical simulation results revealing an experimental bending rigidity of the sandwich of $D = 34 \, N/mm$. A relative error of 13%, 11% and 9% is thus recorded between the 3D, homogenised and 1d/2D FEM model respectively.

Aerospace Science and Engineering - III Aerospace PhD-Days   Materials Research Forum LLC
Materials Research Proceedings 33 (2023) 177-184   https://doi.org/10.21741/9781644902677-26

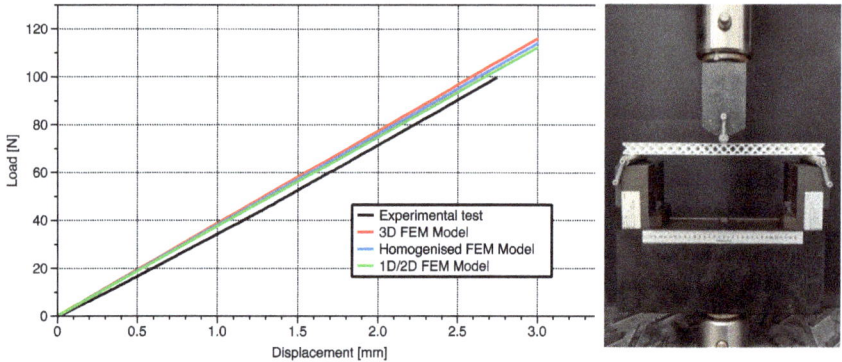

Figure 12 – Experimental bending test. Load-Displacement diagram and experimental test setup.

References

[1] J. J. Sobczak and L. Drenchev, "Metallic Functionally Graded Materials: A Specific Class of Advanced Composites," *J Mater Sci Technol*, vol. 29, no. 4, pp. 297–316, Feb. 2013. https://doi.org/10.1016/j.jmst.2013.02.006

[2] A. Bandyopadhyay and B. Heer, "Additive manufacturing of multi-material structures," *Materials Science and Engineering R: Reports*, vol. 129. Elsevier Ltd, pp. 1–16, Feb. 2018. https://doi.org/10.1016/j.mser.2018.04.001

[3] L. Bai, C. Yi, X. Chen, Y. Sun, and J. Zhang, "Effective design of the graded strut of BCC lattice structure for improving mechanical properties," *Materials*, vol. 12, no. 13, 2019. https://doi.org/10.3390/ma12132192

[4] D. Tumino, A. Alaimo, C. Orlando, S. Valvano, and C. R. Vindigni, "Lattice Core FEM Simulation with a Modified-Beam Approach," *Lecture Notes in Mechanical Engineering*, pp. 946–954, 2023. https://doi.org/10.1007/978-3-031-15928-2_83

[5] R. Gümrük and R. A. W. Mines, "Compressive behaviour of stainless steel micro-lattice structures," *Int J Mech Sci*, vol. 68, 2013. https://doi.org/10.1016/j.ijmecsci.2013.01.006

[6] E. Ptochos and G. Labeas, "Elastic modulus and Poisson's ratio determination of micro-lattice cellular structures by analytical, numerical and homogenisation methods," *Journal of Sandwich Structures and Materials*, vol. 14, no. 5, 2012. https://doi.org/10.1177/1099636212444285

[7] G. Mantegna *et al.*, "Representative volume element homogenisation approach to characterise additively manufactured porous metals, pp. 1–10, Feb. 2022. https://doi.org/10.1080/15376494.2022.2124002

[8] H. Lei *et al.*, "Evaluation of compressive properties of SLM-fabricated multi-layer lattice structures by experimental test and μ-CT-based finite element analysis," *Mater Des*, vol. 169, 2019. https://doi.org/10.1016/j.matdes.2019.107685

Aerospace Science and Engineering - III Aerospace PhD-Days
Materials Research Proceedings 33 (2023) 185-192

Materials Research Forum LLC
https://doi.org/10.21741/9781644902677-27

Interplanetary trajectory design in high-fidelity model: Application to deep-space CubeSats' cruises

Claudio Toquinho Campana[1,a] *

[1]Politecnico di Milano, Department of Aerospace Science and Technology, Via La Masa 34, 20156, Milan, Italy

[a]claudiotoquinho.campana@polimi.it

Keywords: Highly Nonlinear Astrodynamics, Phase Space Analysis, Autonomous Interplanetary Cubesats, High-Fidelity Trajectory Design

Abstract. This paper tackles the problem of first guess trajectory generation for interplanetary missions flying in chaotic environments. Simplified dynamical models are first exploited to perform the preliminary design of deep-space trajectories which leverage orbital perturbations. A real trajectory is then obtained by a refinement procedure in a high-fidelity model. A description of tools and methodologies which will be developed during this PhD research is provided.

Introduction

Just few decades ago, only big companies and international agencies owned the resources necessary to operate in the space sector. It is undoubtful that nowadays the space environment is, instead, getting always more and more accessible and affordable to everyone: the so-called phenomenon of new space economy [1]. This is the result of a slow process started with an increasing interest in space applications by the public, which brought to the spreading of accepted scientific knowledge and to the foundation of smaller businesses active in the space sector [2].

CubeSat technology is an emblematic by-product of this new paradigm. In fact, the competitivity in development and manufacturing costs of CubeSats, which are satellites of reduced size, is attracting an always wider sector of the space community. In particular, the option to adopt CubeSats for travelling interplanetary missions is nowadays extensively investigated. If successful, this research field would open to a world of possibilities.

Despite the relatively low development costs of a CubeSat mission, it is however true that the current paradigm of operating it, once in orbit, from ground, weakens the overall conveniency [3]. With the development of proper technologies, autonomous CubeSats will perform missions in the outer space with a small, if not absent, supervision from ground. CubeSats would be, therefore, an interesting alternative to traditional spacecrafts in the framework of interplanetary cruises. Nevertheless, current technological limits, such as limited on-board computational resources and propulsive capabilities, significantly constraint the possible applicative scenarios. The scientific community, space agencies, and industries are working to design missions and develop technologies which could enable stand-alone travels (e.g. M-ARGO mission [4]).

This PhD project is framed in the context of highly nonlinear astrodynamics applied to assist autonomous interplanetary CubeSat missions. Overall, the purpose is to develop new methodologies for trajectories design and optimization. Particularly, this research investigates how to consciously exploit the dynamics of a chaotic environment for the design of deep-space cruises. The techniques are expected to both improve the on-ground trajectory design phase and make the on-board autonomous generation of a reference trajectory more effective and efficient.

Statement of the problem and research questions

It is well known that the design of an interplanetary cruise is a challenging task because of the intrinsically chaotic dynamics that governs the motion of a spacecraft [5]. Multiple attractors and

Aerospace Science and Engineering - III Aerospace PhD-Days   Materials Research Forum LLC
Materials Research Proceedings 33 (2023) 185-192   https://doi.org/10.21741/9781644902677-27

orbital perturbations, in fact, act together in a relevant way making the phase space highly nonlinear and rather complex to be well characterized. This causes the dynamics to be extremely sensitive to small variations in the states and, therefore, the generation of optimal trajectories in such environments is especially difficult. On the other hand, this intrinsic complexity, if properly managed and accounted for, enables to design trajectories which appear only in this framework [6, 7]. To mention some, dynamical structures like periodic orbits [8], and invariant manifolds [9] can be found and exploited only in autonomous multibody models. A trajectory designed in these simplified dynamical environments may enable for more fuel-efficient transfers, this usually at the expenses of a longer travel time [10].

The design of an interplanetary trajectory is a rather complex activity. Creativity and experience are essential to obtain efficient transfers which respect mission constraints and reach the objectives. When designing interplanetary trajectories in multibody dynamics, there are strategies and methodologies which are commonly used with the aim of simplifying the procedures and obtain better results. As it happens often in many scientific fields, it is useful to begin the analysis considering only simplified models which, nonetheless, try to retain all the important features of the complete phenomenon. Once the design has been accomplished in these reduced environments, it is of paramount importance to assess whether what achieved is reasonable even when contextualized in the real framework.

This procedure is at the base of what is usually done also in the perspective of interplanetary trajectory design [11]. In this regard, as previously mentioned, peculiar dynamical structures appear only in simplified models which try to provide a first approximation and an accurate description of what really happens. A preliminary design of the trajectory is therefore commonly performed in models of reduced complexity to exploit their intrinsic characteristics. In other words, at a first iteration of the design process, the trajectory is developed in models which account only for principal dynamics, while completely neglecting secondary ones. However, the mission will eventually be flown in the real scenario, thus the nominal trajectory is, by definition, the one that exists in the real solar system model [5], which accounts for all possible perturbations and dynamical contributors. The next step in the process, therefore, consists in refining the seeding orbit in the complete dynamical framework. At this aim, the trajectory is corrected using direct transcription and the related optimization problem is solved though a multiple shooting method [12]. The objective is to enforce the resulting trajectory to retain the characteristic features of the initial seed when moving to the real environment, which translates in minimizing the corrections applied by the multiple shooting optimization process [5]. This final step is usually executed using "brute force" since the real dynamics is simply corrected for and not accounted for in the first place. Since the dynamics in such environments is extremely sensitive, this very last step may lead to the loss of optimality and the effectiveness of the resulting trajectory may be compromised. As mentioned in [13], the transition from the Circular Restricted Three-Body Problem (CR3BP) [14] to a realistic model is usually very sharp. In that work the authors proposed a gradual refinement passing through intermediate models of increasing fidelity (for example the elliptic circular restricted three-body problem, and the restricted four-body problem). On the contrary, this project tries to identify methodologies to account for the presence of perturbations in advance, so to make the transition smoother. Techniques are foreseen to be developed facing the problem from different prospectives, so that to make the study more rigorous.

This premise brings to our research questions, here summarized for a better visualization.

RQ1.   *To what extent can the dynamic information be exploited to improve the refinement of trajectories initially designed in simplified models?*

RQ2.   *How much more effective is a methodology that explicitly exploits this enhanced awareness of the dynamical environment, compared to traditional refinement methods?*

Aerospace Science and Engineering - III Aerospace PhD-Days
Materials Research Proceedings 33 (2023) 185-192

Materials Research Forum LLC
https://doi.org/10.21741/9781644902677-27

RQ3. *To what extent can we beforehand account for and exploit perturbations to design more efficient trajectories, yet carrying out their preliminary design in simplified models?*

RQ4. *To what extent can the developed methodologies contribute to the stand-alone efficient and effective generation of first guess trajectories when an interplanetary CubeSat has to design its own journey on board?*

Fig. 1 - Downstream approach

RQ1 and *RQ2* are strictly related to each other. In this case, the problem is approached downstream, which means finding a way to smooth the "brute force" of the refinement process. Fig. 1 schematically represents this. The question mark identifies the contribution yielded by answering to *RQ1* and *RQ2*.

We then ask ourselves if, instead, the problem could be faced upstream, as depicted in Fig. 2 following the same notation of before. This can be rigorously formulated by introducing *RQ3*, the answer of which may be even of more scientific interest and, possibly, bring to relevant technical outcomes. Finally, the analysis accomplished to give answers to all these questions will contribute during the investigation of the last one, which may be regarded as the technical application of this scientific research. The limited computational resources and propulsive capabilities of a CubeSat

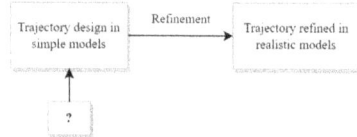

Fig. 2 - Upstream approach

require specific solutions to make up for these problematics. It is clear, however, that these two technological constraints necessitate solutions which are in contrast each other since very fuel-efficient journeys can be calculated only at the expenses of a more extensive processing. If successful, the answer to *RQ4* would contribute to the design of more effective trajectories in a more efficient way.

Expected outcomes

The way it has been formulated makes this scientific research suitable for an incremental approach. In fact, the outcomes of each individual question pose the bases for the successive one. In general, the phases of the project can be subdivided respecting the order with which the research questions have been formulated. For each phase, therefore, some relevant outcomes are expected. In particular, the answers to *RQ1* and *RQ2* may be regarded as a preparatory work in view of *RQ3* and *RQ4*. In this section, each research question is associated to its related expected outcomes. In the next one, the methodologies adopted to tackle the problem will be explained.

Outcomes from RQ1

A deep understanding of the effects of relevant perturbative phenomena, such as, for example, third-body attractions, Solar Radiation Pressure (SRP), and bodies' oblateness, on the refinement of relevant interplanetary trajectories is sought after. This should allow to identify some regions in the dynamical phase space where these effects are more pronounced (sensitive regions) and how / how strongly they play a role in the refinement process. The effect of each individual perturbation relevant for the dynamical system under examination is expected to be better characterized. Furthermore, it is also investigated their impact on the refinement of important dynamical structures such as periodic orbits, manifolds, resonant orbits, etc.. In examining how these evolve, common behaviours may be identified. This preliminary analysis should enhance our overall confidence of the dynamics of notable environments. A modified refinement method is therefore developed with the aim of explicitly accounting for the different sensitivity regions of the phase space.

Outcomes from RQ2
Some real case scenarios will be used as playground to test and validate the developed refinement technique. Its effectiveness in the refinement process will be assessed by comparing the results against those obtained with traditional refinement methodologies.

Outcomes from RQ3
Exploiting the enhanced awareness of the dynamical environment, result of the previous points, a procedure is sought after which would allow to perform a more aware trajectory design. Still working with simplified models, the understanding of how perturbations would affect the designed trajectory is expected to produce positive effects on the results. Firstly, the awareness of how relevant dynamical structures would evolve in a real flyable model may suggest different design strategies. Secondly, a trajectory modelled following this approach is expected to be more robust and, therefore, deviate less during the successive refinement. This is foreseen to be beneficial for the convergency of the refinement process. Some relevant mission analyses will be re-computed to understand whether any improvement is obtained in terms of transfer efficiency if adopting the designed methodology.

Outcomes from RQ4
The outcomes of each previous points are eventually adapted and applied in a unified methodology. This one would enable an autonomous CubeSat to perform a more rigorous and efficient on-board interplanetary trajectory design. The developed technique will be tested to assess how much the computational performances and the effectiveness of the solutions are improved thanks to its adoption.

Methodology
This section describes how the project will be developed. Fig. 3 represents its summary.

Preliminary analysis
The project, expected to last three years, begin with a literature review of modern techniques for design of interplanetary trajectories in multibody environments. Because of the increasing interest on the topic for strategical applications (e.g. ARTEMIS mission [15]), many mission analysis for various kinds of transfers can be found in the literature. It is of paramount importance, therefore, to identify those dynamical environments that describe interesting scenarios for the purpose of this project. These should span a wide enough spectrum of representative cases, so that to make the analysis as complete as possible and applicable in different contexts. Following a preliminary study of the literature, some possible candidates have been identified and are here reported. Still, during the development of the study, this list is likely to be adapted.

- Cislunar environment
- Earth-to-Moon transfer by intersection of Sun–Earth and Earth–Moon manifolds
- Mars and Phobos system
- Jupiter and its moons
- Binary asteroid systems

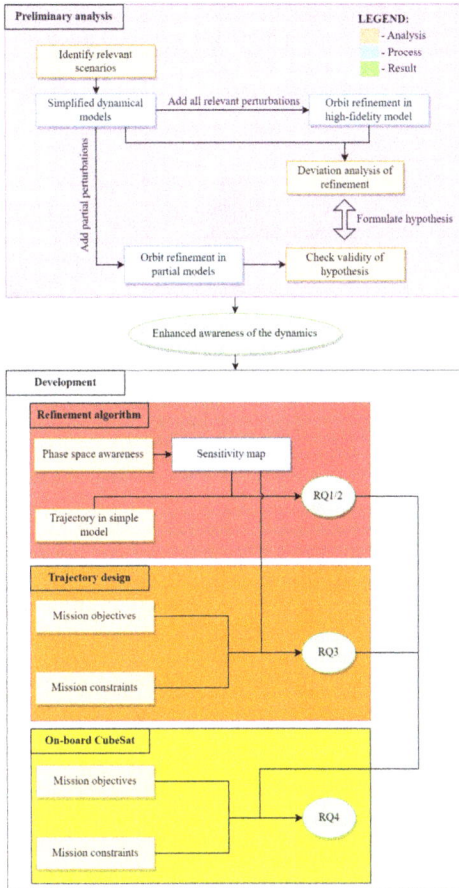

Fig. 3 - Methodology

Simplified multibody models are exploited to perform the preliminary design of trajectories. These are then refined in the real model. Nevertheless, the last two scenarios are much more challenging than the others; the dynamical environment is richer, so much that the simplified model themselves may not be suitable in these frameworks. In fact, in those cases, the effect of perturbations is much more prominent. Constraining the dynamics in a too simplistic model would neglect effects that cannot be anymore regarded as perturbations. It is planned, therefore, to begin the analysis considering more simple environments and then adapt and test the methodologies in those more critical.

The Earth–Moon system is firstly analysed in the CR3BP. Assuming that the design of transfer trajectories is done exploiting dynamical structures such as periodic orbits and manifolds, their refinement in a realistic model is investigated (purple box in Fig. 3). First of all, families of Lyapunov and Halo orbits, which are important structures in modern trajectory design, are refined in the high-fidelity Roto-Pulsating Restricted n-Body Problem (RPRnBP) [5]. This model includes the orbital eccentricity of the primaries, the attraction of all other planets, the Sun, the oblateness of celestial bodies, and the SRP in the description of the dynamics. In this high-fidelity system, periodicity properties are lost. After refinement, sections of the new orbits which most / least deviated from the seeding ones are identified. The analysis on the results should suggest relations between deviations and perturbations. Regions of the phase space more prone to changes are also identified. From this procedure, hypotheses are formulated about the effects that each individual perturbation has on the overall refinement process. To prove what predicted, the refinement process is repeated, this time injecting in the system single or a combination of more perturbations. Finally, the entire process is redone to study how homoclinic and heteroclinic trajectories evolve. In this regard, Poincaré sections are generated to investigate how manifolds change in this new non-autonomous dynamical framework. At this purpose, it may be interesting to introduce in the analysis chaos indicators to better characterize time-varying features of the phase space. This may ease in revealing correlations between natural flow structures and perturbations in the high-fidelity model. The procedure is repeated for all scenarios. A preliminary analysis of the results focuses on understanding, for each specific framework, the following points:

Aerospace Science and Engineering - III Aerospace PhD-Days   Materials Research Forum LLC
Materials Research Proceedings 33 (2023) 185-192   https://doi.org/10.21741/9781644902677-27

- recognize characteristic influence of individual perturbations in the refinement process;
- discretize the time-varying phase space in regions depending on the dynamical contribution introduced by each perturbation;
- perform a sensitivity analysis and identify sensitive regions;
- investigate the effects of perturbations in reference structures (e.g. understand how manifolds and Poincaré intersections evolve in the process);
- define parameters and methodologies that ease the understanding of the relations between perturbations and refinement (e.g. frequency analysis of the dynamical model [16], phase space regions discretization, selection of suitable chaos indicators).

Development

As a result of the previous analysis, the relation between orbital perturbations and the effects they have on the phase space is more intelligible. This enhanced awareness must be exploited to make practical improvements in the field of efficient low-energy transfers in multibody dynamics. First of all, time-dependent sensitivity maps are derived as consequence of the analysis of the previous part. Each applicative scenario will have its own map representing the correlation between perturbations and natural flows as function of time. To answer to *RQ1* and *RQ2*, a traditional multiple shooting technique is modified so that to use the information of the sensitivity map during trajectory refinement (red box in Fig. 3). This is supposed to be done by introducing a variable time step for the discretization of the seeding trajectory designed in simplified models. The time interval between discrete states is therefore adjusted to adapt for the rapidity with which the dynamical behaviour of the system changes. Thanks to the sensitivity map, we already know in advance where this would happen. The reference trajectory is then more densely discretized in the correspondence of sharp variations in the dynamical behaviour of the phase space. Furthermore, the objective function is modified such that diverse weights are introduced to allow different deviations in regions with different dynamical stability. The purpose of this is to take more advantage of the perturbative effects, thus exploiting the natural flow. As result of these two modifications to the standard multiple shooting method, the following benefits are expected:

- the algorithm is efficient since a finer discretization is introduced only where necessary;
- the algorithm is more likely to converge because the dynamics is indulged;
- the result better exploits perturbative phenomena, fostering more efficient transfers.

To test how well the algorithm perform, previous relevant mission analyses are recomputed. This should prove whether improvements are obtained in terms of convergency rate, efficiency of the process, and effectiveness of the obtained solutions.

The core part of the project is now discussed (orange box in Fig. 3). For each simulated scenario, some relevant deep-space missions are hypothesized. If available, some transfers already available in the literature may be adopted for comparison purposes. In any case, clear mission objectives and constraints are formulated to give a physical meaning to the problems. The sensitivity map plays now a fundamental role. From the mission objectives, we know indicatively which transfer we need to perform. Simplified models are adopted to formulate a first guess solution. This one has to be computed trying to exploit as much as possible the natural features enabled by the adoption of a reduced dynamics. At the same time, however, the knowledge of how perturbations will act on it in a high-fidelity representation shall not be overlooked. On the contrary, since we now have this kind of awareness, we need to exploit it. This can be done following two different approaches. Either we try to avoid all those regions, in the reduced phase space, whose counterpart in the sensitivity map would show strong deviations in the dynamical behaviour; or we explicitly try to use them. In the first case there is, therefore, the attempt to suppress perturbations, such that the dynamics of simplified models is preserved. The second approach is, instead, more interesting. In this case, the seeding trajectory, which is the one

Aerospace Science and Engineering - III Aerospace PhD-Days   Materials Research Forum LLC
Materials Research Proceedings 33 (2023) 185-192   https://doi.org/10.21741/9781644902677-27

computed in the simplified model, will be designed such that perturbative regions are crossed with judicious. This increases somehow the uncertainty because, during refinement, the trajectory is expected to deviate more. However, this produces the beneficial effect that mission constraints can be relaxed more, hence increasing the freedom of the system. Many different seeding trajectories can then be designed since they will eventually adhere to the mission requirements after refinement. This process is foreseen to bring benefits in terms of transfer costs.

The last point introduces the developed techniques in the computing loop that allows a multi-mission autonomous CubeSat to perform its own trajectory design. The CubeSat is supposed to receive from ground both its mission requirements and a proper sensitivity map. It is now just a matter of connecting the answers to the previous research questions (yellow box in Fig. 3). In particular, the CubeSat will first generate preliminary trajectories following the procedure described in the previous paragraph. Then, the trajectories are refined thanks to the modified multiple shooting technique. For validation purposes, the algorithm will be tested in processor-in-the-loop simulations using hardware representative for a CubeSat interplanetary mission. The efficiency of the algorithm and its effectiveness, in terms of computing low-cost transfers, is evaluated by comparing the results with those obtained with traditional procedures.

Conclusions

This PhD project tries to give practical answers to the problem of designing interplanetary trajectories in highly nonlinear environments in autonomy. The contribution of perturbation in generating more effective trajectories in more efficient ways is investigated. Starting from an accurate analysis of how perturbations play a role in the dynamical description of the phase space, practical methodologies are then developed. A modified multiple shooting algorithm and a new trajectory design procedure are the outcomes of this research. Their efficiency and effectiveness are tested in relevant simulated scenarios.

Acknowledgments

The author desire to thank Prof. Francesco Topputo and Dr. Gianmario Merisio for their support in writing this abstract.

This research is funded by the Italian Ministry of University and Research, 38th PhD cycle.

References

[1] H. C. Alewine, «Space accounting,» Accounting, Auditing & Accountability Journal, vol. 33, p. 991-1018, 2020.https://doi.org/10.1108/AAAJ-06-2019-4040

[2] E. Kulu, «In-Space Economy in 2021 - Statistical overview and classification of commercial entities,» in 72nd International Astronautical Congress (IAC 2021), Dubai, United Arab Emirates, 2021.

[3] G. Di Domenico, E. Andreis, A. C. Morelli, G. Merisio, V. Franzese, C. Giordano, A. Morselli, P. Panicucci, F. Ferrari e F. Topputo, «Toward self-driving interplanetary CubeSats: the ERC-funded project EXTREMA,» in International Astronautical Congress: IAC proceedings, 2021.

[4] R. Walker, D. Koschny, C. Bramanti e I. Carnelli, «Miniaturised Asteroid Remote Geophysical Observer (M-ARGO): a stand-alone deep space CubeSat system for low-cost science and exploration missions,» in 6th Interplanetary CubeSat Workshop, Cambridge, UK, 2017.

[5] D. A. Dei Tos, «Trajectory optimization of limited control authority spacecraft in high-fidelity models,» 2018.

[6] F. Bernelli-Zazzera, F. Topputo e M. Massari, «Assessment of mission design including utilization of libration points and Weak Stability Boundaries,» 2004.

[7] W. S. Koon, M. W. Lo, J. E. Marsden e S. D. Ross, Dynamical systems, the three-body problem and space mission design, Marsden Books, 2011.

[8] E. J. Doedel, V. A. Romanov, R. C. Paffenroth, H. B. Keller, D. J. Dichmann, J. Galán-Vioque e A. Vanderbauwhede, «Elemental periodic orbits associated with the libration points in the circular restricted 3-body problem,» International Journal of Bifurcation and Chaos, vol. 17, p. 2625-2677, August 2007.https://doi.org/10.1142/S0218127407018671

[9] F. Topputo, M. Vasile e F. Bernelli-Zazzera, «Low energy interplanetary transfers exploiting invariant manifolds of the restricted three-body problem,» The Journal of the Astronautical Sciences, vol. 53, p. 353-372, December 2005.https://doi.org/10.1007/BF03546358

[10] F. Topputo, «On optimal two-impulse Earth-Moon transfers in a four-body model,» Celestial Mechanics and Dynamical Astronomy, vol. 117, p. 279-313, August 2013.https://doi.org/10.1007/s10569-013-9513-8

[11] D. A. Dei Tos, «Automated trajectory refinement of three body orbits in the real solar system model. Dynamical substitutes of Lagrangian points and quasi-periodic orbits about them,» 2014.

[12] K. Oshima, F. Topputo e T. Yanao, «Low-energy transfers to the Moon with long transfer time,» Celestial Mechanics and Dynamical Astronomy, vol. 131, January 2019.https://doi.org/10.1007/s10569-019-9883-7

[13] D. A. Dei Tos e F. Topputo, «On the advantages of exploiting the hierarchical structure of astrodynamical models,» Acta Astronautica, vol. 136, p. 236-247, July 2017.https://doi.org/10.1016/j.actaastro.2017.02.025

[14] V. G. Szebehely, Theory of orbit: the restricted problem of three bodies, Academic Press, 1967, p. 668.https://doi.org/10.1016/B978-0-12-395732-0.50007-6

[15] D. C. Folta, M. Woodard, K. C. Howell, C. Patterson e W. Schlei, «Applications of multi-body dynamical environments: The ARTEMIS transfer trajectory design,» Acta Astronautica, vol. 73, p. 237-249, April 2012.https://doi.org/10.1016/j.actaastro.2011.11.007

[16] G. Gómez, J. J. Masdemont e J. M. Mondelo, «Solar system models with a selected set of frequencies,» Astronomy & Astrophysics, vol. 390, p. 733-749, July 2002.https://doi.org/10.1051/0004-6361:20020625

Aerospace Science and Engineering - III Aerospace PhD-Days   Materials Research Forum LLC
Materials Research Proceedings 33 (2023) 193-199   https://doi.org/10.21741/9781644902677-28

Micromechanical analysis for evaluation of voids effect on thermoelastic properties of composites via 1D higher-order theories

Rebecca Masia[1, a *]

[1]Department of Mechanical and Aerospace Engineering, Politecnico di Torino, 10129, Torino, Italy

[a] rebecca.masia@polito.it

Keywords: Micromechanics, Voids, CUF, RVE, Composite Materials, Statistical Analysis

Abstract. The present work presents statistical results of numerical simulations to investigate the effect of different void percentages on composite materials' coefficient of thermal expansion (CTE). A random distribution of voids is simulated over the Representative Volume Element (RVE) matrix. The use of a high-order beam model within the framework of Carrera Unified Formulation (CUF) leads to a Component Wise (CW) description of the model cross-section. Numerical results for different fiber volume fractions and void concentration percentages are carried out, and the comparison with references from literature demonstrates the agreement in the average CTE trend.

Introduction

Fiber-reinforced composites have proven to have excellent thermo-mechanical properties such as lightweight, mechanical performance at high temperature and good thermal stability. The coefficient of thermal expansion (CTE) plays a critical role in designing composite materials since it can directly determine the dimensional stability of the structure and thermal stress distribution. Moreover, the multiscale nature of composites requires the development of accurate multiscale models. Another critical aspect is the growth of defects that inevitably occur during the manufacturing process. Specifically, voids are the primary manufacturing defects that significantly influence thermal properties, and this influence needs to be investigated.

Most of the numerical simulations for the homogenization analysis usually employ Representative Volume Element (RVE) models, which represent the smallest geometric entity containing all the information about material properties and volume fraction of the constituents. Many numerical and analytical approaches are available for obtaining homogenized properties. For instance, in [1], the Mechanics of Structure Genome (MSG) coupled with Carrera Unified Formulation (CUF) and Hierarchical Legendre Expansions (HLE) for the discretization of the cross-section allows the computation of thermo-elastic properties of a Repeatable Unit Cell (RUC) model. Another MSG-based analysis computes the CTE of solid models [2]. Finally, in [3], semi-analytical models for the computation of homogenized CTE are presented, such as the High Fidelity Generalized Method of Cells (HFGMC).

The present article adopts the CUF to obtain a one-dimensional (1D) high-order Finite Element Method (FEM) analysis reducing the computational costs while maintaining 3D accuracy of the model. Hence, shape functions act along the RVE axis, and expansion functions enhance the cross-section kinematics.

The aim of the present work is to investigate the influence of microscale matrix voids on the prediction of effective CTEs of RVE models with different percentages of void and fiber volume fractions.

This work is organized as follows: Section 2 presents the high-order theory formulation; Section 3 introduces the micromechanics model; Section 4 provides numerical results and their discussion; and the conclusions are given in Section 5.

Aerospace Science and Engineering - III Aerospace PhD-Days   Materials Research Forum LLC
Materials Research Proceedings 33 (2023) 193-199   https://doi.org/10.21741/9781644902677-28

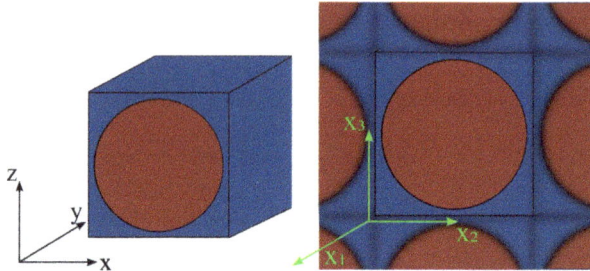

Figure 1 Square-pack RVE model with reference system high-order theory (left), and reference frame in micromechanics (right).

High-order beam theory

The present work exploits the capabilities of a refined 1D kinematic model. The three components of the displacement field are defined according to the reference system in Fig. 1

$$u(x, y, z) = \{u_x \, u_y \, u_z\}^T \tag{1}$$

The governing equations of the problem have been derived within the CUF framework [4]. Therefore, the displacement field becomes:

$$\boldsymbol{u}(x, y, z) = F_\tau(x, z)\boldsymbol{u}_\tau(y), \qquad \tau = 1, 2, \dots, M \tag{2}$$

where $F_\tau(x, z)$ are the expansion functions employed over the cross-section, $\boldsymbol{u}_\tau$ is the vector of general displacement and M is the number of terms in the expansion. In addition, τ denotes the summation of the expansion terms. Lagrange polynomials have been employed as expansion functions, leading to a Component Wise (CW) approach. By describing the longitudinal direction according to the finite element method (FEM), shape functions are introduced:

$$\boldsymbol{u}(x, y, z) = F_\tau(x, z)N_i(y)\boldsymbol{u}_{\tau i}, \quad \tau = 1, 2, \dots M, \quad i = 1, 2, \dots, p+1 \tag{3}$$

where $N_i(y)$ represents the shape functions of p order and $\boldsymbol{u}_{\tau i}$ is the vector of nodal displacements. The stress vector is defined as:

$$\boldsymbol{\sigma}^T = \{\sigma_{xx} \, \sigma_{yy} \, \sigma_{zz} \, \sigma_{xy} \, \sigma_{xz} \, \sigma_{yz}\} \tag{4}$$

The geometrical equation is involved as relation between strains and displacements:

$$\boldsymbol{\varepsilon} = \boldsymbol{Du} \tag{5}$$

where $\boldsymbol{D}$ is the 6×3 differential operator. Then, the Hooke's law allows the relation between strain and stress to be defined as:

$$\boldsymbol{\sigma} = \boldsymbol{C\varepsilon} \tag{6}$$

where $\boldsymbol{C}$ is the stiffness matrix of the material. The governing equations are derived from the Principles of Virtual Displacement (PVD). For the static analysis the PVD can be expressed as:

Aerospace Science and Engineering - III Aerospace PhD-Days   Materials Research Forum LLC
Materials Research Proceedings 33 (2023) 193-199   https://doi.org/10.21741/9781644902677-28

$$\delta L_{int} = \delta L_{ext} \tag{7}$$

where δL_{int} is the strain energy and L_{ext} is the work of the external forces. δ is used to indicate the virtual variation of the quantity. After some mathematical steps, the strain energy can be expressed as:

$$\delta L_{int} = \delta \boldsymbol{u}_{sj}^T \boldsymbol{k}_{ij\tau s} \boldsymbol{u}_{\tau i} \tag{8}$$

where the explicit form of the 3×3 Fundamental Nucleus (FN) $\boldsymbol{k}_{ij\tau s}$ is:

$$\boldsymbol{k}_{ij\tau s} = \int_V N_j F_s \boldsymbol{D}^T \boldsymbol{C} \boldsymbol{D} N_i F_\tau dV \tag{9}$$

The external load can be expressed as:

$$\delta L_{ext} = \delta \boldsymbol{u}_{sj}^T \boldsymbol{P}_{sj} \tag{10}$$

where $\boldsymbol{P}_{sj}$ is the 3×1 external load vector and is referred to as the FN of the load vector. The global stiffness matrix and the external load vector are assembled by iterating the indices i, j, τ, s. The first two indexes i, j denote the loop on the FEM nodes, while τ, s are the loop indices on the cross-section nodes.

Micromechanics formulation
The present work uses a micromechanical analysis to compute the effective thermo-elastic properties, particularly the Coefficient of Thermal Expansion (CTE) of composite materials.

The study makes use of different RVE models. The cross-section of the 1D models, illustrated in Fig.1, is discretized with 9-node bi-quadratic elements (L9), whereas along x_1 direction two 4-nodes quadratic elements are employed. The repeatability of the model in space is guaranteed by the application of Periodic Boundary Condition (PBC), according to the reference micromechanics frame in Fig. 1.

$$u_i^{j+}(x,y,z) - u_i^{j-}(x,y,z) = \bar{\varepsilon}_{ik} \Delta x_k^j, \qquad \Delta x_k^j = x_k^{j+} - x_k^{j-} \tag{11}$$

j+ and j- indicate the positive and negative direction of the x_k and $\bar{\varepsilon}_{ik}$ is the macroscopic strain vector. The thermo-elastic analysis considers the stress field given both from the elastic and thermal contributions:

$$\sigma_{ij} = \sigma_{ij}^E + \sigma_{ij}^T \tag{12}$$

where E stands for elastic and T the thermal. The present homogenization procedure first involves the resolution of the static problem in order to obtain the effective stiffness matrix, which is then employed in the effective CTE vector.

In the micromechanics framework, the local solutions, such as the strain and stress vectors, have an average value over the RVE volume. The local elastic stress field σ_{ij}^E is defined in Eq. (6), and its macroscopic value $\bar{\sigma}_{ij}^E$ can be expressed as follows:

$$\bar{\sigma}_{ij}^E = \frac{1}{V} \int_V \sigma_{ij}^E dV \tag{13}$$

Aerospace Science and Engineering - III Aerospace PhD-Days   Materials Research Forum LLC
Materials Research Proceedings 33 (2023) 193-199   https://doi.org/10.21741/9781644902677-28

in the same way, the macroscopic strain vector is:

$$\bar{\varepsilon}_{ij}^E = \frac{1}{V}\int_V \varepsilon_{ij}^E dV \tag{14}$$

The micromechanics formulation for the elastic analysis can be deepened in [5].
Now, the effective elastic coefficients matrix $\bar{C}_{ijkl}$ can be easily retrieved from:

$$\bar{\sigma}_{ij}^E = \bar{C}_{ijkl}\bar{\varepsilon}_{ij}^E \tag{15}$$

The local stress field given by the thermal contribution due to the application of a temperature variation θ is:

$$\sigma_{ij}^T = \beta_{ij}\theta \tag{16}$$

where $\beta_{ij} = -C_{ijkl}\alpha_{ij}$ is the local thermal induced stress vector.
Note that the temperature field applied to the RVE, unlike the stress and strain, is assumed to be uniform in the entire heterogeneous material.
By applying the integral over the volume of the thermal stress vector, it is possible to find the homogenized value $\bar{\sigma}_{ij}^T$ as done in Eq. 13. Then, by explicating Eq. 16 for $\bar{\beta}_{ij}$ and exploiting the relationship between $\bar{\beta}_{ij}$ $\bar{\alpha}_{ij}$ and $\bar{C}_{ijkl}$, the effective CTE vector can be obtained by rearranging the following relation:

$$\bar{\sigma}_{ij}^T = -\bar{C}_{ijkl}\bar{\alpha}_{ij}\theta \tag{17}$$

More details about the CUF micromechanics framework are available in [6]. The voids are modeled as randomly distributed over the RVE matrix phase, according to the void volume selected for the analysis. It is considered an implicit presence of voids by assigning negligible isotropic elastic properties and a CTE equal to zero to the Gauss points associated with the single void.

Numerical results
The numerical assessments of the current work are conducted to demonstrate the reliability of the numerical simulations in computing the homogenized thermo-elastic properties of composite materials employing RVE models. In detail, the variation of the effective CTE due to the gradually enhance of the void fraction is investigated. For the sake of clarity, for each void percentage within the matrix, 100 random distributions are generated, thus requiring a statistical investigation of the results.

Influence of fiber volume fraction and void percentage on the effective CTE
An hexagonal-pack model, showed in Fig. 2, is employed for the current assessment, and two values of fiber volume fraction are evaluated. The results are compared with those reported in [7]. The material is a C/C composite in which the fiber is assumed orthotropic with $\alpha_{11} = -1 \cdot 10^{-6}[\text{K}^{-1}]$ and $\alpha_{22} = 1.8 \cdot 10^{-5}[\text{K}^{-1}]$, while the isotropic matrix has a CTE of $4 \cdot 10^{-6}[\text{K}^{-1}]$. First, the results with 0 % of voids are produced and listed in Table 1.

Aerospace Science and Engineering - III Aerospace PhD-Days
Materials Research Proceedings 33 (2023) 193-199

Materials Research Forum LLC
https://doi.org/10.21741/9781644902677-28

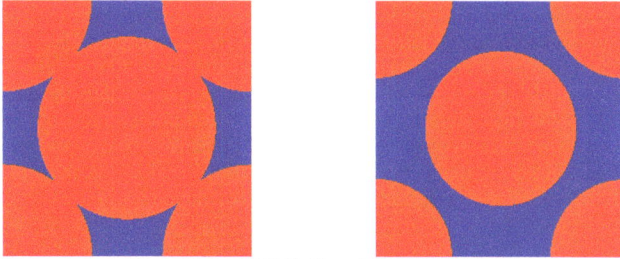

Figure 2 Hexagonal-pack model for 80% (left) and 57% (right) of fiber volume fraction.

Table 1 Effective CTEs of an hexagonal-pack model with 80% and 57% of fiber volume fraction and 0% of voids.

Model	$\alpha_{11} \times 10^6 [K^{-1}]$	$\alpha_{22} = \alpha_{33} \times 10^6 [K^{-1}]$
	$V_f = 57\%$	
Reference [7]	-0.900	12.720
Current CUF simulation	-0.909	13.082
	$V_f = 80\%$	
Reference [7]	-0.962	16.450
Current CUF simulation	-0.972	15.820

As expected, the longitudinal CTE decreases when the V_f rises from 57% to 80% since the fiber becomes the prevailing phase. At the same time, the transverse CTE increases with a higher value of V_f.

The following numerical instances examine the effects of void volume fraction and the fiber reinforcement phase on the effective CTE of composites. In order to understand the way the existence of voids within the material matrix affects the thermo-elastic property, void percentages between 2 and 8% are considered with 80% of fiber volume fraction. The analysis results are presented in Fig. 3.

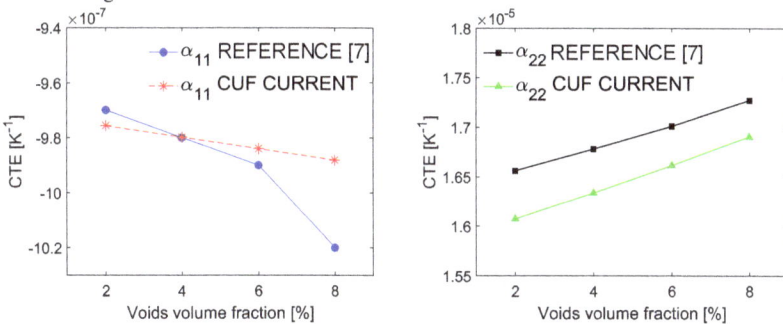

Figure 3 Variation of the longitudinal (left) and transverse (right) CTE according to the percentage voids volume fraction for the present and the reference results in [7], with 80% of fiber volume fraction.

The longitudinal CTE gradually weakens when the void volume fraction rises, while the transverse CTE tends to increase. This behavior is in agreement with evidence existing in the literature. It may be explained by considering that when the void inclusions occur within the matrix, the contribution of a low fiber CTE and the zero value of void CTE make obvious the collapse of the effective CTE along the longitudinal direction. Conversely, when the RVE transverse direction is considered, the rise in void percentage within the matrix makes the transverse CTE of the fiber the most impactful contribution in that direction. Indeed, it can be noticed that as the void fraction increases, the overall CTE tends to be closer to the cross-sectional CTE of the fiber. In Fig. 4, the percentage variation of α_{11} and α_{22} is shown. The calculation assumes the relative CTE with 0% voids of the reference solution in [7].

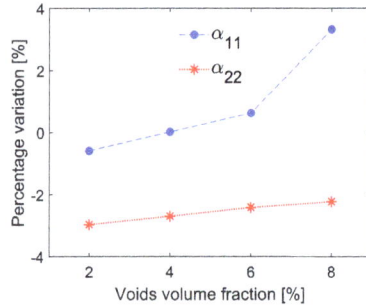

Figure 4 Percentage of the relative variation of longitudinal and transverse CTE between the present and the reference results in [7].

Fig. 4 highlights the discrepancies between the two methods with the decreasing of CTE value. Considering the transversal CTE, the values tend to be closer when higher void concentration occurs.

Conclusions

The present study investigated the effect of random distribution of voids within the matrix phase of composites on the effective coefficient of thermal expansion (CTE) for different Representative Volume Elements (RVE) models and fiber volume fractions. The homogenization analysis is conducted in the micromechanics framework, in which the use of the Carrera Unified Formulation (CUF) allows a Component Wise (CW) approach. The statistical analysis led to detect mean values of outcomes, thus minimizing the influence of the void distribution. A square-pack and a hexagonal-pack RVE with different fiber volume fractions are involved in the investigation. Periodic Boundary Conditions (PBC) ensured the periodicity constraint imposed by micromechanics analyses. The results demonstrated that manufacturing defects might increase or decrease the homogenized CTEs. Future investigations will focus on extending the study to the dehomogenization procedure to investigate localized stress and strain fields over the RVE volume for different void percentages and cluster distribution.

Acknowledgements

This work is part of a project that has received funding from the European Research Council (ERC) under the European Union's Horizon 2020 research and innovation programme (Grant agreement No. 850437). The author would also like to thank Mattia Trombini for the collaboration with the analysis of the outcomes.

References

[1] A. R. Sanchez-Majano, R. Masia, A. Pagani, E. Carrera. Microscale thermo-elastic analysis of composite materials by high-order geometrically accurate finite elements, Composite Structures, 2022, Vol. 300, p. 116105. https://doi.org/10.1016/j.compstruct.2022.116105

[2] X. Liu, W. Yu, F. Gasco, J. Goodsell. A unified approach for thermoelastic constitutive model of composite structures, Composites Part B, 2019, Vols. 172:649-59. https://doi.org/10.1016/j.compositesb.2019.05.083

[3] J. Aboudi, S. M. Arnold, and B. A. Bednarcyk. Practical Micromechanics of Composite Materials, 1st Edition, Buttherworth-Heinemann, Oxford Spires Business Park, Kidlington, Oxforshire, United Kingdom, 2021.

[4] E. Carrera, M. Cinefra, M. Petrolo, E. Zappino. Finite element analysis of structures through unified formulation. John Wiley & Sons, 2014. https://doi.org/10.1002/9781118536643

[5] E. Carrera, M. Petrolo, M.H. Nagaraj, M. Delicata. Evaluation of the influence of voids on 3D representative volume elements of fiber-reinforced polymer composites using CUF micromechanics. Composite Structures, 2020, Vol. 254. https://doi.org/10.1016/j.compstruct.2020.112833

[6] I. Kaleel, M. Petrolo, E. Carrera, A. M. Waas, Computationally Efficient Concurrent Multiscale Framework for the Linear Analysis of Composite Structures. AIAA Journal, 2019, Vol. 57(9). https://doi.org/10.2514/1.J057880

[7] K.L. Wei, J. Li, H.B. Shi, M. Tang. Numerical evaluation on the influence of void defects and interphase on the thermal expansion coefficients of three-dimensional woven carbon/carbon composites. 2020, Composite Interfaces, Vol. 27, pp. 873-892. https://doi.org/10.1080/09276440.2019.1707586

Aerospace Science and Engineering - III Aerospace PhD-Days   Materials Research Forum LLC
Materials Research Proceedings 33 (2023) 200-204   https://doi.org/10.21741/9781644902677-29

Freeform Offner spectrometer for space applications

Doria Chiara[1] *, Guerri Irene[2], Pekala Grzegorz[2], Cremonese Gabriele[3], Naletto Giampiero[4,1], Taiti Alessio[2]

[1]Center of Studies and Activities for Space "Giuseppe Colombo" (CISAS), University of Padova, via Venezia 15, 35131 Padova, Italy

[2]Leonardo S.p.A., via delle Officine Galileo 1, 50013 Campi Bisenzio (FI), Italy

[3]INAF- Astronomical Observatory of Padova, Vicolo dell'Osservatorio 5, 35122, Padova, Italy

[4]Department of Physics and Astronomy, University of Padova, Via Marzolo 8, 35131, Padova, Italy

*chiara.doria.1@phd.unipd.it

Keywords: Optics, Space Instruments, Freeform, Spectrometer, Design

Abstract. The performance in terms of image quality and spatial resolution plays a key role for imaging space instruments. Modern advancement in manufacturing and testing introduced freeform optics to the scene. Thanks to a higher number of degrees of freedoms with respect to the classical optical surfaces, freeform technology is a great opportunity to improve the instrument performance. Here a freeform Offner spectrometer is presented: it has been studied for the PRISMA second generation (SG) instrument, which is dedicated to space application, and it is now at the design phase A at Leonardo S.p.A.

Introduction

In the last few years, freeform optics started to be highly used for imaging and non-imaging systems thanks to the advancement in design, fabrication, and testing fields. Even if requiring an expert knowledge in cross- disciplinary fields, freeform optics can give a huge improvement in terms of performance and instrument compactness. In this first part of the study the focus will be the improvement of the spectrometer performance thanks to the use of freeform optics. Higher compactness design adopting freeform optics will be analyzed in future works, specifically for CubeSat applications.

Background

The geometry of a freeform optical surface is characterized by an absence of an axis of rotational symmetry, thus allowing many more degrees of freedom with respect to standard (also aspherical) optics. Different studies show that there are significant advantages in using those kinds of surfaces in the optic fields, both in terms of optical performance and instrument compactness. These are two major points for space instrumentation, making their application very promising for the next generation of space optical systems.

For what concerns the fabrication of freeform optics, we can presently use numerically controlled machines for grinding, polishing, and diamond turning. The challenge in these techniques includes building knowledge of how materials behave and defining suitable procedures for the surface shaping process. Some exciting developments are the recent advances in additive manufacturing that enable the 3D printing of freeform optics substrates.

Despite all those difficulties to face, freeform optics have already been applied on many different areas such as optical transformation (e.g., quantum cryptography, art forms), lighting and illumination (e.g., luminance, architecture lighting, automotive), manufacturing (e.g., EUV lithography, laser materials processing, machine vision and inspection), mobile displays (e.g.,

Aerospace Science and Engineering - III Aerospace PhD-Days Materials Research Forum LLC

Materials Research Proceedings 33 (2023) 200-204   https://doi.org/10.21741/9781644902677-29

near-eye, head worn handhelds, smart glasses), remote sensing (e.g., down-looking satellite, ubiquitous data collection, astronomical instrumentation, CubeSat), infrared and military instruments (e.g., UAVs and drones, conformal optics, and intelligence surveillance and reconnaissance systems) [4].

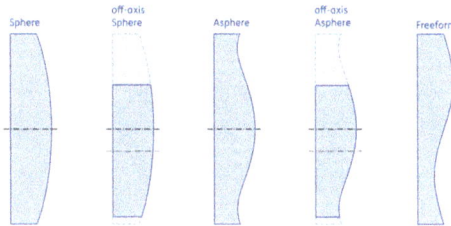

Fig. 1 Evolution of optical surfaces.

In order to introduce freeform optics in space instruments, one of the main issues is to find the best way of mathematically describe these surfaces: not only for designing and optimizing the optical surface by suitable ray-tracing codes, but also to have the correct input to the machines/tools with which these optics will be manufactured and tested.

There are many methodologies for the mathematical description of freeform surfaces, either local, also called non-orthogonal polynomials (XY polynomials, spline surfaces, radial basis function) or global, also called orthogonal polynomials (Zernike, Fringe Zernike, Q-type, Chebyshev) [1][2]. At this kind of analytical functions, we can add many new representation techniques, those employed to obtain a surface from discrete data points fitting, and the so-called hybrid or combined methods (usually applied when the surface presents big slopes).

For example, when using a Zernike polynomial, from the literature we know that the normalized expansion can be used to describe aberrations and the coefficient values of each mode represent the root mean square (RMS) wavefront error attributable to that mode [3]. The main coefficient order is described in Fig. 2.

A key point for the freeform description and analysis is the study of how the different terms of the polynomials affect aberrations, distortions and the main constraints of an optical design.
In addition, it is important to understand which the best design is to develop systems with great performance but not so much sensitive to manufacturing, mounting and stability tolerances.

From the many design methods presented in literature, here two methods are under study, both working on the minimization of the freeform system sensibility to a perturbation of the freeform system during the design phase.
The first one analyses the variation of the optical path difference due to a tilt perturbation: after the definition of a suitable function called error sensitivity function (ESF), it uses the Non-dominated Sorting Genetic Algorithm (NSGA-II) to minimize the ESF function and obtain a set of instrument design solutions. Then the ESF is exploited again to find the best polynomial description for the freeform surfaces [6].
The second method has been proposed to find the optimal design employing freeform surfaces less sensitive not only to tilt errors but to any perturbation applied to all optical elements [7].

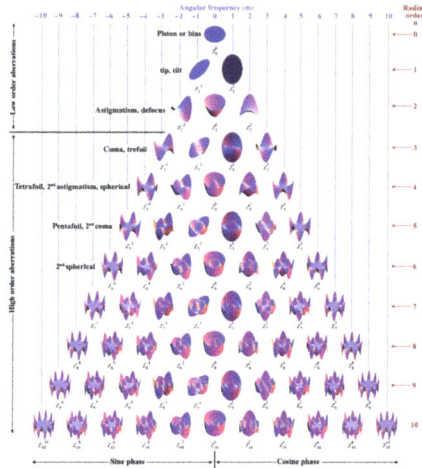

Fig.2 Surface plots of the Zernike polynomials up to 10 orders. The name of the main aberration is associated to the respective order. [3]

The spectrometer of PRISMA SG

High spatial and spectral resolution instruments changed our way of considering and understanding the environment in which we live and its main phenomena and characteristics. In fact, Earth surface observation is one of the most common and important space applications.

On this respect, the Italian Space Agency launched the PRISMA (PRecursore IperSpettrale della Missione Applicativa), a satellite designed and realized by Leonardo S.p.A. dedicated to the observation of the Earth surface, its natural resources and relevant natural processes. The satellite is composed by a hyperspectral sensor that acquires both VNIR (Visible and Near-Infrared) and SWIR (Short-Wave Infra-Red), and a panchromatic camera allowing to capture not only the geometry of the observed object but also the chemical-physical composition of the surface.

PRISMA is now observing the Earth surface from a LEO orbit at 615 km with a spatial scale of 30 meter/pixel over a field of view of 30 km [5]. In Fig. 3a the optical layout of the spectrometer is represented.

Now, entering more in depth of what this project concerns, we are working on the design and manufacturing of the first freeform optical element in Leonardo S.p.A., that will be integrated in the PRISMA SG, which yields a spatial scale improved of a factor 3 respect to the original PRISMA instrument: 10 meter/pixel over a view of 30 km (from an orbit of 520 km).

These very high performances will be reached thanks to the use of two identical Offner spectrometers using freeform optics (preliminary layout shown in Fig. 3b). The introduction of freeform mirrors to the Offner configuration enables to increase the FOV maintaining a good distortions correction thanks to the great flexibility offered by the increased number of degrees of freedom.

Presently, the spectrometer provides a good correction for smile and keystone distortions maintaining a good optical quality over the whole FOV and spectral range, as expected.

During the optimization phase we also studied different kinds of polynomials (Zernike, Chebyshev, XY) in order to understand which should be the best compromise in terms of computational weight and performance. From literature, we understood that the application of different polynomials is related to the pupil shape [1]. In our case, even if the spectrometer pupil

Aerospace Science and Engineering - III Aerospace PhD-Days
Materials Research Proceedings 33 (2023) 200-204

Materials Research Forum LLC
https://doi.org/10.21741/9781644902677-29

will be annular (due to the pupil of the instrument telescope) the footprint of the different fields on the mirrors are overlapped enough to cover all the mirror surface, so we compared only the surface description with the Zernike Standards and the Chebyshev polynomials. We concluded that there are not many differences in terms of performances and number of variables, so we decided to adopt the Zernike ones.

Moreover, we are working on understanding the correlation between a perturbation of one of the instrument parameters (starting from position and tilts of the freeform mirrors) and the variation of the optical quality of the spectrometers.

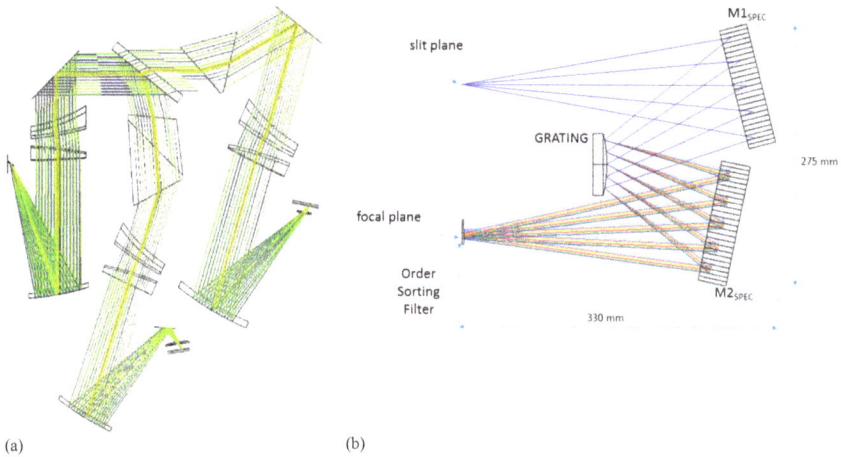

(a)   (b)

Fig.3: (a) PRISMA VNIR/SWIR spectrometer layout [5]; (b) PRISMA SG spectrometer layout where M1 and M2 are freeform.

Conclusions

The development of PRISMA SG spectrometer involving freeform mirrors represents a challenge from the theorical point of view. In this work, a preliminary design of the optical layout and the main objective in terms of performances have been presented. A short overview of the complex world of freeform surfaces is presented: a description of the main advantages of this technology in terms of performance and compactness has been introduced as well as the main difficulties that it is necessary to face for their design, manufacturing and testing.

References

[1] 16. J. F. Ye, L. Chen, X. H. Li, Q. Yuan, and Z. S. Gao, "Review of optical freeform surface representation technique and its application," Opt. Eng. 56, 110901 (2017)https://doi.org/10.1117/1.OE.56.11.110901

[2] C. Hou, Y. Ren, Y. Tan, Q. Xin, and Y. Zang, "Compact optical zoom camera module based on Alvarez elements," Opt. Eng. 59, 025104 (2019)https://doi.org/10.1117/1.OE.59.2.025104

[3] V. Lakshminarayanan, and A. Fleck. "Zernike polynomials: A guide", Journal of Modern Optics - J MOD OPTIC. 58. 1678-1678. 10.1080/09500340.2011.633763 (2011)https://doi.org/10.1080/09500340.2011.633763

[4] J. P. Rolland, M. A. Davies, T. J. Suleski, C. Evans, A. Bauer, J. C. Lambropoulos, and K. Falaggis, "Freeform optics for imaging," Optica 8(2), 161 (2021)https://doi.org/10.1364/OPTICA.413762

[5] D. Labate, M. Ceccherini, A. Cisbani, V. De Cosmo, C. Galeazzi, L. Giunti, M. Melozzi, S. Pieraccini, and M. Stagi, "The PRISMA payload optomechanical design, a high performance instrument for a new hyperspectral mission", Acta Astronautica, Volume 65, Issues 9-10,2009, Pages 1429-1436, ISSN 0094-5765https://doi.org/10.1016/j.actaastro.2009.03.077

[6] Z. Qin, Q. Meng, and X. Wang, "Desensitization design method of a freeform optical system based on local curve control," Opt. Lett. 48, 179-182 (2023)https://doi.org/10.1364/OL.480641

[7] X. Liu, T. Gong, G. Jin, and J. Zhu, "Design method for assembly-insensitive freeform reflective optical systems," Opt. Express 26, 27798-27811 (2018)https://doi.org/10.1364/OE.26.027798

[8] J. P. Rolland, M. A. Davies, T. J. Suleski, Chris Evans, A. Bauer, J. C. Lambropoulos, and K. Falaggis, "Freeform optics for imaging," Optica 8, 161-176 (2021)https://doi.org/10.1364/OPTICA.413762

Aerospace Science and Engineering - III Aerospace PhD-Days   Materials Research Forum LLC
Materials Research Proceedings 33 (2023) 205-212   https://doi.org/10.21741/9781644902677-30

A distributed nanosatellite attitude testing laboratory for joint research activities

Curatolo Andrea[1,a*], Manconi Francesco[2,b], Lughi Chiara[2,c]

[1]Department of Industrial Engineering Alma Mater Studiorum Università di Bologna, Via Fontanelle 40, I-47121 Forlì Italy

[2]Department of Mechanical and Aerospace Engineering Politecnico di Torino, Corso Duca degli Abruzzi 24, 10129 Torino Italy

[a]andrea.curatolo4@unibo.it [b]francesco.manconi@polito.it [c]chiara.lunghi@studenti.polito.it

Keywords: CubeSats, Satellite Testing, Laboratory Equipment

Abstract. During the last decades, there has been an increase in the use of satellites of reduced dimensions. Among the others, microsatellites (mass from 11 to 200 kg) and nanosatellites (1 to 10 kg) have been the ones receiving increasing interest from Universities for educational activities [1], [2]. Their reduced cost, complexity and developing time compared to larger satellites make them particularly suitable for student projects. In this regard CubeSats (satellites of standardised dimensions, based on 1 unit, 10x10x10 cm) were developed at Caltech with the goal of having a low cost and fast to be developed satellite [3]. The CubeSat form factor has then been widely used also for scientific and commercial space missions [4]. Alongside the development of nanosatellites, there has been an increase in the need for better CubeSat testing for improving CubeSat reliability [5]. As reported in an extensive study on 855 CubeSats [4], at the year 2018 almost 25% of CubeSats missions failed in their early life stage (*infant mortality*). One of the subsystems more difficult to be tested is the Attitude Determination and Control System (ADCS). This subsystem includes sensors for attitude determination, actuators for attitude control and an onboard controller. Integrated subsystem testing is a challenging task since the device under test should freely rotate under low torque conditions and sensors/actuators should be stimulated. A common way to provide a free rotational environment is to use an air bearing table [6].

The University of Bologna developed its own solution for ADCS testing with a dynamic attitude simulator testbed for nanosatellites [7]. Recently a collaboration between the Microsatellite and Space Microsystem laboratory of the University of Bologna and the STAR laboratory of the Polytechnic University of Turin was established to develop a distributed laboratory for nanosatellite ADCS testing and development. This project enables joint research activities for both research groups in a collaborative framework. The work done included the development of a new Air Bearing Table (ABT) in the Unibo laboratory, the development of a custom ADCS mock-up compliant with the 1U form factor in the Polito laboratory and a joint hardware integration and test campaign.

Aerospace Science and Engineering - III Aerospace PhD-Days   Materials Research Forum LLC
Materials Research Proceedings 33 (2023) 205-212   https://doi.org/10.21741/9781644902677-30

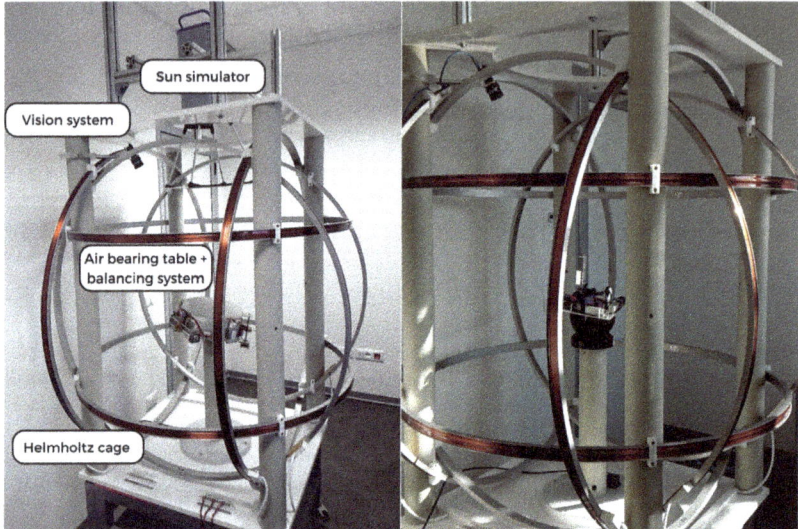

Figure 1: The attitude simulator testbed of the Unibo. The air bearing table for satellites with mass up to 4 kg is shown on the left, the one for satellites with mass up to 1 kg is shown on the right.

The attitude simulator testbed of the University of Bologna integrates several subsystems for the simulation of the Low Earth Orbit (LEO) environment including:

- Air Bearing Table and Automatic Balancing System (ABS) providing three degree of freedom rotational motion in a low torque environment;
- Helmholtz cage for Earth magnetic field simulation;
- Sun simulator;
- Metrology system for ground-truth attitude generation

The facility is described in detail in [7], here its main features are reported. The ABT and ABS are used to reproduce the attitude dynamics of a satellite in space. The external torques acting on a nanosatellite in LEO can be as low as $10^{\wedge}$-6 Nm [8]. For ABT facilities, the gravity torque caused by the misalignment between the Centre of Mass (CM) and Centre of Rotation (CR) of the platform is the biggest disturbance torque and needs to be minimised. This can be done by acting on the relative position between CR and CM through a balancing procedure. A coarse balancing can be obtained by evenly distributing the platform components' mass while fine balancing is achieved using sliding masses actuated by stepper motors in an automatic balancing procedure. The fine balancing includes both a part of feedback control and a part of system identification and is described in [9]. The Helmholtz cage is used to reproduce the magnetic field in LEO, where satellites are often equipped with magnetic field sensors and magnetic actuation. A magnetorquer and a closed loop control system are used to accurately track a desired variable magnetic field. The sun sensor consists of a LED lamp able to reproduce the visible part of the Sun emission spectrum. The metrology system consists of a calibrated monocular system. Two ABT have been developed: one for satellites with mass up to 4 kg and the other for satellites with mass up to 1 kg (see Figure 1). The latter has been upgraded to allow the housing of a CubeSat of 1U standard dimension

Aerospace Science and Engineering - III Aerospace PhD-Days   Materials Research Forum LLC
Materials Research Proceedings 33 (2023) 205-212   https://doi.org/10.21741/9781644902677-30

(provided by the Polito) with a dedicated locking mechanism and to host new models of stepper motors.

The facility of Turin has been used for initial basic tests and components implementation in the ADCS mock-up. The PoliTo Clean Room, which is used in the assembly, integration, and functional verification of CubeSats and products that require a controlled environment and special cleanliness, was used in the integration process. The Clean Room is a secluded volume of 60 m^3 and is a 20 m^2, equipped with appropriate equipment. It is suitable for hosting up to three operators while maintaining an ISO7 cleanliness grade. The vertical laminar flow bench is a carefully enclosed cabinet designed to prevent contamination of any particle-sensitive product. Air is drawn through a HEPA filter and blown in a very smooth, laminar flow towards the user. The cabinet is made of stainless steel and other special materials, with no gaps or joints where spores might collect. The equipment is certified for a cleanliness level of ISO5 (Class 100).

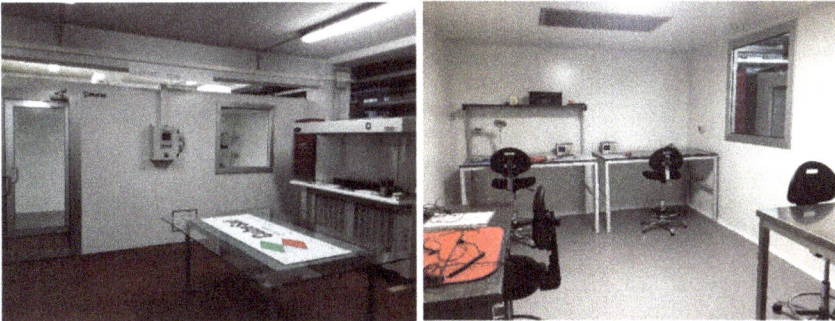

Figure 2: Polito clean room present in Turin

Once the interface with bologna had been studied, a preliminary identification of the possible tests that could be carried out using the elements of the facility was made. An initial analysis identified the following testing activities as compliant with the available instrumentation:

1. Verification of the interfaces of the basic board by Electrical Ground Support Equipment (EGSE);
2. Verification of the interfaces of the ADCS board by EGSE;
3. Verification of the correct functioning of the magnetic torquers by means of Helmholtz cage;
4. Verification of attitude maintenance downstream of the test facility's intrinsic disturbances;
5. Verification of correct operation of the magnetometers by comparison with outputs supplied by the IMU;
6. Verification of correct operation of the magnetometers by comparison with outputs supplied by the visual camera;
7. Verification of correct direction of magnetometer dipole actuation;
8. Stabilisation tests downstream of random input around the friction-free axis determined by the air bearing;
9. Verification of the correct operation of the sun sensors using visual camera and sun simulator;

10. Assuming the equatorial plane as the orbital plane, it is possible to test the maintenance of nadir pointing by the actuators of the subsystem, without restrictions on the number of orbits.

11. Once the testing activities were established, the development of the Polito testbed began.

The development process included two main aspects: mechanical interface development and functional board development. For the mechanical interface, a universal and interchangeable locking system was developed to keep the testbed within standardised dimensions for the CubeSat format. Structural integrity was also taken into consideration, along with the weight of the structure, which needed to be as low as possible. For the functional (base) board, a solution was developed for power supply and data exchange with the subsystem (ADCS) board. A board was designed with the required components for power supply, data communication and data processing, including two batteries, a Raspberry PI0-W processor, a hardline connector, ADCs, 5 and 3.3 V regulator circuits, and a 32GB microSD memory. The chosen microprocessor was equipped with a Wi-Fi communication capability, and four types of bus were used for data transmission to the microprocessor. The ADCS board implements two single-axis control laws under test: PD and Y-dot.

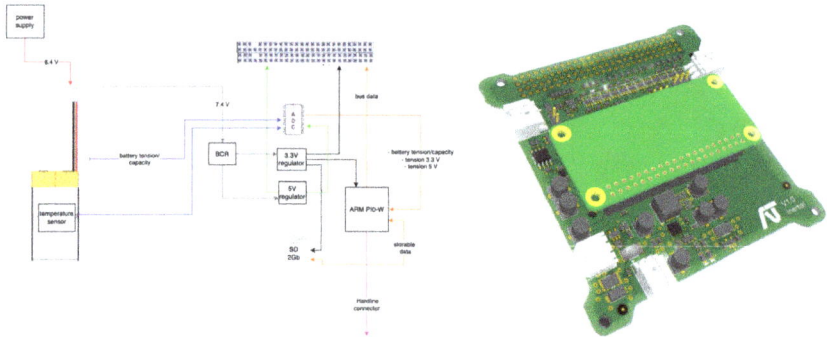

Figure 3: OBC functional board schematic and PCB

Using a CAD software, a 3D model of the mechanical interface was created, which was designed to contain the functional board and the subsystem board and 3D printed in PLA material. A mass mock-up was assembled with the functional and ADCS board models for fine-tuning and verification of correct functioning during balancing procedures in Bologna.

Aerospace Science and Engineering - III Aerospace PhD-Days
Materials Research Proceedings 33 (2023) 205-212

Materials Research Forum LLC
https://doi.org/10.21741/9781644902677-30

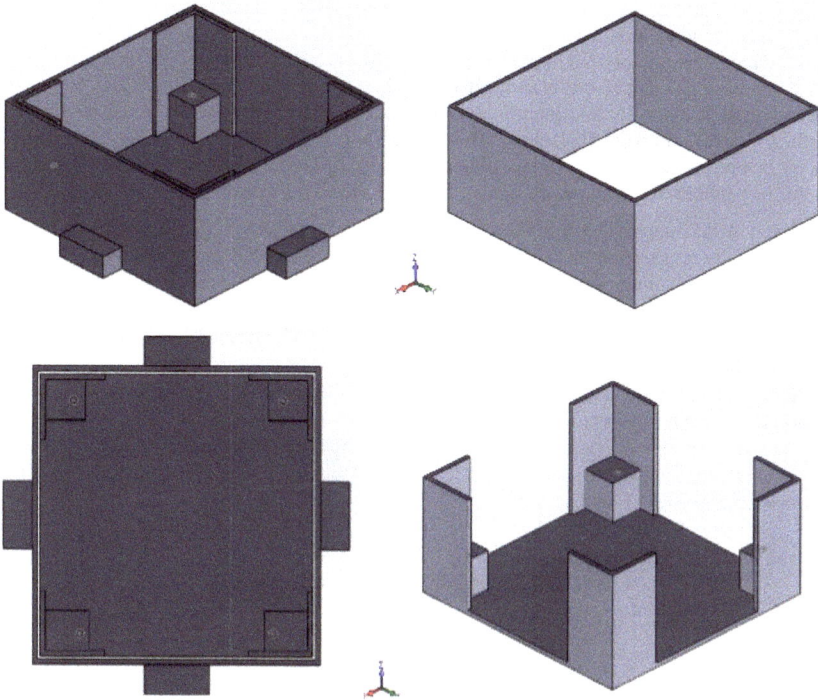

Figure 4: 3D model of the mechanical interface of the mock up

The upgrade on the Unibo facility took into consideration different constrains/requirements:

- Possibility to host and fix a CubeSat with 1U form factor and mass up to 350 g (specifications provided by Polito);
- Minimisation of the disturbance torques with particular attention to the gravity torque;
- Minimisation of the overall mass and dimensions to keep inertia characteristics as similar as possible to the ones of the 1U CubeSat under test;
- Provide hosting for electronic equipment used by the ABS (includes stepper motors, onboard microprocessor, IMU, other electronics and wiring).

The onboard electronics was mainly inherited from the existing platform while new stepper motors with integrated linear guides have been used. The design phase included the exploration of different possibilities for hosting the CubeSat and components and meeting the other requirements. After some iterations between the two teams a solution employing 3D printed L-shaped blocks and pods was adopted as a mechanical interface between the Air Bearing Table and the CubeSat (visible in Figure 5). Although during the realisation of the CAD model we tried to be as accurate as possible, the actual platform differed from the model for some aspects:

Aerospace Science and Engineering - III Aerospace PhD-Days   Materials Research Forum LLC
Materials Research Proceedings 33 (2023) 205-212   https://doi.org/10.21741/9781644902677-30

- The plate was 2 mm thicker due to limited plate thickness options from the supplier, this resulted in an increase in overall mass;
- The stepper motors and linear guides were 250 g heavier with respect to the mass indicated on the CAD model, this also resulted in an increase in overall mass.

The increase in mass, located mainly above the CR, could have jeopardized the possibility of perform the platform coarse balancing. The problem was solved by putting counterweights inside the air bearing hollow. This solution, although not ideal since it increased the overall mass and inertia, was preferred over others solutions being the most time effective one.

The components integration process involved both Unibo team and Polito team. Prior to the integration, the stepper motors were tested at different motor stepped and acceleration. The goal of the tests was the minimisation of the vibration due to motor resonance. Once all the electronic components have been integrated on the platform, the automatic balancing system was tested. This step required a considerable transfer of information from Unibo to Polito, for this reason documentation explaining in detail the balancing process and best practice has been produced. The test of the automatic balancing system showed that the disturbance torque acting on the new platform after the balancing process is lower than $5 \times 10\text{-}5$ Nm. Figure 6 shows the results of the tests in terms of residual torque. Solutions for further improving this result are under study today. An experimental campaign to test the Polito CubeSat Attitude Control System is foreseen for the period of March 2023.

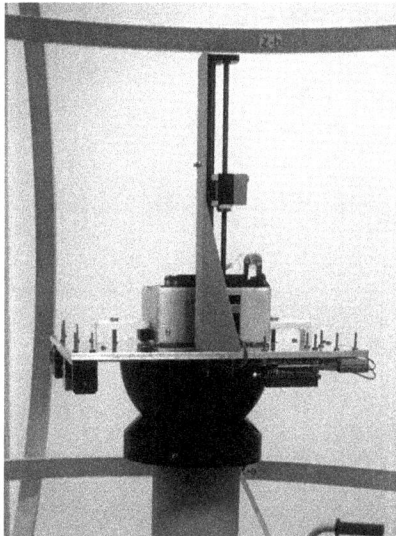

Figure 5: The air bearing table with the Polito mock-up mounted on it

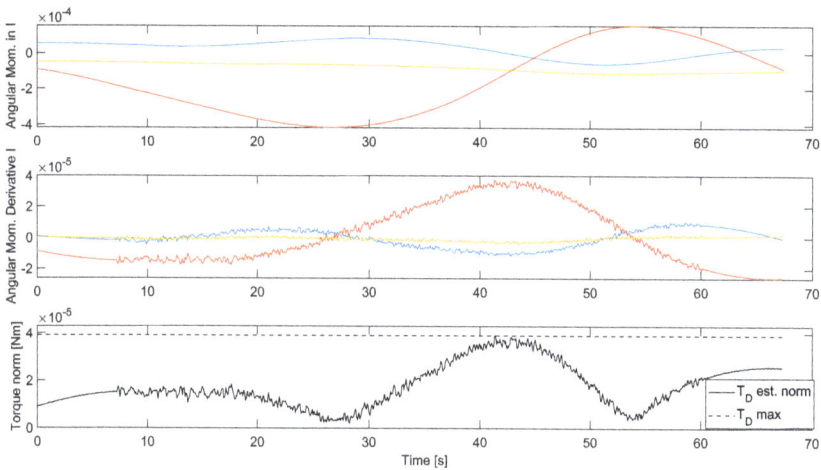

Figure 6: Result of the balancing process. From top to bottom: 1) angular momentum evaluation from filtered angular velocity data; 2) angular momentum derivative; 3) estimated residual torque

References

[1] J. Straub and D. Whalen, 'Evaluation of the Educational Impact of Participation Time in a Small Spacecraft Development Program', *Education Sciences*, vol. 4, no. 1, pp. 141–154, Mar. 2014. https://doi.org/10.3390/educsci4010141

[2] J. A. Larsen and J. D. Nielsen, 'Development of cubesats in an educational context', in *Proceedings of 5th International Conference on Recent Advances in Space Technologies - RAST2011*, Istanbul, Turkey, Jun. 2011, pp. 777–782. doi: 10.1109/RAST.2011.5966948

[3] P. Puig-suari, C. Turner, and R. Twiggs, 'CubeSat: The Development and Launch Support Infrastructure for Eighteen Different Satellite Customers on One Launch', Jan. 2001.

[4] T. Villela, C. A. Costa, A. M. Brandão, F. T. Bueno, and R. Leonardi, 'Towards the Thousandth CubeSat: A Statistical Overview', *International Journal of Aerospace Engineering*, vol. 2019, pp. 1–13, Jan. 2019. https://doi.org/10.1155/2019/5063145

[5] J. Bouwmeester, A. Menicucci, and E. K. A. Gill, 'Improving CubeSat reliability: Subsystem redundancy or improved testing?', *Reliability Engineering & System Safety*, vol. 220, p. 108288, Apr. 2022. https://doi.org/10.1016/j.ress.2021.108288

[6] J. Schwartz, M. Peck, and C. Hall, 'Historical Review of Air-Bearing Spacecraft Simulators', *Journal of Guidance, Control, and Dynamics*, vol. 26, May 2003. https://doi.org/10.2514/1.1035

[7] D. Modenini, A. Bahu, G. Curzi, and A. Togni, 'A Dynamic Testbed for Nanosatellites Attitude Verification', *Aerospace*, vol. 7, p. 31, Mar. 2020. https://doi.org/10.3390/aerospace7030031

[8] A. Cortiella *et al.*, '3 CAT-2: Attitude Determination and Control System for a GNSS-R Earth Observation 6U CubeSat Mission', *European Journal of Remote Sensing*, vol. 49, no. 1, pp. 759–776, Jan. 2016. https://doi.org/10.5721/EuJRS20164940

[9] A. Bahu and D. Modenini, 'Automatic mass balancing system for a dynamic CubeSat attitude simulator: development and experimental validation', *CEAS Space Journal*, vol. 12, no. 4, pp. 597–611, Dec. 2020. https://doi.org/10.1007/s12567-020-00309-5

Aerospace Science and Engineering - III Aerospace PhD-Days
Materials Research Proceedings 33 (2023) 213-218

Materials Research Forum LLC
https://doi.org/10.21741/9781644902677-31

Time domain aeroelastic analysis of wing structures by means of an alternative aeroelastic beam approach

Carmelo Rosario Vindigni[1,a*]

[1]University of Enna Kore, Faculty of Engineering and Architecture, 94100, Cittadella Universitaria, Enna, Italy

[a]carmelorosario.vindigni@unikore.it

Keywords: Aeroelastic Beam, Wing Stick Model, Wing-Aileron Flutter

Abstract. In this work an alternative beam finite element for rapid time-domain flutter analysis of wings equipped with trailing edge control surfaces, generally distributed along the span, is presented. The aeroelastic beam finite element proposed is based on Euler-Bernoulli beam theory, De St. Venant torsion theory and two-dimensional time-domain unsteady aerodynamics. The developed finite element model is attractive for preliminary aero-servo-elastic analyses and flutter suppression systems design purposes; moreover, the finite element matrices obtained could be easily included in existing aeroelastic optimization codes that already use beam modelling of lifting structures to carry out aeroelastic tailoring studies.

Introduction

In modern design of wing structures energy and cost saving considerations have become more important leading to the preference of lightweight structural configurations. Anyway, lightweight structures suffer of susceptibility to aeroelastic instabilities. The simplest aeroelastic model that can be used for preliminary aeroelastic analysis is the three degrees of freedom 3DOF model proposed by Theodorsen that reduces the wing structure to its representative mean airfoil and implements the structural stiffnesses by means of heave, torsional, and flap hinge springs [1]. Anyway, though the 3DOF system is representative of high aspect ratio HAR wings with full-span trailing edge control surfaces, it is not best suited for aero-servo-elastic analysis of wings with considerable variations in stiffness and mass balance along the span or wings that presents locally distributed control surfaces. Nowadays the most popular design and verification tool for aeroelastic analysis of wing structures is the combination of the Finite Element Method FEM, used to model the structural system, coupled with the Doublet Lattice Method DLM, used to model the aerodynamic system, and the connection between structure and aerodynamics is performed by means of splines [2]. Unfortunately, this kind of aeroelastic model presents high computational costs and it is better suitable for the verification phase of the project than for the preliminary assessment phase of the design. A widely used method for preliminary aeroelastic analysis of wings is based on the realization of equivalent beam models of the structure, that can be coupled with different aerodynamic models such as the strip theory and the unsteady vortex lattice method. In this work an alternative modelling approach based on an aeroelastic beam finite element formulation is presented and used for preliminary aero-servo-elastic time domain analyses of wings with trailing edge control surfaces. The equivalent beam modelling of the wing is based on Euler-Bernoulli beam assumptions, De St. Venant torsion theory, and 2D unsteady aerodynamics.

Methods

The structural model considered in this work represents a cantilever wing with thin symmetric airfoil, straight elastic axis EA, and a trailing edge aileron-like control surface. The wing is inextensible and with an infinite chordwise bending stiffness. The elastic axis of the wing lies on the x axis, identifying the spanwise direction, while the cross section lies in the y-z plane, where

Aerospace Science and Engineering - III Aerospace PhD-Days   Materials Research Forum LLC
Materials Research Proceedings 33 (2023) 213-318   https://doi.org/10.21741/9781644902677-31

y and z are oriented in the positive flow and upwards directions, respectively. The aileron control surface is hinged to the wing frame and connected to the actuators that provide a local stiffness k_{act}, it is also considered aerodynamically unbalanced; moreover, the aileron is considered flexible in torsion and with elastic axis close enough to the hinge line such that they could be considered coincident. The wing is modelled as a Euler-Bernoulli beam that also behave in accordance with De Saint Venant torsion theory and presents bending-torsion inertial coupling. The structural degrees of freedom of the wing-stick model are the vertical displacement due to bending w (positive upwards), the elastic torsional rotation ϕ around the elastic axis (positive nose-up), and the control surface rotation δ about its hinge (positive flap down). The wing's geometric parameters of interest are given in Fig. 1.

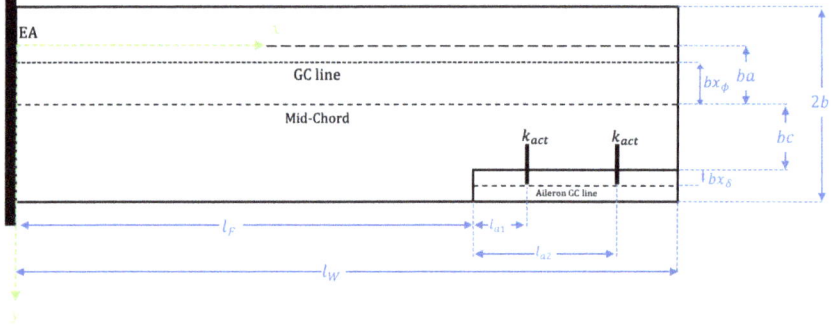

Figure 1 - Wing schematic

The equations of motion of the wing are the ones of a beam with EI bending stiffness and GJ torsional stiffness, while the torsional governing equation of the flap is the one of a beam with a combined torsional stiffness given by the sum of its elastic stiffness $G_f J_f$ and the hinge stiffness K_δ. The loads acting on the beam are given by the unsteady aerodynamic modelling provided by Theodorsen for thin symmetric airfoils undergoing harmonic motion that are written in time domain taking advantage of an indicial function approach [3]. The governing equations of the wing can be written in compact form as [4]

$$[M_s + M_{aer}]\ddot{q}(t) + [C_s + C_{aer}]\dot{q}(t) + [K_s \mathcal{D}^2 + K_\delta + K_{aer}]q(t) = 0. \tag{1}$$

where $\mathbf{q(t)} = [\text{w} \quad \phi \quad \delta \quad \bar{x}]^T$ is the generalized displacement vector, $\mathbf{M}_s$ and $\mathbf{M}_{aer}$ are the structural and aerodynamics mass matrices, $\mathbf{C}_s$ and $\mathbf{C}_{aer}$ are the damping matrices, $\mathbf{K}_s$ and $\mathbf{K}_{aer}$ are the stiffness matrices. Moreover, $\mathcal{D}^2$ and $\mathbf{K}_\delta$ are the the the differential operator that takes into account the derivatives in x (the beam element axis) and the linear stiffness matrix taking into account the actuator stiffness, respectively, that are defined as follows

$$\mathcal{D} = diag\left(\frac{\partial^2}{\partial x^2}, \frac{\partial^2}{\partial x^2}, \frac{\partial}{\partial x}, 0\right); \; \mathbf{K}_\delta = diag\,(0, 0, K_\delta, 0). \tag{2}$$

where

$$K_\delta = k_\delta + k_{act}\delta_F. \tag{3}$$

being k_δ the hinge stiffness per unit length and δ_F an expression of the Dirac delta function $\delta_F = \delta_d(x - l_F - l_{a1}) + \delta_d(x - l_F - l_{a2})$ that identifies the actuators position along the aileron span.

A third order Hermite interpolation function is introduced for the bending displacement w(x) and a linear interpolation function is used for the others generalized displacements; thus, the beam displacement interpolation can be written as

$$\mathbf{q}(x) = \mathbf{N}(x)\Delta. \tag{4}$$

where $\mathbf{N}(x)$ is the shape functions matrix and $\Delta = [w_i, \theta_i, \phi_i, \delta_i, \bar{x}_i, w_j, \theta_j, \phi_j, \delta_j, \bar{x}_j]^T$ is the generalized nodal displacement vector of the $i-j$ beam finite element. The generalized interpolation is then substituted into the governing equations and their weak form is introduced obtaining the following compact form expression, where $\tilde{q} = [\tilde{w} \quad \tilde{\phi} \quad \tilde{\delta} \quad \tilde{\bar{x}}]$ is the virtual displacement vector.

$$\int_L \tilde{q}[\mathbf{M}_s + \mathbf{M}_{aer}]\ddot{\mathbf{q}}(t)dx + \int_L \tilde{q}[\mathbf{C}_s + \mathbf{C}_{aer}]\dot{\mathbf{q}}(t)dx + \int_L \tilde{q}[\mathbf{K}_s\mathcal{D}^2 + \mathbf{K}_\delta + \mathbf{K}_{aer}]\mathbf{q}(\mathbf{t})\mathbf{dx} = \mathbf{0}. \tag{5}$$

From the governing equations weak form the elemental mass, damping, and stiffness matrices can be obtained and are defined as follows

$$\mathbf{M}_{el} = \int_L \mathbf{N}^T[\mathbf{M}_s + \mathbf{M}_{aer}]\mathbf{N}dx$$

$$\mathbf{C}_{el} = \int_L \mathbf{N}^T[\mathbf{C}_s + \mathbf{C}_{aer}]\mathbf{N}dx \tag{6}$$

$$\mathbf{K}_{el} = \int_L [(\mathbf{DN})^T\mathbf{K}_s\mathbf{D} + \mathbf{N}^T\mathbf{K}_\delta + \mathbf{N}^T\mathbf{K}_{aer}]\mathbf{N}dx$$

Once the aeroelastic beam elemental matrices have been obtained the equivalent wing structural model can be realized in a finite element fashion assembling the matrices opportunely and according with the structural discretization. In this way the FEM governing equations are obtained and read as

$$\mathbf{M}_G\ddot{\Delta}_G + \mathbf{C}_G\dot{\Delta}_G + \mathbf{K}_G\Delta_G = 0. \tag{7}$$

where the subscript G stands for "global" (referring to the whole structure) and Δ_G is the unknown displacements vector. Finally, Eq. 6 can be cast in state-space form defining the dynamic matrix of the system as follows

$$\mathbf{A} = \begin{bmatrix} 0 & I \\ -\mathbf{M}_G^{-1}\mathbf{K}_G & -\mathbf{M}_G^{-1}\mathbf{C}_G \end{bmatrix}. \tag{8}$$

Thus, the stability analysis of the wing stick model can be carried out studying the eigenvalues of $\mathbf{A}$ for increasing speed values in order to identify the free-stream velocity for which the eigenvalues real part becomes positive valued.

Results
Validation analyses have been carried out on both clean Goland wing model and its modified wing-aileron configuration [5]. A preliminary convergence study has revealed that a subdivision of the

Aerospace Science and Engineering - III Aerospace PhD-Days   Materials Research Forum LLC
Materials Research Proceedings 33 (2023) 213-318   https://doi.org/10.21741/9781644902677-31

wing span in ten aeroelastic beam elements was enough for the flutter analysis of the clean Goland wing. Fig. 4 shows the Goland wing eigenvalues real part and frequency evolution with the airstream speed at sea level density, from which it can be seen that the classical coupling between the first fundamental bending and torsion modes is present for a flutter speed and frequency of $v_f = 137.4 \, m/s$ and $\omega_f = 11.2 \, Hz$, respectively. These results are in good agreement with the exact solution given by Goland and Luke and with other literature results that use similar approaches, as shown in Table 1.

Table 1 - Goland wing flutter results comparison

REFERENCE	MODELING APPROACH	$v_f[m/s]$	$\%E(v_f)$	$\omega_f[Hz]$	$\%E(\omega_f)$
GOLAND AND LUKE [6]	Analytical	137.25	-	11.25	-
PROPOSED APPROACH	Aeroelastic beam FE	137.40	0.1	11.10	1.33
[5]	Euler – Bernoulli beam + Strip theory (p-k method)	137.01	0.2	11.14	0.99
[7]	Intrinsic beam + strip theory	135.60	1.2	11.17	0.07
[8]	Slender beam+ modified strip theory	137.40	0.1	11.10	1.33
[9]	Thin walled beam + strip theory	137.40	0.1	11.04	1.86

For further validation the so called "heavy" version of the Goland wing has been considered [10]. In detail a structural model made up from 2D shell elements that model a box structure of spars, ribs and skin panels and 1D bar elements that model spars and ribs caps, has been implemented in Patran/Nastran, while its equivalent beam model has been realized with the aeroelastic beam approach here proposed. From the aerodynamic point of view, the heavy Goland wing structural model realized in Patran/Nastran has been completed with a DLM lifting surface with a discretization of 75 panel spanwise and 30 panels chordwise that ensure convergence of flutter results; thus, the structure/aerodynamics coupling has been provided by means of a Finite Plate Spline FPS. From the analysis carried out in Nastran it has been noted that, at sea level air density, the heavy Goland wing also presents the bending torsion modes coupling at the flutter speed $v_{f\ Nastran}^{HG} = 119 \, m/s$, shown in Fig.2.

Figure 2 – Heavy Goland wing flutter mode $\omega_{f\ Nastran}^{HG} = 2.6 \, Hz$

The aeroelastic analysis of the heavy Goland wing model has been then carried out with the proposed aeroelastic beam approach considering a discretization of ten beam finite elements and the results obtained are shown in Fig. 3 where it can be seen that the predicter flutter boundary is $v_f^{HG} = 129 \, m/s$ and the flutter frequency is $\omega_f^{HG} = 3 \, Hz$. Due to strip theory assumptions the

Aerospace Science and Engineering - III Aerospace PhD-Days   Materials Research Forum LLC
Materials Research Proceedings 33 (2023) 213-318   https://doi.org/10.21741/9781644902677-31

predicted flutter speed slightly overestimates the one computed with the more accurate DLM model, but it is worth to be said that the aeroelastic beam approach has allowed to significantly reduce the problem degrees of freedom with respect to the Nastran model. In fact, the Nastran model has 4 896 DoF while its equivalent stick model presents 40 DoF only.

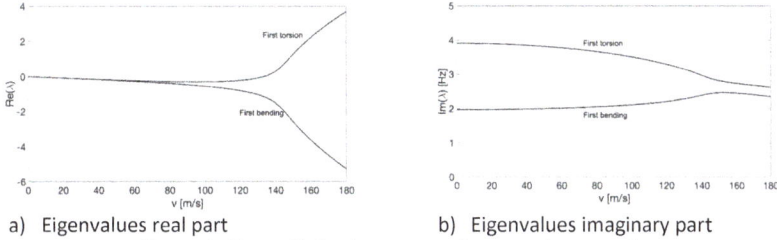

a) Eigenvalues real part   b) Eigenvalues imaginary part

Figure 3. Heavy Goland wing aeroelastic analysis results

Last, the validation for a wing-aileron configuration has been carried out considering the Goland wing equipped with an aileron-type control surface extending from the 60% of the span to the wing tip. The wing-aileron stick model has been implemented considering four aeroelastic beam elements for the clean wing portion, where the flap degree of freedom has been suppressed, and six elements for the wing portion equipped with the aileron. Flutter results, in terms of eigenvalues real part and frequencies, are reported in Fig. 4 where it can be seen that, in accordance with literature results [5], the instability arises for the first wing torsion mode at the flutter speed of $v_f = 109.5\ m/s$ and frequency of $\omega_f = 10.3\ Hz$.

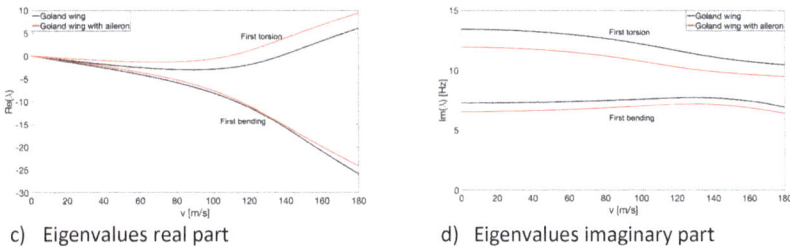

c) Eigenvalues real part   d) Eigenvalues imaginary part

Figure 4. Goland wing aeroelastic analysis results

Conclusions

An alternative aeroelastic beam approach for time domain aeroelastic analysis of wing structures has been presented in this work. The flexural-torsional equivalent beam model governing equations of a wing equipped with an aileron-like control surface have been used to compute the finite element matrices by means of a weak formulation approach. Then, the obtained elemental matrices have been implemented in a beam finite element code to realize the wing numerical model and to carry out the stability analysis. Validation analyses have been carried out for clean wing configurations and for a wing with a trailing edge control surface; a comparison of the results obtained using the aeroelastic beam approach with literature and commercial code results has been provided.

References

[1] Theodorsen, T. (1949). General theory of aerodynamic instability and the mechanism of flutter (No. NACA-TR-496).

[2] Albano, E., & Rodden, W. P. (1969). A doublet-lattice method for calculating lift distributions on oscillating surfaces in subsonic flows. AIAA journal, 7(2), 279-285. https://doi.org/10.2514/3.5086

[3] Vindigni, C. R., & Orlando, C. (2022). Simple adaptive v-stack piezoelectric based airfoil flutter suppression system. Journal of Vibration and Control, 10775463221085854. https://doi.org/10.1177/10775463221085854

[4] Vindigni, C. R., Mantegna, G., Esposito, A., Orlando, C., & Alaimo, A. (2022). An aeroelastic beam finite element for time domain preliminary aeroelastic analysis. Mechanics of Advanced Materials and Structures, 1-9. https://doi.org/10.1080/15376494.2022.2124333

[5] Mozaffari-Jovin, S., Firouz-Abadi, R. D., & Roshanian, J. (2015). Flutter of wings involving a locally distributed flexible control surface. Journal of Sound and Vibration, 357, 377-408. https://doi.org/10.1016/j.jsv.2015.03.044

[6] Goland, M., & Luke, Y. L. (1948). The flutter of a uniform wing with tip weights. J. Appl. Mech. Mar 1948, 15(1),pp. 13-20. https://doi.org/10.1115/1.4009753

[7] Patil, M. J., Hodges, D. H., & Cesnik, C. E. (2000). Nonlinear aeroelastic analysis of complete aircraft in subsonic flow. *Journal of Aircraft*, *37*(5), 753-760. https://doi.org/10.2514/2.2685

[8] Berci, M., & Cavallaro, R. (2018). A hybrid reduced-order model for the aeroelastic analysis of flexible subsonic wings—A parametric assessment. *Aerospace*, *5*(3), 76. https://doi.org/10.3390/aerospace5030076

[9] Z. Qin and L. Librescu, Dynamic aeroelastic response of aircraft wings modeled as anisotropic thin-walled beams, Journal of Aircraft 40 (2003), pp. 532–543. https://doi.org/10.2514/2.3127

[10] Beran, P. S., Khot, N. S., Eastep, F. E., Snyder, R. D., & Zweber, J. V. (2004). Numerical analysis of store-induced limit-cycle oscillation. *Journal of Aircraft*, *41*(6), 1315-1326. https://doi.org/10.2514/1.404

Aerospace Science and Engineering - III Aerospace PhD-Days   Materials Research Forum LLC
Materials Research Proceedings 33 (2023) 219-225   https://doi.org/10.21741/9781644902677-32

Development of accurate fluid-structure interaction models for aerospace problems

Andrea Rubino[1,a] *, Marco Donato de Tullio[1], Dario De Marinis[1], Maria Cinefra[1]

[1]Polythecnic of Bari, Italy

[a]andrea.rubino@poliba.it

Keywords: Fluid-Structure Interaction, Finite Element Method, Carrera Unified Formulation, Adaptive Mesh

Abstract. The research program deals with fluid-structure interaction (FSI), a challenging field of engineering, with a practical application on aerospace problems such as the flutter, an instability problem due to aeroelastic excitations. The research goal is to find a suitable way to deal with flutter in such a way the two macro fields, fluid and structure, are modelled with mid to high-level accuracy. Moreover, the creation of an interface could be useful to study other problems, such as the cabin comfort for an aircraft. To do that, the research activity is firstly split in two to study in depth the structural and the fluid problem, then they are merged together through the interface analysis: the main problem is that each field has an own scale and the union of both requests a suitable modelling and analysis. After a preliminary study on the state of the art of fluid-structure interaction in literature, which forms the pillar of the research project, the structural field is analysed first. The study of the structure system is carried out on the Carrera Unified Formulation (CUF), which allows a reduced degrees of freedom model with the same accuracy of the classical Finite Element Method (FEM). Analysis on possible adaptive mesh methods is needed in order to match the proper scale at interface with fluid dynamics system. This work could be a milestone for future investigation in problems that need a mesh refinement, beyond the aeroelastic field. Then, the fluid system is studied and analysed through the Navier-Stokes equations. In particular, a Dual Time Stepping model for non-stationary Favre Average Navier-Stokes is used. More in general, considering different order of magnitude for the Reynold's number, three different analysis could be done: Reynolds Average Navier-Stokes (RANS) for only large-scale eddies resolved and other components modelled, Large Eddy Simulation (LES) which adds the resolution of the flux of energy with respect to the previous simulation and Direct Numerical Simulation (DNS) which resolves also the dissipating eddies, so representing the best performing simulation but with very high computational cost. Finally, a complete simulation of fluid-structure interaction is performed to find the flutter velocity and study the induced vibrations from turbulent boundary layer to the aircraft cabin and/or to the rocket nose.

State of the art

The state of the art for an accurate fluid-structure interaction model is the following: considering commercial codes and analysing the literature, there is a large number of models where the fluid part is accurate and studied in depth and the structural part is dealt in a simpler manner, or there is a considerable number of models in which the situation is reversed, so the structural part is deepened and the fluid part is approximated with less accuracy. The problem is not simple to deal with, because different scales are involved and combining them is a challenging process, as ell indicated in Fig. 1.

Aerospace Science and Engineering - III Aerospace PhD-Days   Materials Research Forum LLC
Materials Research Proceedings 33 (2023) 219-225   https://doi.org/10.21741/9781644902677-32

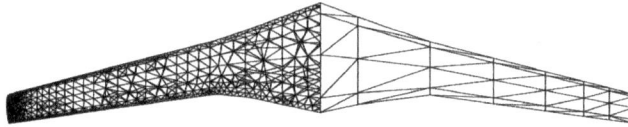

Fig. 1 – Fluid vs. Structural modelling. Credits: CCTech®

In particular, several cases are present in the literature, distinguishing between the models used: linear and non-linear. Among the former, we mention the Classical Aerodynamic Theory (CAT), the Classical Hydrodynamic Stability Theory (CHST) [1], the Parallel Shear Flow with Dynamic Inviscid Perturbation (PSF with DIP) [2] and the Time-Linearized General Analysis (TLGA). Regarding non-linear models, the Harmonic Equilibrium Method (HEM) [3] and the System Identification Method (SIM) [4,5] are mentioned. These are added to the classic methods of studying the modes of vibration of the structure in the presence of external forces that simulate the effect of aerodynamic forces. What emerges from this first research is that there is a trend to widely use Reduced Order Models (ROMs) [6] to create a compromise between the number of degrees of freedom required by the fluid-structure interaction model and the available computational resources.

Methods of analysis

To develop an accurate FSI model, the idea is to use the Carrera Unified Formulation for the structural part and the Dual Time Stepping (DTS) model for non-stationary Favre Average Navier-Stokes for the fluid part.

Structural modelling - CUF

The choice on the CUF for the structural part is driven by several advantages: in fact, this formulation allows to describe the kinematic filed in unified manner, it allows to derive the governing equations in compact way and it gives accurate results with a low number of Degrees Of Freedom (DOFs). Let us consider a generic beam structure whose longitudinal axis, with respect to a Cartesian coordinate system, lays on the coordinate y, being its cross-section defined in the xz-plane. The displacement field of one-dimensional models in CUF framework is described as a generic expansion of the generalized displacements (in the case of displacement-based theories) by arbitrary functions of the cross-section coordinates:

$$\mathbf{u}(x, y, z) = F_\tau(x, z)\mathbf{u}_\tau(y) \qquad \tau = 1, \dots, M \tag{1}$$

where $u = \{u_x, u_y, u_z\}$ is the vector of 3D displacements and $u_\tau = \{u_{x\tau}, u_{y\tau}, u_{z\tau}\}$ is the vector of general displacements, M is the number of terms in the expansion, τ denotes summation and the functions $F_\tau(x, z)$ define the 1D model to be used. In the framework of plate theories, by considering the mid-plane of the plate laying in the xy-plane, CUF can be formulated in an analogous manner:

$$\mathbf{u}(x, y, z) = F_\tau(z)\mathbf{u}_\tau(x, y) \qquad \tau = 1, \dots, M \tag{2}$$

In the equation above, the generalized displacements are function of the mid-plane coordinates of the plate and the expansion is conducted in the thickness direction z. The main advantage of CUF is that it allows to write the governing equations and the related finite element arrays in a

220

Aerospace Science and Engineering - III Aerospace PhD-Days
Materials Research Proceedings 33 (2023) 219-225

Materials Research Forum LLC
https://doi.org/10.21741/9781644902677-32

compact and unified manner, which is formally an invariant with respect to the F_τ functions [7,8].
Let the 3D displacement vector be defined as:

$$\mathbf{u}(x, y, z) = \begin{cases} u_x(x, y, z) \\ u_y(x, y, z). \\ u_z(x, y, z) \end{cases} \tag{3}$$

According to classical elasticity, stress and strain tensors can be organized in six-term vectors with no lack of generality. They read, respectively:

$$\boldsymbol{\sigma}^T = \{\sigma_{yy} \ \sigma_{xx} \ \sigma_{zz} \ \sigma_{xz} \ \sigma_{yz} \ \sigma_{xy}\} \tag{4}$$

$$\boldsymbol{\varepsilon}^T = \{\varepsilon_{yy} \ \varepsilon_{xx} \ \varepsilon_{zz} \ \gamma_{xz} \ \gamma_{yz} \ \gamma_{xy}\} \tag{5}$$

Regarding to this expression, the geometrical relations between strains and displacements with the compact vectorial notation can be defined as:

$$\boldsymbol{\varepsilon} = \boldsymbol{Du} \tag{6}$$

where, in the case of small deformations and angles of rotations, D is the following linear differential operator:

$$D = \begin{bmatrix} 0 & \dfrac{\partial}{\partial y} & 0 \\ \dfrac{\partial}{\partial x} & 0 & 0 \\ 0 & 0 & \dfrac{\partial}{\partial z} \\ \dfrac{\partial}{\partial z} & 0 & \dfrac{\partial}{\partial x} \\ 0 & \dfrac{\partial}{\partial z} & \dfrac{\partial}{\partial y} \\ \dfrac{\partial}{\partial y} & \dfrac{\partial}{\partial x} & 0 \end{bmatrix} \tag{7}$$

On the other hand, or isotropic materials the relation between stresses and strains is obtained through the well-known Hooke's law:

$$\boldsymbol{\sigma} = \boldsymbol{C\varepsilon} \tag{8}$$

where **C** is the isotropic stiffness matrix:

$$C = \begin{bmatrix} C_{22} & C_{21} & C_{23} & 0 & 0 & 0 \\ C_{21} & C_{11} & C_{13} & 0 & 0 & 0 \\ C_{23} & C_{13} & C_{33} & 0 & 0 & 0 \\ 0 & 0 & 0 & C_{55} & 0 & 0 \\ 0 & 0 & 0 & 0 & C_{44} & 0 \\ 0 & 0 & 0 & 0 & 0 & C_{66} \end{bmatrix} \tag{9}$$

The coefficients of the stiffness matrix depend only on the Young's modulus, E, and the Poisson ratio, v, and they are:

$$C_{11} = C_{22} = C_{33} = \frac{(1-v)E}{(1+v)(1-2v)}$$
$$C_{21} = C_{13} = C_{23} = \frac{vE}{(1+v)(1-2v)} \qquad (10)$$
$$C_{44} = C_{55} = C_{66} = \frac{E}{2(1+v)}$$

In the case of 1D models, the discretization along the longitudinal axis of the beam is made by means of the finite element method. The generalized displacements are in this way described as functions of the unknown nodal vector, $\mathbf{q}_{\tau i}$, and the 1D shape functions, $N_i(y)$:

$$\mathbf{u}_\tau(y) = N_i(y)\mathbf{q}_{\tau i} \qquad i = 1, \dots, n_{elem} \qquad (11)$$

where n_{elem} is the number of nodes per element and the unknown nodal vector is defined as:

$$\mathbf{q}_{\tau i} = \left\{ q_{ux_{\tau i}} \quad q_{uy_{\tau i}} \quad q_{uz_{\tau i}} \right\}^T \qquad (12)$$

Similarly, the FEM discretization of generalized displacements on the mid-surface of the plate can be written as follows:

$$\mathbf{u}_\tau(x,y) = N_i(x,y)\mathbf{q}_{\tau i} \qquad i = 1, \dots, n_{elem} \qquad (13)$$

where 2D shape functions $N_i(x,y)$ are employed. Different sets of polynomials can be used to define FEM elements. Lagrange interpolating polynomials have been chosen in this work to generate both one-dimensional and two-dimensional elements. For the sake of brevity, their expression is not provided, but it can be found in the book by Carrera et. al [7], in which two-nodes (B2), three-nodes (B3) and four-nodes (B4) beam elements and four-nodes (Q4), nine-nodes (Q9) and sixteen-nodes (Q16) plate elements are described. By combining the FEM approximation with the kinematic assumptions of the Carrera Unified Formulation, the 3D displacement field can be written as:

$$u = F_\tau N_i \mathbf{q}_{\tau i} \qquad i = 1, \dots, n_{elem} \qquad (14)$$

where the functions F_τ and N_i are defined according to the type of element (beam or plate). Note that in Eq. 14 the shape functions N_i and the expanding functions F_τ are independent. A novel approach [9] is introduced considering a coupling by relating the expanding functions F_τ to the shape functions N_i by means of the following formalism:

$$u = F_\tau^i N_i \mathbf{q}_{\tau i} \qquad (15)$$

The difference of Eq. 15 from Eq. 14 is the additional superscript i of N_i, which is now an index also of the function F_τ. This definition introduces the dependency of the kinematic assumptions to the FE nodes, namely the Node-Dependent Kinematics (NDK) [10].

Fluid modelling - DTS

The choice on the DTS for non-stationary Favre Average Navier-Stokes for the fluid part is driven by the following considerations. The DTS scheme is similar to the approximate-Newton method, but has an additional sink term that can be controlled to optimize convergence. In fact, the iterative method is treated as a time-marching method based on a pseudo-time, and in this formulation, the physical-time derivative becomes a sink in the pseudo-time frame. The total system can be viewed as a steady-state calculation with a sink term dependent on the physical-time step. For large

Aerospace Science and Engineering - III Aerospace PhD-Days   Materials Research Forum LLC
Materials Research Proceedings 33 (2023) 219-225   https://doi.org/10.21741/9781644902677-32

physical-time steps, the sink term is small, the problem behaves like a steady-state problem and the Courant-Friedrichs-Lewy number based on the pseudo-time step (pseudo-CFL) should be set corresponding to the steady-state optimal values. For small time steps, the problem is sink dominated and the pseudo-CFL can be increased. Unlike the physical-time step, the pseudo-time step can be set locally depending on the local flow conditions and physical-time steps. The Reynolds Averaged Navier-Stokes (RANS) equations, in terms of Favre mass-averaged quantities, using the $k - \omega$ turbulence model, in a Cartesian coordinate system can be rewritten:

$$\frac{\partial \rho}{\partial t} + \frac{\partial}{\partial x_j}\left(\rho u_j\right) = 0 \tag{16}$$

$$\frac{\partial(\rho u_i)}{\partial t} + \frac{\partial}{\partial x_j}\left(\rho u_j u_i\right) = -\frac{\partial p_t}{\partial x_j} + \frac{\partial \hat{\tau}_{ji}}{\partial x_j} \tag{17}$$

$$\frac{\partial(\rho \widetilde{H} - p_t)}{\partial t} + \frac{\partial}{\partial x_j}\left(\rho u_j \widetilde{H}\right) = \frac{\partial}{\partial x_j}\left[u_i \hat{\tau}_{ji} + (\mu + \sigma^* \mu_t)\frac{\partial k}{\partial x_j} - q_j\right] \tag{18}$$

$$\frac{\partial(\rho k)}{\partial t} + \frac{\partial}{\partial x_j}\left(\rho u_j k\right) = S_k + \frac{\partial}{\partial x_j}\left[(\mu + \sigma^* \mu_t)\frac{\partial k}{\partial x_j}\right] \tag{19}$$

$$\frac{\partial(\rho \omega)}{\partial t} + \frac{\partial}{\partial x_j}\left(\rho u_j \omega\right) = S_\omega + \frac{\partial}{\partial x_j}\left[(\mu + \sigma \mu_t)\frac{\partial \omega}{\partial x_j}\right] \tag{20}$$

where

$$\widetilde{H} = h + \frac{1}{2}(u^2 + v^2 + w^2) + \frac{5}{3}k \tag{21}$$

$$p_t = p + \frac{2}{3}\rho k = \rho RT + \frac{2}{3}\rho k = \rho\left(R + \frac{2k}{3T}\right) = \rho \tilde{R} T \tag{22}$$

$$S_k = \hat{\tau}_{ji}\frac{\partial u_i}{\partial x_j} - \beta^* \rho \omega k \tag{23}$$

$$S_\omega = \rho \alpha \frac{\omega}{k} \hat{\tau}_{ji}\frac{\partial u_i}{\partial x_j} - \beta \rho \omega^2 \tag{24}$$

and $\hat{\tau}_{ji}$ indicate the sum of the molecular and Reynolds stress tensor components:

$$\hat{\tau}_{ji} = (\mu + \mu_t)\left[\frac{\partial u_i}{\partial x_j} + \frac{\partial u_j}{\partial x_i} - \frac{2}{3}\frac{\partial u_k}{\partial x_k}\delta_{ij}\right] - \frac{2}{3}\rho k \delta_{ij} \tag{25}$$

The heat flux vector components q_j are rewritten as:

$$q_j = -\left(\frac{\mu}{Pr} + \frac{\mu_t}{Pr_t}\right)\frac{\partial h}{\partial x_j} \tag{26}$$

where Pr_t is the turbulent Prandtl number.

Innovation and significant results
In the first part of the PhD programme, the focus was on the structural modelling. In particular, the NDK approach was exploited to investigate the behaviour of simple geometries through static and dynamic analyses. These analyses are conducted using an academic code. Through the Node-Dependent Kinematic approach, a combination between mono- and bi-dimensional elements was

Aerospace Science and Engineering - III Aerospace PhD-Days   Materials Research Forum LLC
Materials Research Proceedings 33 (2023) 219-225   https://doi.org/10.21741/9781644902677-32

possible, obtaining a complete 3D field of displacements. The investigation highlighted the possibility to use *adapted* elements (see Fig. 3) to obtain *adaptive* meshing (see Fig. 2).

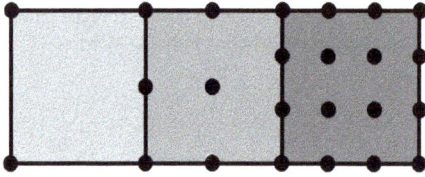

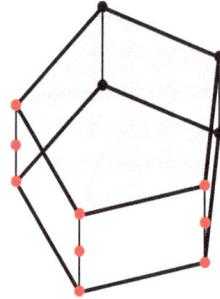

Fig. 2 – Adaptive meshing   Fig. 3 – Adapted elements

Note that in Fig. 2 is possible to connect elements with different number of nodes using a combination of 1D and 2D elements, exploiting the CUF. Moreover, in Fig. 3 is highlighted the fact that the 3D element can be *distorted*, maintaining a different discretization along the thickness. Several benchmark problems are tested using these elements, showing results in accordance with those tabulated in literature or obtained with commercial codes (Patran/Nastran®). These are important achievements considering that the fluid-structure interface have to connect a finer mesh for fluid analysis with a coarse one for structural analysis (*adaptive* mesh), also in region with complicated geometries (*adapted* and *distorted* elements) [11].

Some important results are obtained considering a square plate with a concentrated load in the middle and a Razzaque's skew plate with a set of concentrated loads at the mid-line of the top surface. In Fig. 4 is reported an example of mesh and in Fig. 5 is shown a magnification of an element between two zones with different number of nodes.

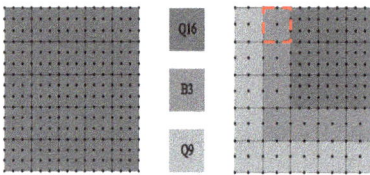

Fig. 4 – Square plate   Fig. 5 – Adapted 1D element

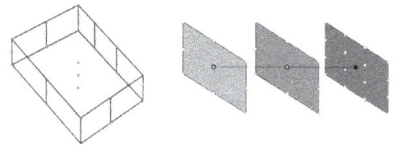

The analysis with the reduced mesh allows to save DOFs:

Table 1 – Comparison between DOFs of full and reduced mesh

Square Plate	Razzaque's Plate
DOFs Full: 11532	DOFs Full: 7500
DOFs Reduced: 6600	DOFs Reduced: 4593
Saving: 42.77%	Saving: 38.76%

Moreover, different distorted meshes are analysed and compared in terms of displacements and frequencies, as shown in Fig. 6.

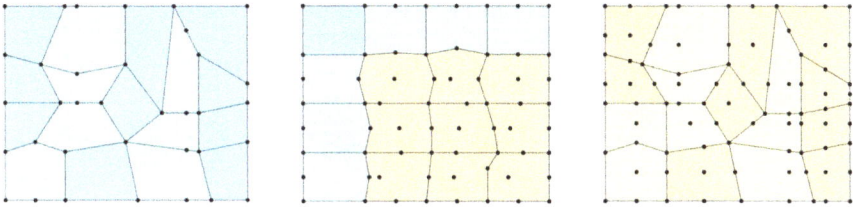

Fig. 6 – Different configurations of distorted meshes

In all the cases considered, the percentage errors are below the 1% almost always. The next steps are to implement the information exchange between the fluid part and the structural part.

Conclusion

In conclusion, the work done by far has demonstrated the potentiality of the method and represents an important milestone for the achievement of the final target: the development of accurate fluid-structure interaction models for aerospace problems.

References

[1] Lin CC. 1955. The Theory of Hydrodynamic Stability. Cambridge, UK: Cambridge, Univ. Press. 155 pp.

[2] Dowell EH. 1971. Generalized aerodynamic forces on a flexible plate undergoing transient motion in a shear flow with an application to panel flutter. AIAA. J. 9 (5):834-41 https://doi.org/10.2514/3.6283

[3] Hall KC, Thomas JP, Clark WS. 2000. Computation of unsteady nonlinear flows in cascades using a harmonic balance technique. Presented at the Int. Symp. on Unsteady Aerodyn., Aeroacoust. and Aeroelast. of Turbomach., 9th, Ecole Centrale de Lyon, Lyon, France, Sept. 4-7, 2000.

[4] Silva WA. 1993. Application of nonlinear systems theory to transonic unsteady aerodynamic responses. J. Aircraft 30(5):660-68 https://doi.org/10.2514/3.46395

[5] Silva WA. 1997. Discrete-time linear and nonlinear aerodynamic impulse responses for efficient (CFD) analyses. PhD thesis. Coll. William Mary. 159 pp.

[6] Dowell, E. H., & Hall, K. C. (2001). Modeling of fluid-structure interaction. Annual review of fluid mechanics, 33(1), 445-490. https://doi.org/10.1146/annurev.fluid.33.1.445

[7] Carrera, E., Giunta, G., Petrolo, M., 2011. Beam Structures: Classical and Advanced Theories. John Wiley and Sons, pp. 45-63. https://doi.org/10.1002/9781119978565

[8] Carrera, E., 2003. Theories and finite elements for multilayered plates and shells: a unified compact formulation with numerical assessment and benchmarking. Arch. Comput. Methods Eng. 10 (3), 216-296. https://doi.org/10.1007/BF02736224

[9] Carrera, E., Giunta, G., 2010. Refined beam theories based on Carrera's Unified Formulation. Int. J. Appl. Mech. 2 (1), 117-143. https://doi.org/10.1142/S1758825110000500

[10] Cinefra, M. (2021). Non-conventional 1d and 2d finite elements based on cuf for the analysis of non-orthogonal geometries. European Journal of Mechanics-A/Solids, 88, 104273. https://doi.org/10.1016/j.euromechsol.2021.104273

[11] Cinefra, M., & Rubino, A. (2022). Adaptive mesh using non-conventional 1D and 2D finite elements based on CUF. Mechanics of Advanced Materials and Structures, 1-11. https://doi.org/10.1080/15376494.2022.2126039

Aerospace Science and Engineering - III Aerospace PhD-Days   Materials Research Forum LLC
Materials Research Proceedings 33 (2023) 226-232   https://doi.org/10.21741/9781644902677-33

A study of characteristic element length for higher-order finite elements

Jiahui Shen[1,a] *

[1] Politecnico di Torino, Corso Duca degli Abruzzi 24, 10129 Torino, Italy

[a]jiahui.shen@polito.it

Keywords: Fracture Energy Regularization, Characteristic Element Length, Damage Analysis, Higher-Order Beam Theories, Carrera Unified Formulation

Abstract. The utilization of a fracture energy regularization technique, based on the crack band model, can effectively resolve the issue of mesh-size dependency in the finite element modelling of quasi-brittle structures. However, achieving accurate results requires proper estimation of the characteristic element length in the finite element method. This study presents practical calculation methods for the characteristic element length, particularly for higher-order finite elements based on the Carrera Unified Formulation (CUF). Additionally, a modified Mazars damage model that incorporates fracture energy regularization is employed for damage analysis in quasi-brittle materials. An experimental benchmark is adopted then for validation, and the result shows that the proposed methods ensure accurate regularization of fracture energy and provide mesh-independent structural behaviors.

Introduction

Higher-order beam finite element method based on Carrera unified formulation (CUF) [1] allows for one-dimensional analysis along the beam direction, with three-dimensional results obtained by expanding the cross-section using different polynomials such as Taylor [2] and Lagrange [3, 4]. The Lagrange expansion is a popular choice for many engineering analyses due to its ability to fit well with arbitrary cross-sections. However, when considering the strain-softening behavior of quasi-brittle materials, where stress decreases as strain increases, achieving mesh objectivity in the numerical results poses a significant challenge in finite element analyses.

The objectivity of numerical results in the framework of continuum mechanics can be restored by various regularization methods, such as integral-type or gradient-type nonlocal models [5, 6], viscous or rate-dependent methods [7], and the crack band approach [8] (or fracture energy regularization method). Among them, the crack band approach is a popular analytical tool due to its simplicity and efficiency. It can give mesh-independent finite element (FE) results by regularizing the softening curves of Gaussian points according to the crack bandwidth during calculation. Therefore, the key parameter in this method is the crack bandwidth (or characteristic element length).

The crack bandwidth is influenced by many characteristics of finite elements such as the shape, the order, the dimension, the interpolation function and scheme, and so on [9]. Over the past decades, several estimations for crack bandwidth have been proposed and can be categorized into three groups: (1) methods based on the square root of the element area or cubic root of the element volume [10]; (2) projection methods or methods proposed by Govindjee [11]; and (3) methods proposed based on Oliver [12]. However, all the above methods are applicable to two-dimensional or linear elements, which is not suitable for higher-order finite elements. Some researchers have made many modifications based on the previous three methods such as adding influencing factors to consider the element order [13, 14], which will also inspire this work. In this work, two methods will be proposed and discussed by comparing results from a numerical benchmark.

Finite elements based on CUF

This work is based on the higher-order beam theories within the framework of CUF, as described in [1]. To maintain brevity and simplicity, only the essential framework is presented here. The 1D unified formulation can be expressed as follows:

$$u(x,y,z) = F_\tau(x,z)u_\tau(y), \quad \tau = 1,2,\ldots\ldots,M \tag{1}$$

where y is set as the axial direction; (x,z) creates the cross-section plane; $F_\tau(x,z)$ varies within the cross-section; $u_\tau(y)$ is the generalized displacement vector; τ represents the summation; M stands for the number of terms in the expansion.

In this work, Lagrange-like polynomials are adopted as cross-section expanding functions F_τ, which is known as the Lagrange expansion (LE). The quadrilateral element with different orders such as four-node bilinear (L4), nine-node quadratic (L9) and sixteen-node cubic (L16) is mainly utilized. As an illustration, the interpolation function for L4 is shown as an example:

$$F_\tau = \frac{1}{4}(1 + rr_\tau)(1 + ss_\tau), \quad \tau = 1,2,3,4 \tag{2}$$

where (r,s) are natural coordinates that vary from -1 to 1 and r_τ and s_τ are the actual coordinates.

The classical beam shape functions can be adopted to approximate the generalized displacement field u_τ and Eq. 1 can be rewritten as:

$$u(x,y,z) = F_\tau(x,z)N_i(y)u_{\tau i}, \quad i = 1,2,\ldots\ldots,N_{NE} \tag{3}$$

where N_i is the shape functions of classical beam elements including two nodes (B2), three nodes (B3), and four nodes (B4) for choice; N_{NE} is the number of nodes per element; $u_{\tau i}$ is the nodal displacement vector.

Modified Mazars damage model

Mazars [15] proposed a simple isotropic damage model considering a scalar damage variable which can be written as:

$$\sigma = (1 - d)E_0\varepsilon \tag{4}$$

where σ and ε represent stress and strain, respectively; E_0 is the stiffness matrix of undamaged material; d is the damage variable.

More details on modified Mazars damage models can be found in [16]. For explaining fracture energy regularization, the tensile damage evolution law is presented here:

$$d_t = \begin{cases} 1 - \dfrac{\varepsilon_{d0}}{\kappa_t}\exp\left(\dfrac{\varepsilon_{d0}-\kappa_t}{\varepsilon_{tu}-\varepsilon_{d0}}\right) & \text{if } \kappa_t \leq \varepsilon_{tres} \\ 1 - \dfrac{p_t \times \varepsilon_{d0}}{\kappa_t} & \text{if } \kappa_t > \varepsilon_{tres} \end{cases} \tag{5}$$

where ε_{d0} is the maximum strain of elastic period; κ_t is a internal variable for loading function which is the equivalent strain in Mazars damage model; p_t is defined as the ratio between the residual tensile stress and uniaxial tensile strength which ensures the damage is infinitely close to 1.0 but not equals 1.0; ε_{tres} is the residual strain corresponding to the residual stress; ε_{tu} is the equivalent ultimate strain for bilinear softening, which is relevant to the constitutive law.

Aerospace Science and Engineering - III Aerospace PhD-Days   Materials Research Forum LLC
Materials Research Proceedings 33 (2023) 226-232   https://doi.org/10.21741/9781644902677-33

The fracture energy regularization technique is employed for controlling the slope of the softening diagram which is related to ε_{tu}. It can be realized by the following equation:

$$\frac{G_f}{l_c} = f_{ctm}(\varepsilon_{tu} - \varepsilon_{d0}) \tag{6}$$

where G_f is the fracture energy of the material; f_{ctm} is the tensile strength; l_c is the crack bandwidth or characteristic element length.

Crack bandwidths estimation

The first method developed for higher-order finite elements was proposed in [10], which is based on the cubic root of the element volume. The volume consists of one beam element and one Lagrange element, as shown in Fig. 1 for further clarification. The element volume is the product of beam element length l_e and Lagrange element area A_e, and can be divided into smaller elements based on the order of the beam and Lagrange elements. In Fig. 1, the element volume is divided into eight smaller elements, and the crack bandwidth can be estimated by taking the cubic root of the smaller volume. Therefore, the crack bandwidth can be estimated using the following equation:

$$l_{p1} = \sqrt[3]{\frac{l_e \times A_e}{\left(\sqrt{M}-1\right)^2 \times (N_{NE}-1)}} \tag{7}$$

However, this method requires that the smaller volume shown in Fig. 1 should be cubical. Therefore, the following equation should be approximately satisfied:

$$\frac{\sqrt{A_e}}{l_e} \approx \frac{\sqrt{M}-1}{N_{NE}-1} \tag{8}$$

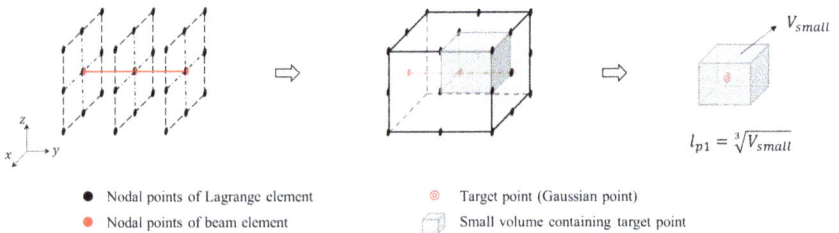

● Nodal points of Lagrange element   ⊚ Target point (Gaussian point)
● Nodal points of beam element   ▱ Small volume containing target point

Figure 1. Description of method 1

The other method is inspired by [9, 13] which reported that strain localization with softening only occurs on some Gaussian points within a single higher-order element, rather than all of them. This phenomenon is attributed to the influence of the element order. The method estimates the crack bandwidths as follows:

$$l_{p2} = l_g \times \alpha \tag{9}$$

where l_g is the estimated length from [11] and α can be considered as a correction factor for the strain localization.

For a clear explanation, Fig. 2 illustrates the calculation process after assembling a cube element and conducting a 3D Govindjee's projection. Then the Gaussian points are divided into three layers, as L9 is adopted, with 9 Gaussian points in each layer due to the use of B3. In the first layer, two-thirds of the Gaussian points undergo softening, suggesting a value of α as $13/18$. In the second layer, all Gaussian points undergo softening, so α is 1.0. In the third layer, all Gaussian points are still in elastic linear period and no damage is detected. It is worth noting that the suggested value of α is from [9, 13], which is actually from the weight ratio of softening Gaussian points to all Gaussian points on one element. But in this CUF based higher-order beam elements, α is suggested as the weight ratio of softening Gaussian points to all Gaussian points on one specific layer, illustrating α is not always the same for all Gaussian points in this higher-order beam theory.

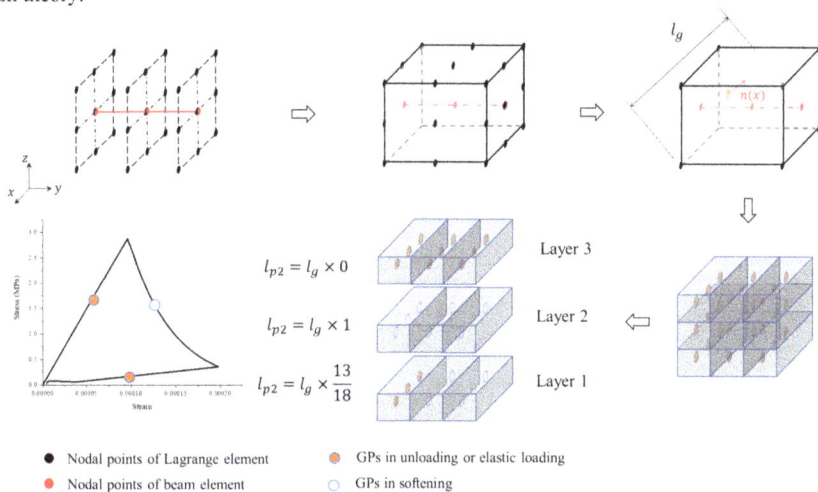

Figure 2. Description of method 2

Numerical examples

The present study utilizes the Hassanzadeh test [17], which is a benchmark direct tension test for quasi-brittle materials, to compare and validate proposed methods. Fig. 3 shows the dimension of a cube sample with four edges notched in the middle. The bottom is fixed and a tension is imposed on the top via displacement control. The material is concrete, with an elastic modulus of 36 GPa, a tensile strength of 3.5 MPa, and a Poisson's ratio of 0.2.

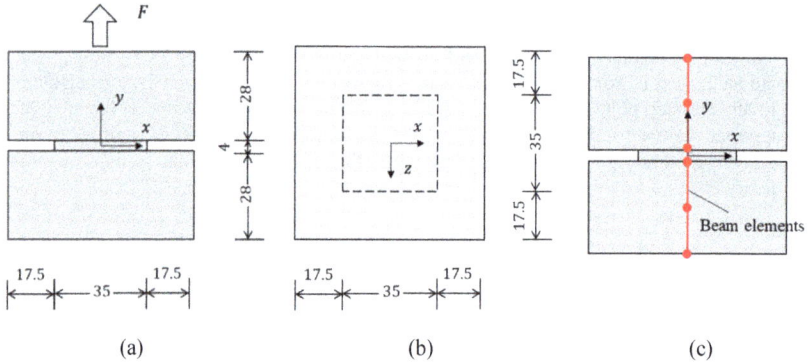

Figure 3. Information of test sample (unit: mm): (a) Front view, (b) Top view, and (c) Assignment of beam elements

Figure. 3(c) illustrates the assignment of beam elements, which are used in the present study. Various numbers and orders of beam elements are considered, as shown in Table 1. The configuration of Lagrange elements for all models will be consistent, using L9 elements with the same size.

Table 1: Model information

Model No.	Model 1	Model 2	Model 3	Model 4	Model 5	Model 6
Beam Configuration	1B2+4B2	2B2+4B2	1B3+4B3	2B3+4B3	1B4+2B4	2B4+2B4
Total DoFs	6498	6681	11193	11919	9390	10479

All load-displacement curves of different models obtained from different methods are plotted in Fig. 4. When two beam elements are adopted in the middle notch, Method 1 fails to provide results due to non-convergence. Additionally, when B4 elements are adopted, the peak load and softening part of the curve are both lower than other models, suggesting that Method 1 is unable to regularise the fracture energy because Eq. 8 is not satisfied. From Fig. 4(b), Method 2 provides great results that all models producing similar curves except for Model 6 which has a slightly lower curve. This indicates Method 2 provides a reliable estimation of crack bandwidths, successfully preserving the fracture energy.

The damage distribution of all models obtained using different methods are shown in Fig. 5. For models with Method 1, the damage distributions that limited to the notched part are reasonable. However, when B3 or B4 elements are adopted for Method 2, some issues arise, resulting in the fracture area extending beyond the middle notched part. Nevertheless, the damage intensity is not too significant.

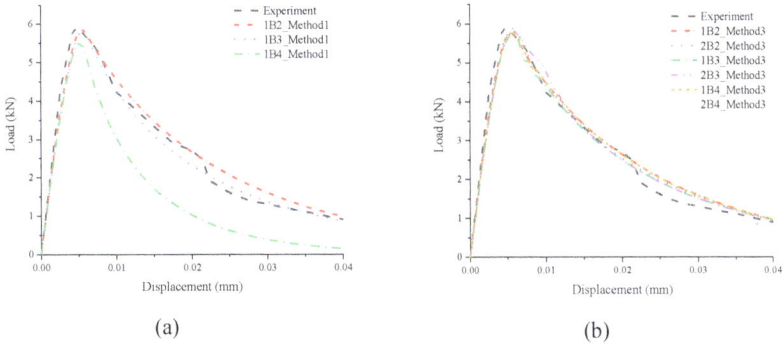

(a)   (b)

Figure 4: Load-displacement curves from: (a) Method 1 and (b) Method 2

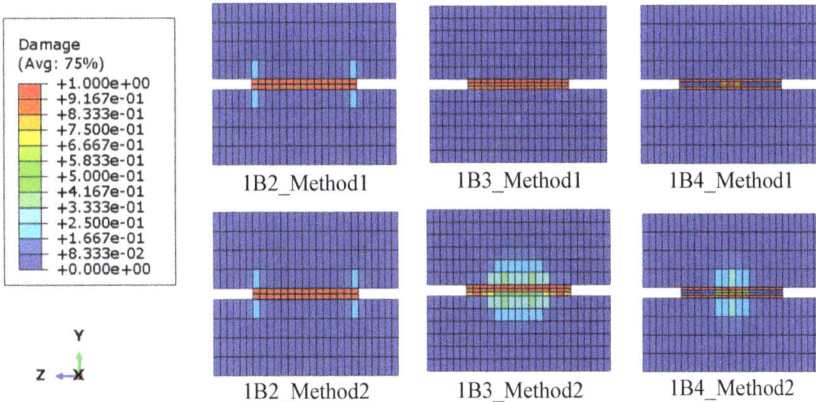

Figure 5: Damage distribution of different models from different methods

Conclusions

This work shows the damage analysis of quasi-brittle materials using CUF-based higher-order finite elements. A modified Mazars damage model with fracture energy regularization is utilized. Two methods are proposed for estimating the crack bandwidths or characteristic element length, which is a crucial parameter for fracture energy regularization in higher-order beam elements. While Method 2 yielded the best performance in the previous analysis, both methods provide objective structural behaviors that are mesh independent and can help preserve fracture energy to some extent.

Reference

[1] M Petrolo, E Carrera, M Cinefra, E Zappino. Finite element analysis of structures through unified formulation. John Wiley & Sons, 2014. https://doi.org/10.1002/9781118536643.index

[2] E Carrera, M Filippi, E Zappino. Laminated beam analysis by polynomial, trigonometric, exponential and zig-zag theories. European Journal of Mechanics-A/Solids 41 (2013): 58-69. https://doi.org/10.1016/j.euromechsol.2013.02.006

[3] A Pagani, E Carrera, R Augello, D Scano. Use of Lagrange polynomials to build refined theories for laminated beams, plates and shells. Composite Structures 276 (2021): 114505. https://doi.org/10.1016/j.compstruct.2021.114505

[4] J Shen, A Pagani, MRT Arruda, E Carrera. Exact component-wise solutions for 3D free vibration and stress analysis of hybrid steel–concrete composite beams. Thin-Walled Structures 174 (2022): 109094. https://doi.org/10.1016/j.tws.2022.109094

[5] G Pijaudier-Cabot, ZP Bažant. Nonlocal damage theory. Journal of engineering mechanics 113.10 (1987): 1512-1533. https://doi.org/10.1061/(ASCE)0733-9399(1987)113:10(1512)

[6] EC Aifantis. On the role of gradients in the localization of deformation and fracture. International Journal of Engineering Science 30.10 (1992): 1279-1299. https://doi.org/10.1016/0020-7225(92)90141-3

[7] G Duvant, JL Lions. Inequalities in mechanics and physics. Vol. 219. Springer Science & Business Media, 2012. http://dx.doi.org/10.1007/978-3-642-66165-5

[8] ZP Bažant, BH Oh. Crack band theory for fracture of concrete. Matériaux et construction 16 (1983): 155-177. https://doi.org/10.1007/BF02486267

[9] M Jirásek, M Bauer. Numerical aspects of the crack band approach. Computers & structures 110 (2012): 60-78. https://doi.org/10.1016/j.compstruc.2012.06.006

[10] J Shen, MRT Arruda, A Pagani. Concrete damage analysis based on higher-order beam theories using fracture energy regularization. Mechanics of Advanced Materials and Structures (2022): 1-15. https://doi.org/10.1080/15376494.2022.2098430

[11] S Govindjee, GJ Kay, JC Simo. Anisotropic modelling and numerical simulation of brittle damage in concrete. International journal for numerical methods in engineering 38.21 (1995): 3611-3633. https://doi.org/10.1002/nme.1620382105

[12] J Oliver. A consistent characteristic length for smeared cracking models. International Journal for Numerical Methods in Engineering 28.2 (1989): 461-474. https://doi.org/10.1002/nme.1620280214

[13] AT Slobbe, MAN Hendriks, JG Rots. Systematic assessment of directional mesh bias with periodic boundary conditions: Applied to the crack band model. Engineering Fracture Mechanics 109 (2013): 186-208. https://doi.org/10.1016/j.engfracmech.2013.06.005

[14] W He, Y Xu, Y Cheng, PF Jia, TT Fu. Tension-compression damage model with consistent crack bandwidths for concrete materials. Advances in Civil Engineering 2019 (2019). https://doi.org/10.1155/2019/2810108

[15] J Mazars. Application de la mécanique de l'endommagement au comportement non linéaire et à la rupture du béton de structure. THESE DE DOCTEUR ES SCIENCES PRESENTEE A L'UNIVERSITE PIERRE ET MARIE CURIE-PARIS 6 (1984).

[16] MRT Arruda, J Pacheco, LMS Castro, E Julio. A modified mazars damage model with energy regularization. Engineering Fracture Mechanics 259 (2022): 108129. https://doi.org/10.1016/j.engfracmech.2021.108129

[17] M Hassanzadeh. Behaviour of fracture process zones in concrete influenced by simultaneously applied normal and shear displacements. Division of Building Materials, Lund Institute of Technology, 1992.

Aerospace Science and Engineering - III Aerospace PhD-Days
Materials Research Proceedings 33 (2023) 233-238

Materials Research Forum LLC
https://doi.org/10.21741/9781644902677-34

On modelling damage in composite laminates using the Ritz method and continuum damage mechanics

Dario Campagna[1,a] *, Alberto Milazzo[1,b], Vincenzo Oliveri[2,c], Ivano Benedetti[1,d]

[1] Department of Engineering, University of Palermo, Viale delle Scienze, Edificio 8, 90128, Palermo, Italy

[2] School of Engineering and Bernal Institute, University of Limerick, V94 T9PX, Limerick, Ireland

[a,]*dario.campagna@unipa.it, [b]alberto.milazzo@unipa.it [c]vincenzo.oliveri@ul.ie, [d]ivano.benedetti@unipa.it

Keywords: Ritz Method, Continuum Damage Mechanics, Gibbs Effect, Variable Angle Tow Laminates

Abstract. In this work, a Ritz formulation for the analysis of damage initiation and evolution in composite plates under progressive loading is presented. The proposed model assumes a first order shear deformation theory and considers geometric non-linearities through the von Karman assumptions. The damage is modelled through Continuum Damage Mechanics. A set of results is presented to show the potential of the method and highlight some issues to be addressed by suitable developments of the method.

Introduction

Composite materials are widely used in several engineering fields such as automotive, naval, and aerospace, due to their superior strength-to-weight ratio and other desirable properties. To further enhance their performance, new manufacturing techniques have been developed, including automated fiber placement and additive manufacturing, which have enabled the design of composite structures with variable fiber angle placement, known as variable angle tow (VAT) laminates. These composites offer numerous advantages and have been extensively studied [1,2].

Due to the wider design space enabled by such innovative manufacturing methods, there is the need, by designers and engineers, of modeling and developing computational tools capable of predicting the structural behavior of components with a satisfactory level of accuracy, including the prediction of their damaging and failure modes. Modeling progressive damage in composite materials is challenging due to the different damage mechanisms that may arise. Damage can be modeled at different scales, from the micro- to macro-scale, depending on the idealization's scale. For example, at the micromechanical level, representative volume elements (RVEs) are used to represent damage in the form of matrix softening or fiber breaking [3]. Instead, at the macro-scale level, damage is often modeled as a hard discontinuity.

Continuum Damage Mechanics (CDM) is one of the most popular frameworks used to handle the damage process at the meso-scale level, where individual plies are represented as homogenous. CDM models are based on the works of Matzenmiller and Ladeveze [4,5] among others, which describe degradation as a progressive loss of material stiffness. Several models have been developed and used in finite element (FE) approaches within the CDM framework [6,7]. In addition to FE-based analysis methods, single domain meshless approaches, such as the Ritz method, have been shown to be competitive, especially when dealing with smeared damaged zones [8].

However, in some cases, damage tends to localize in a narrow band due to loading conditions or initial imperfections [9,10]. This phenomenon can cause the numerical model to suffer from spurious effects when reconstructing the damage state, known as the Gibbs effect, which leads

Aerospace Science and Engineering - III Aerospace PhD-Days   Materials Research Forum LLC
Materials Research Proceedings 33 (2023) 233-238   https://doi.org/10.21741/9781644902677-34

towards a non-physical response. Thus, designers and engineers must carefully consider the limitations of different modeling and computational tools when predicting the structural behavior of composite materials and components.

This work aims at developing an adaptive multi-domain Ritz method to avoid spurious numerical effects and maintain the advantages of the Ritz method in terms of reducing the degree of freedom compared to FE models, while still achieving objective physical response.

Despite the benefits granted by the Ritz method, there are only few works in the literature addressing damage initiation and evolution using the Ritz approach. Few works address the topic [11,12], where the damage models are often overly simplified and offer an on/off representation of damage, more suitable for identifying damage initiation than damage evolution, which tend to be excessively conservative. These considerations highlight the novelty of the proposed method.

Methods

In the present formulation, the kinematic is based on the First order Shear Deformation Theory with von Kármán assumptions accounting for geometric non-linearities. Following a CDM approach, the onset of the damage is predicted using the Hashin's criteria [13] and four damage indices are computed and used for the degradation of the constitutive relations at the ply level, in fiber and matrix directions, both for tension and compression. The evolution of damage is based on a linear softening law, as shown in Fig. 1, and it is assessed by means of equivalent strains.

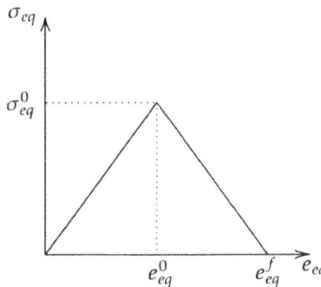

Figure 1. Linear strain softening constitutive law.

To ensure consistent energy dissipation with respect to the Ritz kinematic approximation during the failure process, a smeared approach is adopted by distributing the fracture energy over an area associated with the gauss points used in the numerical integration [14]. Referring to Fig. 1, the fracture energy dissipated per unit area can be computed as

$$G_c = \int_0^w \int_0^{e_{eq}^f} \sigma_{eq} \, de_{eq} \, dx = w \frac{1}{2} \sigma_{eq}^0 e_{eq}^f, \tag{1}$$

where w is the square root of the area associated with a generic gauss point. Using orthogonal polynomials for the approximation of the primary variables [15] and applying the principle of the minimum potential energy, the governing equation can be written as,

$$(K_0 + \bar{K}_0 + K_1 + K_2 + \bar{K}_1 + R)X = F_D + F_L \tag{2}$$

where, X is the vector collecting the unknown Ritz coefficients, $K_0, K_1, K_2, \bar{K}_0, \bar{K}_1$ are the stiffness matrices of the problem, where the subscripts 1 and 2 refer to the geometric non-linear terms, and the over-bar refers to prescribed initial imperfections. Furthermore, in Eq. (2) R is the penalty

Aerospace Science and Engineering - III Aerospace PhD-Days   Materials Research Forum LLC
Materials Research Proceedings 33 (2023) 233-238   https://doi.org/10.21741/9781644902677-34

matrix related to the enforcement of the BCs thorough a penalty approach and the vectors $\boldsymbol{F}_D$ and $\boldsymbol{F}_L$ collect the discrete terms associated with the external loads. To solve the non-linear problem given in Eq.(2) the incremental form of the governing equation may be expressed as

$$R\Delta X + \Delta\left[\left(\boldsymbol{K}_t^g + \boldsymbol{K}_t^d\right)X\right] = \Delta\boldsymbol{F}_D + \Delta\boldsymbol{F}_L. \tag{3}$$

where, $\boldsymbol{K}_t^g$ and $\boldsymbol{K}_t^d$ are the tangent stiffness matrix contributions related to geometric non-linearities and initial imperfections as well as the damage evolution, respectively. The developed semi-analytical model has been implemented in an efficient analysis tool, where Eq.(2) is solved through the Newton-Raphson numerical scheme.
To a better understanding of the matrices appearing in Eqs. (2) and (3) the reader is referred to Ref. [8].
Some remarks on the adaptive multi-domain approach are worthwhile [16]. The onset of the damage is monitored at specific sampling points corresponding to the domain integration points. When damage initiation is triggered, the adaptive multi-domain is activated to refine the discretization in the vicinity of the damaged zone. This refinement allows to represent the global response of the structure, in terms of strains, as a piecewise continuous function removing the Gibbs effect arising in single domain discretization. Some tests regarding the suppression of the Gibbs effect are presented in the next section.

Results
To assess the capability offered by the developed analysis tool, some preliminary results are herein reported and discussed.
Fig. 2 shows the obtained post-buckling results of different simply supported VAT composite plates subjected to compressive in-plane loading, using a single domain approach. The applied load and out-of-plane displacement are normalized with respect to the critical buckling load of a quasi-isotropic layup and the plate thickness, respectively. These results show that, for the layups considered, the laminate achieving the greatest buckling load is also the one that fails at the lowest out-of-plane displacement. This shows an example on how the developed tool may be used to efficiently investigate damage characteristics of VAT laminates and find trade-off design solutions.

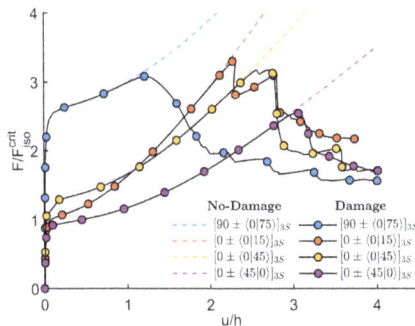

Figure 2. Post-buckling result for different VAT laminates under compression loading.

After considering a test case where the damage is distributed over a well-defined area, some tests of damage localization are reported, to assess the occurrence of the Gibbs phenomenon. To illustrate the problem, a rectangular unidirectional composite plate subjected to uniaxial tension is considered, Fig. 3. Moreover, a narrow band of material along the x_2 axis has strength lower than the rest of the bar, in order to artificially induce the onset of the damage.

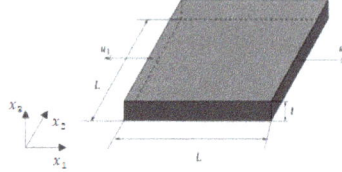

Figure 3. Schematic representation of rectangular plate under uniaxial tension.

Fig. 4 shows the contour plot of the damage state in fiber direction of the plate obtained with four different polynomial expansions, where the emergence of the Gibbs phenomenon is clearly highlighted.

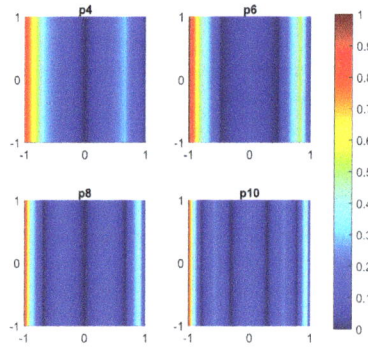

Figure 4. Gibbs effect on damage contour plots of the plate under uniaxial tension.

To address this issue, a possible approach involves discretizing the entire domain using a finer multi-patch representation, which is employed in this study, after having considered also alternative ideas, e.g. the use of filters to smooth out the Gibbs ripples.

Figs. (5a–c) show three different multi-domain discretization of the plate subjected to uniaxial tension. The generalized displacement of the patches in the narrow band, where the damage spread, are approximated using first order polynomials, whilst in the bigger patch a higher number of polynomial degree is maintained to avoid losing accuracy. The corresponding damage plots are reported in Figs. (5d–f), where it is shown how the Gibbs effect is completely removed. Moreover, Fig. 5(g) shows the results in term of force-vs-displacements, which confirms the independence from the level of discretization used, i.e. the objectivity of the response.

Finally, the solution of the present method is compared with FE results obtained with the ABAQUS built-in CDM model, Fig. (5h). The results comparison shows excellent agreement with established FE analysis methods, with the great advantage of a noticeable reduction of computational effort in terms of degrees of freedom.

Conclusions

In this work, a Ritz approach for the analysis of progressive failure of composite plates has been developed.

Aerospace Science and Engineering - III Aerospace PhD-Days   Materials Research Forum LLC
Materials Research Proceedings 33 (2023) 233-238   https://doi.org/10.21741/9781644902677-34

The present study adopts the first order shear deformation theory with non-linear von Karman strains assumptions for representing geometrically non-linear deformations and Continuum Damage Mechanics framework for capturing the initiation and evolution of damage. The method, its capabilities and limits have been presented and discussed. A multi-domain approach has been seen as a good candidate for removing the Gibbs effect that arises when the damage tends to localize in narrows band in single domain discretization.

The developed analysis tool may be useful for identifying the operational limits and providing valuable insights to the designer for the analysis of VAT layups that, with standard FE analysis, require a high number of elements to properly describe the fiber angle variation of a VAT lamina.

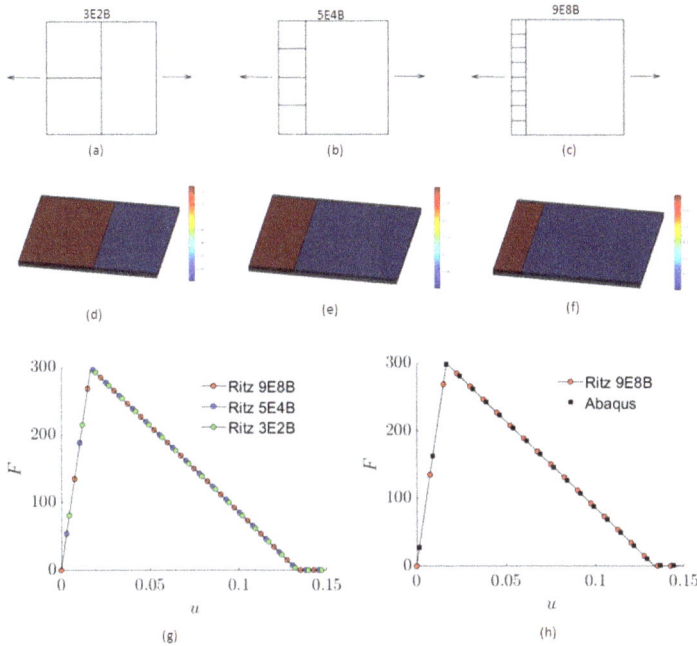

Figure 5. (a)-(c) Multi-patch discretization used. (d)-(f) damage plot of each discretization. (g) Comparison of result for different discretizations. (h) Comparison of results with Abaqus.

References

[1] Z. Gürdal and R. Olmedo, In-plane response of laminates with spatially varying fiber orientations - Variable stiffness concept, AIAA J. 31, 4 (1993) 751–758. https://doi.org/10.2514/3.11613

[2] Z. Gürdal, B. Tatting, and C. Wu, Variable stiffness composite panels: Effects of stiffness variation on the in-plane and buckling response, Compos. A: Appl. Sci. Manuf 39, 5 (2008) 911–922. https://doi.org/10.1016/j.compositesa.2007.11.015

Aerospace Science and Engineering - III Aerospace PhD-Days   Materials Research Forum LLC
Materials Research Proceedings 33 (2023) 233-238   https://doi.org/10.21741/9781644902677-34

[3] M. Lo Cascio and I. Benedetti, Coupling BEM and VEM for the analysis of composite materials with damage, J. Multiscale Modell. 13, 01 (2022) 2144001. https://doi.org/10.1142/S1756973721440017

[4] A. Matzenmiller, J. Lubliner, and R. Taylor, A constitutive model for anisotropic damage in fiber-composites, Mech. Mater. 20, 2 (1995) 125–152. https://doi.org/10.1016/0167-6636(94)00053-0

[5] P. Ladeveze and E. LeDantec, Damage modelling of the elementary ply for laminated composites, Compos. Sci. Technol., 43, 3 (1992) 257–267. https://doi.org/10.1016/0266-3538(92)90097-M

[6] P. Maimí, P. Camanho, J. Mayugo, and C. Dávila, A continuum damage model for composite laminates: Part I – constitutive model, Mech. Mater. 39, 10 (2007) 897–908. https://doi.org/10.1016/j.mechmat.2007.03.005

[7] P. Maimí, P. Camanho, J. Mayugo, and C. Dávila, A continuum damage model for composite laminates: Part II – computational implementation and validation, Mech. Mater. 39, 10 (2007) 909–919. https://doi.org/10.1016/j.mechmat.2007.03.006

[8] D. Campagna, A. Milazzo, I. Benedetti, and V. Oliveri, A non-linear Ritz method for progressive failure analysis of variable angle tow composite laminates, Mech. Adv. Mater. Struc. (2022). https://doi.org/10.1080/15376494.2022.2134951

[9] I. Benedetti, H. Nguyen, R. A. Soler-Crespo, W. Gao, L. Mao, A. Ghasemi, ... and H. D. Espinosa. Formulation and validation of a reduced order model of 2D materials exhibiting a two-phase microstructure as applied to graphene oxide. J Mech Phys Solids 112 (2018) 66-88. https://doi.org/10.1016/j.jmps.2017.11.012

[10] M. Jirásek and D. Rodrigue, Localization analysis of nonlocal models with damage-dependent nonlocal interaction. Int J Solids Struct 174 (2019) 1-17. https://doi.org/10.1016/j.ijsolstr.2015.06.001

[11] Q. J. Yang and B. Hayman, Simplified ultimate strength analysis of compressed composite plates with linear material degradation, Compos. B: Eng. 69 (2015) 13–21 https://doi.org/10.1016/j.compositesb.2014.09.016

[12] Q. J. Yang and B. Hayman, Prediction of post-buckling and ultimate compressive strength of composite plates by semi-analytical methods, Eng. Struct. 84 (2015) 42–53. https://doi.org/10.1016/j.engstruct.2014.11.013

[13] Z. Hashin and A. Rotem, A fatigue failure criterion for fiber reinforced materials, J. Compos. Mater. 7, 4 (1973) 448–464. https://doi.org/10.1177/002199837300700404

[14] M.H. Nagaraj, J. Reiner, R. Vaziri, E. Carrera, and M. Petrolo, Progressive damage analysis of composite structures using higher-order layer-wise elements, Compos Part B-Eng 190 (2020) 107921. https://doi.org/10.1016/j.compositesb.2020.107921

[15] A. Milazzo and V. Oliveri, Buckling and Postbuckling of Stiffened Composite Panels with Cracks and Delaminations by Ritz Approach, AIAA J. 53, 3 (2017) 965-980. https://doi.org/10.1016/j.ijsolstr.2018.10.002

[16] V. Oliveri and A. Milazzo, A Rayleigh-Ritz approach for postbuckling analysis of variable angle tow composite stiffened panels, Comput. Struct. 196 (2018) 263-276. https://doi.org/10.1016/j.compstruc.2017.10.009

Aerospace Science and Engineering - III Aerospace PhD-Days   Materials Research Forum LLC
Materials Research Proceedings 33 (2023) 239-246   https://doi.org/10.21741/9781644902677-35

Development of a flat-sat software for deep-space autonomous GNC hardware-in-the-loop testing

Davide Perico[1,a] *

[1]Department of Aerospace Science and Technology, Politecnico di Milano, Via La Masa 34, 20156, Milano, Italy

[a]davide.perico@polimi.it

Keywords: Autonomous Interplanetary CubeSats, Hardware-In-the-Loop, Flat-Sat On-Board Computer, Software Development Workflow

Abstract. This paper presents the development of software to integrate autonomous Guidance, Navigation, and Control algorithms and perform Hardware-In-the-Loop testing in the EXTREMA Simulation Hub facility to leverage the technology for self-driving deep-space CubeSats. Firstly, the design drivers are derived, and a multi-layered modular architecture for real-time execution is justified accordingly. Moreover, the combination of software and target computing hardware is identified by presenting the adoption of a board that satisfactorily represents the limited computational power typical of CubeSats and drawing the path towards reconfigurable computing.

Introduction

At the present time, the space sector is thriving, and in the next future an important growth in the number of space missions is forecasted [1]. CubeSats participate as one of the major actors in this trend, thanks to their concept that inferences cost-effective and relatively fast access to space. Moreover, the combination of these features with the recent advancements in on-board technology have pushed the adoption of CubeSats also in deep-space missions [2].

However, the classical paradigm for which missions strongly rely on ground-based operations could halter this course. The reliability that ground support guarantees comes at the price of high costs that usually characterize a big portion of the total budget required to design, launch, and operate a mission. Moreover, the high distances characterizing interplanetary scenarios and a forecasted growth of missions data rates ask for large ground antennas, but nowadays the NASA Deep Space Network (DSN) and the ESA Estrack stations are the only ones able to provide performant enough equipment to satisfy these requirements [3]. The result could be a saturation of the existing stations. Finally, another issue is represented by the complexity of robotic operations required for the probe: as that is increasing, the need for real-time commanding becomes paramount.

EXTREMA (Engineering Extremely Rare Events in Astrodynamics for Deep-Space Missions in Autonomy) challenges the current paradigm by enabling self-driving interplanetary spacecraft. Deep-space Guidance, Navigation, and Control (GNC) applied in a complex scenario is the core subject of EXTREMA. EXTREMA finds its foundations on three pillars. Pillar 1 regards autonomous navigation. Pillar 2 concerns autonomous guidance and control. Pillar 3 deals with autonomous ballistic capture. The project has been awarded a European Research Council (ERC) Consolidator Grant in 2019.

The product of each Pillar targets the integration in the EXTREMA Simulation Hub (ESH), a Hardware-In-the-Loop (HIL) facility that enables testing to verify and validate autonomous Guidance, Navigation and Control (GNC) algorithms and architectures [4].

The HIL simulations framework is composed of three main elements: Realistic Experimental faciliTy for vision-based NAvigation (RETINA) [5], EXTREMA THruster In the Loop

Aerospace Science and Engineering - III Aerospace PhD-Days   Materials Research Forum LLC
Materials Research Proceedings 33 (2023) 239-246   https://doi.org/10.21741/9781644902677-35

Experiment (ETHILE) [6], and SpacecrafT Attitude SImulation System (STASIS) [7]. Finally, SPESI oversees the process since it simulates the space dynamics and environment [8].

Moreover, all those systems interface with the EXTREMA flat-sat that reproduces the autonomous spacecraft functionalities. The flat-sat On-Board Computer (OBC) is its computing core and dwells in the STASIS air-bearing platform, which simulates the probe's attitude dynamics. The whole simulation framework is reported in Figure 1.

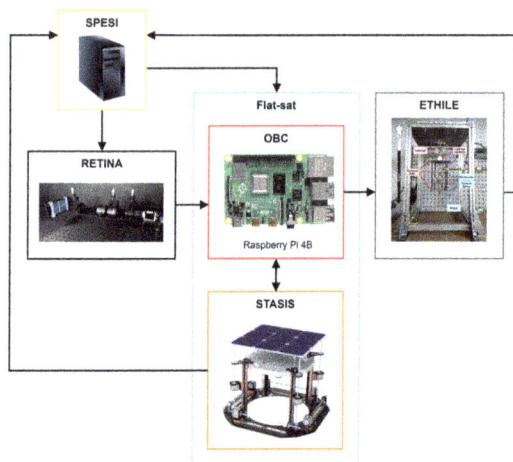

Figure 1: ESH functional breakdown structure including the flat-sat [4,5,6].

The research work focuses on the development of the OBC of the EXTREMA ESH flat-sat to enable closed-loop HIL testing for autonomous GNC algorithms. Therefore, the following research question is shaped:

Which combination of embedded software and hardware architecture is effective in managing the execution of autonomous guidance, navigation, and control algorithms in a Hardware-In-the-Loop facility?

Methodology

The approach followed to develop the flat-sat OBC software is the use of C++ as the main programming language because it offers a deeper level of optimization, and portability to different processor architectures, and it is object-oriented, which makes the code portable, reusable, and flexible.

On the other hand, a branched line of development of the software with Rust is foreseen because even if the latter language is still quite limited in libraries due to its relatively new introduction, it has interesting features for embedded programming such as direct control over running time and memory usage [9].

Furthermore, Figure 2 illustrates the workflow followed to develop the flat-sat OBC, highlighting the verification and validation sides.

The process started with the reception of the requirements from the EXTREMA Pillars and the simulation environment to translate them into the architecture design. Three main requirements were identified considering the simulation objectives of the ESH: software modularity, event-driven state change, and real-time task execution [4].

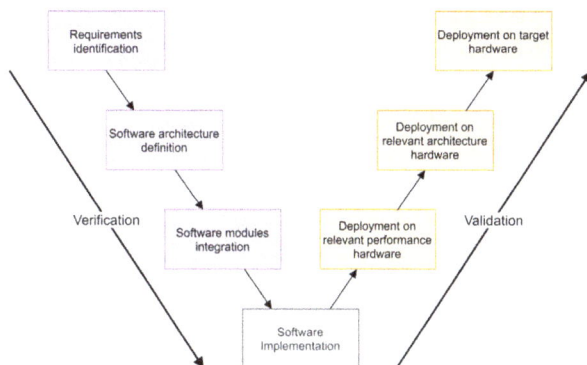

Figure 2: Flat-sat OBC development workflow.

Modularity. It guarantees an agile development of the different modules and an eased integration into the OBC. As the Pillar algorithms are designed using MATLAB and undergo continuous modifications and improvements, the MATLAB Coder is used to generate the C/C++ libraries required for their integration into the core software. The latter is based on wrappers and linking interfaces.

Event-Driven. The autonomy of the GNC algorithms has to be coordinated at the system-level. For instance, ESA definitions of execution autonomy are considered. The current trend is to target levels E3 and E4 which foresee event-based autonomous operations, execution of on-board operations control procedures, and goal-oriented mission re-planning[1]. Considering this framework, processes execution based on events moves towards these directions.

Real-Time. The simulation must happen in real-time as the software Pillar units are characterized by hard time constraints [7], therefore the solution identified was the adoption of a Real-Time Operating System (RTOS) as the operative system on top of which the flight hardware runs [10].

The second step aimed to translate the requirements into the software architecture. From literature, different architectures have been evaluated, such as MEXEC [11], FRESCO [12], Basilisk [13], and SAVOIR Onboard Software Reference Architecture (OSRA)[2].

The comparison is synthesized into the architecture reported in Figure 3, which conceives three layers: Middleware, Application and Hardware Driver.

---

[1] ECSS-E-ST-70-11C Space Segment Operability
[2] OSRA Onboard Software Reference Architecture

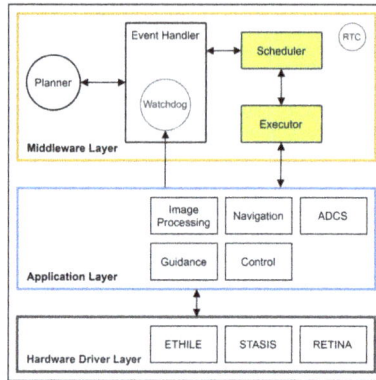

Figure 3: Flat-sat OBC software architecture.

Middleware layer. It is an abstraction layer that hosts the Planner, generating what procedures the spacecraft shall follow and holding its state. This is again beneficial for spacecraft autonomy and can be based on plans generated by a decision-making process or time-scheduled [14]. Consequentially, the plan is sent to the Event Handler, which processes the events associated with it and matches them with the corresponding tasks. Also, the Modules in the Application Layer can generate events, and the same logic is applied. The Event Handler is also responsible for generating a state transition condition to be sent to the planner. Furthermore, the scheduler receives the tasks to be executed from the Event Handler and adds them to a priority-dependent row of processes. A demonstrative scheme of a sequence of actions performed in the middleware layer is reported in Figure 4 where the change of state in the Planner is triggered by the event handled.

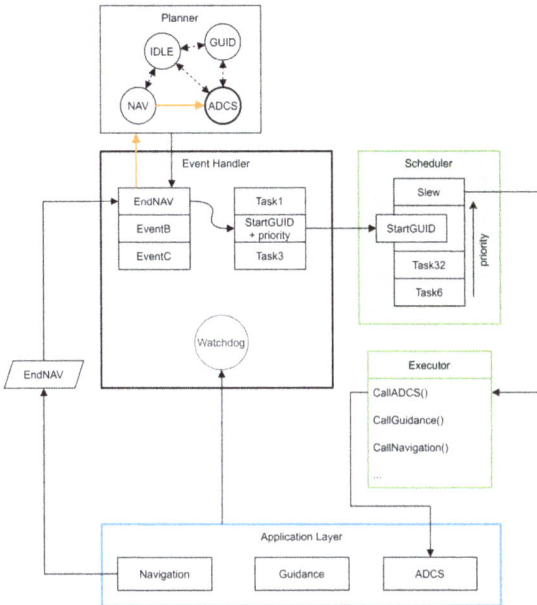

Figure 4: Demonstrative example of the functioning of the Middleware Layer.

Application layer. The application layer collects all the modules developed within the EXTREMA Pillars, as well as the ADCS algorithms. Moreover, specific manually coded wrappers and binding functions are also included in this layer.

Hardware driver layer. This last layer hosts the drivers/interfaces to communicate with the ESH hardware units to send commands and receive data such as hardware state or physical measurements if available from the sensing units.

Once the software core is prototyped, the following steps focus on the development of the remaining software modules and the deployment of the target hardware.

The process moved on to the coding of the required atomic modules of the application layer, together with the agents of the middleware layer. Object-oriented programming is widely used to define the agents and their methods. The execution policy grows as the integration of the functions goes on. The autocoded GNC modules undergo software unit tests to ensure the proper functioning and then communicate with the software middleware through wrappers that allow to eventually convert Coder's data types into standard C++ ones.

At this point the software is deployed on a target embedded hardware that has limited resources and is representative of a real on-board computer. For this purpose, a Raspberry Pi 4 Model B aided with an external Real-Time Clock (RTC) is used because it offers a relatively low-cost and a widely supported hardware. The clock is synchronized at the beginning of each simulation by communicating with SPESI such that initial time synchronization between the environment model and the spacecraft is achieved. The deployment of codes to the board is achieved in two principal ways, targeting the Linux-based operating system installed: direct compilation on the board connected through an SSH tunnel to the host computer and cross-compilation with a toolchain.

Aerospace Science and Engineering - III Aerospace PhD-Days   Materials Research Forum LLC
Materials Research Proceedings 33 (2023) 239-246   https://doi.org/10.21741/9781644902677-35

The Raspberry Pi 4B board model preliminary considered has a quad-core ARM processor clocking at 1.5GHz and 4Gb of RAM. These characteristics, as well as the power absorbed, are not typical for real CubeSat hardware which at the current state of the art has processor clock frequencies up to hundreds of MHz and can rely on low-power computing hardware [15].

Considering these limited on-board resources, having flexible and power-efficient hardware and performant software would be beneficial but hard to achieve with an independent design of the two. Hence, making the computer architecture reconfigurable, that is, rearrangeable on demand at run-time, can provide a flexible, performant, and low-power solution for autonomy because the architecture can be adapted to the algorithms that are executed [16]. An architecture of this type, promising for space applications, is System-on-a-Chip (SoC), whose reconfigurable part is often a Field Programmable Gate Array (FPGA). The different modules of the Application Layer can be optimized for execution using a different configuration of the hardware, even if paying the price of reconfigurability delay and data storage management during the process.

The current SoC under evaluation is the ZedBoard, a development board based on the Zynq-7000 architecture that is shown in Table 1, which compares different boards, some of which are used in deep-space CubeSats OBCs. The comparison highlights that the ZedBoard is a good representative of flown hardware, at relatively low cost. Moreover, the deployment of codes for reconfigurable computing relies on a High-Level-Synthesis Tool (HLS) [17].

Table 1: Relevant examples of SoC boards.

SoC	Processor	Device	Frequency	Power
Gaisler GR712RC[3]	Dual-Core LEON3-FT	Spaceteq OBC-GR712	80 MHz	5 W
Microsemi Smart Fusion-2[4]	ARM Cortex-M3	AAA Clyde Space Kryten-M3	166 MHz	2 W
Xilinx Zynq 7020[5]	Dual-Core ARM Cortex-A9	Xiphos Q7	766 MHz	2 W
Xilinx Zynq UltraScale+[6]	Quad-Core ARM Cortex-A53	Xiphos Q8	1.2 GHz	25W
Xilinx Zynq 7030[7]	Dual-Core ARM Cortex-A9	GomSpace NanoMind Z7000	800 MHz	2.3 W
Xilinx Zynq XC7Z020-1CLG484C[8]	Dual-Core ARM Cortex-A9	Avnet ZedBoard	100 MHz	N/A
Xilinx Zynq XC7Z020-CLG484-1[9]	Dual-Core ARM Cortex-A9	Xilinx ZC702 Evaluation Board	200 MHz	N/A

The final step in the design of the EXTREMA flat-sat OBC software is its deployment to a processor architecture that would normally be used on a deep-space CubeSat that would also include a reconfigurable part and more than one core. The family of processors matching these characteristics is the LEON, also adopted by ESA[10]. Specifically, a LEON3 SoC will be targeted

[3] Spaceteq OBC-GR712 datasheet
[4] AAC Coyde Space Kryten M3 datasheet
[5] Xiphos Q7 datasheet
[6] Xiphos Q8 datasheet
[7] GomSpace NnaoMind Z7000 datasheet
[8] Avnet Zedboard datasheet
[9] Xilinx ZC702 Evaluation Board datasheet
[10]ESA Onboard Computers and Data Handling Microprocessors

Aerospace Science and Engineering - III Aerospace PhD-Days   Materials Research Forum LLC
Materials Research Proceedings 33 (2023) 239-246   https://doi.org/10.21741/9781644902677-35

as it comes with debugging tools, compilation toolchains for RTEMS real-time operative system, and its LEON3-FT fault-tolerant version has already got flight heritage (e.g., on the OBC FERMI by Argotec [18]).

Conclusions

In this work, the architecture and development flow of the EXTREMA flat-sat OBC is presented. The software is used to execute extensive closed-loop Hardware-In-the-Loop simulation campaigns at the EXTREMA Simulation Hub with the objective of verifying and validate Guidance Navigation and Control algorithms for deep-space missions employing autonomous stand-alone CubeSats.

The procedure that is followed is illustrated starting from the definition of the coding language and proceeding with the explanation of the different elements composing the architecture layers. Finally, the different solutions for the target hardware are presented, as well as the novelty of the reconfigurable computing paradigm for an improvement in the implementation and execution of the on-board software.

Acknowledgments

I would like to thank my supervisor, Prof. Francesco Topputo (Politecnico di Milano), Dr. Gianmario Merisio (Politecnico di Milano), and Dr. Gianfranco Di Domenico (Politecnico di Milano) for their assistance.

This research is part of EXTREMA, a project that has received funding from the European Research Council (ERC) under the European Union's Horizon 2020 research and innovation programme (Grant Agreement No. 864697).

References

[1] A. Freeman and C. Norton. Exploring our Solar System with Cubesats and Nanosats. In: Proceedings of the 13th Reinventing Space Conference. Oxford, UK, 2018. https://doi.org/10.1007/978-3-319-32817-11

[2] E. Kulu. Nanosatellite Launch Forecasts - Track Record and Latest Prediction. In: Small Satellite Conference 2022. Logan, Utah, USA, 2022.

[3] L. J. Deutsch, S. A. Townes, P. E. Liebrecht, P. A. Vrotsos and D. M. Cornwell. Deep Space network: The Next 50 Years. In: 14th International Conference on Space Operations. Daejeon, Korea, 2016. https://doi.org/10.2514/6.2016-2373

[4] A. Morselli, G. Di Domenico, E. Andreis, A. C. Morelli, G. Merisio, V. Franzese, C. Giordano, F. Ferrari and F. Topputo. The EXTREMA Orbital Simulation Hub: a Facility for GNC Testing of Autonomous Interplanetary CubeSat. In: 4S Symposium Proceedings. Vilamoura, Portugal, 2022.

[5] P. Panicucci, E. Andreis, F. Franzese and F. Topputo. An overview of the EXTREMA deep-space optical navigation experiment. In: 3rd Space Imaging Workshop. Atlanta, Georgia, USA, 2022.

[6] A. Morselli, A. C. Morelli and F. Topputo. ETHILE: A Thruster-In-the-Loop Facility to Enable Autonomous Guidance and Control for Autonomous Interplanetary Cubesats. In: 73rd International Astronautical Congress (IAC 2022). Paris, France, 2022.

[7] G. Di Domenico and F. Topputo. STASIS: An Attitude Testbed for Hardware-in-the-Loop Simulations of Autonomous Guidance, Navigation, and Control Systems. In: 73rd International Astronautical Congress (IAC 2022). Paris, France, 2022.

[8] C. Giordano and F. Topputo. SPESI: A Real-Time Space Environment Simulator for the EXTREMA Project. In: 33rd AAS/AIAA Space Flight Mechanics Meeting. Austin, Texas, USA, 2023.

[9] N. D. Matsakis and F. S. Klock. The Rust Language. ACM SIGda Ada Letters, 34 (3) 103-104. 2014. https://doi.org/10.1145/2663171.2663188

[10] P. Hambarde, R. Varma and S. Jha. The Survey of Real Time Operating System: RTOS. In: 2014 International Conference on Electronic Systems, Signal Processing and Computing Technologies. Nagpur, India, 2014. doi: https://doi.org/10.1109/ICESC.2014.15

[11] M. Troesch, F. Mirza, K. Hughes, A. Rothstein-Dowden, A. Donner, R. Bocchino, M. Feather, B. Smith, L. Fesq, B. Barker and B. Campuzano. MEXEC: An Onboard Integrated Planning and Exécution Approach for Spacecraft Commanding. In: 30th International Conference on Automated Planning and Scheduling. Nancy, France, 2020.

[12] R. Amini, L. Fesq, R. Mackey, F. Mirza, R. Rasmussen, M. Troesch and K. Kolcio. FRESCO: A Framework for Spacecraft Systems Autonomy. In: 2021 IEEE Aerospace Conference (50100). Big Sky, Montana, USA, 2021. https://doi.org/10.1109/AERO50100.2021.9438470

[13] M. Cols Margenet, H. Schaub and S. Piggott. Flight Software Development, Migration, and Testing in Desktop and Embedded Environments. Journal of Aerospace Information Systems, 18 (4) 157-174, 2021. https://doi.org/10.2514/1.I010820

[14] E. Andreis, P. Panicucci, V. Franzese and F. Topputo. A Robust Image Processing Pipeline for Planets Line-Of-sign Extraction for Deep-Space Autonomous Cubesats Navigation. In: 44th AAS Guidance, Navigation and Control Conference. Breckenridge, Colorado, USA, 2022.

[15] A. D. George and C. M. Wilson. Onboard Processing With Hybrid and Reconfigurable Computing on Small Satellites. Proceedings of the IEEE 106(3) 458 – 470. 2018. doi: https://doi.org/10.1109/JPROC.2018.2802438

[16] Z. Wan, A. Lele, B. Yu, S. Liu, Y. Wang, V. J. Reddi and A. Raychowdhury. Robotic Computing on FPGAs: Current Progress, Research Challenges, and Opportunities. arXiv preprint. 2022. https://doi.org/10.48550/arXiv.2205.07149

[17] G. W. Donohoe and J. C. Lyke. Reconfigurable Computing for Space. In T. A. Thawar, editor, Aerospace Technologies Advancements, chapter 3. InTech, 2010. ISBN: 978-953-7619-96-1. https://doi.org/10.5772/117

[18] V. Marchese, N. R. Benigno, S. Simonetti, E. Fazzoletto, F. Miglioretti, V. Di Tana, S. Pirrotta, M. Amoroso, E. Dotto and V. Della Corte. LICIACube Mission: The Fastest Fly-By Ever Done by a CubeSat. In: Small Satellite Conference 21. Logan, USA, 2021.

Aerospace Science and Engineering - III Aerospace PhD-Days   Materials Research Forum LLC
Materials Research Proceedings 33 (2023) 247-253   https://doi.org/10.21741/9781644902677-36

How space technologies can address the impact of climate change on aeronautic and the aviation

Marianna Valente[1a]

[1]Department of Production and Management, Politecnico di Torino, Torino

[a]marianna.valente@polito.it

Keywords: Aviation, Climate change, Space Technologies

Abstract. Aviation and flight management are strictly influenced by environmental conditions. Greater variability of the climate and the atmospheric phenomena carries greater risks for air navigation and increase cancellations and delay. An increase in the frequency and intensity of storms, torrential rains, hailstorms, fog, wind gusts, and other meteorological events may lead to a reduction in flight times and from a design point of view an increase in the weight of the aircraft, to counteract higher loads in operation. All of this could produce rising management and operational costs, as well as a reduction in profit. The instruments for remote monitoring as satellites, ground stations, and other technologies allow to collect data about several climate parameters such as temperature, humidity, rainfalls, wind patterns, etc. This data could be analyzed to obtain useful information to improve flights managements and airport services. Moreover, Artificial intelligence (AI) can be used to elaborate this information in real-time. This it makes possible to have more accurate forecasts and update the mathematical model for weather forecasting.

Introduction

Climate change is an urgent and growing global crisis. It is the result of human activities such as burning fossil fuels, deforestation, and agricultural production that emit greenhouse gases into the atmosphere, trapping heat and leading to extreme weather events such as floods, droughts, and hurricanes. Climate change is already having an impact on communities around the globe, and its effects are expected to worsen as temperatures continue to rise.

The effects of climate change are widespread and cannot be ignored. It is an interdisciplinary issue, and its effects are felt in almost every aspect of our lives. From public health to energy security, from food security to water availability. Climate change is leading to a wide range of consequences for the environment, society, and the economy. It is essential to understand the science behind this problem, take steps to reduce our contributions to it, and support policies to help mitigate its effects.

Climate change is also a major concern in the aviation industry. Air travel is one of the fastest-growing sources of greenhouse gas emissions, and international aviation emissions are expected to rise massively in the next decades. To reduce the environmental impact of air travel, the aviation industry is exploring new technologies and alternative fuels, as well as implementing measures such as carbon offsets and fuel efficiency standards to reduce emissions. It is also important to adopt policies that incentivize airlines to reduce their emissions and invest in more sustainable practices. Moreover, aircraft performances are strictly conditioned by environmental phenomena, therefore climate changes have a heavy impact on it.

One area that is increasingly becoming important in the fight against climate change is the Space sector. Space technology has the potential to help us better understand climate change and develop new solutions to reduce our emissions. Satellites provide us with data on global temperatures, sea level rise, and other factors that can support us to learn the effects of climate change. They can also track changes in the atmosphere, such as the amount of carbon dioxide, ozone, etc.

Aerospace Science and Engineering - III Aerospace PhD-Days   Materials Research Forum LLC
Materials Research Proceedings 33 (2023) 247-253   https://doi.org/10.21741/9781644902677-36

Observation satellites are the most suitable to collect data and monitor climate events. For example, the NASA-led Orbiting Carbon Observatory-2 mission uses a satellite to measure carbon dioxide levels in the atmosphere [1]. This data can then be used to inform policy decisions and help us identify areas where emissions need to be reduced.

In addition, space technology can also be used to develop new solutions to reduce our emissions. For example, solar panels on spacecraft can be used to generate electricity, and technologies developed for human exploration could be applied to research new forms of energy on Earth, such as new, smaller nuclear fission reactors [2]. These technologies can be crucial on the ground to reduce our reliance on fossil fuels and reduce the number of polluting gases.

Aviation and climate change

Climate change impacts aeronautics and aviation in all aspects, from the design phases to traffic and airport management. Eurocontrol has analyzed in detail the consequences of various meteorological changes on the civil aviation sector.

Observing the sea level rise in Europe (Fig.1), from 1993 to 2019 the trend is rising across the whole mainland.

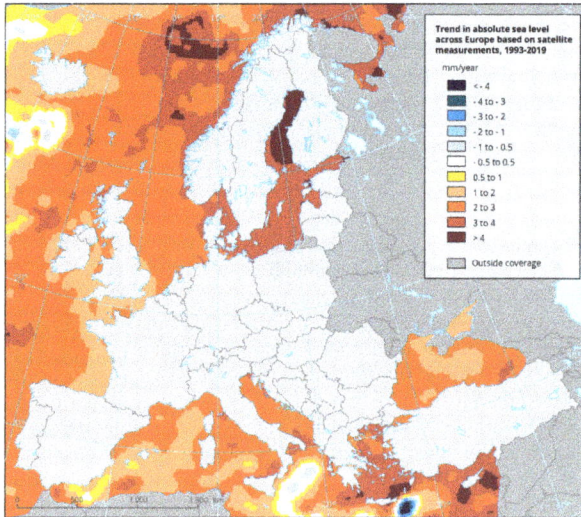

Figure 1. Absolute Sea level trend in Europe from 1993 to 2019 (Source: European Environment Agency https://www.eea.europa.eu/data-and-maps/find/global#c12=sea%20level%20&b_start=0)

The sea level rise increases the risk of flooding along the coastlines. This could be also enlarged by an increase in storm frequency and intensity. The ECAC region (the widest grouping of Member States of any European organization dealing with civil aviation, being currently composed of 44 Member States) is dense with coastal airports, and many of these had already adopted more efficient water drainage measures. A study [3], conducted using the Geographic Information System (GIS) simulations, revealed that, in the ECAC region, two-thirds of the airports are expected to be at risk of marine flooding in the event of a storm surge during the period up to 2090. 91% of these are small airports with less than 10,000 flights for years. Nevertheless, they are crucial for civil aviation e not only, because they have a key role in the development of local tourism and the economy in general.

Aerospace Science and Engineering - III Aerospace PhD-Days   Materials Research Forum LLC
Materials Research Proceedings 33 (2023) 247-253   https://doi.org/10.21741/9781644902677-36

Convective cells, unexpected storms, and wind shear are among the most frequent causes of en-route and arrival delays. These phenomena are easily predictable and more common in specific geographic areas. For instance, in the Mediterranean zone, convective events are the most widespread reason for en-route delays. In the next years, these events are projected to be increased in intensity. Moreover, extreme rainfall days are predicted to rise across northern Europe but decrease across southern Europe, causing an increment in the duration of delays. It is difficult to predict, in the long period, the effects of this on traffic management, taking into account both the variability of wind intensity and the interaction with other meteorological conditions. Nevertheless, it is clear that the ATM (air traffic management) will have to face an increase in the delay and will have to take action in time for reprogramming the routes [4,5].

From a structural point of view, if the gust intensity will increase, also the load on the aircraft rises. As a consequence, the airframe should be stiffer with an increase in the empty weight. This could lead to a reduced fuel storage capacity or a smaller number of passengers on board.

In addition, changes in wind patterns could result in major route changes. Some of them may be congested or less feasible than others, overloading the work of the ATM. Furthermore, changes in routes and the flight envelope, for safety reasons, imply the need to obtain a new flight certification for the aircraft, with huge costs for the airlines, both for the documents themselves, but also because the vehicle should remain in the hangar for a long time.

Climate change with the temperature increasing, alterations in weather patterns, and the environment of locations worldwide, can also affect the tourism demand, geographically or as time shifts or a combination of both [6]. Considering the ECAC region, in central Europe, a rise in tourists is projected in the autumn months (September, October, and November). While in the south of Europe, there will be a small decrease in summer tourism, but an increase both in the autumn and spring months. Other regions such as north Spain and Scandinavia will become more favorable all year. Overall, the mountain areas will suffer most from the temperature rise, especially the ski resorts that will increasingly have to employ artificial snow, while the bathing areas will be available for a longer period, with a reduction in demand in the warmer months (June, July, August) where there is the hottest temperature peak.

According to a study conducted by Eurocontrol [7] Stakeholders increasingly recognize that both the European aviation sector and European ATM have taken steps to adjust to climate change, but more must be done. A majority of respondents to this survey believe the sector has implemented some adaptation measures (Figure 2). Just 6% of participants think the sector has not analyzed adaptation.

Aerospace Science and Engineering - III Aerospace PhD-Days
Materials Research Proceedings 33 (2023) 247-253

Materials Research Forum LLC
https://doi.org/10.21741/9781644902677-36

The European Aviation Sector

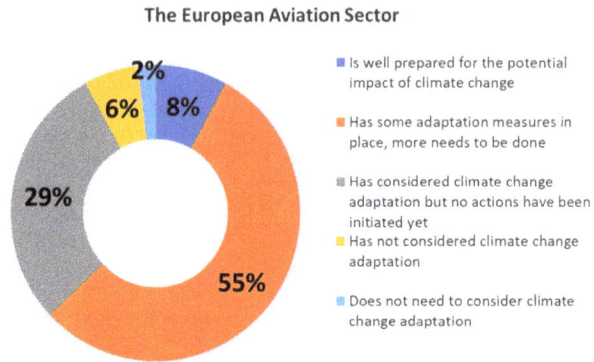

Figure 2. (N = 63) Stakeholder perceptions of the level of preparedness for the potential impacts of climate change for the entire European Aviation sector. Data from [7].

Regarding European ATM, 60% of respondents concluded that certain adaptation measures are in existence (Figure 3). Just a little part of the participants claimed that European ATM has not contemplated adjustment. Awareness is increasing, and more organizations are undertaking action. However, it must be determined if the appropriate measures are being implemented promptly or if we need to expedite action.

The European ATM

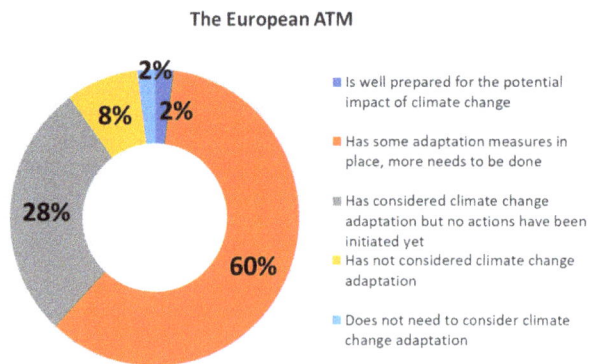

Figure 3. (N = 50) Stakeholder perceptions of the level of preparedness for the potential impacts of climate change for the European Air Traffic Management (ATM). Data from [7].

Space technologies to mitigate the impact.

There are many space technologies and Earth observation missions to collect and provide climate data to support aviation. Considerable efforts are being made to monitor the as best as possible the meteorological events to improve our knowledge and forecasting models. Not only aviation could have benefited, from these data, but also agriculture or meteorological stations could be the end

Aerospace Science and Engineering - III Aerospace PhD-Days   Materials Research Forum LLC
Materials Research Proceedings 33 (2023) 247-253   https://doi.org/10.21741/9781644902677-36

user of the information. Moreover, real-time monitoring helps to manage potential catastrophic events, such as tsunamis, floods, and landslides, to take action on time, limiting the victims.

Among the most dangerous and difficult to observe phenomena, there are convective cells. These systems are characterized by cumulonimbus clouds and occur frequently in tropical and subtropical zones. The convective cell takes place often with severe wind gusts, heavy rainfall, and thunderstorms. These events are sudden and short-lived, and could potentially damage the aircraft compromising flight safety.

Recent missions are being developed to study convective systems, using C-band or Ka-band radar, radiometers, or spectrometers. One of the most innovative is Raincube, a 3U CubeSat equipped with a Ka-band nadir pointing radar, with a deployable antenna [8]. The satellite, measuring the brightness temperature, detects the presence and size of the convective cells. As a negative aspect, it does not have a system to search and track the cells, so the radar could only spot the cell if the satellite flew over. Moreover, Raincube operated in LEO, therefore it is not possible to study the evolution of the system but only the instantaneous situation.

Another important mission, similar to the previous one, was TEMPEST-D, launched in 2018 and operational until 2021. The satellite was a CubeSat 6U, equipped with a microwave radiometer to observe the evolution of storm clouds [9]. To evaluate the storm intensity and the hazard of meteorological events, it measured the brightness temperature, with less accuracy than Ka-band radar but with a wider swath of 825km [10]. The final goal of the mission was to validate the performance of the payload, so a TEMPEST constellation could be developed, in LEO, to examine not only the intensity but also the time evolution of short events as convective cells.

Over the years, many other missions have been developed for the study of these phenomena, with different payload performances related to the aspect that scientists wanted to measure. La and Messager (2022) presented several instruments in LEO that acquire data and estimate ocean surface convective winds [11]. Sentinel-1 Synthetic Aperture Radar (SAR), for instance, offers a high spatial resolution to observe the gust wind, while with ASCAT scatterometer and SMAP radiometer, it is possible to observe the phenomenon at a larger scale.

Thanks to this data collected by satellites over the years, it is possible to develop and update the mathematical models of forecasting, which implemented in the cockpit can prove to be a valuable resource. An innovative method for storm forecasting was introduced [12], based on the global GLD360 lighting data and information from Meteosat, a geostationary satellite. They defined three levels of severity for the cumulonimbus, related to the brightness temperature of the water vapor. Then, the method evaluates the atmospheric motion vectors by analyzing two consecutive images captured by satellite and extrapolating the data by comparing them. This method was implemented on board a few aircraft and the pilots reported that they had the forecasting data 15-20 minutes later than the scans satellite, (usually this period is 30-60 minutes). However, some improvements are necessary to reduce the period between the two scans, delete the parallax errors, etc.

Many precision mathematical models have been developed and are being updated also to support air traffic management (ATM). A mathematical model that uses statistical data collected from satellites to model storms stochastically was developed [13]. This model is combined with an optimal trajectory planning algorithm. The resulting aircraft trajectories maximize the likelihood of reaching a waypoint while avoiding hazards, given the potential uncertainties due to wind disturbances and in those not-safe flight patterns caused by uncertainties in the data of thunderstorms.

A new study [14] explored the potential of lightning data assimilation with the Weather Research and Forecasting model (WRF) from an air traffic management (ATM) perspective for the first time in Europe, using Milano Malpensa as a case study. The authors highlighted the positive impact of lightning using three different methods for convective and stratiform

Aerospace Science and Engineering - III Aerospace PhD-Days   Materials Research Forum LLC
Materials Research Proceedings 33 (2023) 247-253   https://doi.org/10.21741/9781644902677-36

precipitation and emphasized that the assimilation of different data types can provide a reliable forecast in terms of spatial and temporal accuracy which meets ATM requirements.

Moreover, according to the SESAR European ATM Master plans [15] a crucial concept for the future ATM is trajectory-based operations (TBO). This system allows the aircraft to update autonomously the trajectory both to satisfy business needs, such as less fuel consumption and time-saving and to avoid potentially hazardous events. To adopt successfully the TBO it is necessary to reduce the uncertainty associated with the aircraft motion and weather development. Nowadays, the aircraft tracking system is based on the Automatic Depending Surveillance-Broadcast system (ADS-B), which adopts terrestrial antennae to provide information about the vehicle, except in the polar and ocean zones. To address this gap, a satellite constellation is necessary. A parametric study [16] was conducted into the design of a custom ADS-B Satellite constellation, to obtain a low-cost satellite constellation that provides ADS-B coverage equal to or better than the ground antenna. From the first analyses of the work, emerged that using a higher number of satellites at a higher altitude, a more reliable ADS-B coverage of the trans-oceanic flight route is achieved.

Methods like this, which exploit satellite aircraft communication, are increasingly being implemented, making safer flights and more dangerous weather events more easily predictable.

Conclusion

The effects of climate change on the aviation industry are expected to worsen in the future, leading to more delays and cancellations, increased fuel costs, and more frequent and disruptive extreme weather events. To mitigate the effects of climate change on the aviation industry, airlines must take proactive measures such as improving aircraft efficiency, investing in renewable energy sources, and implementing sustainable operational practices. Governments should also recognize that the aviation industry is vulnerable to climate change and provide incentives for airlines to reduce their emissions. By taking proactive steps to mitigate the impacts of climate change on the aviation industry, airlines can protect their operations and help to reduce the overall emissions of the aviation industry. On the other hand, space technologies provide a lot of information about environmental changes. Nevertheless, huge efforts are necessary to improve the forecasting abilities of sudden and strong events as convective cells. Satellite constellations in low Earth orbit could be a trade-off solution to achieve high accuracy and to measure the phenomenon's local aspects. At the same time, using a "train" of satellites allows one to obtain an image sequence, to observe the temporal evolution of the events. As a negative aspect, managing a constellation of small satellites it is not that simple, a more complex instrument on board is necessary to synchronize the "train", and the cost of the satellite increases.

Finally, communication in real time between satellites and aircraft could play a crucial resource to address sudden events, responsively and avoiding possible injury.

References

[1] A. Eldering et al., The Orbiting Carbon Observatory-2 early science investigations of regional carbon dioxide fluxes, Science, vol. 358, fasc. 6360, p. eaam5745, ott. 2017, doi: 10.1126/science.aam5745. https://doi.org/10.1126/science.aam5745

[2] M. A. Gibson et al., Development of NASA's Small Fission Power System for Science and Human Exploration, 2015. https://doi.org/10.2514/6.2014-3458

[3] European Commission, Eurocontrol. Impact of Sea Level Rise on European Airport Operations; Eurocontrol: Bruxelles, Belgium, 2021.

[4] European Commission, Eurocontrol. Impact of Changes in Storm Patterns and Intensity of Flight Operations; Eurocontrol: Bruxelles, Belgium, 2021.

Aerospace Science and Engineering - III Aerospace PhD-Days   Materials Research Forum LLC
Materials Research Proceedings 33 (2023) 247-253   https://doi.org/10.21741/9781644902677-36

[5] European Commission, Eurocontrol Impact of Climate Changes in Wind Patterns On Flight Operations; Eurocontrol: Bruxelles, Belgium, 2021.

[6] European Commission, Eurocontrol. Impact of Climate Change on Tourism Demand; Eurocontrol: Bruxelles, Belgium, 2021.

[7] European Commission, Eurocontrol. European aviation in 2040 - challenges of growth - adapting aviation to a changing climate; Eurocontrol: Bruxelles, Belgium, 2018.

[8] S. Tanelli, E. Peral, E. Im, M. Sanchez-Barbetty, R. M. Beauchamp, e R. R. Monje, RainCube and its legacy for the next generation of spaceborne cloud and precipitation radars, in 2020 IEEE Radar Conference (RadarConf20), Florence, Italy, set. 2020, pp. 1-4. doi: 10.1109/RadarConf2043947.2020.9266437. https://doi.org/10.1109/RadarConf2043947.2020.9266437

[9] C. Radhakrishnan et al., Cross Validation of TEMPEST-D and RainCube Observations Over Precipitation Systems, IEEE J. Sel. Top. Appl. Earth Obs. Remote Sens., vol. 15, pp. 7826-7838, 2022, doi: 10.1109/JSTARS.2022.3199402. https://doi.org/10.1109/JSTARS.2022.3199402

[10] R. M. Schulte et al., A Passive Microwave Retrieval Algorithm with Minimal View-Angle Bias: Application to the TEMPEST-D CubeSat Mission, J. Atmospheric Ocean. Technol., vol. 37, fasc. 2, pp. 197-210, feb. 2020, doi: 10.1175/JTECH-D-19-0163.1. https://doi.org/10.1175/JTECH-D-19-0163.1

[11] T. V. La e C. Messager, Different Observations of Sea Surface Wind Pattern Under Deep Convection by Sentinel-1 SARs, Scatterometers, and Radiometers in Collocation, IEEE J. Sel. Top. Appl. Earth Obs. Remote Sens., vol. 15, pp. 3686-3696, 2022, doi: 10.1109/JSTARS.2022.3172375. https://doi.org/10.1109/JSTARS.2022.3172375

[12] R. Müller, A. Barleben, S. Haussler, e M. Jerg, A Novel Approach for the Global Detection and Nowcasting of Deep Convection and Thunderstorms, Remote Sens., vol. 14, fasc. 14, p. 3372, lug. 2022, doi: 10.3390/rs14143372. https://doi.org/10.3390/rs14143372

[13] D. Hentzen, M. Kamgarpour, M. Soler, e D. González-Arribas, On maximizing safety in stochastic aircraft trajectory planning with uncertain thunderstorm development, Aerosp. Sci. Technol., vol. 79, pp. 543-553, ago. 2018, doi: 10.1016/j.ast.2018.06.006. https://doi.org/10.1016/j.ast.2018.06.006

[14] V. Mazzarella et al., Is an NWP-Based Nowcasting System Suitable for Aviation Operations?, Remote Sens., vol. 14, fasc. 18, p. 4440, set. 2022, doi: 10.3390/rs14184440. https://doi.org/10.3390/rs14184440

[15] European Commission, Eurocontrol, SESAR European ATM Master Plan, Edition 2020.

[16] T. H. Nguyen, N. Tsafnat, E. Cetin, B. Osborne, e T. F. Dixon, Low-Earth orbit satellite constellation for ADS-B based in-flight aircraft tracking, Adv. Aircr. Spacecr. Sci., vol. 2, fasc. 1, pp. 95-108, gen. 2015, doi: 10.12989/AAS.2015.2.1.095. https://doi.org/10.12989/aas.2015.2.1.095

Aerospace Science and Engineering - III Aerospace PhD-Days   Materials Research Forum LLC
Materials Research Proceedings 33 (2023) 254-261   https://doi.org/10.21741/9781644902677-37

Design and challenges of an IOD/IOV 12U Cubesat mission

Matteo Gemignani[1,a] *

[1]Dept. of Civil and Industrial Engineering, University di Pisa, Italy

[a]matteo.gemignani@phd.unipi.it

Keywords: CubeSats, Propulsion, Demonstration, Validation

Abstract. EXCITE ("EXtended Cubesat for Innovative Technology Experiments") is a technology demonstration mission selected by ASI in 2021 in the frame of the "Future Cubesat Missions" call. Based on a custom-designed 12U CubeSat platform featuring a full-composite structure, EXCITE is aimed at in-orbit demonstration / in-orbit validation of a number of innovative small spacecraft technologies in the domains of chemical and electric onboard propulsion, thermal management of significant heat loads in limited volumes, COTS GPU computing for IoT applications, and steerable, integrated S-band antennas. In this paper we describe the EXCITE platform design, outline the main expected technological innovations, and discuss the possible methodologies for a multi-disciplinary optimization approach.

Introduction and Background

Recent years have seen an exponential increase in the quantity of small satellites and CubeSats launched into orbit each year. As a natural consequence, the variety of types of missions that can be performed with these spacecrafts, such as in-orbit demonstrations, remote sensing, or scientific experimentation, has also grown [1]. This variety led to an increasing demand for enhanced performance, for instance, relatively to pointing accuracy, power generation, and data downloading. Thanks to this increased performance requirements, CubeSat project design is gradually detaching from its original attributes, such as simplicity, low-cost and high-risk, going towards a new philosophy which is more performance oriented, but still cost-efficient. With this background and this demands, new challenges emerged for the design and project strategies for

Figure 1: The EXCITE 12U Cubesat

CubeSats, in particular regarding optimization techniques, which until now were mostly devoted to large scale projects [2].

In the frame of this context, in this paper a 12U CubeSat mission named EXCITE ("EXtended Cubesat for Innovative Technology Experiments") (Fig. 1) for in orbit demonstration/validation is presented, illustrating its main technological challenges, with a focus on optimization strategies that can be adopted in order to maximize the performance of this platform.

The EXCITE mission is jointly developed by a team including the University of Pisa as team leader. The proposal was submitted in 2020 in response to the Italian Space Agency's "Call for Future CubeSat Missions" and was selected for funding. The mission proposal was prepared in late 2020 in response to ASI's call for Future Cubesat Missions and was selected in 2021. Four SMEs based in the Tuscany region act as industrial partners: Aerospazio Tecnologie, a small company of the Siena/Livorno area with a strong background in space propulsion and testing; CRM Compositi, a structural workshop specialized in composite materials in Livorno; IngeniArs, a spin-off company of UniPi, dedicated to space electronics; and MBI, a telecom company located in Pisa, active in satellite telecommunications and networking. The consortium is a working example of a regional-scale initiative leveraging on Cubesat technology for local development. Activities will start in mid-2023 in the frame of ASI's Alcor programme [3], with flight planned for 2026.

Mission Overview

The project concerns the IOD/IOV of five innovative technologies:

- *Green monopropellant thruster*: a hydrogen peroxide propulsion system capable of performing moderate delta-V manoeuvres on a Cubesat, under development at UniPi [4].
- *Pulsed plasma thruster*: a miniaturized electric thruster from Aerospazio Tecnologie S.r.l., capable of delivering very low impulse bits, for proximity operations or fine attitude control.
- *Reconfigurable integrated S-band antenna*: an electronically steerable antenna based on exciters distributed on suitable spacecraft surfaces, developed at UniPi [5].
- *Internet-of-things GPU demodulator*: a technology by MBI S.r.l, based on the utilization of a COTS GPU for on-board processing of advanced IoT waveforms and VDES protocol.
- *Pulsating heat pipes*: high throughput heat pipes under development at UniPi [6], based on unsteady fluid flow, especially suited for high heat flux applications (e.g., thermal management of high-power microsatellites).

The mission has the following additional objectives:

- To develop a high-performance 12U bus equipped with deployable solar panels.
- To integrate a commercial optical sensor as an onboard source of a data stream representative of real applications.
- To set-up a ground station at UniPi to be used for satellite operations but also as a valuable teaching tool.

Two of the experiments hosted on EXCITE belong to the IOV category (chemical and electric propulsion); the other three are extended scope IOD experiments (pulsating heat pipes, IoT GPU demodulator, reconfigurable integrated antenna).

Aerospace Science and Engineering - III Aerospace PhD-Days   Materials Research Forum LLC
Materials Research Proceedings 33 (2023) 254-261   https://doi.org/10.21741/9781644902677-37

The mission is based on a 12U CubeSat platform featuring a full-composite structure, deployable solar panels, 3-axes attitude control, and S-band telecommunications. Some of the subsystems (Fig. 2) will be procured on the commercial market as COTS, while some high performance, critical elements (power generation, thermal control, structure and deployables) will be entirely designed and manufactured by the EXCITE team.

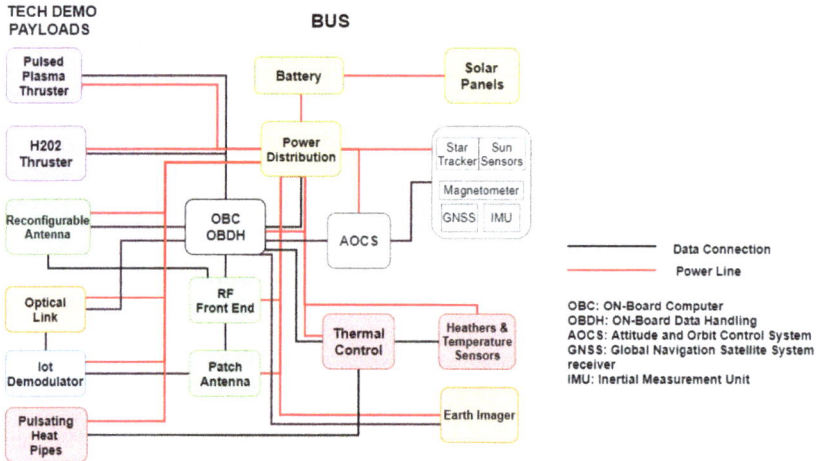

Figure 2: EXCITE Functional Block Diagram

A Sun-synchronous, 550 km Earth orbit is assumed as the baseline operational environment of EXCITE. The mission can however easily be adapted to different LEO locations, should a convenient flight opportunity arise. This provides ample flexibility in the choice of launch opportunities. Considering the relatively low ballistic coefficient of the platform when solar panels are deployed, de-orbit will occur naturally form the chosen orbit within the 25 years term mandated by the international debris mitigation guidelines. It is additionally foreseen to perform the last chemical thruster burn in such a way as to lower the orbital attitude and further accelerate re-entry. Space-qualified COTS will be used for the platform subsystems wherever possible, but the EXCITE mission will also provide an opportunity to develop and use operationally a number of advanced microsatellite bus technologies that are already under development at UniPi and partners:

- The structure will be made of carbon composite materials. This will allow for mass reduction with respect to metallic frames and to the possibility to build custom layers into the composite structure, e.g. for the Reconfigurable Antenna experiment.
- Deployable solar panels (Fig. 4) will be designed, manufactured, and integrated using the same composite manufacturing techniques as for the spacecraft body, integrating Shape Memory Alloy actuators developed at UniPi.
- Thermal control, a critical issue in such a high-power, small volume bus, will integrate with the PHP experiment.

Aerospace Science and Engineering - III Aerospace PhD-Days   Materials Research Forum LLC
Materials Research Proceedings 33 (2023) 254-261   https://doi.org/10.21741/9781644902677-37

Payload Demonstrations

The H2O2 propulsion system occupies a volume of about 2U, including a propellant mass of about 2 kg. The demonstration maneuvers are designed to change different orbital parameters in a sensible way, demonstrating both in-plane and out-of-plane maneuvers, while deviating minimally from the nominal flight trajectory. Starting from the nominal orbit at 550 km, the maneuvers envisaged are as follows:

1. apogee lowering to 500 km. The required consumption is about 22% of the initial propellant mass equivalent to 27 m/s of ΔV.
2. apogee raising back to 550 km; propellant consumption and ΔV approximately as before.
3. change of RAAN by 0.5 degrees; the maneuver is best performed providing a 65 m/s ΔV by firing at argument of latitude almost equal to 90 degrees, for a 50% propellant consumption.
4. the remaining propellant is used for the final de-orbiting maneuvering phase.

The PPT tests are designed to demonstrate the capability to carry out precision maneuvers. Four independent thrusters will be installed on EXCITE (Fig. 3), so to have the possibility to generate either pure thrust or pure torque by selecting thrusters in pairs.

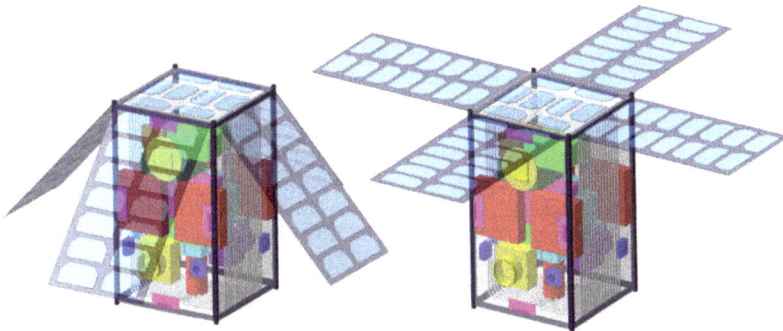

Figure 3: Solar panels deployment sequence

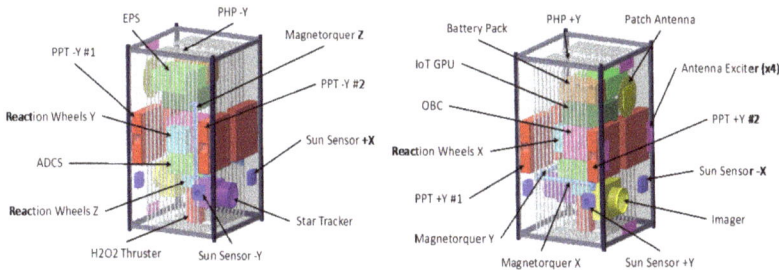

Figure 4: Schematics of the subsystem arrangement inside EXCITE

Aerospace Science and Engineering - III Aerospace PhD-Days   Materials Research Forum LLC
Materials Research Proceedings 33 (2023) 254-261   https://doi.org/10.21741/9781644902677-37

It is envisioned to perform three experiments:

1. validate the use of PPTs to control the attitude of the satellite in one axis , by spinning up the satellite using a pair of thrusters pairs on opposite sides of the spacecraft body. The burns last 5 minutes in order to reach a spin rate of about 0.8 deg/s. The resulting angular rate measured by the AOCS sensors provides a measure of the impulse delivered. The opposite pairs of thrusters are then fired to spin the spacecraft back to zero angular speed.
2. provide off-loading capabilities of reaction wheels of the on-board 3-axis AOCS, by momentarily taking over the task of the magnetorquers to provide de-saturation torque to the reaction wheel assembly.
3. provide translational thrust, needed e.g. for precise proximity operations (rendez-vous, close formation control), by firing thrusters on the same side of the spacecraft. In the latter case, the spacecraft orientation will be set so to have the thrusters oriented along the orbital velocity direction. Firing for a few minutes at nominal rate (1 Hz) and impulse bit (40 μN s) will provide enough delta-V to change orbital altitude by a few tens of meters, a change that can be easily detected by the onboard GNSS sensors.

The PHP experiment is totally passive; it will automatically start as soon as proper thermal conditions are established, which will occur almost immediately after acquisition of the nominal Sun pointing attitude and deployment of the solar panels.

Regarding the GPU demodulator, various demonstration activities will be performed:

1. generic utilization of the GPU for onboard data processing, anytime during flight.
2. a designed IoT signal is transmitted by MBI's IoT ground terminal when EXCITE is passing within its visibility area. In particular, an innovative spread-spectrum waveform named IURA (IoT Universal Radio Access) will be used. The signal collected by on-board telecommunication hardware is demodulated and stored by the GPU.
3. during a generic orbit, S-Band I+Q samples from different world areas are collected by the RF front-end radio, recorded by the GPU and processed onboard.
4. when the spacecraft is in view of the mission's ground station, or of another suitable receiving station located elsewhere, the GPU generates a Continuous Wave (CW) signal in S-band that is broadcast by the RF front-end. The receiving ground terminals can perform a detailed analysis of the received beacon for performance assessment and for additional auxiliary scientific purposes.

The reconfigurable antenna will be tested by rotating the spacecraft to different attitude angles with respect to the ground station and measuring the strength of the received signal. The experiment can be performed anytime, provided visibility with the GS is ensured and the instantaneous power balance on board allows for the RF front end operation. On basis of future risk analysis results, the telecommunication activities can be arranged to perform combined IoT GPU - ReconfAntenna experiments.

Challenges and Optimization

EXCITE's mission philosophy is to provide affordable, effective in-orbit technology demonstrations by making use of COTS, wherever possible, but also dedicated developed hardware, while maintaining adequate best practices in documentation, product/mission assurance, verification, and testing, balancing the need for rigorous space project management with the limited resources of a Cubesat programme. In order to achieve such an ambitious goal, some points in the "traditional" approach to CubeSat need to be abandoned: for example, for this mission design we are going to give up the modularity of the CubeSats, preserving the 12U form factor but using

Aerospace Science and Engineering - III Aerospace PhD-Days   Materials Research Forum LLC
Materials Research Proceedings 33 (2023) 254-261   https://doi.org/10.21741/9781644902677-37

the entire internal volume of the satellite for an integration that maximizes the space available for payloads for the eventual reuse of this platform as a test bed for innovative satellite technologies. The final goal is to develop new methodologies by merging optimization techniques in order to achieve an optimal design for the EXCITE mission. But first, it is important to clearly state which mission aspects has to be optimized.

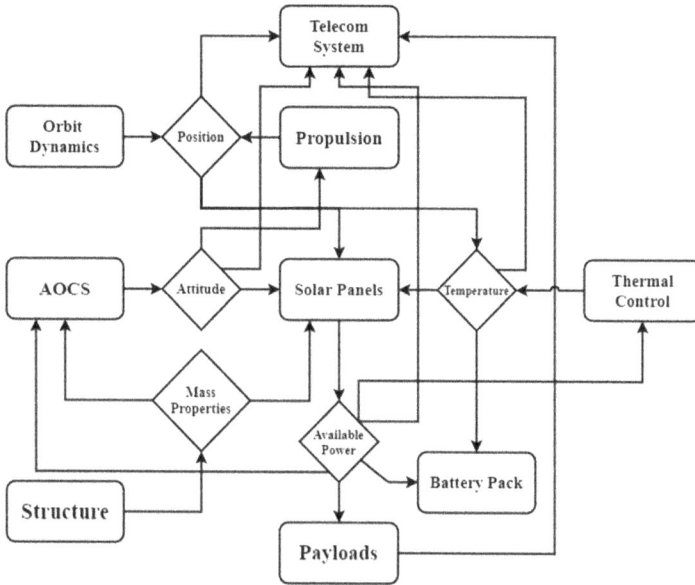

Figure 5: Schematics of the coupling between subsystems and parameters

For a space mission in general, the problem of optimization is never driven by a single objective but, rather, by a set of objectives that concur to achieve the mission statement most efficiently. In this case, validate the 5 embarked experiments so as to elevate their TRL. For this reason, the first step, as part of the design process will be to develop an opportune objective function. This objective function will be a sum of subfunctions each specifically tailored to the type of experimentation to be done on the relevant payload technology. For example, for the hydrogen peroxide engine, the success of the mission will depend crucially on the accuracy with which it can perform the scheduled maneuvers, while for the PhPs it will be the degree of efficiency in heat dissipation. Moreover, the success of the mission will depend on the total duration of the payload functionality. This objective function definition will be definitely a complex task to undergo, and will be affected also by the schedule of the mission operations that will essentially define the sequence of experiments.

Of course, the performance of the technologies will be affected by the degree of optimization of the overall platform, therefore in the sense of maximum available power, maximum achievable data rate, spacecraft agility and precise pointing. EXCITE is expected to generate a considerable amount of data, but the quantity that the communication system can send to the ground station is dependent on the amount of power this system has available to use. In general, the more power provided by the photovoltaic array and battery, the more data can be transmitted. Furthermore, if the ground station antenna is aligned with the satellite antenna, more data can be transferred with

Aerospace Science and Engineering - III Aerospace PhD-Days   Materials Research Forum LLC
Materials Research Proceedings 33 (2023) 254-261   https://doi.org/10.21741/9781644902677-37

less power. However, to align the satellite antenna, attitude operation is required and, sometimes, the actuator power spent in aligning the antenna is more than the power used by communicating without the alignment. The attitude with respect to the Sun during an orbit will affect the Cubesat temperature, that in turn will affect the solar panels' efficiency and battery duration.

Is therefore self-evident (Fig. 5) how the coupling between disciplines in a spacecraft leads to the necessity of using multi-disciplinary optimization to maximize performances.

There are many optimization techniques, widely used in spacecraft engineering, that account for single discipline optimization (SDO). One typical characteristic of SDO is the usage of gradient-free optimization with indifferentiable design variables that originate from detailed modelling. The detailed and complex modelling often includes discrete or indifferentiable variables that cannot be handled with gradient-based optimization. These variables can only be handled with gradient-free optimization such as Genetic Algorithm (GA) or Particle Swarm Optimization (PSO). Hence, many single discipline optimization researches focus on gradient-free optimization Another characteristic is that most SDO applications have a small number of design variables because the gradient-free optimizer cannot handle a large number of design variables. Even with the gradient-based optimization, the number of design variables is less than 100 due to the limited characteristic posed by a "single" discipline. Many aerospace system designers have applied multidisciplinary approaches to their system design projects. However, most MDO efforts have focused on the design of aircraft structures and space launch vehicles, and very little work has considered the MDO application to the space system with complex constraints, such as a small satellite. Gradient-based methodologies will be probably most suited for this particular case, because for these methods the number of function evaluations is nearly constant with respect of an increase of design variables, while for gradient-free methods the number of function evaluations grows exponentially. Therefore one of the biggest challenges of such an approach will be the development of differentiable models that would allow the usage for derivative-based optimization, in order to have reasonable computational times.

Conclusions

An introduction and an overview of the EXCITE mission has been presented. The main technical challenges related to the demonstration of 5 technological demonstration/validation experiments have been reported together with a possible schedule for the operations. The optimization problem for such a platform has been discussed, with a focus on how disciplines related to different spacecraft systems are strongly coupled in space missions and will be of special importance in a high integrated, compact, high power CubeSat such as EXCITE. Moreover, the performance of the mission will be strictly related to the In-orbit demonstration of the payloads, therefore to the spacecraft operations schedule which will be considered in the optimization formulation.

For this reason, in order to maximize the overall performance, a multi-disciplinary approach that unifies the design and the operations will be adopted. On the other hand, the complexity and the computational cost of such an approach will have to be evaluated and justified in terms of the expected gains with respect to more traditional design philosophies.

References

[1] Bryce and Space Technology, SmallSat by the Numbers, 2022, https://brycetech.com/reports/report-documents/Bryce_Smallsats_2022.pdf

[2] B. Yost, S. Weston, State-of-the-Art: Small Spacecraft Technology, 2022. https://ntrs.nasa.gov/citations/20220018058

[3] EXCITE (EXtended Cubesat for Innovative Technology Experiments) https://www.asi.it/tecnologia-ingegneria-micro-e-nanosatelliti/micro-e-nanosatelliti/il-programma-alcor/excite/

Aerospace Science and Engineering - III Aerospace PhD-Days
Materials Research Proceedings 33 (2023) 254-261

Materials Research Forum LLC
https://doi.org/10.21741/9781644902677-37

[4] A. Pasini, L. Sales, E. Puccinelli, L.Lin, A. Apollonio, R. Simi, G. Brotini, L. d'Agostino, Design of an Affordable Hydrogen Peroxide Propulsion System for CubeSats, AIAA 2021-3690. *AIAA Propulsion and Energy 2021 Forum*, 2021. https://doi.org/10.2514/6.2021-3690

[5] S. Genovesi, F. A. Dicandia, Characteristic Mode Analysis for the Design of Nanosatellite Reconfigurable Antennas, 14th European Conference on Antennas and Propagation (EuCAP), Copenhagen, Denmark, 2020, pp. 1-3. https://doi.org/10.23919/EuCAP48036.2020.9135534

[6] M, Mameli, A. Catarsi, D. Mangini, L. Pietrasanta, N. Michè, M. Marengo, P. Di Marco, S. Filippeschi, Start-up in microgravity and local thermodynamic states of a hybrid loop thermosyphon/pulsating heat pipe, Applied Thermal Engineering, Volume 158, 2019, 113771,ISSN 1359-4311. https://doi.org/10.1016/j.applthermaleng.2019.113771

Aerospace Science and Engineering - III Aerospace PhD-Days
Materials Research Proceedings 33 (2023) 262-268

Materials Research Forum LLC
https://doi.org/10.21741/9781644902677-38

Design of an orbit determination computer for AI autonomous navigation

Aurel Zeqaj[1,a] *

[1]Dipartimento di Ingegneria Industriale, Alma Mater Studiorum – Università di Bologna, Forlì, Italy

[a]aurel.zeqaj2@unibo.it

Keywords: EKF, DeepNav, Navigation, Autonomous

Abstract. In the context of the growing demand for autonomous navigation solutions able to reduce the cost of a space mission, the DeepNav project, financed by ASI, has the objective to develop a navigation subsystem relaying solely on the use of onboard assets, e.g. optical images, artificial intelligence algorithms and an Extended Kalman Filter, to perform the navigation of a CubeSat around minor celestial bodies. This manuscript describes the work performed at University of Bologna in the context of the project, in particular for the development of an orbit determination computer, which uses an estimation filter to reconstruct the trajectory of the spacecraft taking as input the optical observables previously processed by the artificial intelligence algorithms.

Introduction

In recent years, the use of micro- and nano-satellites for space exploration has increased rapidly, mainly due to the reduction in launch and space segment costs that this type of technology entails [1,2,3,4]. Unfortunately, this does not (at least for now) correspond to a reduction in the ground segment. For example, the current approach for deep space navigation is to use radiometric tracking, obtained from ground antennas [5,6], which is still very demanding in terms of human and technical resources. In that respect, autonomous navigation in the vicinity of a celestial body can benefit greatly from the application of artificial intelligence approaches.

The 'DeepNav' [7] project, financed by the Italian Space Agency (ASI) and developed by a consortium consisting of AIKO[1] (prime), Politecnico di Milano[2], and Università di Bologna[3], has the aim of designing and testing an innovative navigation technique based on images and the use of artificial intelligence algorithms, for the exploration of asteroids [8,9,10], without any dependence on the Earth ground segment.

In the following sections, the work performed at the University of Bologna on the navigation computer (Figure 1) will be described.

Dynamical Model

For the final test campaign of the project, a set of case studies were designed, including circular and hyperbolic trajectories around the two targeted bodies, i.e. asteroids 101955 Bennu and 25143 Itokawa, varying the following characteristics:

- o Radius at the pericenter;
- o Inclination of the orbit;
- o Rotation angle of the asteroids;

The dynamic model [11,12,13] was developed accordingly, taking into account contributions due to the following accelerations:

---

[1] https://www.aikospace.com/, last access: 08/03/2023
[2] https://dart.polimi.it/, last access: 08/03/2023
[3] https://site.unibo.it/radioscience-and-planetary-exploration-lab/en, last access 08/03/2023

Aerospace Science and Engineering - III Aerospace PhD-Days   Materials Research Forum LLC
Materials Research Proceedings 33 (2023) 262-268   https://doi.org/10.21741/9781644902677-38

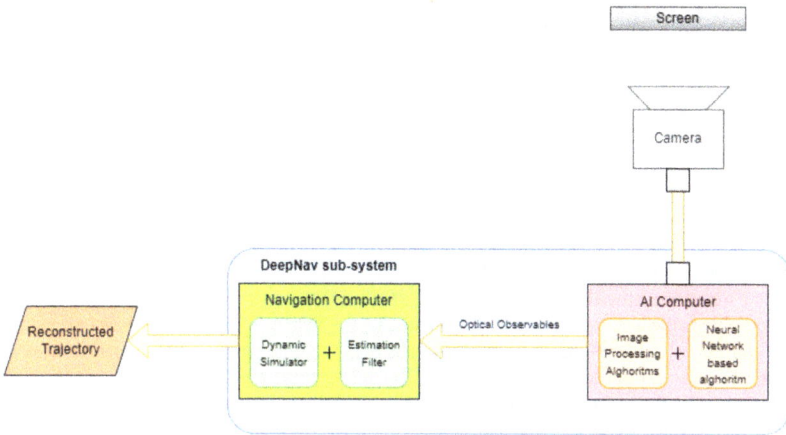

Figure 1: Structure of the DeepNav sub-system

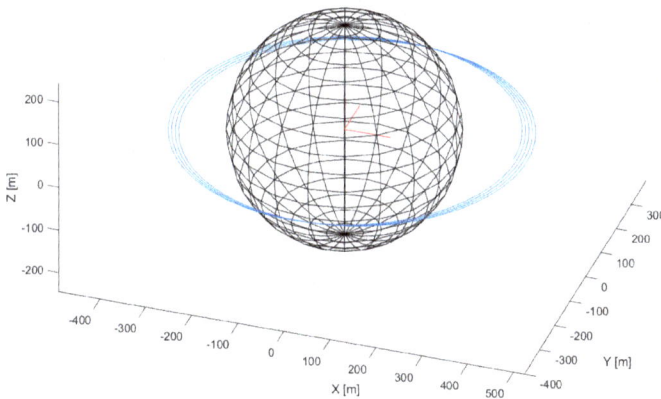

Figure 2: An example of simulator output. A quasi-circular orbit propagated using the point-mass model of the asteroid Bennu and the contribution due to the SRP. The S/C is at 5/3 of the mean radius of Bennu and at an inclination of 0°.

Aerospace Science and Engineering - III Aerospace PhD-Days   Materials Research Forum LLC
Materials Research Proceedings 33 (2023) 262-268   https://doi.org/10.21741/9781644902677-38

- Point-mass gravitational acceleration generated by the asteroids[4];
- Third-body gravitational acceleration produced by the Sun and Jupiter[5];
- Non-spherical gravitational acceleration, modeled using an expansion in spherical harmonics;
- Solar radiation pressure (SRP) acceleration using a simplified box-wing model made by a collection of flat plates.

Figure 2 shows an example of a quasi-circular orbit[6] of the DeepNav spacecraft around asteroid Bennu[7] obtained from the simulator, taking into consideration only the point-mass gravity and the SRP accelerations.

Orbit Determination Filter
The filter developed for the simulator is an Extended Kalman Filter (EKF) [11,13] (Figure 3), which combines the a-priori information provided by the dynamic model with measurements from the AI computer (representing the position of the S/C obtained from the processed images of the camera) in a sub-optimal sequential algorithm. The non-optimality is given by the fact the Kalman filter is designed for linear systems that are only affected by Gaussian noise both in the process and the measurements, while the extension to non-linear system implies a certain degree of approximation, which in the case of EKF is a first-order approximation. In fact, the computation of the Kalman gain and the covariance matrix associated with the estimation is performed by linearizing the system around the last estimated state point. For weakly non-linear systems, such as the orbit model with time updates in the order of seconds, the use of EKF is fully justified, and no definitive evidence is given in the literature on the eventual benefits coming from different types of filters.

The equation that realizes the Kalman based estimation (the combination between a priori propagation and the measurements) is given by:

$$\hat{x}_k = \hat{x}_x^- + K_k(z_k - H_k\hat{x}_k^-). \qquad (1)$$

Where $\hat{x}_x^-$ is the a priori propagated sate, K_k is the Kalman gain, z_k is the vector of measurements and H_k is the matrix that traduces the a priori state estimation into the related a priori measurement. The provided filter is based on a linear discrete system assumption with additive white Gaussian noise, which can be generally expressed as

$$x_{k+1} = \phi_k x_k + u_k. \qquad (2)$$

$$z_k = H_k x_k + v_k \qquad (3)$$

where u_k is the process noise, white Gaussian with zero mean, characterized by the covariance matrix Q_k and v_k is the measurements noise, white Gaussian with zero mean, characterized by the

---

[4] Asteroid Bennu and Itokawa were taken as a study cases
[5] These are the only celestial bodies that give a "relevant" contribution, although several order of magnitude lower than the other contributions.
[6] The reference frame used for the simulation is EME2000
[7] In the picture asteroid Bennu is represented as a sphere for simplicity

Aerospace Science and Engineering - III Aerospace PhD-Days   Materials Research Forum LLC
Materials Research Proceedings 33 (2023) 262-268   https://doi.org/10.21741/9781644902677-38

covariance matrix R_k. The filter is then completed with an equation that provides estimation of the covariance matrix associated with the estimation error, given by

$$P_{k+1}^- = \phi_k P_k \phi_k^T + Q_k. \tag{4}$$

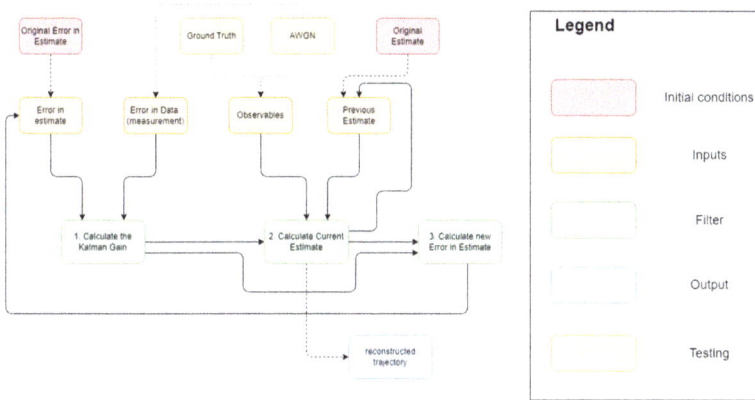

Figure 3: *Flowchart* of the Kalman Filter

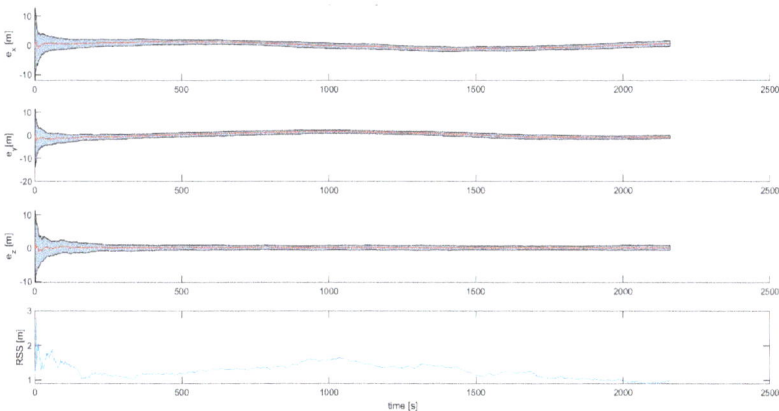

Figure 4: Error in trajectory reconstruction. From top to bottom, the error on the x, y and z axes and the quadratic sum of the RMS in the three coordinates.

In order to test the filter, a "ground truth" trajectory was generated using the dynamic model of the propagator, and it was used as a reference to compare with the reconstructed one generated by the filter. The synthetic observables that were fed to the filter were created from the "true" trajectory, adding white gaussian noise with different values of standard deviation and zero mean. Then, two different sets of Monte Carlo simulations were performed, the first one with different

Aerospace Science and Engineering - III Aerospace PhD-Days   Materials Research Forum LLC
Materials Research Proceedings 33 (2023) 262-268   https://doi.org/10.21741/9781644902677-38

values of the process noise covariance matrix, i.e. Q^8, in order to find the optimal value/s. These values were then used in the second simulation to test the filter performances. As can be seen in Figure 4, after a short initial transient, the filter is able to reconstruct the trajectory, minimizing the difference with the reference one.

Design of the Navigation Computer

The Navigation computer (Figure 5 and Figure 6) was designed with the software Altium Designer, and is composed of the following elements:

- *Microcontroller*, STM32F407, a 32-bit ARM Cortex-M4 based device which is the core of the breadboard;
- *Connections*:
  - UART/USB protocols, used to communicate with the AI computer;
  - JTAG protocol, used to communicate with the Programmer (which is needed to upload the SW);
- *Memory*:
  - EEPROM;
- *Debug elements*:
  - LED, used to verify the correct functioning of the power alimentation and of the microcontroller;
  - Test pad, used to test the correct alimentation of some components of the breadboard.

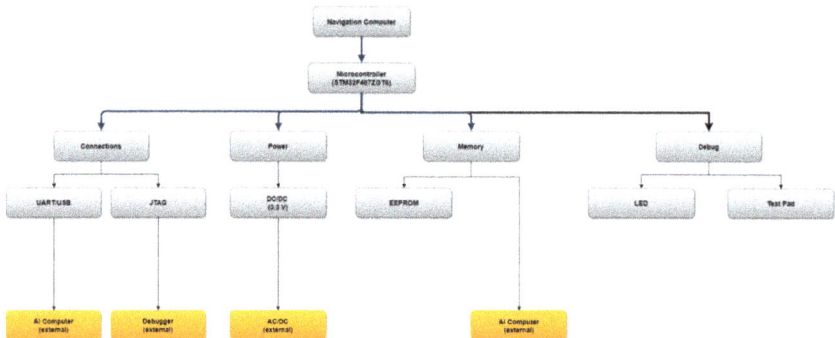

Figure 5: Tree structure of the navigation computer.

Future work

The orbit determination simulator, together with the Extended Kalman Filter, will be integrated within the navigation computer, while the artificial intelligence algorithms will be implemented in a dedicated AI computer. Both of the aforementioned computers will be integrated into the same board and will have data interfaces that will allow internal data exchange between the two logic units and communication with the outside world. Specifically, the internal data exchange will mainly consist of sending information extracted from the images from the AI computer to the navigation computer. The latter will integrate this information with the dynamics of the system through an estimation filter, thus allowing the orbit of the CubeSat to be precisely estimated and

---

[8] The process noise covariance Q is a nxn weighting matrix for the system process, where n is the number of state variables.

propagated over time. The last part of the project will consist in a campaign test to demonstrate the capabilities of the sub-system.

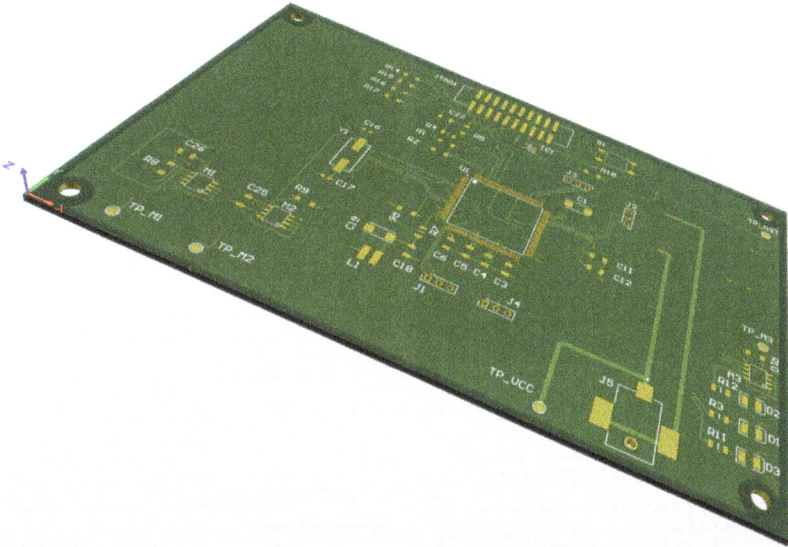

Figure 6: Design of the Navigation Computer. The PCB is composed of 4 layers, with the GND and PWR layers in the middle.

Acknowledgment

The DeepNav (Deep Learning for Navigation of Small Satellites about Asteroids) project has been financed under ASI Contract N. 2021-16-E.0.

References

[1] R. Lasagni Manghi, M. Zannoni, and P. Tortora, An autonomous optical navigation filter for a CubeSat mission to a binary asteroid system, presented at the IEEE International Workshop on Metrology for AeroSpace, MetroAeroSpace, 2019, pp. 116-120.

[2] E. Turan, S. Speretta, and E. Gill, Autonomous navigation for deep space small satellites: Scientific and technological advances, Acta Astronautica, vol. 193, pp. 56-74, Apr. 2022, doi: 10.1016/j.actaastro.2021.12.030. https://doi.org/10.1016/j.actaastro.2021.12.030

[3] V. Franzese, F. Topputo, F. Ankersen, and R. Walker, Deep-Space Optical Navigation for M-ARGO Mission, J Astronaut Sci, vol. 68, no. 4, pp. 1034-1055, Dec. 2021, doi: 10.1007/s40295-021-00286-9. https://doi.org/10.1007/s40295-021-00286-9

[4] E. Dotto et al., LICIACube - The Light Italian Cubesat for Imaging of Asteroids In support of the NASA DART mission towards asteroid (65803) Didymos, Planetary and Space Science, vol. 199, p. 105185, May 2021, doi: 10.1016/j.pss.2021.105185. https://doi.org/10.1016/j.pss.2021.105185

[5] R. Lasagni Manghi, D. Modenini, M. Zannoni, and P. Tortora, Measuring the mass of a main belt comet: Proteus mission, in Proceedings of the International Astronautical Congress, Oct. 2018. https://doi.org/10.1109/MetroAeroSpace.2019.8869616

[6] E. Gramigna et al., Hera Inter-Satellite link Doppler characterization for Didymos Gravity Science experiments, presented at the 2022 IEEE 9th International Workshop on Metrology for AeroSpace, 2022, pp. 430-435. https://doi.org/10.1109/MetroAeroSpace54187.2022.9856049

[7] C. Buonagura et al., Deep Learning for Navigation of Small Satellites about Asteroids: an Introduction to the DeepNav Project, presented at the 2nd International Conference on Applied Intelligence and Informatics, Sep. 2022. https://doi.org/10.1007/978-3-031-25755-1_17

[8] H. Demura et al., Pole and Global Shape of 25143 Itokawa, Science, vol. 312, no. 5778, pp. 1347-1349, Jun. 2006, doi: 10.1126/science.1126574. https://doi.org/10.1126/science.1126574

[9] D. J. Scheeres et al., Heterogeneous mass distribution of the rubble-pile asteroid (101955) Bennu, Sci. Adv., vol. 6, no. 41, p. eabc3350, Oct. 2020, doi: 10.1126/sciadv.abc3350. https://doi.org/10.1126/sciadv.abc3350

[10] D. J. Scheeres et al., The dynamic geophysical environment of (101955) Bennu based on OSIRIS-REx measurements, Nat Astron, vol. 3, no. 4, pp. 352-361, Mar. 2019, doi: 10.1038/s41550-019-0721-3. https://doi.org/10.1038/s41550-019-0721-3

[11] D. A. Vallado and W. D. McClain, Fundamentals of astrodynamics and applications, 3. ed., 1. printing. Hawthorne, Calif.: Microcosm Press [u.a.], 2007. https://doi.org/10.1016/S1874-9305(07)80003-3

[12] O. Montenbruck and E. Gill, Satellite orbits: models, methods, and applications: with ... 47 tables, 1. ed., corr. 3. printing. Berlin Heidelberg: Springer, 2005.

[13] B. D. Tapley, B. E. Schutz, and G. H. Born, Statistical orbit determination. Amsterdam ; Boston: Elsevier Academic Press, 2004. https://doi.org/10.1016/B978-012683630-1/50020-5

Aerospace Science and Engineering - III Aerospace PhD-Days   Materials Research Forum LLC
Materials Research Proceedings 33 (2023) 269-276   https://doi.org/10.21741/9781644902677-39

Simulations of vertical sloshing in a partially filled rectangular tank subjected to time-periodic excitation

Daniele Rossi [1,a] *

[1]Dipartimento di Ingegneria Meccanica e Aerospaziale, Sapienza Università di Roma, via Eudossiana 18, Rome, 00184, Italy

[a]daniele.rossi@uniroma1.it

Keywords: Sloshing, Multiphase Flows, VOF Solver, Energy Dissipation

Abstract. An in-house Navier-Stokes multiphase VOF (Volume-Of-Fluid) solver is used to study sloshing phenomena in a partially filled rectangular tank subjected to vertical sinusoidal excitation. The flow dynamics is found to be significantly affected by vertical acceleration and forcing frequency. Specifically, when the acceleration is strong and the excitation frequency is low, the flow exhibits a more chaotic and three-dimensional nature. Consequently, certain properties such as the energy dissipation and the mixing efficiency of the system are poorly predicted by two-dimensional simulations in that range of parameters, making more expensive three-dimensional simulations necessary. The time history of the sloshing force and instantaneous flow visualizations are used to analyze the effects of liquid impacting on the walls on energy exchanges between the fluid and the tank. Finally, the evolution of mixing efficiency and its influence on the energy losses are discussed.

Introduction

The term sloshing generally refers to motion of free liquid inside a partially filled container, resulting from any form of disturbance or imparted excitation. Depending on the type of disturbance and container shape, the liquid free surface can experience various types of motion. Vertical sloshing is a typical phenomenon for fluid stowed in aircraft tanks. The recent studies of Saltari et al. [1] showed that this type of sloshing yields substantial increase of the overall structural damping, hence its accurate prediction could make it possible to design less conservative aircraft configurations, thus enabling lighter structures.

Whereas lateral and rotational sloshing have been popular research topics for scientists and engineers [2], the literature concerning vertical motion is comparatively scarcer, probably because much less amenable to analytical treatment. Some insight can however be gained from experiments and numerical simulations [1,2,3,4] and with nonlinear reduced-order model [5, 6,7,8].

Numerical studies were initially based on potential flow models, which however require ad-hoc corrections to correctly predict energy dissipation caused by viscous effects [9]. Navier-Stokes simulations have only become available in recent times [10]. Canonical models for interface tracking based on the Volume-Of-Fluid (VOF) [11] or the Level-Set method [12] are computationally quite expensive, hence, sloshing usually has been studied with either the Particle Finite Element Method [13] and the Material Point Method [14]. A powerful alternative resides in the Smoothed Particle Hydrodynamics (SPH) approach [4,15], which is a fully Lagrangian method particularly suitable for solving fluid problems in the presence of large deformation of the free surface. However, it has been reported that in phenomena such as high fragmentation and violent free surface impacts, the predictive power of SPH is significantly reduced [16].

In this study, a high-fidelity Navier-Stokes solver based on the VOF method is used to examine liquid sloshing inside a rectangular tank set into vertical harmonic motion, with particular focus on the sloshing-induced dissipation. The equations are solved by using an energy-preserving numerical algorithm, without resorting to any turbulence model. The main reference is the recent comprehensive experimental study of Saltari et al. [1], covering a wide range of excitation frequencies and acceleration amplitudes. Numerical simulations are used to get additional insight into the sloshing dynamics, especially in terms of three-dimensional effects and water/air mixing efficiency. In the experimental works mentioned above, the main attention is generally given to the flow dynamics and the resulting energy exchange between fluid and structure, but the mixing process is hardly discussed. The color function in the VOF method is here exploited to analyze the volume fractions of the two fluids across the tank, and to quantify the overall mixing efficiency. In addition, the color function allows to precisely track the position of the fluids interface, thus allowing detailed flow visualizations of the three-dimensional sloshing dynamics, even in highly chaotic configurations.

Numerical formulation

The two-phase flow of air and water taking place during the sloshing event is numerically simulated using an incompressible gas/liquid Navier-Stokes solver. The fluids are assumed to be immiscible, and the interface is implicitly tracked by means of an indicator function. The relevant governing equations are as follows

$$\nabla \cdot \mathbf{u} = 0 \,, \tag{1}$$

$$\frac{\partial u}{\partial t} + \nabla \cdot (uu) = \frac{1}{\rho}\left[-\nabla p + \nabla \cdot \Sigma + \mathbf{f}_\sigma\right] + \left(-g + a(t)\right)\mathbf{k} \,, \tag{2}$$

where $\mathbf{u} = \mathbf{u}\,(\mathbf{x}, t)$ is the fluid velocity, $p = p\,(\mathbf{x}, t)$ is the pressure, $\rho = \rho\,(\mathbf{x}, t)$ is the density, $\Sigma = 2\mu S$ is the viscous stress tensor, where $\mu = \mu\,(\mathbf{x}, t)$ is the dynamic viscosity and $S = (\nabla u + \nabla u^T)$ is the rate of strain or deformation tensor, $a(t)$ is the acceleration imposed to the tank, g is the gravity acceleration and $\mathbf{k}$ is the unit vector oriented upwards. Although surface tension only acts at the interface, its effects are modelled as a distributed volumetric force $\mathbf{f}_\sigma = \mathbf{f}_\sigma(\mathbf{x}, t)$, with:

$$\mathbf{f}_\sigma = \sigma \kappa \delta(x - x_s)n \,, \tag{3}$$

where σ is the surface tension coefficient, κ is the local curvature of the interface between the two fluids, $\mathbf{n}$ is the unit normal to the interface, and δ is the Dirac function concentrated at the interface points x_s [17]. The Navier-Stokes equations are solved with a classical projection method [18]. Time integration is carried out by means of the Adams-Bashforth explicit scheme for the convective terms and for the off-diagonal part of the viscous terms, and the Crank-Nicolson scheme for the diagonal diffusion terms. The advection of the material interface is carried out by means of an algebraic VOF method. The numerical scheme for the advection is based on a total variation diminishing (TVD) discretization with flux limiter tailored for effective representation of binary functions [19].

Let χ be a passive tracer advected by a continuous divergence-free velocity field $\mathbf{u}$, it shall satisfy the scalar advection equation

$$\frac{\partial \chi}{\partial t} + \nabla \cdot (\chi \mathbf{u}) = 0 \,, \tag{4}$$

For the case of two immiscible fluids under scrutiny χ is in principle either 1 or 0. In the VOF method a color function C is introduced to approximate the cell average of χ, which in the illustrative case of one space dimension is defined as

$$C_i = \frac{1}{\Delta x_i} \int_{x_{i-1/2}}^{x_{i+1/2}} \chi(x, t^n)dx \,, \tag{5}$$

Aerospace Science and Engineering - III Aerospace PhD-Days   Materials Research Forum LLC
Materials Research Proceedings 33 (2023) 269-276   https://doi.org/10.21741/9781644902677-39

where $\Delta x_i = x_{i+1/2} - x_{i-1/2}$ is the cell size, while t^n is the n-th discrete time step. Equation (4) is then discretized in time yielding

$$C_i^{n+\frac{1}{2}} = C_i^{n-\frac{1}{2}} - \frac{1}{\Delta x_i}\left(\hat{f}_{i+\frac{1}{2}} - \hat{f}_{i-\frac{1}{2}}\right),$$ (6)

where the numerical flux $\hat{f}_{i+1/2}$ is an approximation for χ transported through the cell interface $x_{i+1/2}$ during the time interval $(t^{n-1/2}, t^{n+1/2})$. Details of the algorithm are provided in the original reference [20]. After computing the color function C, density and viscosity are determined by $\rho = \tilde{C}\rho_1 + (1 - \tilde{C})\rho_2$ and $\mu = \tilde{C}\mu_1 + (1 - \tilde{C})\mu_2$, where $\tilde{C}$ is a smoothed version of the color function, evaluated by averaging C over twenty-seven neighboring cells [20], the subscript 1 is used to denote water properties, while the subscript 2 is used for air.

The momentum equations are discretized in the finite difference framework with a staggered grid layout, where the pressure, the color function and the material properties are defined at the cell centers, whereas the velocity components are stored at the middle of the cell faces [21]. A centered second-order discretization is used for the convective terms [18,21] and diffusive terms [17], which is generally best suitable for fully resolved flows, yielding discrete preservation of the total kinetic energy in the case of uniform density. The Poisson equation is discretized by a finite-difference scheme and the resulting sparse linear system is solved by iterative methods. In particular, the HYPRE library [22] is found to be very efficient in massively parallel computations. The surface tension effects are modelled through the continuum surface force method [23], which is used to determine the equivalent local body force, and the interface curvature is evaluated through a modified version of the height function technique [20,24,25]. All details regarding the VOF solver can be found in [26].

Results and discussion

Hereafter we report the results of numerical simulations of vertical sloshing of water in air for a partially filled rectangular tank subjected to vertical sinusoidal excitation (see Fig. 1). We have carried out many two-dimensional simulations to characterize the effects of acceleration and excitation frequency on the sloshing dynamics. A tank with size $L \times H = 0.1172m \times 0.0272m$ is considered, replicating the experimental setup of Saltari et al. [1], and covered with a uniform Cartesian mesh of 128×64 grid nodes. A limited number of three-dimensional simulations has also been conducted at selected conditions to verify possible limitations of two-dimensional simulations. The governing equations are solved in a non-inertial reference frame attached to the tank, with the origin placed at the middle of its base. External excitation is modeled through imposed harmonic acceleration added to the right-hand side of the momentum equation (2), $a(t) = a_0 sin(\omega t)$, with $\omega = 2\pi f$. The initial filling level $\alpha = h/H$, expressing the quiescent fluid level as a fraction of the total depth of the tank, is set to $\alpha = 1/2$. The densities of water and air were set to $998\, Kg/m^3$ and $1.225\, Kg/m^3$, respectively, and their dynamic viscosities to $8.899 \times 10^{-2}\, Kg/(ms)$, $1.831 \times 10^{-3}\, Kg/(ms)$, respectively. The surface tension coefficient was set to $0.07199 N/m$.

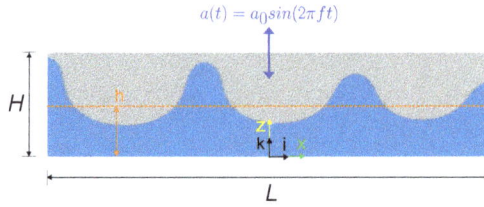

Figure 1: Computational setup for numerical simulations of vertical sloshing. L and H are the width and height of the computational domain, respectively, h is the initial level of the liquid, $a(t)$ is the vertical acceleration in vertical sloshing, f is the forcing frequency.

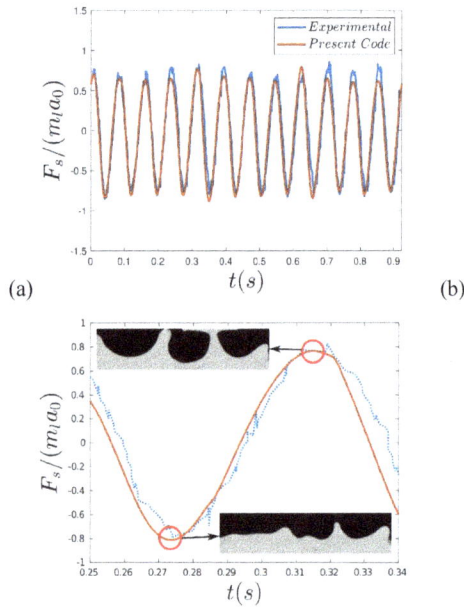

Figure 2: Time history of vertical sloshing force, compared with measurements of Saltari et al. [1] for $a_0 = 1.5g$ and $f = 13$ Hz (a) and detail of the sloshing force trend (b). The orange line and the blue line denote the numerical and experimental results, respectively. Instantaneous snapshots from the numerical simulation are shown at the time instants at which the sloshing force is minimum/maximum.

The sloshing vertical force is determined as

$$F_s = -\int_{walls} p\, \boldsymbol{n} \cdot \boldsymbol{k}\ dx, \tag{7}$$

where integration is carried out over the top and bottom walls of the tank, p is the pressure applied to the walls, and $\boldsymbol{n}$ and $\mathbf{k}$ are the wall-normal and vertical unit vector, respectively. F_s is normalized with respect to the maximum of the inertial force of the liquid, namely $m_l a_0$, with m_l the fluid mass. A quantitative comparison with the experiment carried out by Saltari et al. [1] is reported in Fig.2a, where the time history of the sloshing force over twelve forcing cycles is shown for $a_0 = 1.5g$ and $f = 13\ Hz$. We can see that the phase of the sloshing force is correctly predicted and very good agreement between simulations and experiments is found. In Fig.2b, a detail of the sloshing force in the time interval $0.25s - 0.34\ s$ is considered. Snapshots of numerical simulation illustrate the flow configuration at time instants at which the sloshing force has a (positive or negative) peak. The figure well highlights that positive peaks of the sloshing force are related to fluid impacting the ceiling of the tank, whereas negative peaks occur when the fluid retreats towards the bottom wall. This explains that a large fraction of the energy transfer between structure and fluid occurs during impacts, and that the primary mechanism for dissipation is directly associated with these fluid-structure collisions and interface fragmentation [27].

Once the vertical forces are calculated, the average work exchanged between fluid and shaker can be evaluated at any instant time t_n and over an arbitrary number $(2P)$ of periods T as

$$W(\omega, a_0) = \frac{1}{2P} \int_{t_n - PT}^{t_n + PT} F_s(\omega, a_0) \dot{s}(\tau)\ d\tau, \tag{8}$$

which can be expressed in nondimensional form as $\overline{W} = W/(m_l s_0^2 \omega^2)$, with s_0 displacement amplitude. The average work evaluated from numerical simulations for $a_0 = 1.5g$ and $a_0 = 3g$ is reported in Fig.3 as a function of the forcing frequency $f(Hz)$. The figure shows that for $a_0 = 1.5g$ and high forcing frequency, two-dimensional numerical simulations replicate the experimental behavior very well, whereas at $a_0 = 3g$ and/or low forcing frequency the energy dissipation is largely underestimated. These findings suggest that 2D simulations may not be appropriate to accurately characterize the energy exchanges under high forcing acceleration, where the dynamics of the air/water interface is more chaotic and three-dimensional. The numerical results obtained from 3D simulations in fact generally show improved prediction as compared to the 2D simulations. In this respect, we note that the predicted energy dissipation approaches the experimental value as the grid resolution is improved, although convergence is clearly far from being achieved, for both acceleration amplitudes. In order to accurately reproduce the fluid impacts responsible for the exchange of energy, a much finer mesh is required especially for the case with $a_0 = 3g$, where discrepancies with experimental data are more significant.

Finally, the air/water mixing process in vertical sloshing is quantified by considering the vertical profiles of the time-averaged color function, $\bar{C}(z)$. As previously pointed out, C indicates the volumetric fraction of liquid inside a computational cell, so that $C = 1$ means that the cell is filled with water, whereas if $C = 0$ the cell is filled with air. $\bar{C}$ is defined as

$$\bar{C}(z) = \frac{1}{2\Delta T} \int_{t_k - \Delta T}^{t_k + \Delta T} \left(\frac{1}{L} \int_{-L/2}^{L/2} C(x, z, t) dx \right) dt, \tag{9}$$

where the averaging interval ΔT is taken to be twenty-four forcing cycles. A global index is then introduced to quantify the overall mixing efficiency,

$$\eta = 1 - 2\sqrt{\frac{1}{H} \int_0^H (\bar{C}(z) - 0.5)^2\ dz}, \tag{10}$$

such that $\eta = 1$ in the case of perfect mixing of water and air ($\bar{C} = 0.5$), whereas $\eta = 0$ if water and air were completely segregated. Figure 4 shows the profile of η as a function of frequency for cases with $a_0 = 1.5g$ and $a_0 = 3g$. The figure shows that increasing the forcing frequency and reducing the acceleration results in less efficient mixing between fluid and air. A relatively large jump occurs between $f = 11\ Hz$ and $f = 13\ Hz$, for $a_0 = 1.5g$, which indicates the transition to the standing waves regime, implying a significant reduction of the mixing efficiency. It should also be noted that the three-dimensional simulations are in good agreement with the corresponding two-dimensional results for $a_0 = 1.5g$, whereas for $a_0 = 3g$, the 3D simulations suggest higher mixing. Higher mixing efficiency is found in cases with higher energy dissipation, i.e. lower frequency and higher acceleration. Such behavior confirms that energy dissipation is associated with fluid-structure impacts and fragmentation of the liquid-gas interface [27].

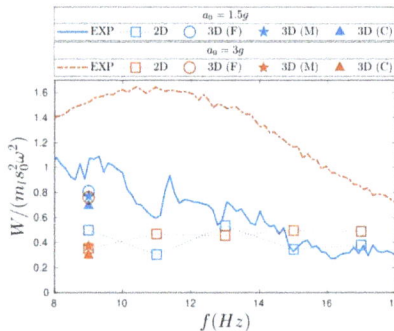

Figure 3: normalized dissipated energy $W / (m_l s_0^2 \omega^2)$ as a function of forcing frequency $f(Hz)$ for $a_0 = 1.5g$ and $a_0 = 3g$. The numerical results obtained from 2D and 3D simulations are compared with the experimental data (EXP) of Saltari et al. [1]. 3D simulations are performed with the three different grids: grid $F = 256 \times 128 \times 192$, grid $M = 192 \times 96 \times 128$, grid $C = 128 \times 64 \times 96$.

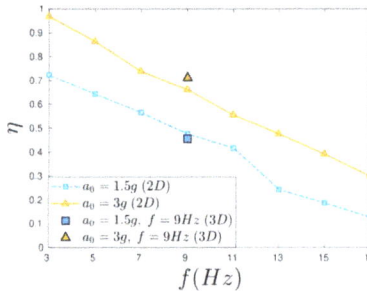

Figure 4: Mixing parameter η as a function of frequency, for $a_0 = 1.5g$ and $a_0 = 3g$.

Conclusion

By analyzing the trend of the vertical sloshing force, together with instantaneous snapshots from the numerical simulations, we concluded that positive and negative peaks of $F_s(t)$ correspond to strong liquid impacts at the top and bottom walls, which occur slightly after the inversion of the

Aerospace Science and Engineering - III Aerospace PhD-Days
Materials Research Proceedings 33 (2023) 269-276

Materials Research Forum LLC
https://doi.org/10.21741/9781644902677-39

tank motion. The characteristic flow dynamics are found to be significantly affected by the forcing parameters and to exhibit more chaotic and three-dimensional behavior in the case of strong accelerations and low forcing frequencies. Therefore, certain characteristics of the flow could not be properly reproduced by resorting only to two-dimensional simulations, and more expensive three-dimensional simulations have been carried out for comparison for the forcing frequency $f = 9\ Hz$. The results show significant improvement with respect to the 2D simulations, although the energy dissipation still exhibits some scatter. Finally, the dependence of liquid-air mixing on acceleration and frequency was quantified in terms of a global mixing efficiency indicator, arriving at the conclusion that mixing can be optimized by enhancing the acceleration amplitude and by reducing the excitation frequency. The numerical results have also highlighted that the increase of mixing corresponds to increase of the energy dissipation, suggesting that power exchanges between fluid and tank walls are closely related with the breaking of the liquid-gas interface and with fluid-wall impacts, which occur as a result of harmonic excitation. Follow-up studies might include deeper assessment of the mechanisms responsible for energy loss (e.g. impacts of liquid on the tank walls or enhanced mixing between the two phases), and assessment of the effect of turbulence models, here intentionally disregarded.

References

[1] Saltari, F., Pizzoli, M., Coppotelli, G., Gambioli, F., Cooper, J.E., Mastroddi, F., 2022a. Experimental characterisation of sloshing tank dissipative behaviour in vertical harmonic excitation. J. Fluids Struct. 109, 103478. https://doi.org/10.1016/j.jfluidstructs.2021.103478

[2] Faltinsen, O., 2017. Sloshing. Adv. Mech. .

[3] Titurus, B., Cooper, J.E., Saltari, F., Mastroddi, F., Gambioli, F., 2019. Analysis of a sloshing beam experiment, in: Proceedings of the International Forum on Aeroelasticity and Structural Dynamics, Savannah, GA, USA, pp. 10–13

[4] Calderon-Sanchez, J., Martinez-Carrascal, J., Gonzalez-Gutierrez, L.M., Colagrossi, A., 2021. A global analysis of a coupled violent vertical sloshing problem using an sph methodology. Eng. Appl. Comput. Fluid Mech. 15, 865–888. https://doi.org/10.1080/19942060.2021.1921849

[5] De Courcy, J.J., Constantin, L., Titurus, B., Rendall, T., Cooper, J.E., 2021. Gust loads alleviation using sloshing fuel, in: AIAA Scitech 2021 Forum, p. 1152. https://doi.org/10.2514/6.2021-1152

[6] Pizzoli, M., Saltari, F., Mastroddi, F., Martinez-Carrascal, J., González-Gutiérrez, L.M., 2022b. Nonlinear reduced-order model for vertical sloshing by employing neural networks. Nonlinear Dyn. 107, 1469– 1478. https://doi.org/10.1007/s11071-021-06668-w

[7] Saltari, F., Pizzoli, M., Gambioli, F., Jetzschmann, C., Mastroddi, F., 2022b. Sloshing reduced-order model based on neural networks for aeroelastic analyses. Aerosp. Sci. Technol. 127, 107708. https://doi.org/10.1016/j.ast.2022.107708

[8] Pizzoli, M., Saltari, F., Coppotelli, G., Mastroddi, F., 2022a. Experimental validation of neural-network-based nonlinear reduced-order model for vertical sloshing, in: AIAA Scitech 2022 Forum, p. 1186. https://doi.org/10.2514/6.2022-1186

[9] Antuono, M., Bouscasse, B., Colagrossi, A., Lugni, C., 2012. Two-dimensional modal method for shallow-water sloshing in rectangular basins. J. Fluid Mech. 700, 419– 440. https://doi.org/10.1017/jfm.2012.140

[10] Demirbilek, Z., 1983. Energy dissipation in sloshing waves in a rolling rectangular tank — iii. results and applications. Ocean Eng. 10, 375– 382. https://doi.org/10.1016/0029-8018(83)90006-9

Aerospace Science and Engineering - III Aerospace PhD-Days   Materials Research Forum LLC
Materials Research Proceedings 33 (2023) 269-276   https://doi.org/10.21741/9781644902677-39

[11] Hirt, C.W., Nichols, B.D., 1981. Volume of fluid (VOF) method for the dynamics of free boundaries. J. Comput. Phys. 39, 201–225. https://doi.org/10.1016/0021-9991(81)90145-5

[12] Sussman, M., Fatemi, E., Smereka, P., Osher, S., 1998. An improved level set method for incompressible two-phase flows. Comput. Fluids 27, 663– 680. https://doi.org/10.1016/S0045-7930(97)00053-4

[13] Gimenez, J.M., González, L.M., 2015. An extended validation of the last generation of particle finite element method for free surface flows. J. Comput. Phys. 284, 186–205. https://doi.org/10.1016/j.jcp.2014.12.025

[14] Li, J., Hamamoto, Y., Liu, Y., Zhang, X., 2014. Sloshing impact simulation with material point method and its experimental validations. Comput. Fluids 103, 86–99. https://doi.org/10.1016/j.compfluid.2014.07.025

[15] Colagrossi, A., Landrini, M., 2003. Numerical simulation of interfacial flows by smoothed particle hydrodynamics. J. Comput. Phys. 191, 448– 475. https://doi.org/10.1016/S0021-9991(03)00324-3

[16] Banim, R., Lamb, R., Bergeon, M., 2006. Smoothed particle hydrodynamics simulation of fuel tank sloshing. Proceedings 25th international congress of the aeronautical sciences .

[17] Tryggvason, G., Scardovelli, R., Zaleski, S., 2011. Direct Numerical Simulations of Gas-Liquid Multiphase Flows. Cambridge University Press.

[18] Orlandi, P., 2012. Fluid flow phenomena: a numerical toolkit. volume 55. Springer Science & Business Media.

[19] Pirozzoli, S., Di Giorgio, S., Iafrati, A., 2019. On algebraic TVD-VOF methods for tracking material interfaces. Comput. Fluids 189, 73–81. https://doi.org/10.1016/j.compfluid.2019.05.013

[20] Popinet, S., 2009. An accurate adaptive solver for surface-tension-driven interfacial flows. J. Comput. Phys. 228, 5838–5866. https://doi.org/10.1016/j.jcp.2009.04.042

[21] Harlow, F.H., Welch, J.E., 1965. Numerical calculation of time-dependent viscous incompressible flow of fluid with free surface. Phys. Fluids 8, 2182–2189. https://doi.org/10.1063/1.1761178

[22] Falgout, R.D., Yang, U.M., 2002. hypre: A Library of High Performance Preconditioners, in: International Conference on computational science, pp. 632–641. https://doi.org/10.1007/3-540-47789-6_66

[23] Brackbill, J.U., Kothe, D.B., Zemach, C., 1992. A continuum method for modeling surface tension. J. Comput. Phys. 100, 335–354. https://doi.org/10.1016/0021-9991(92)90240-Y

[24] Cummins, S.J., Francois, M.M., Kothe, D.B., 2005. Estimating curvature from volume fractions. Comput. Struct. 83, 425–434. https://doi.org/10.1016/j.compstruc.2004.08.017

[25] Francois, M.M., Cummins, S.J., Dendy, E.D., Kothe, D.B., Sicilian, J.M., Williams, M.W., 2006. A balanced-force algorithm for continuous and sharp interfacial surface tension models within a volume tracking framework. J. Comput. Phys. 213, 141–173. https://doi.org/10.1016/j.jcp.2005.08.004

[26] Di Giorgio, S., Pirozzoli, S., Iafrati, A., 2022. On coherent vortical structures in wave breaking. J. Fluid Mech. 947, A44. https://doi.org/10.1017/jfm.2022.674

[27] Marrone, S., Colagrossi, A., Di Mascio, A., Le Touzé, D., 2015. Prediction of energy losses in water impacts using incompressible and weakly compressible models. J. Fluids Struct. 54, 802–822. https://doi.org/10.1016/j.jfluidstructs.2015.01.014

Aerospace Science and Engineering - III Aerospace PhD-Days
Materials Research Proceedings 33 (2023) 277-280

Materials Research Forum LLC
https://doi.org/10.21741/9781644902677-40

Storage and visualization on-the-ground and in near real-time of the data measured by the optical sensors connected to a flying test bench

Antonio Costantino Marceddu[1,a,*], Bartolomeo Montrucchio[1,b]

[1]Politecnico di Torino, Corso Duca degli Abruzzi 24, 10129 Torino, ITALY

[a]antonio.marceddu@polito.it, [b]bartolomeo.montrucchio@polito.it

Keywords: Computer Graphics, Data Visualization, Databases, Graphical User Interfaces, Middleware, Optical Fiber Sensors, Remotely Piloted Aircraft, Wireless Communication

Abstract. Nowadays, sensors are massively used for different types of applications, ranging from the control of production processes to the continuous monitoring of systems of various kinds. Often, due to the large amount of data collected, it can be difficult to fully understand them or extract their inherent value. For this reason, the main research objective discussed in this paper regards the creation of intuitive tools for the visualization of sensor data in a human-readable way so as to facilitate their deep analysis. One of the main cases of study is linked to the PhotoNext Interdepartmental Center and regards the intuitive visualization of near real-time data coming from a flying test bench.

Introduction

The *sensor* is a device that, based on the value assumed by a given input quantity, returns an output signal dependent on it. The term *transducer* is often used interchangeably as it is frequently based on the same working principle, but they differ in their function. The sensor is normally used to control or regulate the system in which it is immersed through the measurement of a certain variable of interest, while the transducer is normally employed for direct use, the recording or measurement of the conversion made on the variation of a physical quantity into a corresponding variation of an alternative physical quantity [1].

The sensors can be differentiated according to the type of:
- Input signals, such as temperature, light, and speed.
- Output signals such as acoustic, optical, and electrical.

They are adopted in various contexts, such as saving electricity through the automatic heating and switching on of the lights based on the presence of people in a room, for emergency situations such as the actuation of automatic braking of vehicles based on the detection of obstacles, or even the control of air quality and the presence of toxic gases [2].

Throughout their entire lifetime, sensors can produce a wide amount of data. Measured variables can sometimes contain more information than is visible directly from the collected data. This information can be brought to light through special tools, which often make use of an intuitive visualization carried out through the use of graphs, tables, or others. The purpose of this paper is therefore to create useful tools for the analysis and visualization of data coming from different types of sensors.

Aerospace Science and Engineering - III Aerospace PhD-Days
Materials Research Proceedings 33 (2023) 277-280

Materials Research Forum LLC
https://doi.org/10.21741/9781644902677-40

Figure 1 - Logo of the PhotoNext Interdepartmental Center. Taken from [1].

Since then, the authors's attention has been posted to the following kind of sensors:

- Fiber Bragg Gratings (FBG) sensors, which are optical sensors fabricated within the core of a single-mode optical fiber through the use of patterns of intense UV laser light that can vary spatially [3].
- Camera sensors, which are commonly used to digitize images for processing, analysis, or recording purposes in cameras, camcorders, scanners, etc. [1].
- Photodiode sensors, which are used to obtain electrical energy by converting light energy [4].

Data from SCADA systems (an acronym for Supervisory Control And Data Acquisition) are also being analyzed. They are systems composed of software and hardware part for monitoring, analyzing, and controlling industrial processes and devices [5].

Research Project on Fiber Bragg Gratings Sensors

Concerning the FBG sensors, the inherent research presented in this paper is carried out in collaboration with the Interdepartmental Center of the Politecnico di Torino dedicated to the study of photonic technologies, called *PhotoNext* [6]. Fig. 1 shows the logo of the PhotoNext Interdepartmental Center. It brings together the work of five different departments:

- Department of Electronics and Telecommunications (DET).
- Department of Mechanical and Aerospace Engineering (DIMEAS).
- Department of Applied Science and Technology (DISAT).
- Department of Environment, Land and Infrastructure Engineering (DIATI).
- Department of Control and Computer Engineering (DAUIN).

Its work combines pure research with applied experimentation activities and mainly focuses on:

- Optical components and sensors for industrial applications.
- High-speed optical fiber communication systems.

The research presented in this paper and located inside the PhotoNext Interdepartmental Center is the result of a collaboration between the DIMEAS and the DAUIN. It concerns the recording and visualization in near real-time and on-the-ground of temperature and strain data coming from a flying test bench [7]. This flying test bench is composed of two interconnected parts, defined *physical system* and *software applications*. The *physical system* encompasses the device to be measured, which is the model aircraft called *Anubi* created by the Icarus student team of the Politecnico di Torino [8], and the FBG sensors needed to perform temperature and strain measurements. It also includes the *SmartScan interrogator* [9], which is the device currently used for reading the values coming from the FBG sensors, a *Raspberry Pi™ 3 Model B+* System-On-Chip (SOC) connected both to the interrogator through an Ethernet connection and to the Internet via a 4G USB dongle, and the batteries needed to power both. Fig. 2 shows the complete physical system.

Aerospace Science and Engineering - III Aerospace PhD-Days   Materials Research Forum LLC
Materials Research Proceedings 33 (2023) 277-280   https://doi.org/10.21741/9781644902677-40

*Figure 2 - The complete physical system that makes up the flying test
bench. In order: (1) Anubi model aircraft made by the Icarus Student
Team, (2) interrogator, (3) Raspberry Pi^{TM}, (4) 4G Internet Dongle.*

The software applications oversee reading, transmitting, storing, and displaying data to the end user. Their description will follow the same order as the data flow. The first application that will be discussed is therefore the *Middleware*, an application written in C and C++ that runs on the Raspberry Pi^{TM} to get sensor data from the interrogator and send them to the *Cloud Database*. The *Cloud Database* is based on *MongoDB®* and stores sensor data and forwards them to any listening *3D Viewer* applications. Finally, the *3D Viewer* application allows the almost real-time visualization of the deformation and temperature data measured during the flight through the use of:

- A *graph visualization*, capable of providing information related to the variation of the value of the sensor with respect to their base value (delta).

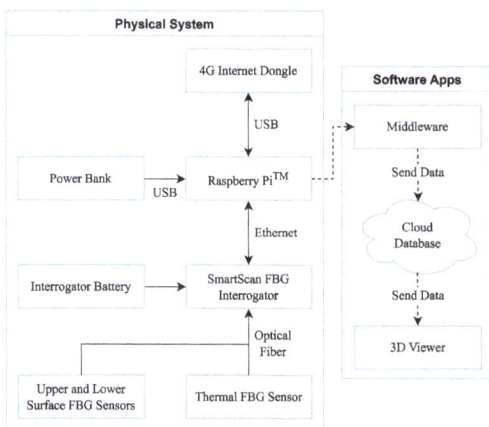

*Figure 3 - A block diagram representing the flying test bench in its
entirety.*

Aerospace Science and Engineering - III Aerospace PhD-Days   Materials Research Forum LLC
Materials Research Proceedings 33 (2023) 277-280   https://doi.org/10.21741/9781644902677-40

- A *3D heat map visualization*, capable of providing more qualitative information than that provided by the graphic visualization since it varies the color gradation of the area in which a particular sensor is positioned based on its current value. However, an indication of the approximate physical location of the sensors in the system is also provided. This is especially useful for monitoring areas of the system under stress.

A block diagram showing the full system is represented in Fig. 2.

To validate the entire system, several tests were carried out at the Tetti Neirotti runway, located near Turin. For the interested reader, please refer to the information contained in [7].

Conclusions

This paper concerns the intuitive visualization of data measured by sensors of different types. In particular, it regards the implementation of a complete flying test bench capable of reading, transmitting, storing, and displaying the data measured in near real-time to the user on the ground. The authors of this paper are currently working on a more complete version of the Desktop version of the *3D Viewer* together with a new Augmented Reality (AR) version. They are also working on adding new sensors to enrich the measurements currently made by the flying test bench.

Funding

This work was carried out under the PhotoNext initiative at Politecnico di Torino (http://www.photonext.polito.it/).

References

[1] Treccani S.p.A., Sensore. Link: https://www.treccani.it/vocabolario/sensore/

[2] Mohd Javaid, Abid Haleem, Shanay Rab, Ravi Pratap Singh, Rajiv Suman, Sensors for daily life: A review, Sensors International, vol. 2, 2021, 100121, ISSN 2666-3511. https://doi.org/10.1016/j.sintl.2021.100121

[3] Smart Fibres Ltd., Our Technology. Link: https://www.smartfibres.com/technology

[4] Ravi Teja, What is a Photodiode? Working, V-I Characteristics, Applications, May 15, 2021. Link: https://www.electronicshub.org/photodiode-working-characteristics-applications/

[5] Scada International A/S, What is SCADA?. Link: https://scada-international.com/what-is-scada/

[6] PhotoNext, PhotoNext Official Site. Link: https://www.photonext.polito.it/it/

[7] A. C. Marceddu et al., Air-to-Ground Transmission and Near Real-Time Visualization of FBG Sensor Data via Cloud Database, IEEE Sensors Journal, vol. 23, no. 2, pp. 1613-1622, Jan. 15, 2023. https://dx.doi.org/10.1109/JSEN.2022.3227463

[8] Icarus Student Team, Icarus Student Team Official Site. Link: https://icarus.polito.it/

[9] Smart Fibres Ltd., SmartScan interrogator. Link: https://www.smartfibres.com/products/smartscan

Aerospace Science and Engineering - III Aerospace PhD-Days   Materials Research Forum LLC
Materials Research Proceedings 33 (2023) 281-287   https://doi.org/10.21741/9781644902677-41

HACK: a Holistic modeling Approach for Cubesat cyberattacKs

Salvatore Borgia[1,a] *, Francesco Topputo[b], Stefano Zanero[c]

[1] Politecnico di Milano, Italy

[a] salvatore.borgia@polimi.it, [b] francesco.topputo@polimi.it, [c] stefano.zanero@polimi.it,

Keywords: Cyberattacks, Digital Model, CubeSat, Cybersecurity, Holistic Modeling, Multiphysics Simulation, Digital Twin

Abstract. In recent years, the threat of cyberattacks has been growing rapidly in numerous industrial sectors that have an impact on our daily life. One of these is the space industry, where the risk of hacking a single satellite can lead to dangerous effects not only for economics but also for Earth critical infrastructure like: transportation systems, water networks, and electric grid. The vulnerability of complex space systems has already been demonstrated in the past. In 1998, for example, hackers took control of the ROSAT X-Ray satellite pointing its solar panels directly to the Sun and causing physical damage. Nowadays, since the attention is moved on small and less sophisticated system, such as CubeSat, the risk of cyber intrusions is even higher as the COTS (Commercial-Off-The-Shelf) technology they use is based on open-source operating systems. In order to counteract this imminent problem, the development of a high-fidelity CubeSat digital model is needed to study and solve related space cybersecurity issues. In fact, thanks to the virtual prototype, what-if simulations can be performed allowing to analyze different cyberattacks scenarios and predict undesirable events on the CubeSat flying on its operative orbit. Moreover, the building of the digital model requires a holistic modeling approach and simulation tools which allows to consider Multiphysics phenomena occurring on the space system itself. Finally, the possibility of connecting the virtual model to a real space system, obtaining the so-called Digital Twin (DT), will help engineers to conduct more accurate actions during the mission.

Introduction

According to G. Falco [1], space assets are sophisticated pieces of equipment with a complex production and operational chain, making space systems vulnerable to cyberattacks. Unlike most critical infrastructure sectors, a space system is not owned by the same organizations that operate the system itself. All of this make the cybersecurity responsibility challenging to be well-defined and assigned. Fig. 1 shows how the cybersecurity risks pathways (colored lines) are accumulated along a typical satellite project. Furthermore, the recent trend to use CubeSat introduces additional cybersecurity risks due to the exploitation of COTS technology which could be an uncontrolled door for external adversary. Considering the growing number of CubeSat orbiting the Earth, malicious organizations can hack only a single unit to provoke collisions phenomena with dangerous effects. This research project is framed within the urgent need of limiting cyberattacks with high-fidelity models and simulation tools. The latter can be designed through the DT and MBSE (Model-Based System Engineering) concepts. The most common definition of DT was formulated by E. Glaessgen et al. [2] as: "An integrated multi-physics, multi-scale, probabilistic simulation of a vehicle or system that uses the best available physical models, sensor updates, fleet history, and so forth, to mirror the life of its flying twin". So, a DT is an emulated version of a real system where mathematical modeling, data-based methods, and hardware are strongly interconnected to have a correct model system identification [3]. NASA was the first association to forge the definition of DT and to build already two identical space vehicles (one on the Earth and the other one in the space) for supporting Apollo program (1961) [4]. Fig. 2 depicts an example of DT application for a bending beam test bench [5]. It is worthy to highlight how the data collected by test bench sensors can be sent to the digital model of the beam and used to refine the virtual model itself.

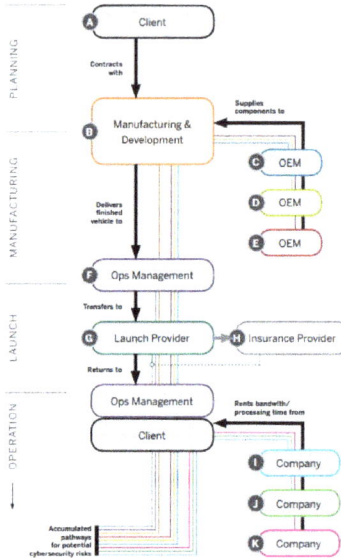

Figure 1: Cybersecurity responsibility landscape for a satellite project [1]

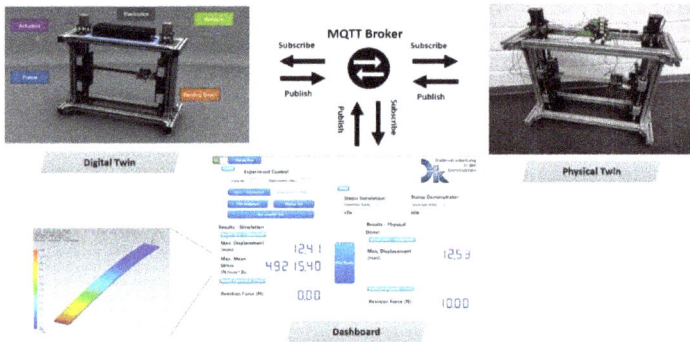

Figure 2: Overview of a DT system [5]

The MBSE method is needed to have a holistic modeling design and to identify the functional relationship between and inside the CubeSat subsystems by defining the relative interfaces, needs, requirements and interconnections. Baker et al. [6] indicates the key MBSE processes (Fig. 3) which shall be applied for each development phases of a project and for each subsystem that composes a space system.

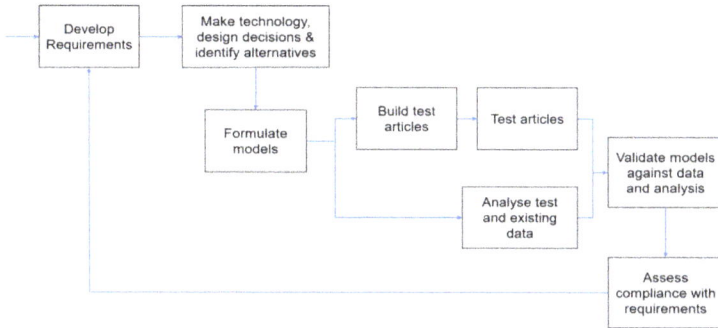

Figure 3: Sub-processes for MBSE method [6]

Methods and tools
In the following sections, the main steps for developing a preliminary digital model of a CubeSat are described starting from the knowledge of the real object to the simulation block which allows to perform system's behavior analyses.

Modeling and Simulation
After the preliminary considerations on space cybersecurity and the importance of simulating cyberattacks on a CubeSat, the main modeling steps [7] to be applied are now presented. First of all, it is important to recall the four stages of the dynamic investigation which leads to system behavior prediction:
- Specify the real system to be examined and deduce a physical model whose behavior is representative of the real one.
- Derive a mathematical model to represent the physical one.
- Solve the differential and algebraic equations of the mathematical model.
- Make design decision or adjust physical parameters to match the virtual space to the real one.

Where the actual system can be the whole CubeSat or a combination of its subsystems (propulsion, electrical, etc.); the physical model is a virtual model obtained through approximations and engineering judgment; the mathematical model can be derived from first principles; the dynamic system can be solved using an appropriate numerical integration scheme; and the refinement model step can be performed adopting the communication real-time interface between the real object and its DT or, for example, collecting on-orbit data to continuously update the physical/mathematical model properties. The physical abstraction step shall be performed in a way that lets to simplify the next simulations considering only dominant dynamics aspects on the system.

The procedure described until now belongs to the causal-modeling approach. Because of the involvement of different CubeSat subsystems in a complex Multiphysics domain, the A-causal method can be useful to simplify the modeling part. In fact, according to this method, the modeling is more focused on the physical connections (connectors or ports) between parts rather than on their interaction in terms of input and output. "There are no input and output variables in real life" (Dr. Yaman Barlas). Fig. 4 shows this difference of modeling approach in the electric domain (capacitor object) using OpenModelica against Simulink language. It is worthy to notice how: in

Aerospace Science and Engineering - III Aerospace PhD-Days   Materials Research Forum LLC
Materials Research Proceedings 33 (2023) 281-287   https://doi.org/10.21741/9781644902677-41

the first case, there is the definition of the elementary port (physical connection), the non-flow variable (voltage), the flow-variable (current), the constitutive and connection equations; on the contrary in the Simulink case, there is the direct translation of the mathematical model where the voltage is assumed as input and the current as output (losing the physical meaning).

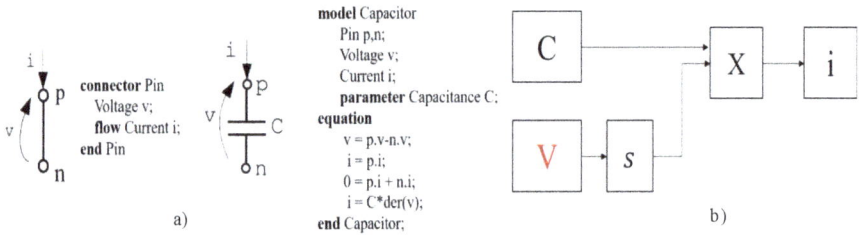

Figure 4: a) Declarative capacitor's model b) Simulink scheme

Within the MBSE framework, the SysML (Systems Modeling Language) can be exploited in order to better define the CubeSat system highlighting its parts or units, and their corresponding requirements. This also allows to better identify: the interconnections between subsystems, and test cases to visualize how a hacker can start an attack and interact with others system blocks. In [8], it is reported an example of SysML application for a crewed Mars mission (exploiting the Cameo Systems Modeler environment), where hierarchically the structure design principle of the heterogenous cyberphysical systems is shown. Finally, the cyberattacks can lead to unpredictable phenomena where the space system is in non-nominal conditions, so the modeling phase shall include also non-linear cases. Fig. 5 shows the pipeline which summarizes what it has been discussed so far.

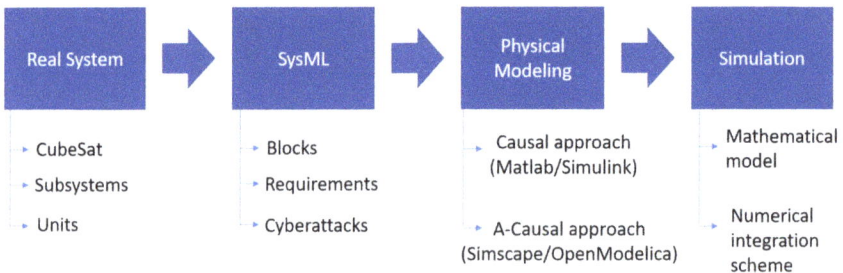

Figure 5: Work pipeline to identify, model, and simulate a CubeSat

Space cybersecurity: threat model

In this section, some of the principal vulnerabilities of a space asset are presented. As suggested by G. Falco [1], using a ground antenna, a black hacker can intercept the IP address from a satellite (CubeSat) internet user and start a TCP/IP connection. From the stolen IP, an attacker can directly uplink false data to the user system connected to that specific IP address. Another cyberattack is the GPS spoofing, where an adversary uses a GPS signal simulator to insert a fake signal behind the true signal and progressively increase the power of the fake signal until the point it is considered real from the satellite's receiver. Moreover, in [9] the technical feasibility of satellite-to-satellite cyberattacks is described. Looking at some past satellite failures (Anik E1, Anik E2, and TDRS 1), we know that they have been occurred because of background radiations or solar flare. The same effects can be obtained with a EMP (Electromagnetic Pulse) attack. In this case, the attacking satellite, equipped by an EMP actuator, shall be oriented towards the victim satellite before to start the attack. Fig. 6 shows graphically the two ways of cyberattack.

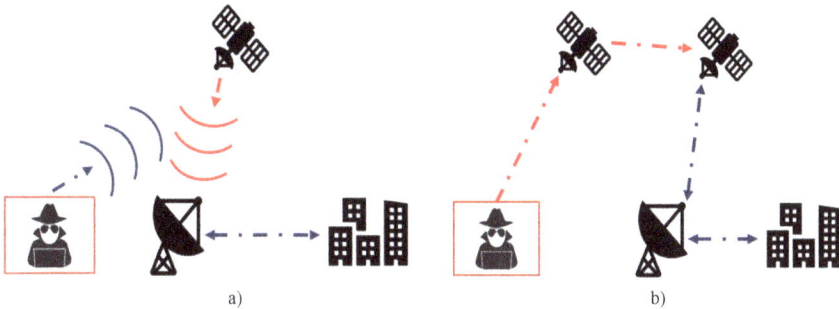

Figure 6: a) Satellite signal spoofing cyberattack b) Satellite-to-Satellite attack scenario

In Fig. 7, the heterogeneous domain of a CubeSat is shown considering the subsystems: OBCS (On-Board Computer Subsystem), AOCS (Attitude and Orbit Control Subsystem), EPS (Electrical Power Subsystem), TT&C (Telemetry, Tracking, and Command), Structure, and Thermal. In real life, each subsystem affects all the others an vice versa. One of the modeling challenges is to keep only the main connections in order to visualize important Multiphysics events for the space mission integrity. In addition, a cyber intrusion path (*1-2-3-4-5-5'*) is highlighted in red. In this scenario, it can be noticed how the insertion of a fake signal received by TT&C subsystem and processed by OBCS, can provoke dangerous internal loops and escalation effects as:

- The fake instructions can change the CubeSat orientation
- The new attitude leads to a different temperature distribution
- The EPS produces new level of power (solar panel exposed area variation to the Sun)
- The OBCS is affected by the new power produced on-board

According to these events, an attacker can send instructions to point solar panel toward the Sun or deep-space, and this can lead to: overheating or structure deformations in the first case, and CubeSat components shut down in the second case (due to insufficient level of power). This kind of failure propagation chain can be analyzed and studied as function of the achieved modeling complexity. The latter allows to reveal more variables and parameters that links a CubeSat physical domain with another.

Aerospace Science and Engineering - III Aerospace PhD-Days   Materials Research Forum LLC
Materials Research Proceedings 33 (2023) 281-287   https://doi.org/10.21741/9781644902677-41

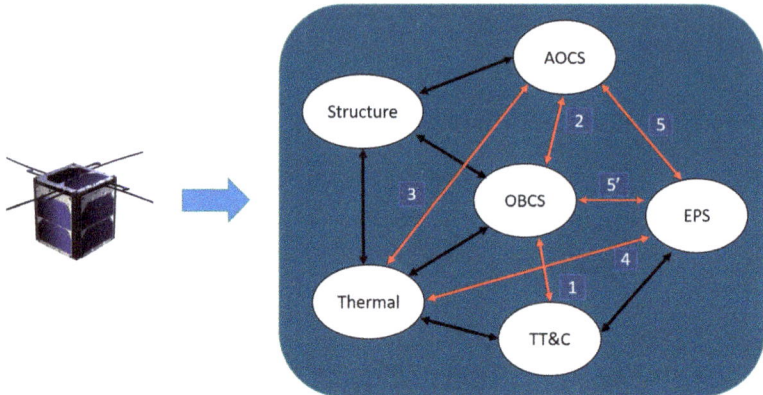

Figure 7: CubeSat domain: the red path shows how a cyber intrusion can spread inside the system

Conclusions

This research project aims to help the realization of a fully-integrated DT of a CubeSat capable to be tested and analyzed under different input conditions (in order to simulate cyberattacks). Thanks to this project, computer science and space engineering will work in synergy, during a space mission development phase, planning defense strategies against cyber threats. Moreover, the high-fidelity digital model allows to optimize the design phase and to reduce the production costs. Finally, some of the candidate research questions, associated with this research work, can be formulated as:

- Using a *holistic design* approach for building a pre-digital twin of a CubeSat, what kind of cross-phenomenona (between subsystems) can be triggered?
- After the simulation campaign, is it possible to define and isolate critical *coupling effects* and relations between a CubeSat subsystem and another?
- Which could be the main escalation events in space (Earth's orbit or deep-space) that lead to a *mission failure*?

References

[1]   G. Falco, Job One for Space Force: Space Asset Cybersecurity, Belfer Center for Science and International Affairs, Harvard Kennedy School, Vol. 79, 2018.

[2]   E. Glaessgen, D. Stargel, The Digital Twin Paradigm for Future NASA and U.S. Air Force Vehicles, 53rd AIAA/ASME/ASCE/AHS/ASC Structures, Structural Dynamics and Materials Conference, 2012. https://doi.org/10.2514/6.2012-1818.

[3]   J. Liu et al., Dynamic Evaluation Method of Machining Process Planning Based on Digital Twin, IEE, 2019. https://doi.org/10.1109/ACCESS.2019.2893309.

[4]   R. Rosen, G. von Wichert, G. Lo, K.D. Bettenhausen, About The Importance of Autonomy and Digital Twins for the Future of Manufacturing, IFAC-PapersOnLine 48-3 (2015) 567-572. https://doi.org/10.1016/j.ifacol.2015.06.141.

[5]   S. Haag, R. Anderl, Digital twin – Proof of concept, Manufacturing Letters, Vol. 15, Part B, pp. 64-66, 2018. https://doi.org/10.1016/j.mfglet.2018.02.006.

[6]   J. A. Estefan, Survey of Model-Based Systems Engineering (MBSE) Methodologies, INCOSE MBSE Initiative, 2008.

[7]   A. Maria, Introduction to Modeling and Simulation, Proceedings of the 1997 Winter Simulation Conference. https://dl.acm.org/doi/pdf/10.1145/268437.268440.

[8]   M. Kirshner, Model-Based Systems Engineering Cybersecurity for   Space Systems, Aerospace 2023, 10, 116. https://doi.org/10.3390/ aerospace10020116.

[9]   G. Falco, When Satellites Attack: Satellite-to-Satellite Cyber Attack, Defense and Resilience, ASCEND November 16-18, 2020, Virtual Event. https://doi.org/10.2514/6.2020-4014.

Aerospace Science and Engineering - III Aerospace PhD-Days   Materials Research Forum LLC
Materials Research Proceedings 33 (2023) 288-293   https://doi.org/10.21741/9781644902677-42

Optical fiber sensor fusion for aerospace systems lifecycle management

Alessandro Aimasso [1,a]

[1]Department of Mechanical and Aerospace Engineering, Politecnico di Torino, Turin, Italy

[a]alessandro.aimasso@polito.it

Keywords: Optical Fiber, Fiber Bragg Gratings, Sensors, Systems, Aerospace, Prognostics and Diagnostics

Abstract. Optical fiber is a material that can transport light signals, so resulting useful for data transmission and sensing applications. Fiber Bragg Gratings (FBG) are a specific type of optical sensors that can measure parameters like temperature, strain, and vibration. The PhD program focuses on developing a sensing and monitoring strategy for aerospace systems using FBG sensors networks. The study will include material selection, optical fiber manufacturing, sensors packaging and integration, calibration and interrogation techniques and smart logics development for acquiring and controlling phenomena affecting the equipment under test. Some experimental activities have already been conducted to analyse thermal and mechanical sensing and to define a reliable methodology for integrating sensors into various systems. During the tests, FBGs were found to have high accuracy and sensitivity for thermal variations, mechanical strain and short-term thermal transients. The crucial role of bounding technique was also enhanced. Additionally, more complex tests have been conducted for sensor more realistic systems, both for space and aeronautic environments. The results gained in this first period are positive and encouraging, suggesting further developments during PhD program.

Introduction

The optical fiber is a glass and polymeric material that can conduct a light signal inside itself. This particular feature has revolutionised the methodology of transporting data and information. The advent of optical fiber has therefore transformed communications technology, starting with the Internet connection. However, its unique physical characteristics have allowed it to be applied to a much wider spectrum in numerous and varied technical applications, and also in aerospace [1].

Optical fiber guarantees a high lightness and a small cross-section of its cables, combined with relatively low production costs. Moreover, it is immune to electromagnetic disturbances (a really crucial aspect for aerospace), while ensuring chemical and electrical passivity: consequently, it can guarantee high performances even in particularly hostile or potentially explosive environments. But the most interesting feature for engineering research is linked to the possibility of creating, directly within the fiber itself, optical structures that can act as sensors, measuring different parameters such as temperature, strain, vibration, humidity, etc... In fact, in the most frontier engineering projects the role of sensors is becoming increasingly important. Nowadays, they are required to not only provide a large amount of data with high accuracy, but also to correctly work in hostile environmental conditions. In aerospace, guaranteeing high performances in hostile environments is a typical requirement and optical-based sensors can meet it [2].

Generalities about optical fiber and FBG sensors

The optical fiber has a cylindric section with several concentric layers, from the inside to the outside, called *core*, *cladding* and *coating*. The external coating (together with eventual additional outer layers) is composed of a typical polymeric material and can vary depending on the application. Its purpose is solely to increase the mechanical features of the fiber, due to its extreme

Aerospace Science and Engineering - III Aerospace PhD-Days   Materials Research Forum LLC
Materials Research Proceedings 33 (2023) 288-293   https://doi.org/10.21741/9781644902677-42

fragility. The core and cladding, on the other hand, are the two innermost glass layers that allow the fiber to transmit the light signal. In fact, if the light is entered in the core with a correct orientation, when it reaches the interface with the cladding it undergoes a total reflection, thus remaining confined to the fiber itself and so propagating the information. In general, core and cladding reach a diameter of 125 µm, which becomes 250 µm with the addition of a typical coating layer.

Optical sensors are created directly inside the fiber itself with techniques that vary depending on the type of sensors and their final application [3]. In this work only Bragg gratings (FBG) are considered. Although there are different techniques, they are generally produced by laser photoengraving. In this way, a periodic remodulation of the core's refractive index is created. This periodic remodulation acts as a filter against the electromagnetic radiation that passes through the fiber. All frequencies, in fact, pass through the grating, except for a specific frequency that is reflected. This frequency, called *Bragg frequency*, is quantified in terms of length as:

$$\lambda_b = 2n_{eff}\Lambda \tag{1}$$

Where n_{eff} is the refractive index of the section of the core containing the grating and Λ the physical distance between one remodulation and the following, called *grating pitch*. It is precisely the dependence of the Bragg frequency on the grating pitch that allows to associate the wavelength variation detected by the optical instrument to the measurement of a physical parameter acting on the sensor itself. On a general level, considering the temperature and the mechanical strain, the two quantities able to physically deform the sensor, the general equation of the FBG sensor results to be:

$$\Delta\lambda = K_T\Delta T + K_\varepsilon\Delta\varepsilon \tag{2}$$

where $\Delta\lambda$ is the variation of the reflected wavelength, K_T and K_ε are the correlation coefficients that allow the optical output to be converted into a temperature/strain value.

Optical fiber and FBG sensors for aerospace
The physical characteristics of the optical fiber, together with the possibility of embedding sensors inside itself, has aroused a significant interest in research also for the aerospace industry. Several studies are available in the literature on this subject, covering different application areas. A first example is represented by structural monitoring: thanks to the minimal fiber's diameter, it results suitable to create "smart" components, integrating the cable into the material during the manufacturing process [4], [5]. This technique can be used for both metallic and composite pieces, albeit with a different methodology. More in details, optical fiber can provide surface information either directly adhering to the structure, by acting as an "*optical strain gauge*" [6], [7]. Another intriguing aspect is the possibility of using FBGs to monitor temperature and structural integrity in remote locations or explosive environments (such as tanks). Because of the fiber's electrical passivity and inability to generate sparks, it is particularly well suited for this type of application, in which traditional sensors cannot be used.

Finally, thermal sensing is a significant field of interest for FBGs. The small fiber cross-section reduces the disturbance introduced by the sensor, which is a significant advantage over a traditional sensor. This could be especially important for thermal testing in space applications, where very thin layers must be tested. More broadly, many aerospace systems require temperature or other information in remote locations (i.e. insight servo-mechanism [8]), for which FBGs may be the best solution [9]. Furthermore, they appear to be suitable for cryogenic as well as high temperature applications. Nowadays, thermocouples with high thermal resistance are primarily used for this

Aerospace Science and Engineering - III Aerospace PhD-Days   Materials Research Forum LLC
Materials Research Proceedings 33 (2023) 288-293   https://doi.org/10.21741/9781644902677-42

application. However, they degrade due to high-temperature oxidation, erosion, and contaminant intrusion into the probes and wiring, resulting in inaccuracies of up to 50°C. Because silica fibers can now withstand thermal cycling at temperatures close to 1000°C with a life time far superior to traditional sensors, optical sensors hold great promise for the development of new high-temperature resistant measuring systems [10].

This brief overview of possible FBGs aerospace applications explains solutions that appear to be extremely strategic. However, they are primarily laboratory-based studies and testing campaigns that have yet to be implemented in real-world systems. Furthermore, there is currently no standard procedure for integrating FBG and optical fibers in aerospace systems and/or components.

Optical sensor networks for aerospace systems lifecycle management

The overall activity about PhD program (Fig. 1) is focused on the study of optical fiber sensors networks, integrated with other traditional sensors, to realize a sensing and monitoring strategy for aerospace systems along the whole development lifecycle. To do this, at first, a methodology to define a network optical multi-sensor chain will be developed for the acquisition of different quantities on the basis of the application (space or aeronautics). In this phase, special attention will be dedicated to the material selection, the optical fiber manufacturing, and sensors packaging and integration, taking into account the environmental characteristics: level of vibration, temperature, atmosphere pressure, etc…

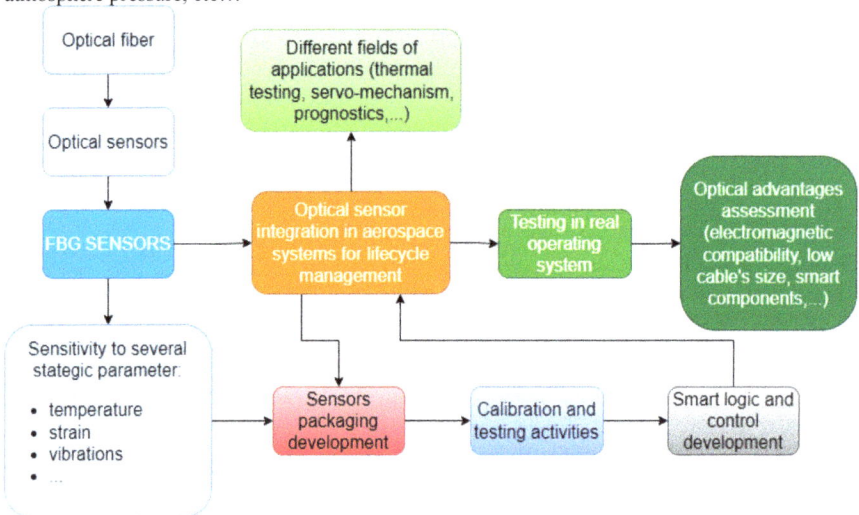

Figure 1. Scheme resuming the overall approach to research activities

Then, the study of the calibration and the interrogation techniques (data storage) of the sensing network will be performed, also using the correlation of data produced by different type of sensors, not necessarily based on the optical technology, that supports the measures and increase the confidence level. Finally, the definition of a set of techniques to interpret the data drawn from the sensing network will be developed. Depending on the applications (servomechanisms, on board systems, thermal testing, etc..) smart logics will be developed to acquire quantities the are useful to capture and control some phenomena affecting the equipment under test. Special emphasis will

Aerospace Science and Engineering - III Aerospace PhD-Days   Materials Research Forum LLC
Materials Research Proceedings 33 (2023) 288-293   https://doi.org/10.21741/9781644902677-42

be addressed to diagnostic and prognostic strategies for a defined aerospace system along the whole development lifecycle.

Some experimental activities have already been carried out, in order to have enough know-how to define a standard and reliable methodology for sensors integration in the systems of interest. In particular, the experimental activities currently conducted are divided into three main areas, in which FBGs were bounded on:

- *samples* for laboratory tests about thermal and mechanical sensing;
- a *model aircraft* for structural monitoring and real time data transmission;
- *space MLI* for thermal testing in vacuum.

Laboratory tests covered sensitivity analysis of FBG sensors to temperature variations and mechanical strain when applied to typical aerospace materials. In particular, great attention has been paid on a *gluing methodology* and to the *sensor reaction time*.

As previously stated, the Bragg wavelength, which is nominally defined during the manufacturing phase by the grating etching process, can vary as a result of physical changes in the sensor itself or environmental conditions (physical stresses applied to the grating or variations of temperature). To ensure satisfactory levels of accuracy and reliability, the correlation of the Bragg wavelength variation with these variations is required. At first, the sensitivity of the Bragg gratings to temperature changes was verified, varying the materials used, the environmental conditions of measurement and the bonding technique. The experimental data collected were used to perform the thermal calibration of the sensors used. The influence of boundary conditions on the sensor calibration process was then defined. Moreover, the FBG sensitivity to environmental temperature conditions is also a fundamental aspect when the same optical sensors are used for other measuring purposes. In fact, the main criticality of FBG is represented by the *cross-sensitivity*: the sensor provides a single optical output, whose variation may depend on multiple factors that the sensor is exposed to. In particular, using FBGs as mechanical strain sensors, it is crucial to estimate with proper accuracy the disturbance generated by other factors (such as temperature) and conceive an effective compensation method. Tests highlighted a very precise linear relationships between optical output and thermal/mechanical stresses applied on the sensor. In this way, the correlation coefficients have been experimentally calculated and, thanks to them, the over lapping process could be used to filter data. The technique to decouple the different contributions is still under investigation, due to the significative complexity of the process. The simplest strategy, already applied with great successful results (Fig. 2), consists in three different steps. At first, an optical sensor is thermally calibrated. Then, after it has been placed on the component for mechanical measures, it is flanked by a second sensor. This second sensor is not mechanically stressed and it provides only the temperature value: in this way, data from first sensor can be filtered by the thermal contribute. The second sensor, essential for the decoupling process, could be electronic or optic.

The results highlighted how, once the wavelength trend is filtered from the contribution of the environmental conditions, FBG could read mechanical strain with high level of accuracy. Furthermore, during the test campaign different bounding techniques were compared. In particular, the effects of a fiber pre-tensioning were analysed. The results showed a possible influence on the correlation coefficients, but with minimal difference on the final accuracy.

Finally, it was tested the FBG capability to read short-term thermal transients, by comparing performances with a conventional thermal probe (PT100). The campaign showed that optical sensors have an extremely high sensitivity and a much shorter reaction time. Data collected allow to consider strategic the use of FBG for thermal monitoring, above all considering the fiber minimally invasiveness and high accuracy.

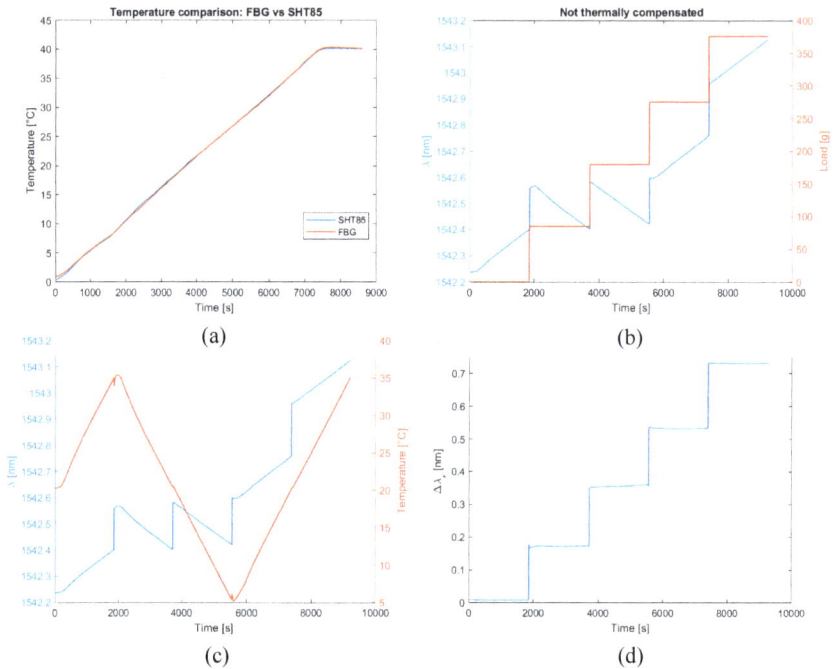

Figure 2. Comparison between temperature sensing of FBG and electronic sensor (a), FBG output when mechanical loads and thermal variation contextually act on the sensor (b,c) and after thermal compensation (d).

After verifying the high sensitivity of FBG sensors, some more complex activities have been conducted for starting to sensor example of aerospace systems.

A first application was to install some FBGs along the wing of a model aircraft made of composite materials. So, it was created a flying experimental test bench for testing performances of monitoring aircraft systems with optical technology and to develop the electronic system for transmitting data to ground in near real-time [11]. Thanks to a great work developed with a multi-disciplinary team, an integrated open-source solution was proposed. Using data detected by FBG, the system is able to display the temperature and displacements of the structure on a heat map arranged on a 3D model and visualized through a computer application on the ground. The methodology can be applied to various scenarios, ranging from maintenance planning activities to performance checks, providing an all-in-one solution for flight data management and structural monitoring.

The last field of application in which some tests have already been conducted is thermal testing for space. In particular, the reliability and sensitivity of FBG sensors for temperature measurements in a thermal protection system (MLI) were tested, comparing performances with thermocouples. From experimental data it was found that optical sensors have several advantages, including lower noise in terms of heat flow and faster reaction time. However, tests have also demonstrated the extreme importance of the methodology of integration of optical sensors, which could be critical and source of inaccuracy above all at low temperatures.

Aerospace Science and Engineering - III Aerospace PhD-Days   Materials Research Forum LLC
Materials Research Proceedings 33 (2023) 288-293   https://doi.org/10.21741/9781644902677-42

Results and further development

Some extremely positive and encouraging results have already been achieved in this first doctoral period. In particular, these refer to the high sensitivity shown by optical sensors, even when applied to materials of typical aerospace environment and in hostile environmental conditions. The physical advantages of fiber, together with the ability to measure different physical parameters with the same instrument, make optical sensors particularly interesting for aerospace.

However, some important issues still need to be properly studied and deepened. First of all, it is necessary to have a standardized procedure of gluing, packaging and calibration of the sensor, in order to fully comply with industry regulations. Secondly, a uniquely tested technique must be developed to overcome the problem of sensor cross sensitivity. Finally, it is necessary to verify on real operating systems what is currently verified by laboratory tests.

References

[1] A. Behbahani, M. Pakmehr, and W. A. Stange, "Optical Communications and Sensing for Avionics," in *Springer Handbooks*, Springer Science and Business Media Deutschland GmbH, 2020, pp. 1125–1150. https://doi.org/10.1007/978-3-030-16250-4_36

[2] S. J. Mihailov, "Fiber Bragg Grating Sensors for Harsh Environments," *Sensors*, vol. 12, pp. 1898–1918, 2012. https://doi.org/10.3390/s120201898

[3] S. J. Mihailov *et al.*, "Ultrafast laser processing of optical fibers for sensing applications," *Sensors*, vol. 21, no. 4. MDPI AG, pp. 1–23, Feb. 02, 2021. https://doi.org/10.3390/s21041447

[4] R. P. Beukema, "Embedding Technologies of FBG Sensors in Composites: Technologies, Applications and Practical Use."

[5] F. Heilmeier *et al.*, "Evaluation of strain transition properties between cast-in fibre bragg gratings and cast aluminium during uniaxial straining," *Sensors (Switzerland)*, vol. 20, no. 21, pp. 1–19, Nov. 2020. https://doi.org/10.3390/s20216276

[6] H. Wang, S. Li, L. Liang, G. Xu, and B. Tu, "Fiber grating-based strain sensor array for health monitoring of pipelines," *SDHM Structural Durability and Health Monitoring*, vol. 13, no. 4, pp. 347–359, 2019. https://doi.org/10.32604/sdhm.2019.05139

[7] Q. Zhang, D. Zhang, J. Li, B. Shui, and Y. Guo, "Strain measurement inside a strong pulsed magnet based on embedded fiber Bragg gratings," in *OFS2012 22nd International Conference on Optical Fiber Sensors*, Oct. 2012, vol. 8421, pp. 84213P-84213P–4. https://doi.org/10.1117/12.966334

[8] D. Belmonte, M. D. L. Dalla Vedova, and P. Maggiore, "Prognostics of onboard electromechanical actuators: A new approach based on spectral analysis techniques," *International Review of Aerospace Engineering*, vol. 11, no. 3, pp. 96–103, 2018. https://doi.org/10.15866/irease.v11i3.13796

[9] E. J. Friebele *et al.*, "Optical fiber sensors for spacecraft applications," *Smart Mater. Struct*, vol. 8, pp. 813–838, 1999, Accessed: Feb. 01, 2023. [Online]. Available: www.iop.org

[10] M. Pakmehr and A. Behbahani, "Optical Exhaust Gas Temperature (EGT) Sensor and Instrumentation for Gas Turbine Engines Advanced Pressure Sensing for Gas Turbine Engines View project Intelligent systems View project". https://doi.org/10.14339/STO-MP-AVT-306-18-PDF

[11] A. C. Marceddu *et al.*, "Air-To-Ground Transmission and Near Real-Time Visualization of FBG Sensor Data Via Cloud Database," *IEEE Sens J*, 2022. https://doi.org/10.1109/JSEN.2022.3227463

Aerospace Science and Engineering - III Aerospace PhD-Days   Materials Research Forum LLC
Materials Research Proceedings 33 (2023) 294-300   https://doi.org/10.21741/9781644902677-43

An extended ordinary state-based peridynamics model for ductile fracture analysis

Jing Zhang[1,2,a], Qingsheng Yang[1*], Xia Liu[1*]

[1] Department of Engineering Mechanics, Beijing University of Technology

[2] MUL2 Group, Politecnico di Torino

[a] zh_jing@emails.bjut.edu.cn

Keywords: Peridynamics, Meshfree Method, Elastoplastic Deformation, Fracture

Abstract. The present research establishes a two-step strategy to incorporate classical elastoplastic constitutive model into ordinary state-based peridynamics (OSB-PD) for ductile fracture analysis. Three length levels are notified, respectively, bond level, material particle level and bulk level. The unified Bodner-Partom theory is incorporated into the OSB-PD framework to define the bond-wise relationship between deformation state and force state. Particle-wise variables indicating plastic deformation state are extracted from connecting bonds to establish the unified ductile damage model at particle level. The damage indicator in turn exerts effects on the following plastic deformation. At present study, the collaboration among PD and unified theories amplifies the theoretical unity of PD in defining material behaviors. Simulations under quasi-static and impact loading conditions are carried out to demonstrate the effectiveness of the present model in reproducing ductile fractures at bulk level.

Introduction

Ductile fracture is a non-trivial issue, accompanied by high plastic deformation, that affects the workability of engineering materials [1]. During plastic deformation, with the proliferation of dislocations and the emergence of voids and cracks, the amplified effects of material inhomogeneity and discontinuity can influence the computational accuracy and efficiency [2]. Therefore, nonlocal meshfree methods like smoothed particle hydrodynamics (SPH) and reproducing kernel particle method (RKPM) are utilized to describe the phenomena where long-range interactions affect the global behavior of the solution. However, it is inevitable to deal with spatial derivatives or their approximations during dynamic crack propagation [3]. Special treatments for fracture and fragmentation are still needed to accommodate the emergence of new free surfaces. To eliminate these concerns, peridynamics (PD) theory reformulates the classical continuum mechanics (CCM) into an integral-differential form with no spatial derivatives required [4].

Originated from molecular dynamics, PD concerns the nonlocal effects of material particles within a certain cutoff distance named horizon. With the newly-proposed "state" instead of tensor, PD establishes a general version of CCM, making it mathematically compatible with crack initiation and propagation without any complementary numerical treatments [5]. PD can then provide a unified formulation of for nonlocal theories including SPH, CSPH, RKPM, G-RKPM and it has been successfully applied on the failure analysis of rocks, metals, polymers and composites [6]. The constitutive models are expressed as the relations between force state and deformation state. They can be carefully divided into two categories concerning ordinary state-based PD (OSB-PD) and non-ordinary state-based PD (NOSB-PD), according to the geometrical relations between force state and deformation state. To describe ductile fracture, PD has been collaborated with numerous constitutive models from different length scales [7], concerning

Aerospace Science and Engineering - III Aerospace PhD-Days   Materials Research Forum LLC
Materials Research Proceedings 33 (2023) 294-300   https://doi.org/10.21741/9781644902677-43

Johnson-Cook theory, Comacho and Ortiz theory, crystal plasticity theory, Gurson-Tvergaard-Needleman theory, Multiscale Micromorphic Molecular Dynamics theory, *etc.*

However, most of the developed PD models concentrate on a certain stage of deformation. The yield surface needs to be defined additionally for plastic deformation, which undermines the theoretical unity of PD theory. The present paper aims to define a unified elastoplastic model coupled with ductile damage to describe the full-course ductile fracture process in a unified manner. The Bodner-Partom (BP) theory [8] is modified to collaborate with OSB-PD theory, while Wang's ductile damage theory [9,10], in a unified manner akin to the BP theory, works along with the modified BP theory and functions as a material deterioration indicator. Final fracture is predicted using a critical bond stretch criterion due to its effectiveness and computational stability. The present elastoplastic model is carefully verified through numerical simulations.

The remainder of this study is threefold. Section 2 introduces the extended elastoplastic OSB-PD theory and the two-step implementation strategy. Multiple verification tests under various loading conditions are conducted in Section 3. Main conclusions and future works are summarized in Section 4.

Theory

A PD body B_0 moves to B_t with a certain deformation. The material body is discretized into numerous material particles. Each central particle possesses a spherical neighborhood H_x with a radius of δ, named horizon. Within the horizon, central particle $\mathbf{x}$ generates bonds with its neighbor particles $\mathbf{x}'$. In OSB-PD, material response at a point depends on the collective deformation of all connecting bonds. PD equations of motion are given by [5]

$$\rho(\mathbf{x})\ddot{\mathbf{u}}(\mathbf{x},t) = \int_{H_x}\{\underline{\mathbf{T}}[\mathbf{x},t]\langle\mathbf{x}'-\mathbf{x}\rangle - \underline{\mathbf{T}}[\mathbf{x}',t]\langle\mathbf{x}-\mathbf{x}'\rangle\}dV_{\mathbf{x}'} + \mathbf{b}(\mathbf{x},t) \qquad (1)$$

where ρ is mass density in the reference configuration; $\mathbf{u}$ is the displacement vector field; $\mathbf{b}$ is the prescribed body force field; $V_{\mathbf{x}'}$ is the volume of particle $\mathbf{x}'$; $\underline{\mathbf{T}}[\mathbf{x}',t]\langle\mathbf{x}'-\mathbf{x}\rangle$ is the force vector state of particle $\mathbf{x}'$ contributed by bond $\mathbf{x}'-\mathbf{x}$ at time t.

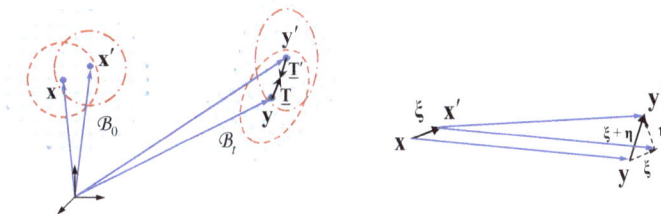

Fig. 1 PD kinematics.

Bond $\boldsymbol{\xi} = \mathbf{x}' - \mathbf{x}$ is also known as the reference position vector state $\underline{\mathbf{X}}[\mathbf{x},t]\langle\boldsymbol{\xi}\rangle$ of particle $\mathbf{x}$ at time t. It is mapped to the deformed configuration as a deformation vector state $\underline{\mathbf{Y}}[\mathbf{x},t]\langle\boldsymbol{\xi}\rangle = \mathbf{y}(\mathbf{x}',t) - \mathbf{y}(\mathbf{x},t) = \boldsymbol{\xi} + \boldsymbol{\eta}$. The displacement vector state $\boldsymbol{\eta}$ is written as $\underline{\mathbf{U}}[\mathbf{x},t]\langle\boldsymbol{\xi}\rangle = \mathbf{u}(\mathbf{x}',t) - \mathbf{u}(\mathbf{x},t)$. The extension scalar state $\underline{e}$ is defined as

$$\underline{e} = \underline{y} - \underline{x} \qquad (2)$$

where $y = |\mathbf{Y}|$, $x = |\mathbf{X}|$. It can be decomposed into an isotropic part $\underline{e}^i = \frac{\theta x}{3}$ and a deviatoric part $\underline{e}^d = \underline{e} - \underline{e}^i$. The volume dilatation is defined as $\theta = \frac{3}{m}(\underline{\omega}\underline{x}) \bullet \underline{e}$, where m is the weighted volume defined as $m = (\underline{\omega}\underline{x}) \bullet \underline{x}$; $\omega(\|\boldsymbol{\xi}\|) = \frac{1}{\|\boldsymbol{\xi}\|}$ is a spherical influence function for isotropic materials.

For plastic deformation, deviatoric extension state $\underline{e}^d$ is further decomposed into an elastic part $\underline{e}^{de}$ and a plastic part $\underline{e}^{dp}$. In consideration of plastic incompressibility, the extension state $\underline{e}$ can be expressed as

$$\underline{e} = \underline{e}^i + \underline{e}^{de} + \underline{e}^{dp} \tag{3}$$

where p is PD pressure defined as $p = -k\theta$. The plastic deformation follows the modified Prandtl-Reuss rule. The rate form of plastic bond extension is expressed as

$$\dot{e}^{dp} = \frac{R_0}{\sqrt{J_{2s}}} \exp\left[-\left(\frac{n_{BP}+1}{2n_{BP}}\right)\left(\frac{Z^2}{3J_{2s}}\right)^{n_{BP}}\right]\underline{t}^d \cdot \underline{x} \tag{4}$$

where J_{2s} is defined as $J_{2s} = \frac{1}{2}\underline{t}^d\underline{t}^d$; Z is a history-dependent variable representing the material hardening state, defined as $Z = Z_1 - (Z_1 - Z_0)e^{-m_{BP}W^p}$. W^p is an equivalent bond-wise plastic work; $R_0, Z_0, Z_1, m_{BP}, n_{BP}$ are material parameters. Then the force scalar state $\underline{t}$ is given by

$$\underline{t} = \underline{t}^i + \underline{t}^d = -\frac{3p}{m}\underline{\omega}\underline{x} + \alpha\omega\left(\underline{e}^d - \underline{e}^{dp}\right) \tag{5}$$

The unified ductile damage is defined at material particle level to measure the isotropic damage degree. A two-step strategy is hereby proposed to connect elastoplastic deformation and ductile damage. The elastoplastic deformation status extracted from bond level can manifest the degree of material damage, while the damage indicator $\bar{D}$ computed from particle level, in turn, exerts impacts on the following development of plastic deformation through bond-wise damage indicator D. The PD equation of motion coupled with damage is then modified as

$$\rho(\mathbf{x})\ddot{\mathbf{u}}(\mathbf{x},t) = \int_{H_\mathbf{x}}(1-D)\{\underline{\mathbf{T}}[\mathbf{x},t]\langle\mathbf{x}'-\mathbf{x}\rangle - \underline{\mathbf{T}}[\mathbf{x}',t]\langle\mathbf{x}-\mathbf{x}'\rangle\}\mathrm{d}V_{\mathbf{x}'} + \mathbf{b}(\mathbf{x},t) \tag{6}$$

where $D = \frac{\bar{D}+\bar{D}'}{2}$ takes an average between the damage values of particle $\mathbf{x}'$ and $\mathbf{x}$. The evolution of D can be described by Wang's unified damage theory

$$\bar{D} = \bar{D}_0 + \frac{\bar{D}_c - \bar{D}_0}{k_D(\bar{s}_c - \bar{s}_0)^{\alpha_D}}\left[(\bar{s}_c - \bar{s}_0)^{\alpha_D} - \left[\bar{s}_c - f^{1/\alpha_D}\left(\frac{\bar{\sigma}^m}{\bar{\sigma}^{eq}}\right)\bar{s}^p\right]^{\alpha_D}\right] \tag{7}$$

$$f\left(\frac{\bar{\sigma}^m}{\bar{\sigma}^{eq}}\right) = \frac{2}{3}(1+v) + 3(1-2v)\left(\frac{\bar{\sigma}^m}{\bar{\sigma}^{eq}}\right)^2 \tag{8}$$

where $\bar{D}_0, \bar{D}_c, \bar{s}_0, \bar{s}_c, \alpha_D, k_D$ are material constants. k_D, among others, is defined for smoothing over the discrepancy between bond-wise damage intensity and bulk damage intensity. $\bar{s}^p$ denotes plastic bond stretch; $\bar{\sigma}^m$ denotes PD pressure, which functions as a measure of isotropic expansion; $\bar{\sigma}^{eq}$ is PD equivalent stress [11]. The complete logic of the present model is illustrated in Fig. 2.

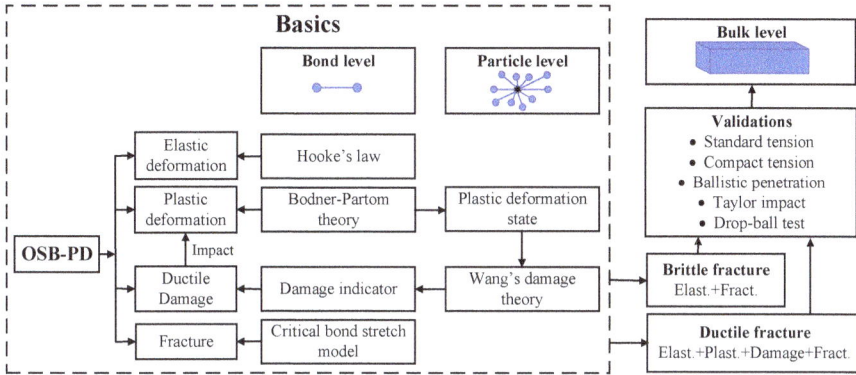

Fig. 2 theory basics.

Main results

This section aims to exploit the effectiveness of the extended OSB-PD constitutive model in predicting the natural development of plastic deformation and material failure. Simulations under quasi-static and high-velocity loading conditions are conducted, including standard tension tests, compact tension (CT) tests and Taylor impact (TI).

A series of standard tension tests are conducted on aluminum alloy 6061(T6) dog-bone specimens for model validation [1,12]. Three aspects are carefully discussed, concerning the effects of model constituents, model applicability on specimens with different degrees of non-uniformity and rate sensitivity effects. The strain-hardening effect is introduced by internal variable Z from the BP theory. The damage model functions as a counterpart to the strain-hardening variable by introducing a material deterioration factor with effect throughout all deformation stages. The bond deterioration further contributes to the collective description of material degradation, leading to the slightly downward trend in simulation curves. By integrating and coordinating all the constituents, satisfactory simulation result can be obtained. Specimens with higher degree of geometrical non-uniformity show better performances than those uniform ones. It indicates that the current computational model cannot deal with massive yielding situations. Numerical instabilities occur near the end of the deformation process, which are translated as severe fluctuations, even collapse, of the simulation curve. Moreover, the modelling results of standard tension tests under different loading rates show a good agreement to experiment data. The fitting degree were satisfactory with little fluctuations at the beginning of plastic flow, which is caused by the stiffness problem of BP theory.

Aerospace Science and Engineering - III Aerospace PhD-Days   Materials Research Forum LLC
Materials Research Proceedings 33 (2023) 294-300   https://doi.org/10.21741/9781644902677-43

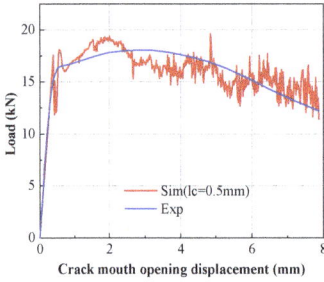

Fig. 3 Load vs. Crack mouth opening
displacement curves

Fig. 4 Crack length vs. Crack mouth
opening displacement curves

The compact tension test [13] is carried out to show the capability of the present model in describing crack propagation behaviors. The test material was DH36 steel with Young's modulus of 210 GPa and the Poisson's ratio of 0.3. The PD model was discretized into 467376 particles in simple cubic structure with the lattice constant (lc) of 0.5 mm. The horizon was 1.5 mm. A quasi-static displacement boundary condition was enforced on two cylinders with a diameter of 6.35 mm along y-axis. The total displacement of 8.0 mm was finished within 89000 steps. According to Fig. 3, the simulation curve obtained by the present PD model (red) follows a similar trend to that of the experiment curve (blue). Oscillations can be noticed during the deformation, which attributes to the effects of sudden bonding and bond breakage. Two sets of synchronic cloud charts, showing the intensity of plastic deformation and crack propagation path, are presented in Table 1. The crack grows steadily and clearly. Around the crack tip there is an enlarging elliptical plastic zone. When the discretization degree attains plane-wise 3904 pixels, corresponding to the lattice constant of 0.5 mm, the simulation results are believed trustful according to Fig. 4.

Table 1 The distribution of plastic deformation and crack propagation path (lc=0.5mm)

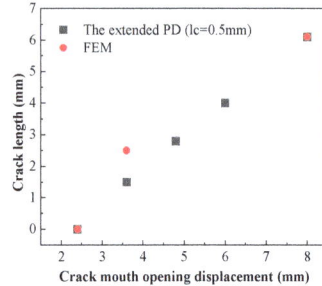

Aerospace Science and Engineering - III Aerospace PhD-Days   Materials Research Forum LLC
Materials Research Proceedings 33 (2023) 294-300   https://doi.org/10.21741/9781644902677-43

In Taylor impact test [14], the specimen was a 31.75 mm long aluminum alloy 6061(T6) cylinder with the diameter of 6.35 mm. The system was discretized into 101988 material particles. The grid size was 0.3 mm and the horizon was 0.54 mm. The impact bar was moving towards a rigid wall at 289 m/s. The time step was chosen as 1×10^{-9}s. The ultimate configuration of the cylinder bar is presented in Fig. 5. Fig. 5(b) exhibits a typical mushroom head with a drum-like section near the impact face, in conformity to the EMU results. Material particles near the impact face are drastically mal-positioned with higher level of bond breakage. The normalized diameter at the impact face for the simulation is around 2.25 and the normalized length is around 0.85, in a satisfying conformity with experimental results.

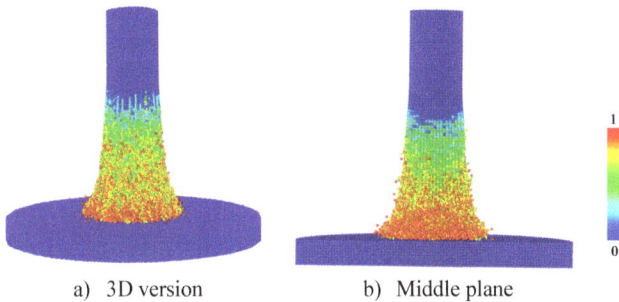

a) 3D version   b) Middle plane
Fig. 5 The degree of bond breakage

Computation tasks are performed on a HP Z8 G4 work station, equipped with two Intel Xeon Gold 6154 CPUs. For all the benchmark tests afore-discussed, 16 processors are invoked in parallel. The CPU time per step per zone ranges from 0.5×10^{-6} s to 3×10^{-6} s. For a model composed of 5×10^5 particles running for 5×10^3 steps, at least 750 seconds are needed. The computational potential can still be further tapped by means of Carrera Unified Formulation and the CUF-based PD-FEM coupling schemes.

Summary

A two-step strategy is proposed to incorporate the BP theory and the Wang's ductile damage theory into OSB-PD framework. A unified elastoplastic peridynamics model is hereby established to simulate the full-course fracture behaviors for ductile materials in consideration of dislocation dynamics, void growth and coalescence. It retains the advantages from both unified constitutive theory and PD theory in predicting the natural development of plastic deformation and material failure. Its capabilities in describing material fractures are verified through a series of tension and impact simulations. The plastic deformation patterns, crack propagation paths and synchronic physical behaviors at different deformation stages have been reproduced to a satisfactory degree.

To optimize the present model, three-dimensional peridynamics can be further coupled with high-order two-dimensional finite elements based on local elasticity [15,16]. The collaboration between PD and CUF-based FEM [17] can retain the salient features of both local and nonlocal theories, which will bring new possibilities for the applications of the present two-step strategy.

References

[1] Li H, Fu MW, Lu J, Yang H. Ductile fracture: Experiments and computations. International Journal of Plasticity 2011;27:147–80. https://doi.org/10.1016/j.ijplas.2010.04.001.

[2] Silling SA, Madenci E. Editorial: The World Is Nonlocal. J Peridyn Nonlocal Model 2019;1:1–2. https://doi.org/10.1007/s42102-019-00009-7

Aerospace Science and Engineering - III Aerospace PhD-Days   Materials Research Forum LLC
Materials Research Proceedings 33 (2023) 294-300   https://doi.org/10.21741/9781644902677-43

[3]   Erdogan Madenci, Atila Barut, Mehmet Dorduncu. Peridynamic Differential Operator for Numerical Analysis. Springer; 2019.

[4]   Silling SA. Reformulation of elasticity theory for discontinuities and long-range forces. Journal of the Mechanics and Physics of Solids 2000;48:175–209. https://doi.org/10.1016/S0022-5096(99)00029-0

[5]   Silling SA, Epton M, Weckner O, Xu J, Askari E. Peridynamic States and Constitutive Modeling. Journal of Elasticity 2007;88:151–84.https://doi.org/10.1007/s10659-007-9125-1

[6]   Gu X, Zhang Q, Madenci E, Xia X. Possible causes of numerical oscillations in non-ordinary state-based peridynamics and a bond-associated higher-order stabilized model. Computer Methods in Applied Mechanics and Engineering 2019;357:112592. https://doi.org/10.1016/j.cma.2019.112592

[7]   Javili A, Morasata R, Oterkus E, Oterkus S. Peridynamics review. Mathematics and Mechanics of Solids 2019;24:3714–39. https://doi.org/10.1177/1081286518803411

[8]   Bodner SR. Review of a Unified Elastic—Viscoplastic Theory. In: Miller AK, editor. Unified Constitutive Equations for Creep and Plasticity, Dordrecht: Springer Netherlands; 1987, p. 273–301. https://doi.org/10.1007/978-94-009-3439-9_6

[9]   Wang T-J. Unified CDM model and local criterion for ductile fracture—I. Unified CDM model for ductile fracture. Engineering Fracture Mechanics 1992;42:177–83. https://doi.org/10.1016/0013-7944(92)90289-Q

[10]   Wang T-J. Unified CDM model and local criterion for ductile fracture—II. Ductile fracture local criterion based on the CDM model. Engineering Fracture Mechanics 1992;42:185–93. https://doi.org/10.1016/0013-7944(92)90290-U

[11]   Madenci E, Oterkus S. Ordinary state-based peridynamics for plastic deformation according to von Mises yield criteria with isotropic hardening. Journal of the Mechanics and Physics of Solids 2016;86:192–219. https://doi.org/10.1016/j.jmps.2015.09.016

[12]   Scapin M, Manes A. Behaviour of Al6061-T6 alloy at different temperatures and strain-rates: Experimental characterization and material modelling. Materials Science and Engineering: A 2018;734:318–28. https://doi.org/10.1016/j.msea.2018.08.011

[13]   Xue Z, Pontin MG, Zok FW, Hutchinson JW. Calibration procedures for a computational model of ductile fracture. Engineering Fracture Mechanics 2010;77:492–509. https://doi.org/10.1016/j.engfracmech.2009.10.007

[14]   Foster J. Dynamic crack initiation toughness: experiments and peridynamic modeling. 2009. https://doi.org/10.2172/1001000

[15]   Pagani A, Carrera E. Coupling three-dimensional peridynamics and high-order one-dimensional finite elements based on local elasticity for the linear static analysis of solid beams and thin-walled reinforced structures. Int J Numer Methods Eng 2020;121:5066–81. https://doi.org/10.1002/nme.6510

[16]   Pagani A, Enea M, Carrera E. Quasi-static fracture analysis by coupled three-dimensional peridynamics and high order one-dimensional finite elements based on local elasticity. Numerical Meth Engineering 2022;123:1098–113. https://doi.org/10.1002/nme.6890

[17]   Carrera E, Cinefra M, Petrolo M, Zappino E. Finite element analysis of structures through unified formulation. Chichester, West Sussex: Wiley; 2014.

Aerospace Science and Engineering - III Aerospace PhD-Days
Materials Research Proceedings 33 (2023) 301-308

Materials Research Forum LLC
https://doi.org/10.21741/9781644902677-44

Structural modeling of manufacture-induced gaps and overlaps by high-order unified finite elements

Alberto Racionero Sánchez-Majano[1,a] *

[1]Politecnico di Torino, Corso Duca degli Abruzzi, 10129, Turin, Italy

[a,]alberto.racionero@polito.it

Keywords: Variable Angle Tow Composites, Defect Modeling, Gaps and Overlaps, Unified Formulation

Abstract. Variable stiffness composites present additional tailoring capabilities if compared to classical straight fiber composites. However, the fabrication of these advanced composites leads inevitably to the presence of manufacturing signatures such as gaps and overlaps, which alter the mechanical behavior of the laminated components. In this work, the fundamental frequency of variable stiffness laminates with the presence of gaps and overlaps is studied. The Defect Layer Method is considered for modeling the defects, while the Carrera Unified Formulation is employed to derive the governing equations which will be solved by the Finite Element Method.

Introduction

Novel manufacturing techniques, such as Automated Fiber Placement (AFP), have permitted to conceive new families of laminated structures, namely Variable Stiffness Composites (VSC) or Variable Angle Tow (VAT), in which fiber tows are steered conforming curvilinear paths. For instance, Olmedo and Gürdal [1] investigated the buckling of VAT plates for different boundary conditions and rotations of the fiber path. They conducted a parametric analysis of how the fiber path parameters affect the non-dimensional buckling load factor. They found improvements up to 19 and 80% with respect to classic laminates. However, in the cases that Olmedo and Gürdal considered, the minimum turning radius, which determines whether a laminate is manufacturable or not, was not taken into account. Therefore, not all the solutions were feasible from a fabrication point of view. Besides, due to the manufacturing features inherent to AFP, imperfections are prone to arise during the fabrication process, namely gaps and/or overlaps, and affect the structural performance [2].

Many authors have proposed numerical methods to investigate the effect of gaps and/or overlaps on the mechanical properties of variable stiffness composites. Blom et al. [2] investigated how gaps affect the strength and stiffness of VAT using the Finite Element Method (FEM). They concluded that increasing the laminate's total gap area deteriorates strength and stiffness. The main drawback of their proposed method was that a very fine mesh was required to capture the gap areas, drastically increasing the computational cost. Fayazbakhsh et al. [3] proposed the Defect Layer Method (DLM), which permits to capture the gap and/or overlap areas that appear in the laminate without incurring into an excessive computational burden as in [2].

In this manuscript, we tackle the influence of gaps and overlaps on VAT laminates by coupling DLM with the Carrera Unified Formulation (CUF) [4]. CUF permits obtaining the governing equations of any structural theory without making *ad hoc* assumptions. So far, CUF has proven to predict accurate stress states [5], as well as capture the influence of multiscale uncertainty defects on the mechanical performance of VAT plates [6].

Variable stiffness composite plates and defect modeling

VAT laminated components are fabricated by steering fiber bands along curvilinear paths. Throughout the years, different variation laws have been investigated. However, two kinds of

Aerospace Science and Engineering - III Aerospace PhD-Days   Materials Research Forum LLC
Materials Research Proceedings 33 (2023) 301-308   https://doi.org/10.21741/9781644902677-44

variations have been the most researched: the constant curvature and linear variation. This work focuses on the latter, in which the local fiber orientation, $\boldsymbol{\theta}$, varies along the $\boldsymbol{x'}$ direction as follows:

$$\theta(x') = \phi + T_0 + \frac{T_1-T_0}{d}|x'|. \tag{1}$$

In the previous equation, T_0 is the fiber angle orientation at $x' = 0$, whereas the fiber angle orientation T_1 is reached at $x' = d$, being d the semi-length or semi-width of the laminate. Last, ϕ is the fiber angle path rotation, which generally equals $0°$ or $90°$. All of these parameters are illustrated in Figure 1.

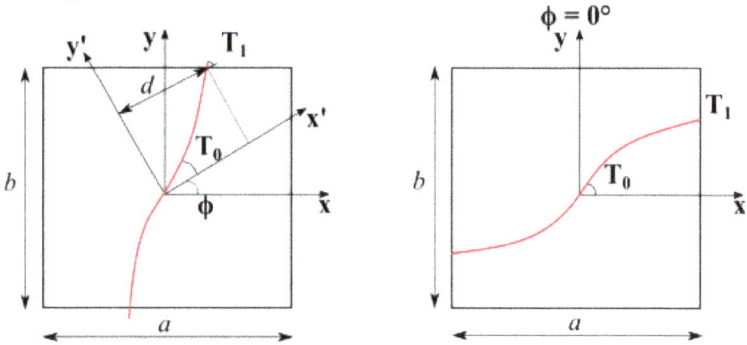

Figure 1. Representation of the fiber parameters involved in the definition of a VAT laminate.

Despite the steering capability of VAT plates, the AFP machines that manufacture them present some limitations, being one of them the curvature of the laid fiber path. In case of exceeding the turning radius, fiber wrinkles and upfolding will generate within the placed tape. An additional manufacturing feature inherent to AFP is the presence of gaps and overlaps in the final part, thereby affecting the structural performance. For instance, gaps reduce the laminate strength, while overlaps can cause an increase in strength. DLM [3] is considered to model these imperfections. As mentioned in [3], the defect area percentage is the only parameter to modify the elastic properties or the thickness associated to the finite element. Note that the elastic properties vary if gaps are considered, whereas an increase in the thickness is provided in the case of overlap. The elastic properties are modified according to Fig. 6 from [3]. The thickness increases proportionally to the defect area within the element and is topped up by 95% of the original thickness of the ply due to the compaction pressure in the autoclave, as reported in [7].

Unified finite elements
2D FE are implemented within the CUF formalism. According to [4], the 3D field of displacement can be expressed in terms of arbitrary through-the-thickness expansion functions, $\boldsymbol{F_\tau(z)}$, of the 2D generalized unknowns laying over the $\boldsymbol{x-y}$ plane. That is,

$$\boldsymbol{u}(x,y,z) = F_\tau(z)\boldsymbol{u}_\tau(x,y). \ \tau = 1,...,M \tag{2}$$

Therein, M denotes the number of expansion terms, and $\boldsymbol{u}_\tau(x,y)$ is the vector containing the generalized displacements. Note that τ denotes summation. The analysis of multi-layered structures is commonly conducted by following an Equivalent-Single-Layer (ESL) and Layer-Wise (LW) approach. In this manuscript, ESL models are built using Taylor polynomials as F_τ in the thickness direction. On the other hand, LW utilizes Lagrange polynomials over the single

Aerospace Science and Engineering - III Aerospace PhD-Days   Materials Research Forum LLC
Materials Research Proceedings 33 (2023) 301-308   https://doi.org/10.21741/9781644902677-44

layers and then imposes the continuity of displacements at the layer interfaces, as in [8]. In this context, TEn denotes a TE of the n-th order, whilst LEn indicates the usage of an LE with n-th order polynomials. Moreover, XLEn means that X Lagrange polynomials of n-th order are used to describe each layer of the laminate.

Utilizing the FE and shape functions $N_i(x, y)$, the displacement field becomes:

$$u(x, y, z) = N_i(x, y)F_\tau(z)q_{\tau i}(x, y). \quad i = 1, \ldots, N_n \tag{3}$$

In Eq. (3), $q_{\tau i}$ denotes the unknown nodal variables, and N_n indicates the number of nodes per element. In this work, 2D nine-node quadratic elements, referred to as Q9, are employed as N_i for the $x - y$ plane discretization.

The governing equations are obtained by means of the Principle of Virtual Displacements (PVD). PVD states that the virtual variation of the internal strain energy, $\delta\mathcal{L}_{int}$, has to be equal to the virtual work of the external forces, $\delta\mathcal{L}_{ext}$, minus that of the inertia forces, $\delta\mathcal{L}_{ine}$. That is:

$$\delta\mathcal{L}_{int} = \delta\mathcal{L}_{ext} - \delta\mathcal{L}_{ine}. \tag{4}$$

In the case of free vibration analyses, Eq. (4) becomes:

$$\delta\mathcal{L}_{int} + \delta\mathcal{L}_{ine} = 0. \tag{5}$$

The virtual variation of the strain energy can be calculated as:

$$\delta\mathcal{L}_{int} = \int_V \delta\boldsymbol{\varepsilon}^T \boldsymbol{\sigma} dV, \tag{6}$$

whereas the virtual work of the inertia forces is computed as:

$$\delta\mathcal{L}_{ine} = \int_V \rho\delta\boldsymbol{u}^T \ddot{\boldsymbol{u}} dV, \tag{7}$$

in which ρ represents of the mass density of the material. Equation (6) can be rewritten using Eq. (3), the constitutive law $\boldsymbol{\sigma} = \boldsymbol{C}\boldsymbol{\varepsilon}$, and the geometrical relations between the strains and displacements, yielding:

$$\delta\mathcal{L}_{int} = \delta\boldsymbol{q}_{sj}^T \left[\int_V \boldsymbol{D}^T(N_j F_s) \, \tilde{\boldsymbol{C}} \boldsymbol{D}(N_i F_\tau) \, dV \right] \boldsymbol{q}_{\tau i} = \delta\boldsymbol{q}_{sj}^T \boldsymbol{k}_0^{ij\tau s} \boldsymbol{q}_{\tau i}, \tag{8}$$

where $\boldsymbol{k}_0^{ij\tau s}$ is the 3×3 Fundamental Nucleus (FN) of the stiffness matrix, which is invariant regardless of the order of the 2D shape function and the through-the-thickness expansion. $\boldsymbol{D}(\cdot)$ denotes the differential operator matrix containing the geometrical relations, and $\tilde{\boldsymbol{C}}$ is the material stiffness matrix in the global reference frame, i.e., $\tilde{\boldsymbol{C}} = \boldsymbol{T}(x, y)^T \boldsymbol{C} \boldsymbol{T}(x, y)$. Since the fiber varies point-wise within the $x - y$ plane, so does the rotation matrix $\boldsymbol{T}$.

The virtual work of the inertia forces can be expressed as:

$$\delta\mathcal{L}_{ine} = \delta\boldsymbol{q}_{sj}^T \left[\int_V \rho I N_i N_j F_\tau F_s \, dV \right] \ddot{\boldsymbol{q}}_{\tau i} = \delta\boldsymbol{q}_{sj}^T \boldsymbol{m}^{ij\tau s} \ddot{\boldsymbol{q}}_{\tau i}, \tag{9}$$

in which $\boldsymbol{I}$ is the 3×3 identity matrix and $\boldsymbol{m}^{ij\tau s}$ is the 3×3 FN of the mass matrix. Hence, the undamped free vibration problem can be written as follows:

Aerospace Science and Engineering - III Aerospace PhD-Days   Materials Research Forum LLC
Materials Research Proceedings 33 (2023) 301-308   https://doi.org/10.21741/9781644902677-44

$$M\ddot{q} + K_0 q = 0. \tag{10}$$

In Eq. (10), M and K_o denote the overall mass and stiffness matrices, respectively. They are obtained by looping over the FN's through the indices i, j, τ and s to obtain the mass and stiffness matrices for the single element, and subsequently assembled to conform that of the whole structure. Last, by imposing harmonic solutions $q = \tilde{q}e^{i\omega t}$, Eq. (10) turns into the following eigenvalue problem:

$$(K_0 - \omega_i^2 M)\tilde{q}_i = 0 \tag{11}$$

where ω_i and $\tilde{q}_i$ are the ith natural frequency and eigenvector, respectively.

Machine simulation: identification of gap and overlap locations.
The steering of fiber bands along a fixed direction, and shifting the AFP head in its perpendicular direction to generate the subsequent fiber course, leads to the presence of gaps and/or overlaps. The location in which they appear depends not only on T_0 and T_1 but also on the steering strategy. That is, one can impose contact between two adjacent courses at the edge or the center of the plate. Figure 2 illustrates the case of a $[\langle 0,45\rangle]$ plate in which the fiber courses touch each other at the edge (Figure 2 left) and at the center of the plate (Figure 2 right). The yellow area indicates a gap area, whereas the green area highlights an overlap area.

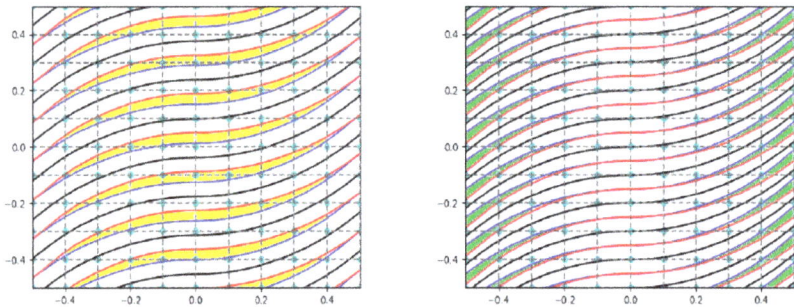

Figure 2. Example of a plate with $[\langle \mathbf{0}, \mathbf{45} \rangle]$ stacking sequence with full gap (left) and full overlap (right) manufacturing strategy.

Table 1 reflects the kind of defects that will arise depending on the fiber path parameters and the contact position of adjacent courses.

Condition	Contact at edge	Contact at center
$|cosT_0| > |cosT_1|$	Gap at center	Overlap at edges
$|cosT_1| > |cosT_0|$	Overlap at center	Gap at edges

Table 1. Defects arising within the plate depending on the fiber parameters and the manufacturing strategy.

The previous imperfections affect such large areas because the course width is kept constant throughout the steering process. In order to reduce the defect area, the course width has to decrease or increase whenever a course intersects the successive one or it does not reach the precedent course's edge, respectively. The increase or decrease of the course width is achieved by cutting an individual tow and restarting its deposition. This comports the generation of small triangular defected regions, as evidenced in Figure 3.

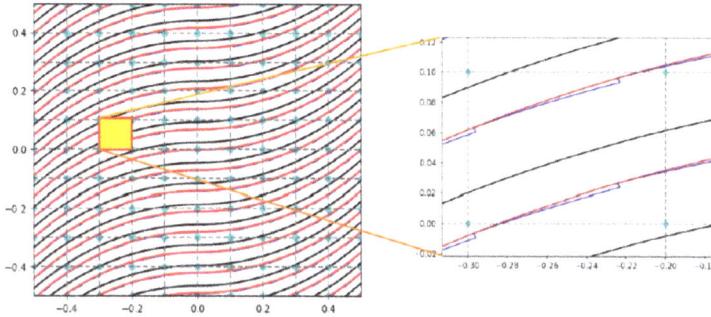

Figure 3. Gap defect correction over a $[\langle 0, 45 \rangle]$ ply The zoomed area shows the triangular gaps that are generated.

As stated in Section 2, the material elastic properties, or element thickness, will be modified according to the defect area within each finite element.

Results

Model verification
The first step towards the maximization of the fundamental frequency of a VAT plate is the model verification. For doing so, the results presented in the work by Akhavan and Ribeiro [9] are used. A three-layered square plate with lamination sequence $[\langle 0, 45 \rangle, \langle -45, -60 \rangle, \langle 0, 45 \rangle]$ is considered. The length and width of the plate are $a = b = 1$ m, and the width-to-thickness ratio is $a/h = 10$, having each ply the same thickness. The plate is clamped on all of its four edges.

First, a convergence analysis is performed employing Q9 FE and LE2 expansion through the thickness direction. The results are enlisted in Table 2. It is observed a good agreement between Ref. [9] and the present approach. Plus, convergence is reached with the 10x10 Q9 mesh, since the relative error between the 10x10 and 12x12 mesh is about 0.1%. Thus, the 10x10 Q9 mesh will be used in the upcoming analyses.

Table 2. Mesh convergence in terms of the first five fundamental frequencies of the $[\langle 0, 45 \rangle, \langle -45, -60 \rangle, \langle 0, 45 \rangle]$ plate. Each discretization uses 1 LE2 expansion per layer through the thickness direction.

Model	DOF	f_1 [Hz]	f_2 [Hz]	f_3 [Hz]	f_4 [Hz]	f_5 [Hz]
Ref.	-	613.79	909.04	1231.65	1337.69	1484.53
6x6 Q9	3549	614.82	916.05	1230.29	1361.72	1492.04
8x8 Q9	6069	611.26	907.03	1219.97	1337.03	1475.20
10x10 Q9	9261	609.91	903.93	1216.18	1328.88	1469.58
12x12 Q9	13125	609.28	902.56	1214.47	1325.46	1467.15

Subsequently, the effect of the structural theory on the fundamental frequency is addressed. The results are available in Table 3.

Table 3. Effect of the structural theory on the first five natural frequencies for the
[⟨$\mathbf{0,45}$⟩, ⟨$\mathbf{-45, -60}$⟩, ⟨$\mathbf{0,45}$⟩] plate. Each model employs a 10x10 Q9 mesh.

Model	DOF	f_1 [Hz]	f_2 [Hz]	f_3 [Hz]	f_4 [Hz]	f_5 [Hz]
Ref.	-	613.79	909.04	1231.65	1337.69	1484.53
TE 1	2646	638.87	955.51	1278.43	1419.71	1553.57
TE 3	5292	611.17	908.11	1218.00	1338.39	1473.10
TE 6	9261	609.49	903.63	1214.24	1328.59	1774.46
1 LE1	5292	621.64	917.66	1244.85	1347.15	1499.66
1 LE 2	9261	609.91	903.93	1216.18	1328.88	1469.58

As expected, TE 1 model provided a higher value of f_1 since it leads to a higher stiffness of the model, whereas the TE 3 provides practically the same value of the reference, where a third-order structural theory was employed. TE 6 calculated a value very similar to 1 LE2, with the exception of the fifth fundamental frequency, where TE 6 provided a higher value. Then, regarding the LW models, the 1 LE1 approach led to the computation of fundamental frequencies that lay between those computed with a TE 3 and 1 LE2 model.

Effect of manufacturing defects: gaps and overlaps
In this section, the effect of manufacturing defects is addressed. As depicted in Section 4, two manufacturing strategies are considered: without and with defect correction. These two strategies are denoted as Type 1 and Type 2, respectively. For the following analyses, the tow paths are conformed by sixteen tows, each with a width of $t_w = 3.125$ mm. The FE mesh comprises 10x10 Q9 elements, as it showed to provide convergence in the previous subsection.

Table 4 presents the effect of gaps on the first five fundamental frequencies of the [⟨0,45⟩, ⟨−45, −60⟩, ⟨0,45⟩] plate. An LW-1 LE2 model is employed since, when gaps are considered, the thickness of the laminate remains unaltered, and no additional computational effort is required when compared to the ideal plate. It is observed that gaps lead to a decrease in the fundamental frequency because resin-rich areas are present within the layers. Type 1 defects present a much lower frequency than the ideal case, whereas Type 2 limits the fundamental frequency reduction.

Table 4. Effect of manufacturing gaps on the [⟨$\mathbf{0,45}$⟩, ⟨$\mathbf{-45, -60}$⟩, ⟨$\mathbf{0,45}$⟩] plate. Each model
uses a 10x10 Q9 mesh and 1 LE2 structural theory.

Model	f_1 [Hz]	f_2 [Hz]	f_3 [Hz]	f_4 [Hz]	f_5 [Hz]
No defects	609.91	903.93	1216.18	1328.88	1469.58
Gap Type 1	547.82	797.10	1083.41	1159.71	1295.57
Gap Type 2	599.10	888.99	1193.21	1306.87	1442.75

The influence of overlap imperfections is gathered in Table 5. An LW-1 LE2 and ESL-TE 3 were considered for the ideal case, whereas only an ESL-TE 3 expansion is considered when modeling overlaps. As expected, the overlaps stiffen the structure, which leads to higher fundamental frequencies. Therefore, overlap Type 1 provides higher frequencies than Type 2, in which defects are reduced by modifying the number of tows while steering them.

An ESL approach is utilized when overlaps are involved due to the vast increase in terms of DOF that an eventual LW model may lead to. In an overlap LW model, the continuity of the interfaces has to be guaranteed through the thickness but also within the ply. In this regard, the number of DOF dramatically increases for a minimal improvement in the prediction of the fundamental frequency.

Table 5. Effect of manufacturing overlaps on the $[\langle 0, 45 \rangle, \langle -45, -60 \rangle, \langle 0, 45 \rangle]$ plate. Each model uses a 10x10 Q9 mesh and the Overlap Types use a TE 3 structural theory.

Model	f_1 [Hz]	f_2 [Hz]	f_3 [Hz]	f_4 [Hz]	f_5 [Hz]
No defects 1 LE2	609.91	903.93	1216.18	1328.88	1469.58
No defects TE 3	611.17	908.11	1218.00	1338.39	1473.10
Overlap Type 1	679.02	980.20	1290.34	1414.51	1558.46
Overlap Type 2	620.72	922.19	1230.51	1356.06	1490.76

Conclusions

This manuscript has discussed the effect of manufacture-induced gaps and overlaps on the fundamental frequency of VSC laminates. First, ideal VAT plates have been modeled using CUF-based FE, and have proven to provide similar results as those found in the literature. Later, the position where gaps and overlaps will appear during manufacturing has been predicted. These defects were incorporated into the FE model by means of the Defect Layer Method. The presence of gaps incurred a decrease in the fundamental frequency, whereas overlaps led to an increase of it due to the stiffening effect they have on the structure.

Future works will be related to the optimization of VAT plates in which the aforementioned manufacturing imperfections are considered.

Acknowledgements

This work is part of a project that has received funding from the European Research Council (ERC) under the European Union's Horizon 2020 research and innovation programme (Grant agreement No. 850437). The author would also like to thank Eugenio Sabatini for providing some of the results shown in this document.

References

[1] R. Olmedo, Z. Gürdal. Buckling response of laminates with spatially varying fiber orientations. In 34th Structures, Structural Dynamics and Materials Conferences (1993), 1567. https://doi.org/10.2514/6.1993-1567

[2] A.W. Blom, C.S. Lopes, P.J. Kromwijk, Z. Gürdal, P.P. Camanho. A theoretical model to study the influence of tow-drop areas on the stiffness and strength of variable-stiffness laminates. Journal of Composite Materials 43(5) (2009), 403-425. https://doi.org/10.1177/0021998308097675

[3] K. Fayazbakhsh, M.A. Nik, D. Pasini, L. Lessard. Defect layer method to capture effect of gaps and overlaps in variable stiffness laminates made by automated fiber placement. Composite Structure 97 (2013), 245-251. https://doi.org/10.1016/j.compstruct.2012.10.031

[4] E. Carrera, M. Cinefra, M. Petrolo, E. Zappino. Finite Element Analysis of Structures through Unified Formulation. Wiley & Sons, Hoboken, New Jersey. 2014.

[5] A.R. Sánchez-Majano, R. Azzara, A. Pagani, E. Carrera. Accurate stress analysis of variable angle tow shells by high-order equivalent-single-layer and layer-wise finite element models. Materials 14(21) (2021), 6486. https://doi.org/10.3390/ma14216486

[6] A. Pagani, M. Petrolo, A.R. Sánchez-Majano. Stochastic characterization of multiscale material uncertainties on the fibre-matrix interface stress state composite variable stiffness plates.

International Journal of Engineering Science 183 (2023), 103787.
https://doi.org/10.1016/j.ijengsci.2022.103787

[7] A.A. Vijayachandran, P. Davidson, A.M. Waas. Optimal fiber paths for robotically manufactured composite structural panels. International Journal of Non-Linear Mechanics 126 (2020), 103567. https://doi.org/10.1016/j.ijnonlinmec.2020.103567

[8] E. Carrera. Theories and finite elements for multi-layered, anisotropic, composite plates and shells. Archives of Computational Methods in Engineering 9(2) (2002), 87-140.

[9] H. Akhavan, P. Ribeiro. Natural modes of vibration of variable stiffness composite laminates with curvilinear fibers. Composite Structures 93(11) (2011), 3040-3047.
https://doi.org/10.1016/j.compstruct.2011.04.027

Aerospace Science and Engineering - III Aerospace PhD-Days   Materials Research Forum LLC
Materials Research Proceedings 33 (2023) 309-314   https://doi.org/10.21741/9781644902677-45

Integrated optical and X-ray pulsar methods for deep-space autonomous navigation based on an adaptive nonlinear filter

Sui Chen[1,a] *

[1]Politecnico di Milano, Department of Aerospace Science and Technology, Via La Masa 34, 20156, Milan, Italy

[a]sui.chen@polimi.it

Keywords: Autonomous Deep-Space Navigation, Optical Navigation, X-Ray Pulsar Navigation, Adaptive Nonlinear Filter

Abstract. Recent technological advancement and the commercialisation of the space sector have led to a significant surge in the development of space missions for deep-space exploration. Currently, deep-space missions mainly rely on ground-based Guidance, Navigation and Control (GNC) operations involving human-in-the-loop processes. Although being reliable, the ground-based navigation approach is prone to prolonged periods of communication delay, lacking real-time capabilities and autonomy. In addition, the booming growth of users in space will unavoidably lead to saturation of ground slots, hindering the progression of space exploration. Reducing the dependence on ground operation by developing on-board autonomous navigation methods represents a potential solution for future deep-space missions. Currently, navigation based on optical and X-ray pulsar measurements represents the two prominent methods for achieving autonomous deep-space navigation and will be investigated in this paper.

Literature Review

NASA Deep Space 1 was the first deep-space mission to use autonomous navigation throughout the entire mission phase, determining its orbit using optical images of distant asteroids [1]. Among the several autonomous navigation methods being proposed and developed, optical navigation represents the most mature and feasible solution for spacecraft autonomous navigation [2]. Using optical sensors, spacecraft's state is estimated by extracting the line-of-sight directions to several known navigation beacons in the Solar System which are subsequently fed into an orbit determination algorithm for state estimation. Optical navigation has been developed and tested extensively in current literature [3,4,5,6]. Several researchers [7,8,9] improved the state estimation accuracy by correcting the planetary light-time and light-aberration effects and by exploiting the optimal beacons selection strategy. Song and Yuan [10] enhanced the numerical stability and computation efficiency of the deep-space orbit determination problem by applying UD factorisation to the state estimator.

The innovative concept of X-ray pulsar navigation (XNAV) was first proposed in 1981 [11], after the first detection of pulsar in 1967. Pulsars are fast rotating neutron stars that can emit regular signals in both low and high energy bands. The pulse time of arrival (TOA) difference between observed pulse profile and standard pulse profile is used to estimate the spacecraft's state. In November 2016, as the first pulsar navigation mission, China launched an X-ray pulsar test satellite XPNAV-1 with a primary objective of detecting the details of X-ray signals emitted by 26 nearby pulsars and subsequently creating a pulsar navigation database. They utilised observation data obtained from the Crab Pulsar to validate the satellite's orbit determination, achieving an average positioning accuracy of 38.4 km [12]. In 2017, NASA demonstrated a fully autonomous, real-time pulsar navigation experiment in space for the first time, attaining an average accuracy of 16 km [13,14]. In general, although XNAV is a less mature and less established method, using pulsar navigation is typically expected to result in more accurate navigation solutions compared to optical navigation [15].

Since XNAV technology is still in the early development stage, it is believed that combining XNAV with other types of navigation methods is the way forward for future deep-space exploration navigation. In fact, a significant amount of work has been dedicated to investigating integrated navigation systems for deep-space exploration. Ning et al. 2017 [16] proposed combining XNAV with traditional celestial navigation, which involves measuring star angles, to create an integrated navigation method for the cruise phase of Mars exploration. Jiao et al. 2016 [17] integrated X-ray pulsar measurement with Mars optical measurement for Mars exploration orbiting phase, yielding an autonomous navigation accuracy of less than 1 km. Xiong et al. 2016 [18] incorporated an ultraviolet optical sensor to enhance the performance of XNAV system and demonstrated that the proposed XNAV/optical integrated navigation scheme can effectively scale down the spacecraft orbit error. However, in their work, an extended Kalman filter (EKF) was used which only works well with linear or slightly nonlinear problems and might produce unsatisfactory estimation results when dealing with highly nonlinear systems. Zhang et al. 2018 developed a novel orbit determination algorithm exploiting both the pulsar timing data and position vector to enhance the performance of XNAV. The Adaptive Divided Difference Filter (ADDF) adopted in their work has demonstrated to have a stronger system noise adaptive capacity and can handle unknown measurement noise and non-additive noise. However, navigation is solely relied on the information from pulsar sources which are still considered to be immature measurement techniques to date and might degrade the overall system reliability and performance. Gu et al. 2019 [19] proposed an optical/radio/pulsars integrated navigation algorithm with an adaptive extended Kalman filter (AEKF) for optimal state estimation, and illustrated that optical/radio/pulsars integrated navigation exhibits a higher level of estimation accuracy compared to both conventional optical navigation and optical/radio integrated navigation.

Objective

Achieving fully autonomous navigation for deep-space applications is a relatively new field of study. As illustrated in the literature review, although optical navigation represents the most mature technology for autonomous navigation, it typically results in low navigation accuracy when relying solely on optical measurements. For this reason, it becomes essential to incorporate other forms of measurement data to improve the performance of optical navigation. Considering the aforementioned limitations found in the current literature, this paper aims to fill in these gaps to advance the current technology for autonomous deep-space navigation. Specifically, information fusion will be performed by integrating optical and pulsar timing measurements and an adaptive nonlinear filter will be adopted to handle unknown measurement noise robustly. It is hoped that the contribution from this work will act as a stepping stone towards achieving fully autonomous deep-space navigation in the future.

Methodology

Measurement Models

Optical Measurement Model The spacecraft inertial position is defined as $r = [x, y, z]^T$ and the beacon inertial position is $r_p = [x_p, y_p, z_p]^T$ which is assumed to be known. The objective is to estimate r by obtaining the line-of-sight (LoS) directions to the beacons via celestial triangulation.

From geometry, the beacon position ρ and the LoS direction $\hat{\rho}$ can be expressed as:

$$\rho = r_p - r = \begin{bmatrix} x_p - x \\ y_p - y \\ z_p - z \end{bmatrix} \quad ; \quad \hat{\rho} = \frac{\rho}{\|\rho\|} = \frac{1}{\|\rho\|} \begin{bmatrix} x_p - x \\ y_p - y \\ z_p - z \end{bmatrix} \tag{1}$$

For optical navigation using the LoS directions to known navigation beacons, the measurements are the azimuth (Az) and elevation (El) to the beacons, defined by the following equation:

$$h = \begin{bmatrix} Az \\ El \end{bmatrix} = \begin{bmatrix} \arctan(\frac{\hat{\rho}_y}{\hat{\rho}_x}) \\ \arcsin(\hat{\rho}_z) \end{bmatrix} = \begin{bmatrix} \arctan(\frac{y_p - y}{x_p - x}) \\ \arcsin(\frac{z_p - z}{\|\rho\|}) \end{bmatrix} \tag{2}$$

Therefore, at each given epoch t_k, the optical measurement model is:

$$y_{optical\_k} = h_{optical}(x_k) + v_{optical\_k} = \begin{bmatrix} \arctan(\frac{y_p - y}{x_p - x}) \\ \arcsin(\frac{z_p - z}{\|\rho\|}) \end{bmatrix}_k + v_{optical\_k} \tag{3}$$

where $v_{optical\_k}$ is the measurement noise of the optical sensor.

X-ray Pulsar Measurement Model For X-ray pulsar-based navigation, the key measurement is the pulsar timing in the form of photon time of arrival (TOA) at the X-ray detector. Position error is corrected by finding the difference between the measured pulse TOA and the predicted pulse TOA. This pulse TOA difference can be determined by comparing the observed pulse profile and the standard pulse profile which is usually constructed in the solar system barycentre (SSB) inertial reference frame. For this reason, the measured photon is first transferred from the spacecraft moving frame to the SSB frame using the following time transfer equation [18]:

$$t_k^{(b)} = t_k^{(sc)} + \frac{1}{c} n \cdot r_{sc,k} + \frac{1}{2cD_0} \left[r_{sc,k}^2 - (n \cdot r_{sc,k})^2 \right] - \frac{2\mu_{Sun}}{c^3} \ln \left| \frac{n \cdot r_{sc,k} + r_{sc,k}}{n \cdot b + b} + 1 \right| \tag{4}$$

where k indicates the discrete time, $t_k^{(sc)}$ is the photon TOA measured at the spacecraft local frame, $t_k^{(b)}$ is the photon TOA at the SSB frame, n is the unit vector from the spacecraft to the pulsar, $r_{sc,k}$ is the spacecraft position vector relative to the SSB origin, c is the speed of light, D_0 is the distance between the pulsar and the SSB, μ_{Sun} is the gravitational constant of the Sun and b is the position vector of the SSB relative to the centre of the Sun.

Therefore, for the X-ray pulsar navigation, the measurement model is defined as:

$$y_{X-ray\_k} = h_{X-ray}(x_k) + v_{X-ray\_k} = \left(t_k^{(b)} - t_k^{(sc)} \right) + v_{X-ray\_k} \tag{5}$$

For an integrated system with both optical and X-ray pulsar measurement information, the combined navigation measurement equation can be written as follow, where v_k is the measurement noise with covariance R:

$$y_k = \begin{bmatrix} y_{optical\_k} \\ y_{X-ray\_k} \end{bmatrix} = h(x_k) + v_k \tag{6}$$

Integrated Navigation Algorithms
System Dynamic Model The spacecraft state x in deep space is defined as $x = [r \quad v]^T$, where r and v are the inertial position and velocity vectors of the spacecraft respectively. The dynamic model is described using the following equation of motion:

$$\dot{x} = f(x) + w = \begin{bmatrix} v \\ -\mu_{Sun}\dfrac{r}{r^3} + C_R\dfrac{P_0R_0^2}{c}\dfrac{A_s}{m_s}\dfrac{r}{r^3} \end{bmatrix} + w \tag{7}$$

where f is the system dynamics equation and w is the process noise with covariance Q.

Adaptive nonlinear Filter For deep-space navigation problems, highly nonlinear models and time-varying measurement noise covariance are likely to be encountered, so using the traditional Kalman filters is likely to lead to unsatisfactory estimation results. In this paper, an adaptive nonlinear filter will be adopted instead in order to improve the overall navigation performance. Derived from the work of [20], a new recursive adaptive unscented Kalman filter (RAUKF) is implemented to estimate the spacecraft's state. Based on the measurement and dynamic models defined in Eq. 6 and Eq. 7, the RAUKF algorithm follows time update and measurement update steps similar to Kalman filters, but with the measurement noise covariance R being scaled by a scaling factor which is derived from current innovation recursively. The full RAUKF algorithm can be found in [20].

Expected Results
The proposed integrated deep-space navigation architecture in this paper has taken inspiration from [18], in which the feasibility of integrated optical and X-ray pulsar navigation systems is assessed and simulation results show that such an integrated system outperforms the ones based solely on either optical or X-ray pulsar measurements. In their work with an extended Kalman filter used for orbit determination, the position estimation mean root-mean-square (RMS) errors of the integrated system are found to be 1.59 km for an Earth-orbiting satellite and 1.73 km for a lunar-orbiting satellite. In Ning's work [20], the implementation of the new RAUKF algorithm in a purely optical-based navigation system has led to a reduction of 15% and 35% in the position and velocity estimation errors respectively as compared to UKF.

The novelty in this paper is the integration of optical and X-ray pulsar navigation methods combined with the implementation of an adaptive nonlinear filter, and this is expected to lead to both robustness to time-varying measurement noise and enhanced state estimation accuracy. Once the proposed algorithms described in this paper are implemented in computer-based codes with subsequent simulations being run, it is expected that the state estimation results will outperform those found in the existing literature given the aforementioned breakthrough of the system, leading to further improvements of the state estimation accuracy. In the long term, the integrated autonomous navigation system will ease the interplanetary exploration, improving our current knowledge of the Solar System and opening the doors to a new era of space exploitation.

Conclusion
This paper investigates an integrated deep-space navigation system using both optical and X-ray pulsar measurements. The proposed system is considered superior and innovative because it combines well-established optical measurement methods with less mature X-ray pulsar measurement techniques. The time-varying measurement noise is handled by the utilisation of an adaptive nonlinear filter instead of the conventional Kalman filters. Once implemented, the results are expected to lead to further improvements of the state estimation accuracy. It is hoped that the contribution from this work will serve as a foundational milestone for achieving fully autonomous deep-space navigation in the future.

Acknowledgement
The author acknowledges the valuable support provided by Prof. Francesco Topputo in the development of this abstract.

References

[1] Shyam Bhaskaran, S Desai, P Dumont, B Kennedy, G Null, W Owen Jr, J Riedel, S Synnott, and R Werner. Orbit determination performance evaluation of the deep space 1 autonomous navigation system. 1998.

[2] William M Owen Jr. Methods of optical navigation. 2011.

[3] S Bella, E Andreis, V Franzese, P Panicucci, F Topputo, et al. Line-of-sight extraction algorithm for deep-space autonomous navigation. In 2021 AAS/AIAA Astrodynamics Specialist Conference, pages 1-18, 2022.

[4] RW Gaskell, OS Barnouin-Jha, Daniel J Scheeres, AS Konopliv, T Mukai, Shinsuke Abe, J Saito, M Ishiguro, Takashi Kubota, T Hashimoto, et al. Characterizing and navigating small bodies with imaging data. Meteoritics & Planetary Science, 43(6):1049-1061, 2008. https://doi.org/10.1111/j.1945-5100.2008.tb00692.x

[5] Hongliang Ma and Shijie Xu. Only feature point line-of-sight relative navigation in asteroid exploration descent stage. Aerospace Science and Technology, 39:628-638, 2014. https://doi.org/10.1016/j.ast.2014.05.017

[6] Shuang Li, Ruikun Lu, Liu Zhang, and Yuming Peng. Image processing algorithms for deep-space autonomous optical navigation. The Journal of Navigation, 66(4):605-623, 2013. https://doi.org/10.1017/S0373463313000131

[7] Reza Raymond Karimi and Daniele Mortari. Interplanetary autonomous navigation using visible planets. Journal of Guidance, Control, and Dynamics, 38(6):1151-1156, 2015. https://doi.org/10.2514/1.G000575

[8] Eleonora Andreis, Vittorio Franzese, and Francesco Topputo. An overview of autonomous optical navigation for deep-space cubesats. In 72nd International Astronautical Congress (IAC 2021), pages 1-11, 2021.

[9] Eleonora Andreis, Vittorio Franzese, and Francesco Topputo. Onboard orbit determination for deep-space cubesats. Journal of guidance, control, and dynamics, pages 1-14, 2022. https://doi.org/10.2514/1.G006294

[10] Min Song and Yunbin Yuan. Autonomous navigation for deep spacecraft based on celestial objects. In 2008 2nd International Symposium on Systems and Control in Aerospace and Astronautics, pages 1-4. IEEE, 2008. https://doi.org/10.1109/ISSCAA.2008.4776308

[11] TJ Chester and SA Butman. Navigation using x-ray pulsars. Jet Propulsion Laboratory, Pasadena, CA, NASA Tech. Rep. 81N27129, 1981.

[12] Liangwei Huang, Ping Shuai, Xinyuan Zhang, and Shaolong Chen. Pulsar-based navigation results: data processing of the x-ray pulsar navigation-i telescope. Journal of Astronomical Telescopes, Instruments, and Systems, 5(1):018003-018003, 2019. https://doi.org/10.1117/1.JATIS.5.1.018003

[13] Alexandra Witze. Nasa test proves pulsars can function as a celestial gps. Nature, 553(7688):261-263, 2018. https://doi.org/10.1038/d41586-018-00478-8

[14] Lori Keesey and Clare Skelly. Nasa team first to demonstrate x-ray navigation in space, 2018.

[15] Erdem Turan, Stefano Speretta, and Eberhard Gill. Autonomous navigation for deep space small satellites: Scientific and technological advances. Acta Astronautica, 2022. https://doi.org/10.1016/j.actaastro.2021.12.030

Aerospace Science and Engineering - III Aerospace PhD-Days   Materials Research Forum LLC
Materials Research Proceedings 33 (2023) 309-314   https://doi.org/10.21741/9781644902677-45

[16] Xiaolin Ning, Mingzhen Gui, Jiancheng Fang, and Gang Liu. Differential x-ray pulsar aided celestial navigation for mars exploration. Aerospace Science and Technology, 62:36-45, 2017. https://doi.org/10.1016/j.ast.2016.10.032

[17] Rong Jiao, Luping Xu, Hua Zhang, and Cong Li. Augmentation method of xpnav in mars orbit based on phobos and deimos observations. Advances in Space Research, 58(9):1864-1878, 2016. https://doi.org/10.1016/j.asr.2016.07.021

[18] Kai Xiong, Chunling Wei, and Liangdong Liu. Performance enhancement of x-ray pulsar navigation using autonomous optical sensor. Acta Astronautica, 128:473-484, 2016. https://doi.org/10.1016/j.actaastro.2016.08.007

[19] Long Gu, Xiuqiang Jiang, Shuang Li, and Wendan Li. Optical/radio/pulsars integrated navigation for mars orbiter. Advances in Space Research, 63(1):512-525, 2019. https://doi.org/10.1016/j.asr.2018.09.005

[20] Xiaolin Ning, Zhuo Li, Weiren Wu, Yuqing Yang, Jiancheng Fang, and Gang Liu. Recursive adaptive filter using current innovation for celestial navigation during the mars approach phase. Science China Information Sciences, 60(3):1-15, 2017 https://doi.org/10.1007/s11432-016-0405-2

Aerospace Science and Engineering - III Aerospace PhD-Days

Materials Research Forum LLC

Materials Research Proceedings 33 (2023) 315-322

https://doi.org/10.21741/9781644902677-46

iDREAM: A multidisciplinary methodology and integrated toolset for flight vehicle engineering

Giuseppe Narducci[1,a*], Nicole Viola[1], Roberta Fusaro[1],
Giuseppe Governale[1], Jasmine Rimani[1], Davide Ferretto[1]

[1]Politecnico di Torino, Torino, Italy

[a]giuseppe.narducci@polito.it

Keywords: Rapid Prototyping, Multidisciplinary Design Tools, Aerospace Vehicle Design, Technology Roadmaps, Cost Estimation

Abstract. For the rapid prototyping of a flight engineering vehicle, data from various building blocks and the required engineering tools for designing critical subsystems are crucial elements. In this context, Politecnico di Torino developed an integrated methodology toolset capable of speeding up the design and validation of various space transportation systems, with a focus on microlaunchers and human landing systems. This research was carried out under the direction of the European Space Agency (ESA).

iDREAM methodology

This methodology not only enables the conceptual design of the new vehicle, but also completes it with an exhaustive analysis of the solution's viability from an economic and technical standpoint. As a result, this methodology can be utilized for three primary purposes, each of which can be used independently or together with automatic connections, as schematically represented in Figure 1:

1. Design and related mission analysis
2. Life Cycle Cost (LCC) assessment
3. Technology roadmap

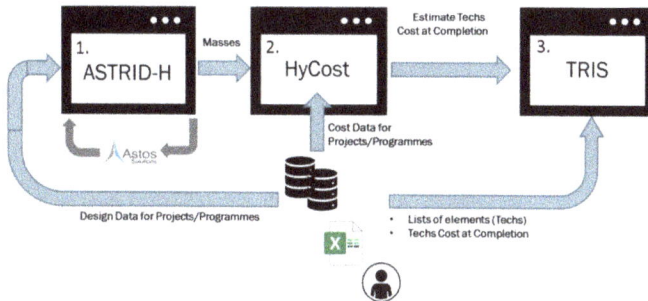

Figure 1: iDREAM path.

ASTRID-H

The first iDREAM capability was exploited using ASTRID-H, as shown in
Figure 1. The main purpose of this tool is to support the rapid prototyping of Micro-Launcher (ML) and Human Landing System (HLS).

Aerospace Science and Engineering - III Aerospace PhD-Days   Materials Research Forum LLC
Materials Research Proceedings 33 (2023) 315-322   https://doi.org/10.21741/9781644902677-46

Building on extensive experience in developing methodologies to support the conceptual and preliminary design of complex aerospace vehicles and integrated subsystems, and leveraging the proprietary software tool ASTRID, Politecnico di Torino has improved ASTRID-H, a methodology and tool that was originally designed for high-speed vehicle applications.

Leveraging on the European Space Agency's interests, two different case studies have been identified: (i) MicroLauncher and (ii) Human Landing System.

The increasing interest in launching small satellites and the limitations associated with launching such payloads have led to the development of MicroLaunchers (ML) as a potential solution to make such missions more affordable. In this context, PoliTO has devised a two-fold methodology. First, it involves using two mass estimation algorithms (Restricted and Optimal Staging) to validate an existing vehicle design, which leads to a redesign and a deeper exploration of subsystems and missions. The second part involves creating a new ML and its associated mission based on a set of high-level requirements. To ensure the accuracy and viability of the design, there is a direct connection with the commercial mission analysis software ASTOS, which enables iterative improvement of the design and mission by working in tandem with the vehicle design and mission analysis routines. Overall, this approach facilitates the development of an optimized Micro-Launcher or verification of the design and mission of an existing one.

The second scenario pertains to a Lunar Lander, also known as a Human Landing System, which is an essential component for exploring the Moon, as outlined in the Global Exploration Roadmap 2018 [1] and Global Exploration Roadmap Supplement – Lunar Surface Exploration Scenario Update 2020 [2] by the International Space Exploration Coordination Group (ISECG). PoliTO has developed a methodology to aid in the conceptual design of a Human Landing System (HLS) for future lunar missions involving astronauts. Depending on the information provided by the Agency, the HLS is intended to be a single-stage vehicle and will be responsible for transportation between the Moon's surface and the Lunar Orbital Platform-Gateway (LOP-G). Based on the requirements specified by the Agency, the Human Landing System (HLS) must be reusable up to five times, host four astronauts, and remain on the moon's surface for three moon nights. With the primary objective of the design process established, the PoliTO team developed a design methodology involving a preparatory conceptual design activity in ASTRID-H and a mission analysis using the commercial software ASTOS. However, the design approach for the HLS differs from that of the MicroLauncher, as it follows a bottom-up approach starting with subsystem design and proceeding to vehicle design. Once the mass, power, and volume budgets of the spacecraft are estimated, they can be utilized as inputs to initialize ASTOS and supplement the design with a proper mission analysis.

HyCost

The second component depicted in Figure 1, is the economic feasibility module.

Benefitting from the long-time experience in developing methodologies to support the assessment of the Life Cycle Cost (LCC) of complex aerospace vehicles, Politecnico di Torino upgraded its proprietary tool HyCost to support a wide range of high-speed vehicles [3] [4] [5], using a different approach for ML and HLS.

For what concern the first case study, the MicroLaunchers (ML), various cost estimation methodologies were evaluated, and those proposed by Drenthe [6] and ESA [7] were chosen. This methodology, which is based on T1 equivalent units and linear factors applied at the subsystem and equipment levels, is focused on small commercial launch vehicles. This method was chosen as the best fit because it provides more flexibility when taking into account new technologies, and subsystem-level considerations give the estimates a high degree of accuracy. The estimates can also be easily refined when new data and updates regarding these innovative launchers are released.

Aerospace Science and Engineering - III Aerospace PhD-Days   Materials Research Forum LLC
Materials Research Proceedings 33 (2023) 315-322   https://doi.org/10.21741/9781644902677-46

The second case study undertaken by Politecnico di Torino focused on assessing the life cycle cost (LCC) of Human Landing Systems (HLS). In order to perform this analysis, a thorough review of the relevant literature on cost-estimating methodologies for space systems and programs was conducted, which identified three primary strategies: (i) Analogy, (ii) Parametric, and (iii) Engineering build-up [8]. Following this, various cost models and tools were evaluated, including the Advanced Missions Cost Model (AMCM) and the Unmanned Space Vehicle Cost Model (USCM).

Based on the specific requirements of this project and the available data, the analogy cost-estimating methodology was determined to be the most suitable. This approach was deemed appropriate due to the early design phase of the system and the limited availability of consistent data on similar systems. Other available models were used for comparison and validation purposes.

TRIS

The third module, TRIS, produces a technology roadmap that assesses the technology readiness and risk for each technology, as well as identifies future activities, missions, and necessary developments. Such technology roadmapping methodologies are designed to pinpoint the essential technologies and activities required for technology development, operational capabilities, and building blocks based on pre-determined performance targets [9].

The third module, TRIS, generates a technology roadmap that estimates the technology readiness and risk assessments for each technology, along with necessary future works, activities, and missions.

Technology roadmapping methodologies are meant to identify the enabling technologies and activities needed for technology development, operational capabilities, and building blocks based on predefined performance targets [9].

Current roadmap activities aim to analyze complex systems and generate an incremental and sustainable technology development plan, or technological roadmap, that must be periodically reviewed by experts involved in strategic decisions. The TRIS methodology developed at Politecnico di Torino is complementary to other approaches found in the literature [10], [11], [12], [13], [14], [15] and generates technology roadmaps capable of supporting strategic decisions in combination with brainstorming sessions with experts' opinions By utilizing a rational, objective, and traceable methodology, TRIS defines gradual paths of technology maturation for new missions, products, or capabilities. TRIS enhances traditional techniques by highlighting feasible incremental pathways to the ultimate goal, by utilizing shared System Engineering tools and processes [16] [17] and purpose-built tools. As intended in TRIS, a technology roadmap is the product of various activities that identify, prioritize, select, and merge elements that fall under the technology roadmap pillars (Operational Capabilities, Technologies, Building Blocks, and Mission Concepts).

Looking at TRIS in detail, the methodology consists of five main steps: Stakeholders' Analysis, Elements' Definition, Prioritization Studies, Planning Definition and Results Evaluation, as detailed in Figure 2:

Figure 2: TRIS methodology.

The process begins with the Stakeholders' Analysis, which plays a crucial role in identifying all the involved parties, determining their roles (Sponsors, Operators, End-users and Customers) and assessing their impact (keep engaged, keep informed, keep satisfied, and monitor) on the final decision. Depending on the category and area of interest, the influence and interest of each actor can be predicted, and their level of influence and interest can be used as a weighting factor for the needs they express.

The second step involves defining and characterizing the various elements or pillars that constitute a roadmap. These elements include (i) Operational Capability (OC), which refers to a high-level function that satisfies a mission statement or research objectives; (ii) Technology Area (TA), which refers to a set of technologies that accomplish one or more OCs and is often further subcategorized into Technology Subject and Technology; (iii) Building Block (BB), which is a physical element that may include several technologies combined to achieve specific functions (OCs); and (iv) Mission Concept (MC), which is defined by a mission statement and comprises BBs to implement several OCs and make use of certain technologies.

The third step of the process entailed a trade-off analysis-based prioritization. In this phase, all criteria derived from the Stakeholders' Analysis can function as figures of merit that contribute to the final ranking of technologies based on their corresponding stakeholder influence and interest. Using the information gathered during the Elements' Definition phase, a list of Operational Capabilities (OCs) and Mission Concepts (MCs) associated with each identified technology was created to facilitate the required Technology Readiness Level (TRL) Transit, as specified by the ESA Standards.

The fourth step in the roadmap process involves Planning Definition, which entails appropriately scheduling the list of Mission Concepts (MCs) within a suitable timeline. To achieve this, a semi-empirical model was developed by conducting a comprehensive literature review of Space Exploration data, which employs a specific time-allocation breakdown. This breakdown is used to determine an initial development timeline for each technology, which is then further refined with the actual list of Activities (ACs) and Mission Concepts (MCs) required to meet each Technology Readiness Level (TRL) Transit, as per the standards set by the European Space Agency (ESA).

Aerospace Science and Engineering - III Aerospace PhD-Days   Materials Research Forum LLC
Materials Research Proceedings 33 (2023) 315-322   https://doi.org/10.21741/9781644902677-46

In cases where a specific AC or MC is associated with multiple technologies, its starting date is fixed only after all relevant technologies have attained the minimum TRL required by the AC/MC. This step facilitates the definition of the final timeline for each technology, and by merging it with the ordered and linked list of ACs/MCs, the final plan is generated, along with the incremental path for each technology's maturation. The expected graphical output includes two Gantt Charts: one displaying the time and budget allocation for each technology on TRL Transits, along with TRL Milestones, and the other focusing on the ordered list of ACs/MCs along the same timeline. This dual visualization is feasible due to the well-established relationship between technologies and ACs/MCs.

During the Results Evaluation step, the roadmapping activities carried out in the previous steps are integrated and risk analysis is performed to assess the level of risk associated with each feasible roadmap, taking into account the anticipated challenges in achieving the target Technology Readiness Level (TRL). The risk analysis aids in exploring different out-of-nominal scenarios and conducting a sensitivity analysis to assess the influence of stakeholders' expectations on the final roadmap.

Database connection
In addition to the aforementioned capabilities, iDREAM rely on two distinct MySQL databases, which are specifically designed to support the other modules of the toolset. This support is made possible by a Database Management Library that ensures a unified connection between the input and output of data. The two databases used were named TREX and HyDat. TREX, the first database, is exclusive to the European Space Agency and was utilized in the Human Landing System case study. The second database, HyDat, was developed internally by Politecnico di Torino and was originally intended for use in the design of hypersonic and reusable vehicles for space access. Subsequently, it was adapted to facilitate the MicroLauncher case study and was also employed in another contract between PoliTo and ESA. Eventually, this database has been modified to accommodate a wide range of vehicles, including the Human Landing System.

Case study: MicroLauncher
The case study presented here is based on Rocket Lab ML Electron [18]. Completely designed and produced by Rocket Lab, the Electron launch vehicle is one of the first ML ever launched.
Table 1 details the main results achieved by comparing the iDREAM ASTRID-H tool with the actual values of the Electron ML, while Table 2 lists the outcomes of the cost-estimating tool. Errors under 10% were obtained, which is entirely consistent with the conceptual design process.

Table 1: Results of iDREAM ASTRID-H routine compared with Electron actual values.

Global Input Variable Name	Electron [iDREAM]	Electron	Percentage differences [%]
Payload Mass [kg]	268.59	280.00	-4.08
Payload Diameter [m]	1.07	1.08	-0.93
MTOM [t]	12.49	12.5	-0.08
1st Stage Inert Mass [t]	0.89	0.90	-1.11
2nd Stage Inert Mass [t]	0.19	0.20	-5.00
Fairing mass [kg]	44.04	44.00	0.09
Fairing Length [m]	2.57	2.40	7.08
Total Length [m]	18.00	18.00	0.00
1st Stage Thrust [kN]	244.97	224.30	9.22

2nd Stage Thrust [kN]	27.79	25.8	7.71
1st Stage engine mass [kg]	35.58	35.00	1.66
2nd Stage engine mass [kg]	38.15	35.00	9.00

Table 2: Results of iDREAM HyCost routine compared with Electron actual values.

	Price per Flight [k€]	Specific Cost [k€/kg]
Electron	16200	54
Electron [iDREAM]	17327	55
Percentage differences [%]	6.96	-3.70

Case study: Human Landing System

This article presents a case study based on the ESAS LSAM spacecraft [19] [20], which is considered one of the most comprehensive projects in lunar surface access. The spacecraft, developed by the ESAS team with NASA's assistance, features state-of-the-art manufacturing technologies and multi-lunar mission capabilities. Although the ESAS LSAM is a 2-stage spacecraft, for this analysis, the lander is considered a single stage that encompasses both the ascent and descent modules, with a focus on the descent mission phases. The ascent module is treated as payload mass in this study. However, due to this approximation, some percentage differences exceed 10% when comparing the results obtained using the iDREAM ASTRID-H methodology with the actual values of the ESAS LSAM, as shown in Table 3. These discrepancies may be attributed to the lack of data about the LSAM subsystem [19] [20] [21].

Table 3: Results of iDREAM ASTRID-H routine compared with ESAS LSAM actual values.

Global Input Variable Name	ESAS LSAM [iDREAM]	ESAS LSAM [15]	Percentage differences [%]
ECLSS Mass [kg]	1177	1312	-10.29
Avionics mass[kg]	678	655	3.51
Propulsion mass [kg]	3810	3905	-2.43
Structure mass [kg]	2965	2841	4.36
EPS mass [kg]	1310	1246	5.14
Other mass [kg]	1155	1022	13.01
Dry mass [kg]	10421.3	11264	-7.48
Wet mass [kg]	40163	45861.6	-12.43
Fuel mass [kg]	25580.7	29820	-14.22

The results achieved by ASTRID-H are used as inputs to run Cost Estimation Routine (HyCost) and to obtain the outcomes detailed in Table 4.

Table 4: Results of iDREAM HyCost routine compared with ESAS LSAM actual values.

	Development and Production Cost [M€]
ESAS LSAM [13]	5500
ESAS LSAM [iDREAM]	5993
Percentage differences	8.23 %

Aerospace Science and Engineering - III Aerospace PhD-Days   Materials Research Forum LLC
Materials Research Proceedings 33 (2023) 315-322   https://doi.org/10.21741/9781644902677-46

Conclusion and Future works

Although the results obtained in this study were in line with the expectations, future work will employ a distinct approach that capitalizes on the significant advancements made possible by AI-based tools. Specifically, a design assistant leveraging a Knowledge Graph (KG) will be developed to evaluate the feasibility of various mission architectures. The KG, which is a semantic network that represents entities and their relationships, will be utilized to collect explicit and implicit knowledge and serve as the backbone of this new methodology. This approach will enhance the design process by enabling quick and effortless access to past design choices and exploring new design alternatives.

References

[1] I. S. E. C. G. (ISEGC), "Global Exploration Roadmap 2018," 2018.

[2] I. S. E. C. G. (ISEGC), "Global Exploration Roadmap - Supplement August 2020 - Lunar Surface Exploration Scenario Update," 2020.

[3] R. Fusaro, N. Viola, D. Ferretto, V. Vercella, V. Fernandez Villace and J. Steelant, "Life cycle cost estimation for high-speed transportation systems," 2020. https://doi.org/10.1007/s12567-019-00291-7

[4] R. Fusaro, V. Vercella, D. Ferretto, N. Viola and J. Steelant, "Economic and environmental sustainability of liquid hydrogen fuel for hypersonic transportation systems," 2020. https://doi.org/10.1007/s12567-020-00311-x

[5] R. Fusaro, N. Viola, D. Ferretto, V. Vercella and J. Steelant, "Life-cycle cost estimation for high-speed vehicles: From engineers' to airlines' perspective," 2020. https://doi.org/10.2514/6.2020-2860

[6] N. Drenthe, "SOLSTICE: Small Orbital Launch Systems, a Tentative Initial Cost Estimate," 2016.

[7] G. Reinbold, "Successful Cost Estimation with T1 Equivalents," in ICEAA Int. Train. Symp, 2016.

[8] NASA, "NASA Cost Estimating Handbook," 2015.

[9] M. Carvalho, A. Fleury and A. Lopes, "An overview of the literature on technology roadmapping (TRM): Contributions and trends," Technol. Forecast. Soc. Change. 80, p. 1418-1437, 2013. https://doi.org/10.1016/j.techfore.2012.11.008

[10] H. Abe, "The Innovation Support Technology (IST) Approach: Integrating Business Modeling and Roadmapping Methods," in M. Moehrle, R. Isenmann, R. Phaal (Eds.), Technol. Roadmapping Strateg. Innov., Berlin, Heidelberg, Springer, 2013, p. 173-188. https://doi.org/10.1007/978-3-642-33923-3_11

[11] H. Geschka and H. Hahnenwald, "Scenario-Based Exploratory Technology Roadmaps - A Method for the Exploration of Technical Trends," in M. Moehrle, R. Isenmann, R. Phaal (Eds.), Technol. Roadmapping Strateg. Innov., Berlin, Heidelberg, Springer, 2013, p. 123-136. https://doi.org/10.1007/978-3-642-33923-3_8

[12] D. Kanama, "Development of Technology Foresight: Integration of Technology Roadmapping and the Delphi Method," in M. Moehrle, R. Isenmann, R. Phaal (Eds.), Technol. Roadmapping Strateg. Innov., Berlin, Heidelberg, Springer, 2013, p. 151-171. https://doi.org/10.1007/978-3-642-33923-3_10

Aerospace Science and Engineering - III Aerospace PhD-Days   Materials Research Forum LLC
Materials Research Proceedings 33 (2023) 315-322   https://doi.org/10.21741/9781644902677-46

[13] M. Moehrle, "TRIZ-based technology roadmapping," in M. Moehrle, R. Isenmann, R. Phaal (Eds.), Technol. Roadmapping Strateg. Innov. Charting Route to Success, Berlin, Heidelberg, Springer, 137-150, p. 2013. https://doi.org/10.1007/978-3-642-33923-3_9

[14] R. Phaal, C. Farrukh and D. Probert, "Fast-Start Roadmapping Workshop Approaches," in M. Moehrle, R. Isenmann, R. Phaal (Eds.), Technol. Roadmapping Strateg. Innov., Berlin, Heidelberg, Springer, 91-106, p. 2013. https://doi.org/10.1007/978-3-642-33923-3_6

[15] D. Knoll, A. Golkar and O. De Weck, "A concurrent design approach for model-based technology roadmapping," 12th Annual IEEE International Systems Conference, SysCon 2018 - Proceedings, p. 1 - 6, 2018. https://doi.org/10.1109/SYSCON.2018.8369527

[16] M. Viscio, N. Viola, R. Fusaro and V. Basso, "Methodology for requirements definition of complex space missions and systems," Acta Astronautica, pp. 79-92, 2015. https://doi.org/10.1016/j.actaastro.2015.04.018

[17] R. Shishko, "NASA systems engineering handbook," 2007.

[18] R. Lab, "Electron, User's Guide," Rocket Lab, [Online]. Available: rocketlabusa.com.

[19] NASA, NASA's exploration system architecture study, 2005.

[20] J. F. Connolly, After LM - NASA LUNAR LANDER CONCEPTS BEYOND APOLLO, 2019.

[21] I. Masafumi, M. Ian and C. Bernd, "A New Sizing Methodology for Lunar Surface Access Systems," AIAA Scitech 2020 Forum, 2020.

Aerospace Science and Engineering - III Aerospace PhD-Days   Materials Research Forum LLC
Materials Research Proceedings 33 (2023) 323-328   https://doi.org/10.21741/9781644902677-47

Could a remotely operated UAV fleet improve emergency response?

A. Avi*, G. Quaranta

Dipartimento di Scienze e Tecnologie Aerospaziali, Politecnico di Milano, via Giuseppe La Masa, 34, 20156 Milano MI

arrigo.avi@polimi.it, giuseppe.quaranta@polimi.it

Keywords: UAV, UAS, Docking Station, Emergency, Firefighter, SA, Civil Protection

Abstract. Unmanned aerial vehicles could reduce risks in emergency response missions. The UAV firefighter team of Provincia autonoma di Trento proposes a new concept of operations. A fleet of drones placed in strategically selected locations and operated from the main headquarters in Trento. The first UAV, with a docking station, was placed on Marmolada glacier for trial.

Introduction and motivation

Unmanned aerial vehicles are changing many aspects of our society. We have examples of constructions, surveys, rescue, etc. One of the most important aspects is the reduction of the risks compared to manned aircraft, other things are lower costs and automation.

The civil protection system of Provincia autonoma di Trento [1] (a region in the north of Italy, known for the Dolomiti mountains) is complex and articulate. There are lots of different actors: more than 5000 volunteer firefighters, a professional brigade, alpine rescue teams, red cross groups, etc. One of the main objectives of this organization is rescuing people and about that, and the coordination between different teams is paramount. Since 2015 a team of the firefighter professional brigade regularly use UAVs as an important tool. They are involved in many activities like searching for missing people, wildfires, notable accidents, surveying landslides and cliffs, and other important activities for prevention management.

In particular, the three most important activities performed are:

- Missing People

  The deployment of UAV teams during the search for or missing person missions is standard [2], coupled with local firefighter volunteer brigades. Response time is one of the more challenging aspects of this type of operation: the team moves from Trento headquarters to the site of the last known position. This could take more than an hour and pose a risk against a successful result. Furthermore, sites are in narrow valleys with few roads, some in a bad condition.

- Hydrogeological risk

  In the Alps region, it is necessary to monitor the environment against climate change, especially hydrogeological risks. UAVs conducted a lot of surveys related to hydrogeological management, assessment of damages, and risk assessment. Monitoring activities are ever more important and require a higher frequency, so UAV could be an effective platform to do the monitoring in a cheaper, safer and more effective way.

- Situational awareness

  The frequency and cost of natural disasters are increasing worldwide [3]. It is paramount to have a fast and reliable way for situational awareness. It is essential to have accurate information from the disaster sites for an effective high-level decision-making process. This is one of the most demanding activities to be performed, it requires lots of personnel (that could be already involved in rescuing people) and equipment. Currently, there are some systems that could give an overview of the situation. There are two systems: a fleet of steerable cameras that cover only 20% of the entire territory, and a fleet of sensors that could give only punctual information. Lacking coverage and punctiform information leads often to poor situational awareness.

Proposed Concept of Operations
Following the previously highlighted problems, a new concept of operations was proposed. A fleet of drones, that remains inside docking stations, placed in strategically selected locations and operated from the main headquarters in Trento. Docking stations guarantee a safe takeoff and landing area, recharging, and storage. In this way, it is possible to have a distributed system that could have a larger coverage area (up to the entire territory) and almost real-time response time. The UAV fleet is composed of a docking station, UAV, a data network, and interfaces.

The *docking station* is the logistical part of the system. It provides a safe takeoff-landing area with precision landing aids, recharging or replacing systems for batteries, and automatic systems for payload change. It is connected to the data network and acts as a datalink repeater for UAVs.
UAV is the frontline object of this system, plays an important role in terms of reliability and availability in marginal weather conditions like often is during natural disasters.

Data networks have a key role in the entire system. It needs to guarantee a safe and reliable way to communicate with the UAV and with the docking station. Another important role is the deployment of information for other people involved in operations: video feed for firefighters, repeater for communications, and redundant datalink with the main operating center.

Interface is the link between pilots and UAV, it should be customized based on the different missions in which will be used.

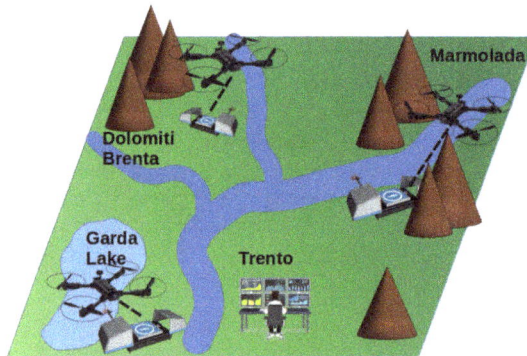

Fig. 1 - concept of operations

Feasibility study and proposed implementation
A remotely located fleet of drones is an intensive initiative to develop. It is important, and mandatory for a public administration, to do a feasibility study to evaluate sustainability in the most general way.

The deployment of many docking stations is expensive, so it is important to evaluate how much the UAV network could be used for different activities. To evaluate the feasibility of the proposed UAV network that, a session to present the project was organized with members of the Civil Protection Department: geological, forest, and water management services. After the meeting a

Aerospace Science and Engineering - III Aerospace PhD-Days Materials Research Forum LLC
Materials Research Proceedings 33 (2023) 323-328 https://doi.org/10.21741/9781644902677-47

brief questionnaire was sent, regarding which services were interested in using the remotely operated UAV network and in particular:

- how many areas interested and extent;
- how many yearly operations;
- distance from villages;
- time of operations (h24/7, seasonal, real time, programmed);
- objectives;
- data needed (RGB, lidar, multispectral, video feed...)

5 forms were collected with the following outcomes:

- a fleet of UAV was an interesting idea;
- most service need programmed operations;
- services are very interested in collecting data, but don't have enough resources to process;
- almost all areas are outside of villages;
- all services need to monitor both small areas and large areas.

The most important outcome of this questionnaire was the number of operations needed by every service: at least 250 operations every year; this information highlights that a sustainable design of the network could be achieved.

Finding a compromise to satisfy all the preceding requirements is not simple. A current solution that involves COTS components already available on the market could be depicted. In particular:

- UAV should have an endurance of at least 2 hours, and range of 15 kilometers;
- it is necessary to have at least 12 docking stations with UAV [4];
- it is necessary to deploy a digital data radio network [5].

With these requirements, sustainability could be reached because the fleet covers the territory of Trento region, every UAV could reach every location inside its working area in less than 10 minutes, and 12 docking stations could guarantee reliability with lots of backup alternatives.

In terms of economic evaluation, the proposed solution requires initial investments of about 2.500.000 €, and yearly costs of 85000 €.

The management of this proposed solution needs some dedicated human resources: there should be a team of 10 people because it is necessary to have at least 2 pilots H24/7 and a team for maintenance.

All these evaluations led to consider this system sustainable and effective for a regional firefighter department.

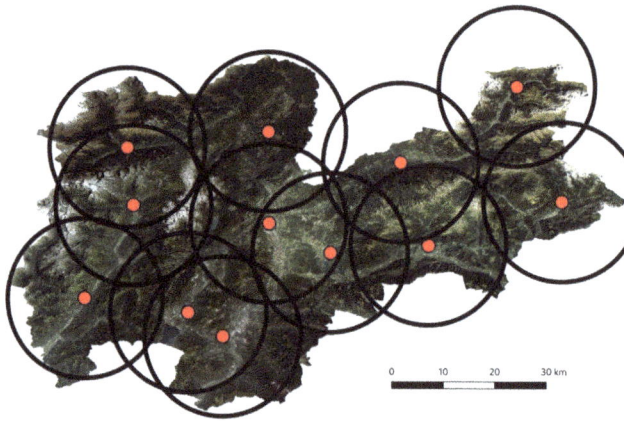

Fig. 2 - proposed docking station network

In a feasibility study [6] it is important to evaluate alternatives. Three options were analyzed:

- option *zero*: do nothing, it isn't an actual solution, all the troubles highlighted before remaining unchanged;
- option *one UA*: deploy a very expensive and performing UAV that could fly directly from main headquarters with endurance of many hours, simplest solution but with no backup (only one UA in one location);
- option *UAV everywhere*: deploy lots of UAVs managed by a local team, it is almost impossible to manage due to the large team and due to people already involved in rescue operation couldn't be available for UAV piloting.

Fig. 3 - Marmolada trial

Trial in Marmolada

On 3 July 2022, a collapse involving the glacier of Marmolada, it took 11 lives. After that, the first UAV docking station was placed near the glacier to perform surveys on the remaining glacier. From 03 September to 03 November, the station was operational, and 37 missions were flown. Most missions were inspections of a crevasse or photogrammetric survey. This first deployment was composed of an agnostic docking station, a DJI drone and a commercial cloud interface. Managed from the headquarters in Trento, it was an important opportunity to test, in a small and simple way, the effectiveness of the proposed network.

Aerospace Science and Engineering - III Aerospace PhD-Days   Materials Research Forum LLC
Materials Research Proceedings 33 (2023) 323-328   https://doi.org/10.21741/9781644902677-47

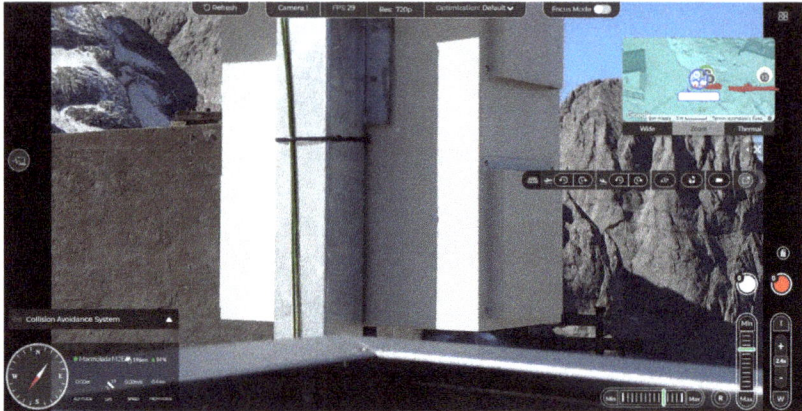

Fig. 4 - Interface

References

[1] Information on http://www.protezionecivile.tn.it/organizzazione/Dipartimento/

[2] Jake N. McRae, Christopher J. Gay, Brandon M. Nielsen, Andrew P. Hunt, Using an Unmanned Aircraft System (Drone) to Conduct a Complex High Altitude Search and Rescue Operation: A Case Study, Wilderness & Environmental Medicine, Volume 30, Issue 3, 2019, Pages 287-290, https://doi.org/10.1016/j.wem.2019.03.004

[3] Information on https://www.eea.europa.eu/ims/economic-losses-from-climate-related

[4] Wankmüller, C., Truden, C., Korzen, C. et al. Optimal allocation of defibrillator drones in mountainous regions, OR Spectrum 42, 785–814, https://doi.org/10.1007/s00291-020-00575-z

[5] R. Bassoli, C. Sacchi, F. Granelli and I. Ashkenazi, "A Virtualized Border Control System based on UAVs: Design and Energy Efficiency Considerations," 2019 IEEE Aerospace Conference, Big Sky, MT, USA, 2019, pp. 1-11, https://doi.org/10.1109/AERO.2019.8742142

[6] Linee guida per la redazione del progetto di fattibilità tecnica ed economica da porre a base dell'affidamento di contratti pubblici di lavori del PNRR e del PNC (Art. 48, comma 7, del decreto-legge 31 maggio 2021, n. 77, convertito nella legge 29 luglio 2021, n. 108)

Aerospace Science and Engineering - III Aerospace PhD-Days   Materials Research Forum LLC
Materials Research Proceedings 33 (2023) 329-333   https://doi.org/10.21741/9781644902677-48

Innovative navigation strategies based on multiple signals for performance improvement of drone-based operations

Verdiana Bottino[1,a] *

[1]Department of Industrial Engineering, University of Naples Federico II, Italy

[a]verdiana.bottino@unina.it

Keywords: UAS, MEMS, Remotely Piloted Aircraft, Drones, Integrated Navigation, Heading, 5G Technology

Abstract. The growing employment of Unmanned Aerial Systems (UASs, commonly referred to as drones) in civilian and military environments, along with the increasing accuracy of avionics components, has led to the definition of more stringent drone-based mission requirements. In this sense, the following work summarizes a series of Ph.D. research activities aimed at improving UAS performance under different aspects, in order to enhance their use in a variety of application contexts and environmental conditions while ensuring a satisfactory level of accuracy.

The last decade experienced a rapid expansion of the UASs market thanks to the significant cost reduction of new technologies. Indeed, compared to fixed-wing aircraft or helicopters, drones are characterized by much lower dimensions and weight, and the substantial performance improvement reached by avionic components made them a cost-effective and versatile solution capable of meeting various mission needs. As a result, the use of drones is no longer limited to military operations but is spreading to a wide range of application domains, such as precision agriculture, monitoring, surveillance, communication, and goods delivery. Moreover, the autonomy and flexibility of drones make their employment highly beneficial for emergency operations, cinematography, and infrastructure inspection, which often require costly equipment, repetitive procedures, and a high level of technical experience [1–5].

The general improvement of UAS performances is also owed to their capability to integrate more sensors at the same time, with the aim of allowing simultaneous data collection from different sources, including electronic devices, high-resolution cameras, thermal cameras, and Micro-electro Mechanical Sensors (MEMS). The latter, in particular, play a fundamental role in the development of aerial vehicles and platforms: in fact, compared to the traditional heavy and expensive navigation units, the inertial units manufactured using MEMS benefit from miniaturized electronic components (such as microphones, gyroscopes, accelerometers, and pressure sensors) that are lighter, more compact, and definitely cheaper than other sensors. Also, their most advanced releases are becoming comparable in terms of accuracy with the bulky and costly fiber-optic technology, which is why MEMS are being massively employed also in the aerospace and defense field [1,6–10].

Despite the increased reliability and ease of integration, the use of MEMS is associated with a series of issues (such as the inertial sensors bias drift with time) whose resolution is still the object of numerous studies. Navigation-related issues must be adequately considered, with the aim of proposing suitable and cost-effective solutions capable of meeting both accuracy and reliability requirements; for this reason, the Ph.D. project outlined in the present work was aimed at improving UAS navigation performances to enhance drone employment in different mission types, also developing innovative navigation solutions to successfully use drones in challenging environments.

Aerospace Science and Engineering - III Aerospace PhD-Days   Materials Research Forum LLC
Materials Research Proceedings 33 (2023) 329-333   https://doi.org/10.21741/9781644902677-48

Since UAS-based operations require accurate and autonomous navigation capabilities exploited in the estimate of attitude parameters, the initial part of the research activities has been focused on the heading angle computation. An innovative gyrocompassing procedure based on the integration of a tactical-grade MEMS Inertial Measurement Unit (IMU), composed of an accelerometer and a gyroscope, and a low-cost MEMS magnetometer was proposed [11]: particularly, a Matlab^{TM} tool was implemented to acquire the IMU data, and the magnetometric measurements allowed to compute the three components of the Earth magnetic field which were used to initialize the gyrocompassing procedure. A Kalman Filtering algorithm composed of a prediction stage and a correction stage was implemented to perform data fusion, and the estimated heading angle was compared with a reference value obtained by means of a certified Attitude and Heading Reference System (AHRS). The results reported in Table 1 showed that the heading angle computed by the proposed system was compliant with the certified value, with a difference of 0.76 deg in Root Mean Square Error, proving that the magnetometer measurements integration allows the designed system to reach remarkable performances in terms of heading estimate.

Table 1: Difference in heading angle ($\Delta\theta$) between the reference and the proposed system in terms of Mean, Root Mean Square Error (RMSE), and Standard Deviation (STD).

	Mean	**RMSE**	**STD**
$\Delta\theta$ [°]	0.76	0.76	0.04

The same goal was reached through an innovative methodology exploiting the Sun polarization properties and designed to support GNSS/IMU integrated solutions [12]. A polarimetric camera was embarked on a commercial drone to acquire different images, which were processed to deduce the Sun Meridian/anti-meridian position in the local reference frame and, subsequently, the heading angle. Ground and flight tests were performed using an 8mm focal length lens, and these measurements were integrated with the ones obtained by the IMU in the filtering phase. Then, the computed attitude parameters were compared with the ones obtained by a certified AHRS for ground tests and by drone telemetry for flight tests: in both cases, an RMSE error of about 1 deg was shown, proving that the use of a polarimetric camera permits adequate compensation of IMU-related bias drift and that the proposed system represents a suitable solution to adopt when traditional GNSS/IMU cannot be employed. An example of the results obtained in a flight test is shown in Table 2.

Table 2: Average values of roll (ϕ), pitch (θ), and heading (ψ) angles computed by the drone telemetry during a flight test along a straight path, using an 8mm focal length lens. The average error is computed as the difference between the heading angle obtained by the drone telemetry and the one measured by the proposed system. The reported values refer to a single group of images acquired at different drone locations.

Parameter	Value
Average ϕ [°]	-1
Average θ [°]	-1.2
Average ψ [°]	199.5
Average ε_ψ [°]	-0.2

Aerospace Science and Engineering - III Aerospace PhD-Days   Materials Research Forum LLC
Materials Research Proceedings 33 (2023) 329-333   https://doi.org/10.21741/9781644902677-48

Present studies are aimed at the elimination of magnetometric measurements to perform the gyrocompassing procedure, despite a Kalman Filter-based algorithm still representing the best option to adopt for data fusion; regarding the visual-based UAS navigation, further studies based on polarimetric camera measurements are being conducted, with particular focus on fisheye lenses.

During the last year, research activities have also been carried out to perform accurate UAS heading estimation in unstructured environments where GNSS satellite positioning is problematic. Indeed, despite working well outdoors, GNSS signal easily suffers from blockage or degradation in urban areas with tall buildings, indoor settings, deserts, and under bridges, having large errors affecting the drone attitude and position estimates consequently [13]. Different options were investigated but, eventually, the most suitable solution was found in exploiting 5G technology, because of the numerous advantages associated with this type of connection (including higher carrier frequencies, insensitivity to magnetic interference, and the capability to ensure a positioning service with sub-meter accuracy) [14–16]. So, a preliminary system architecture for UASs operating in GNSS-challenging scenarios based on an integrated IMU/GNSS solution was proposed [13]: as shown in Figure 1, the possibility to switch to a 5G/GNSS integrated configuration in case of GPS signal unavailability or degradation was included.

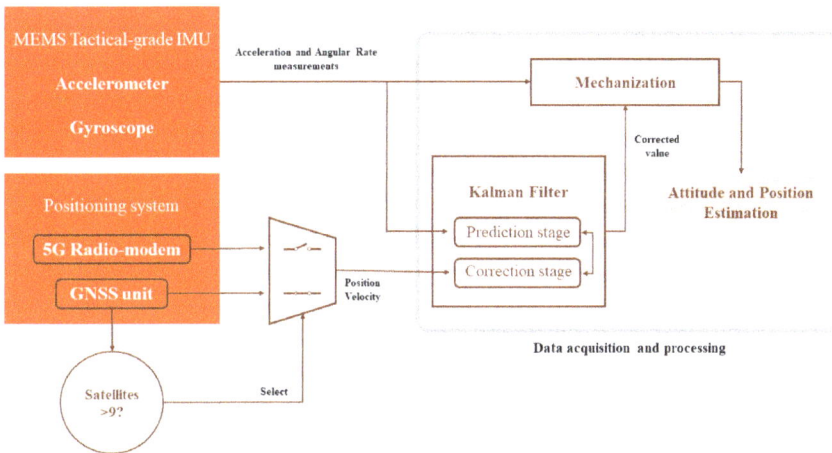

Figure 1: Proposed 5G-GNSS RTK/IMU system architecture. The 5G Radio-modem is integrated onboard the drone to replace the GNSS service during its outage periods or in case the GNSS positioning service doesn't provide sufficiently accurate results. The navigation process is divided into two main steps: a mechanization process that allows to obtain the navigation solutions in their continuous form, and a Kalman Filtering stage that provides position, velocity, attitude, and bias corrected values.

Currently, Ph.D. research activities concerning this topic are proceeding with the elaboration of a Matlab™ tool aimed at simulating 5G communication and computing the drone position using trilateration techniques so that, in the near future, flight tests will be performed embarking a 5G radio modem on a commercial drone with a flying platform composed by a MEMS-based IMU and a GNSS receiver. The results will be compared with the ones obtained through Matlab™

Aerospace Science and Engineering - III Aerospace PhD-Days   Materials Research Forum LLC
Materials Research Proceedings 33 (2023) 329-333   https://doi.org/10.21741/9781644902677-48

simulations, while data fusion will be performed by means of a Kalman Filter algorithm. During the experiment planning and trajectory design phases, particular attention will be paid to the application UASs for automated cinematography, which frequently operates in GNSS-denied or GPS-challenging environments.

References

[1] de Alteriis, G., Silvestri, A. T., Conte, C., Bottino, V., Caputo, E., Squillace, A., Accardo, D., and Schiano Lo Moriello, R. "Innovative Fusion Strategy for MEMS Redundant-IMU Exploiting Custom 3D Components." *Sensors*, Vol. 23, No. 5, 2023, p. 2508. https://doi.org/10.3390/s23052508

[2] Khosravi, M., Enayati, S., Saeedi, H., and Pishro-Nik, H. "Multi-Purpose Drones for Coverage and Transport Applications", *IEEE Transactions on Wireless Communications*, Vol. 20, No. 6, 2021. https://doi.org/10.1109/TWC.2021.3054748

[3] Kangunde, V., Jamisola, R. S., and Theophilus, E. K. "A Review on Drones Controlled in Real-Time". *International Journal of Dynamics and Control*. 4. Volume 9, 1832–1846.

[4] Zhu, P., Wen, L., Du, D., Bian, X., Fan, H., Hu, Q., and Ling, H. "Detection and Tracking Meet Drones Challenge." *IEEE Transactions on Pattern Analysis and Machine Intelligence*, Vol. 44, No. 11, 2022, pp. 7380–7399. https://doi.org/10.1109/TPAMI.2021.3119563

[5] Ayamga, M., Akaba, S., and Nyaaba, A. A. "Multifaceted Applicability of Drones: A Review." *Technological Forecasting and Social Change*, Vol. 167, 2021. https://doi.org/10.1016/J.TECHFORE.2021.120677

[6] Han, S., Meng, Z., Omisore, O., Akinyemi, T., and Yan, Y. "Random Error Reduction Algorithms for MEMS Inertial Sensor Accuracy Improvement—a Review". *Micromachines*. 11. Volume 11, 1–36.

[7] Narasimhappa, M., Mahindrakar, A. D., Guizilini, V. C., Terra, M. H., and Sabat, S. L. "MEMS-Based IMU Drift Minimization: Sage Husa Adaptive Robust Kalman Filtering." *IEEE Sensors Journal*, Vol. 20, No. 1, 2020, pp. 250–260. https://doi.org/10.1109/JSEN.2019.2941273

[8] Fontanella, R., de Alteriis, G., Accardo, D., Schiano, R., Moriello, L., and Angrisani, L. "Advanced Low-Cost Integrated Inertial Systems with Multiple Consumer Grade Sensors", IEEE/AIAA 37th Digital Avionics Systems Conference (DASC). IEEE, 2018.

[9] de Alteriis, G., Accardo, D., Conte, C., and lo Moriello, R. S. "Performance Enhancement of Consumer-Grade Mems Sensors through Geometrical Redundancy." *Sensors*, Vol. 21, No. 14, 2021. https://doi.org/10.3390/s21144851

[10] Navidi, N., Landry, R., Mariani, S., and Gravina, R. "A New Perspective on Low-Cost MEMS-Based AHRS Determination." 2021. https://doi.org/10.3390/s21041383

[11] de Alteriis, G., Bottino, V., Conte, C., Rufino, G., and lo Moriello, R. S. "Accurate Attitude Initialization Procedure Based on MEMS IMU and Magnetometer Integration", *2021 IEEE 8th International Workshop on Metrology for AeroSpace (MetroAeroSpace). IEEE, 2021.*

[12] Bottino, V., de Alteriis, G., Schiano, R., Moriello, L., Rufino, G., and Accardo, D. "Drone Navigation Based on Integrated MEMS Inertial and Polarimetric Camera Measurements", AIAA SCITECH 2023 Forum (p. 2705).

[13] Bottino, V., de Alteriis, G., Accardo, D., Schiano, R., and Moriello, L. "System Architecture Design of a UAV for Automated Cinematography in GNSS-Challenging Scenarios", *2022 IEEE 9th International Workshop on Metrology for AeroSpace (MetroAeroSpace), 2022.*

[14]   Bai, L., Sun, C., Dempster, A. G., Member, S., Zhao, H., Cheong, J. W., and Feng, W. "GNSS-5G Hybrid Positioning Based on Multi-Rate Measurements Fusion and Proactive Measurement Uncertainty Prediction." Vol. 71, 2022.

[15]   Li, F., Tu, R., Hong, J., Zhang, S., Zhang, P., and Lu, X. "Combined Positioning Algorithm Based on BeiDou Navigation Satellite System and Raw 5G Observations." *Measurement: Journal of the International Measurement Confederation*, Vol. 190, No. August 2021, 2022, p. 110763. https://doi.org/10.1016/j.measurement.2022.110763

[16]   Shi, L., Shi, D., Zhang, X., Meunier, B., Zhang, H., Wang, Z., Vladimirescu, A., Li, W., Zhang, Y., Cosmas, J., Ali, K., Jawad, N., Zetik, R., Legale, E., Satta, M., Wang, J., and Song, J. "5G Internet of Radio Light Positioning System for Indoor Broadcasting Service." *IEEE Transactions on Broadcasting*, Vol. 66, No. 2, 2020, pp. 534–544. https://doi.org/10.1109/TBC.2020.2981755

Aerospace Science and Engineering - III Aerospace PhD-Days   Materials Research Forum LLC
Materials Research Proceedings 33 (2023) 334-340   https://doi.org/10.21741/9781644902677-49

Accurate characterization of the noise sources affecting BepiColombo's radio tracking observables

David Bernacchia[1,a] *

[1]Dipartimento di Ingegneria Industriale, Alma Mater Studiorum – Università di Bologna, Forlì, Italy

[a]david.bernacchia@unibo.it

Keywords: BepiColombo, Radio Science, Calibrations, Tropospheric Delay, Antenna Mechanical Noise

Abstract. The ESA mission BepiColombo, nowadays, represents a milestone for the precision levels that can be obtained in the determination of the orbit of a spacecraft. The high accuracy provided by the embedded radio system, together with modern techniques used for the calibration of tropospheric noise, such as the usage of a microwave radiometer, allows to estimate the state of the spacecraft with a very small uncertainty. Doppler measurements collected during the first two Solar Conjunction Experiments performed by BepiColombo have been analyzed in order to understand which are the main sources of residual noise still affecting the data after the calibration process, with the aim of improving the overall calibration procedure itself. The study showed that, even though the usage of radiometers led to a strong reduction of the noise due to the troposphere, the main error source can still be identified in some un-calibrated tropospheric effects. This indicates that the improvement of the overall calibration process strongly relies on the refinement of the method used for tropospheric noise removal.

Introduction

The latest advances in the technology of the radio systems used for deep space navigation led to a great improvement in terms of precision and accuracy in the deep space orbit determination. The strong reliability offered by the present-day radio systems necessarily requires very high and demanding standards in terms of calibration procedures, in order to maximize the precision levels that can be obtained.

BepiColombo spacecraft is equipped with one of the most precise radio systems ever flown in space, composed by two different transponders, the Deep Space Transponder (DST), enabling X-band/X-band and X-band/Ka-band radio links, and the Ka Transponder (KaT), enabling Ka/Ka radio link, both for ranging and Doppler. The two transponders, used simultaneously, allow the establishment of the so called multifrequency link, which plays a crucial role in the calibration process. It has been shown, in fact, that, through the linear combination of triple-link observables, it is possible to remove from the measurements the noise introduced by dispersive media such as solar plasma and Earth's ionosphere [1,2], which usually represent the dominant source of noise affecting the signal.

The radio system described before represents the key instrument of the Mercury Orbiter Radio science Experiment (MORE) which, during the entire duration of the mission, will be performing different investigations in terms of gravity, geodesy and fundamental physics [3]. During cruise the main radio science experiments performed are the Solar Conjunction Experiments (SCE), which are mainly focused on the estimation of the Parametrized Post-Newtonian (PPN) parameters [4].

This work presents an analysis performed on the BepiColombo tracking data collected during the first two solar conjunction experiment (SCE1 in March 2021 and SCE2 in January/February 2022), which have been used as a testbed to verify the effectiveness of different calibration

Aerospace Science and Engineering - III Aerospace PhD-Days   Materials Research Forum LLC
Materials Research Proceedings 33 (2023) 334-340   https://doi.org/10.21741/9781644902677-49

techniques. The data analyzed have been collected at the ESA's Deep Space Antenna in Malargue (DSA3) and comprise a set of Doppler and Range measurements in X- and Ka- bands for both uplink and downlink, registered during a total of 31 tracking passes (17 related to SCE1 and 14 related to SCE2).

Tropospheric delay calibration

Once removed the errors introduced in radiometric observables by the dispersive media through the linear combination of the triple-link measurements, the main noise source affecting the signal is represented by Earth's troposphere. The water vapor layer present in the troposphere introduces a time-variable delay which affects doppler and, to a lesser extent, range observables. The standard procedure used to remove tropospheric effects from the measurements is based on data obtained from the Global Navigation Satellite System (GNSS). However, a new calibration system based on a water vapor radiometer, called Tropospheric Delay Calibration System (TDCS), has been recently installed at the Malargue ground station in order to improve the calibration of the tropospheric effects. A detailed description of the TDCS and the results of a first validation test, performed with the Gaia spacecraft, are given in [5,6].

In the context of the MORE experiment, the first two solar conjunction experiments represented a great opportunity to further validate the TDCS performances, also in correspondence of a solar conjunction. A recent work has been carried out in which the TDCS and the GNSS calibrations have been compared performing an orbit determination analysis on the data collected during SCE1 and SCE2. The main point of the study was the analysis of the noise in the residuals, which are defined as the difference between the observables measured at the ground station and the observables generated by a mathematical model of the spacecraft's dynamics. As described in detail in [7], the TDCS calibrations led to an average reduction of doppler observables noise up to 51% with respect the noise level reachable with GNSS calibration. Concerning range observables, the TDCS didn't show an improvement as strong as the one brought into doppler ones but still a better performance than GNSS. In general, we can say that, nowadays, the TDCS represents the best option for the calibration of tropospheric effects.

At the current state of the art, the calibration procedures performed on the BepiColombo data comprise the removal of the errors introduced by the ground station and spacecraft's electronics (ranging only), the errors introduced by the dispersive media (solar plasma and ionosphere) through the multifrequency link and the errors introduced by the troposphere through the TDCS calibrations. In order to further improve this calibration process, a preliminary analysis, described in the following section, has been conducted on the post-calibration residuals relative to SCE1 and SCE2, with the aim of understanding which are the main residual noise sources still affecting the measurements. Identifying the main sources of un-calibrated noise is the first step to improve the overall calibration process. In order to reach this goal, the Doppler residuals obtained from the analysis of SCE1 and SCE2 data have been studied in terms of Allan standard deviation (ASD) and autocorrelation, relating these functions to two quantities that are suspected of being indicators of possible residual noise sources: the wind speed and the Sun Earth Probe (SEP) angle.

Doppler residuals analysis: SCE1

Starting from SCE1 residuals, the first step of the study concerned the visualization of the ASD and autocorrelation in relation with the wind speed registered at the Malargue station during the different tracking passes. Figure 1 shows the outcome of the analysis.

Aerospace Science and Engineering - III Aerospace PhD-Days   Materials Research Forum LLC
Materials Research Proceedings 33 (2023) 334-340   https://doi.org/10.21741/9781644902677-49

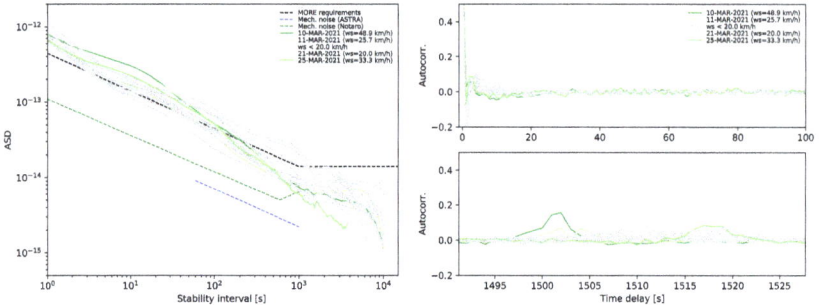

*Figure 1: Allan standard deviation (left) and autocorrelation function (right) of SCE1
Doppler residuals related to the wind speed.*

The graph on the left is showing the residuals' Allan standard deviation relative to each tracking passes. The curves are presented in shades of green in which a darker green represents an higher wind speed registered during that tracking passes. The grey curves, instead, represent passes in which the wind speed registered was below 20 km/h. The same concept applies for the autocorrelation graph. What can be observed from the ASD curves is that the 4 passes in which a moderate/high wind speed was recorded are, for low values of integration time (τ), all well above the MORE's stability requirement for two-way Doppler residuals (represented by the black dashed line) while, at high values of τ, the requirement is fulfilled. This indicates that, in terms of calibration, the residual noise still strongly affects the instrument stability at low values of integration time so a further noise removal is still required. About this point, the autocorrelation curves on the right are showing high autocorrelation peaks around the Round Trip Light Time (RTLT) giving us a strong hint about the nature of the main residual noise source. A peak at the RTLT, in fact, indicates that the noise affecting the residuals recurs at the location of the ground station. Moreover, it is clearly visible that the highest autocorrelation peak is related to the pass in which the highest wind speed value (48.9 km/h) was record, suggesting that the noise sources could be strongly related to the wind at the station. This consideration leads to the assumption that this main noise sources could be residual un-calibrated tropospheric effects or antenna mechanical noise (AMN). Concerning the latter, the ASD graph has been completed with two colored dashed line representing some literature models of antenna mechanical noise: the green one has been extracted from [8] while the blue one is extracted from the noise budget drafted in the frame of the ASTRA studies [9,10]. The comparison between these dashed lines and the residuals Allan deviation curves shows that at values of τ around 1000s the AMN may become the predominant noise sources while for smallest values the biggest noise contribution is given by another source which, as what discussed before, is very likely to be un-calibrated troposphere.

The second step of the study concerned a similar analysis to the one already described but, this time in relation of the Sun Earth Probe angle. The SEP is a parameter of interest for this analysis because, in condition of superior solar conjunction (SEP < 3°), the TDCS radiometer enters the *sun-avoidance mode* which is aimed to avoid the direct intrusion of the Sun into the observation beam of the instrument. This task is accomplished by creating a pointing offset between the pointing direction of the radiometer and the actual position of the spacecraft. However, even if this technique avoids an undesired presence of the Sun into the radiometer's observations it also implies that the observed troposphere portion, used for the calibration, is different from the actual troposphere which interferes with the ground station/spacecraft link.

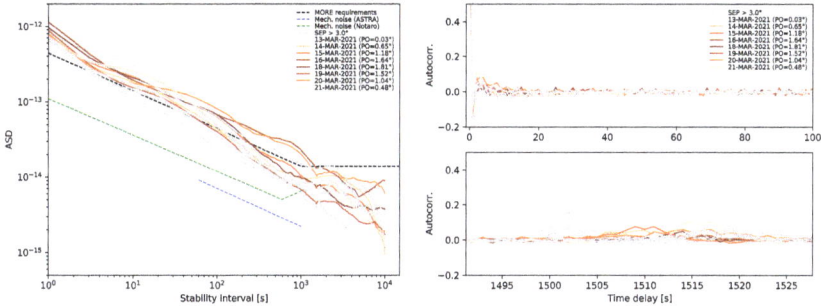

Figure 2: Allan standard deviation (left) and autocorrelation function (right) of SCE1 Doppler residuals related to the SEP.

Figure 2 shows the outcome of the analysis of the residuals in relation with the Sun Earth Probe angle. Similarly to what discussed before, the curves are presented in shades of orange in which a darker orange represents an higher value of radiometer pointing offset. The highest value of pointing offset angle registered during SCE1 is about 1.81°. The grey curves represents all the passes in which the SEP was above 3°. The first difference we can notice between the ASD curves of the two different analysis is that, while the passes with strong wind showed a good behaviour for high values of the stability interval, in this case we can clearly see that the passes which are above the black dashed line at high τ values are characterized by the radiometer working in Sun-avoidance mode, suggesting that the main reason of this divergence from the stability requirement can be found in un-calibrated troposphere due to the reasons explained before. This observation is somehow confirmed by the autocorrelation curves showing the typical feature around the RTLT. However it has to be underlined how these autocorrelation peaks in this case are lower with respect to the one of the pass characterized by high wind speed. Another consideration to be done is that, among the two curves which are consistently above the requirement it is not comprised the pass in which the highest value of pointing offset was registered. This could be due to the fact that during the two critical passes some more unfavorable conditions added to the radiometer working in Sun avoidance. Some more aspects to be discussed concerning the Allan deviation curves are that, as shown in the wind speed analysis, when the curves are under the requirement at 1000s, the antenna mechanical noise becomes the predominant noise source while at low values of τ the two dashed line are still far from the residuals' curves, underlining how the un-calibrated troposphere has to be pointed out as the main source of noise still to be calibrated.

Doppler residuals analysis: SCE2
The same analysis performed for SCE1 was replicated with SCE2 data.

Aerospace Science and Engineering - III Aerospace PhD-Days   Materials Research Forum LLC
Materials Research Proceedings 33 (2023) 334-340   https://doi.org/10.21741/9781644902677-49

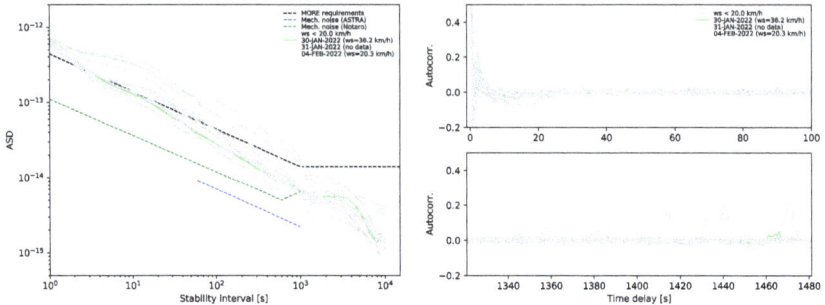

*Figure 3: Allan standard deviation (left) and autocorrelation function (right) of SCE2
Doppler residuals related to the wind speed.*

Unfortunately, as shown in Figure 3, during the second solar conjunction experiments only a couple of passes where characterized by a moderate wind speed so, concerning the wind analysis, no much more consideration to what discussed for SCE1 can be done. We can notice how one of the two passes in question presents an high autocorrelation peak around the RTLT.

Much more interesting, instead, the outcome of the analysis performed in terms of SEP angle.

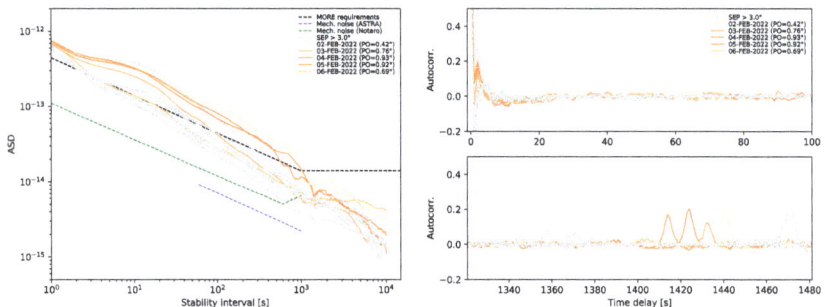

*Figure 4: Allan standard deviation (left) and autocorrelation function (right) of SCE2
Doppler residuals related to the SEP.*

Observing the left graph in Figure 4, we can see that before the stability interval point $\tau = 1000s$, almost all the curves related to passes characterized by the radiometer working in Sun-avoidance mode are well above the requirement curve. On the contrary, a good amount of passes in which the SEP was above 3° reach the stability requirement at low values of integration time. This is a strong indication of how the mismatch between the troposphere observed by the radiometer and the actual troposphere present along the antenna's line of sight is a relevant source of error. Also in this case, the autocorrelation function shows the typical peaks around the round trip light time, reinforcing the idea that the main noise sources can be found close to the antenna location. An aspect that is particularly interesting to notice is the fact that, during SCE2, the pointing offset registered are much smaller than those occurred during SCE1, however the autocorrelation peaks of the former are much higher than those showed by SCE1 data. This could be due to the fact that SCE2 was performed during the period between January and February which in Malargue

Aerospace Science and Engineering - III Aerospace PhD-Days   Materials Research Forum LLC
Materials Research Proceedings 33 (2023) 334-340   https://doi.org/10.21741/9781644902677-49

(Argentina) is comprised in the summer time. The highest temperature during the summer could lead to a stronger turbulence in the atmosphere thus the highest autocorrelation peaks may be a consequence of this particular condition.

Summary

The presented work had been performed in the frame of the first two BepiColombo's solar conjunction experiments with the aim of extracting information about the noise sources that are mainly affecting the residuals after the full calibration process. The high-precision radio system embedded in BepiColombo's probe requires a very accurate calibration of the measurements in order to maximize the accuracy of the spacecraft's state estimation and the identification of the main noise is a crucial first step for the improvement of the calibration process. For this purpose a study on the calibrated data from SCE1 and SCE2 had been conducted by analyzing the behaviour of Doppler residuals' Allan standard deviation and autocorrelation in function of the wind speed and Sun Earth Probe angle.

Conclusions and future work

The study conducted showed that, in all the different cases analyzed, a common behaviour of the residuals were to present a peak of autocorrelation function around the round trip light time. This particular peaks are due to the fact that there is a recurrent noise that enters into the observables at a location close to the ground station, indicating that some residual un-calibrated tropospheric effects and the antenna mechanical noise due to the vibration of the antenna structure induced by the wind are probably the main sources of error in the calibrated data. The fact that both of this noise sources can be located at the antenna location, thus both contributing to the autocorrelation peaks at RTLT, makes the separation of their effects a difficult task to accomplish. However, the analysis of the residuals' Allan standard deviation gives us a strong hint of the fact that between the two, the main component of the overall noise is the un-calibrated troposphere. The comparison, in fact, between the different ASD curves and the literature model of the antenna mechanical noise shows that the former are, especially for low values of integration time, quite far from the AMN models demonstrating that the residual troposphere represents a bigger source of error with respect the mechanical noise. However, at high values of integration time ($\tau > 1000s$) the residuals' Allan deviation match with one of the two AMN literature model, indicating that, once removed the residual troposphere, the antenna mechanical noise may represent the new biggest noise source.

In conclusion, this work had been very useful to individuate the trail to enter in the next future in order to improve the calibration process. It was discussed how the very precise BepiColombo's radio system allows, thanks to the multifrequency link, the cancellation of dispersive media-induced noise and how the recently installed TDCS improves strongly the calibration of the tropospheric effects. However it has been showed how this latter system can still be improved in order to reduce the residual un-calibrated troposphere. Different techniques, aimed to the minimization of the un-calibrated troposphere and described in literature, such as the so called *beam-crossing technique* described in [11], will be the attention of future works and analysis in order to obtain the desired improvement in the noise removal process. Other parallel works will concern the analysis of the antenna mechanical noise which, after the minimization of the residual troposphere, will represent the biggest challenge in the calibration process.

References

[1] Bertotti, B., Comoretto, G., & Iess, L. (1993). Doppler tracking of spacecraft with multi-frequency links. Astronomy and Astrophysics (ISSN 0004-6361), vol. 269, no. 1-2, p. 608-616., 269, 608-616.

[2] Mariotti, G., & Tortora, P. (2013). Experimental validation of a dual uplink multifrequency dispersive noise calibration scheme for Deep Space tracking. Radio Science, 48(2), 111-117. https://doi.org/10.1002/rds.20024

[3] Iess, L., Asmar, S. W., Cappuccio, P., Cascioli, G., De Marchi, F., di Stefano, I., ... & Zannoni, M. (2021). Gravity, geodesy and fundamental physics with BepiColombo's MORE investigation. Space Science Reviews, 217(1), 21. https://doi.org/10.1007/s11214-021-00800-3

[4] Di Stefano, I., Cappuccio, P., & Iess, L. (2020). The BepiColombo solar conjunction experiments revisited. Classical and Quantum Gravity, 38(5), 055002. https://doi.org/10.1088/1361-6382/abd301

[5] Manghi, R. L., Maschwitz, G., Tortora, P., Rose, T., Martellucci, A., de Vicente, J., ... & Quibus, L. (2019). Tropospheric Delay Calibration System (TDCS): design and performances of a new generation of microwave radiomters for ESA deep space ground stations. In TT&C workshop.

[6] Manghi, R. L., Zannoni, M., Tortora, P., Martellucci, A., De Vicente, J., Villalvilla, J., ... & Rose, T. (2021). Performance characterization of ESA's tropospheric delay calibration system for advanced radio science experiments. Radio Science, 56(10), 1-14. https://doi.org/10.1029/2021RS007330

[7] Manghi, R. L., Bernacchia, D., Casajus, L. G., Zannoni, M., Tortora, P., Martellucci, A., ... & Iess, L. Tropospheric Delay Calibration System performance during the first two BepiColombo solar conjunctions. Radio Science, e2022RS007614.

[8] Notaro, V., Mariani, M. J., Di Ruscio, A., Iess, L., Armstrong, J. W., & Asmar, S. W. (2018, March). Feasibility of an innovative technique for noise reduction in spacecraft Doppler tracking. In 2018 IEEE Aerospace Conference (pp. 1-10). IEEE. https://doi.org/10.1109/AERO.2018.8396480

[9] Iess, L., Budnik, F., Colamarino, C., Corbelli, A., Di Benedetto, M., Fabbri, V., et al. (2012). ASTRA: Interdisciplinary study on enhancement of the end-to-end accuracy for spacecraft tracking techniques. In Proceedings of the International Astronautical Congress, IAC, (pp. 53425-53435)

[10] Iess, L., Di Benedetto, M., James, N., Mercolino, M., Simone, L., & Tortora, P. (2014). Astra: Interdisciplinary study on enhancement of the end-to-end accuracy for spacecraft tracking techniques. Acta Astronautica, 94(2), 699-707. https://doi.org/10.1016/j.actaastro.2013.06.011

[11] Graziani, A., Jarlemark, P., Elgered, G., Martellucci, A., Mercolino, M., & Tortora, P. (2014). Assessment of Ground-Based Microwave Radiometry for Calibration of Atmospheric Variability in Spacecraft Tracking. IEEE transactions on antennas and propagation, 62(5), 2634-2641. https://doi.org/10.1109/TAP.2014.2307582

Aerospace Science and Engineering - III Aerospace PhD-Days | Materials Research Forum LLC
Materials Research Proceedings 33 (2023) 341-346 | https://doi.org/10.21741/9781644902677-50

Coupled 3D peridynamics and refined 2D finite elements models embedded in a global\local approach

Marco Enea[1,a] *

[1]Politecnico di Torino, Corso Duca degli Abruzzi 24, 10129, Turin, Italy

[a,]*marco.enea@polito.it

Keywords: Global\local, Peridynamics, Higher-Order Finite Elements, Carrera Unified Formulation

Abstract. The present work proposes a two-step global\local approach for the three-dimensional analysis of structures . In particular, the first step makes use of classical finite element modelling, based on classical 2D elements, whereas a refined higher-order model based on the Carrera Unified Formulation (CUF) is coupled with a 3D peridynamics domain for the local analysis. The solution from the proposed method is compared with a full FEM solution. The objective of the present work is to pave the way to the building of a framework where coupled FE\PD models are embedded into this global\local approach to solve progressive failure problems.

Introduction

During the last century, various models have been proposed to describe the complex mechanism of the failure process. Most of them are based on the Finite Element Method (FEM), such as, for instance, the introduction of the Cohesive Zone Element (CZE) [1] and the eXtended Finite Element Method (XFEM) [2]. However, the main limitation of theories based on continuum mechanics is related to the difficulty in dealing with discontinuities, such as a crack. To overcome this obstacle, non-local theories have been widely used to solve fracture mechanics problems in the past years. For example, a recent non-local theory is Peridynamics (PD), introduced by Silling in [3]. According to PD, a solid body is made of physical particles, which interact in a pairwise manner when their distance is lower than a prescribed value. The main advantage of PD theory is the integro-differential nature of its governing equations, leading to the possibility of dealing with discontinuous displacement fields. Nevertheless, a characteristic of non-local theories is that they involve large, sparse, and not banded matrices, resulting in high computational demand.

In this context, researchers are working towards methodologies to couple FEM models based on classical elasticity with PD regions, intending to exploit the advantages of both methods. Pagani and Carrera [4] have proposed a coupling between the 3D peridynamics domain and 1D refined finite elements based on the Carrera Unified Formulation (CUF) [5]. This technique is based on Lagrange multipliers and has been successfully applied to progressive failure analysis [6].

This paper first proposes a coupling between 3D PD domains and 2D CUF-based higher-order elements. Then, this method is embedded in a two-step global\local approach [7,8]. The first step is a global analysis performed via commercial FEM software. Then, a region to be locally refined is chosen. After some manipulations, the displacements obtained from the global analysis are used as input for the local model, which is built by adopting CUF-based higher-order finite elements. This work wants to represent a first action towards creating a standalone framework, where coupled FE\PD models are embedded into this global\local approach to solve progressive failure problems.

Aerospace Science and Engineering - III Aerospace PhD-Days   Materials Research Forum LLC
Materials Research Proceedings 33 (2023) 341-346   https://doi.org/10.21741/9781644902677-50

Global\local approach

In the present research, a one-way global\local coupling technique is adopted. The procedure is divided into two steps. The first one consists of a global analysis of an entire structure through two-dimensional (2D) classical finite elements. The static analysis is performed through a commercial software based on Finite Element Method (FEM). Then, a region to be locally refined has to be chosen, according to an arbitrary criterion, such as for instance, the maximum stress. The second step of the global\local procedure consists in the creation of a refined local model, built using higher-order 2D plate finite elements, based on the Carrera Unified Formulation (CUF). In the CUF framework, the 3D displacement field $\boldsymbol{u}(x,y,z)$ is defined as a 1D through-the-thickness expansion function of the unknowns, evaluated through the finite element method. The relation can be expressed in the following way:

$$\boldsymbol{u}(x,y,z) = F_\tau(z)N_i(x,y)\boldsymbol{q}_{\tau i} \tag{1}$$

where F_τ is the expansion function, N_i the shape function and $\boldsymbol{q}_{\tau i}$ the nodal unknowns vector. The index τ indicates the number of terms in the thickness expansion, while the subscript i represents the number of the structural finite element nodes. In this work, sixteen-nodes cubic elements (Q16) are used as shape functions for the in-plane discretization, whereas four-node cubic Lagrange Expansion (LE3) are employed as expansion function F_τ over the thickness (see **Figure 1**).

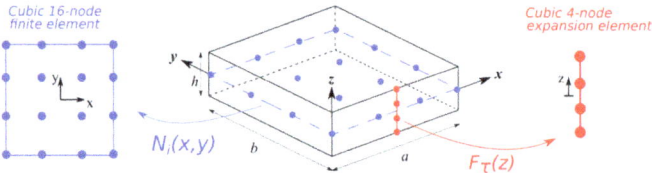

Figure 1. Graphical representation of the cubic interpolation for in-plane and thickness domain.

However, the Degrees of Freedom (DOFs) in the global model are expressed in terms of translational displacements (u_x^0, u_y^0, u_z^0) and rotations $(\theta_x^0, \theta_y^0, \theta_z^0)$,whereas the CUF-based local model makes use of a displacement-based formulation. Thus, a procedure to transform the rotations of the global model into pure displacement DOFs is conducted, by means of a Reissner-Mindlin displacement field. The equations read:

$$
\begin{aligned}
u_x(x,y,z) &= u_x^0(x,y) + z\theta_y^0(x,y) - y\theta_z^0(x,y) \\
u_y(x,y,z) &= u_y^0(x,y) - z\theta_x^0(x,y) - x\theta_y^0(x,y) \\
u_z(x,y,z) &= u_z^0(x,y) - x\theta_y^0(x,y) - y\theta_y^0(x,y)
\end{aligned}
\tag{2}
$$

where quantities with superscript 0 denotes displacements and rotations of global DOFs. The computed pure displacement values from the global nodes are then interpolated and applied as boundary conditions into the refined local model, in order to perform a static analysis.

Coupling of 3D Peridynamics with 2D higher-order finite elements

Peridynamics is a non-local theory based on integro-differential equations. It assumes that a solid body is composed by material particles and each pair of those interacts if their distance is less than a material horizon of radius δ. The physical interaction between two particles at $\boldsymbol{x}$ and $\boldsymbol{x}'$ is called a *bond*, which extends over a finite distance. The non-local nature of peridynamics makes this theory attractive for being successfully applied to solid mechanics and crack propagation

Aerospace Science and Engineering - III Aerospace PhD-Days   Materials Research Forum LLC
Materials Research Proceedings 33 (2023) 341-346   https://doi.org/10.21741/9781644902677-50

problems. Nevertheless, this method can be computationally prohibitive for many applications, because the resulting matrices are sparse, not generally banded, and large.

For this reasons, in previous works a coupling between CUF-based 1D finite elements and peridynamics has been proposed. In the present research, the extension to a coupling with 2D higher-order finite elements is proposed. This technique is based on the application of Lagrange multipliers at the interfaces $\mathcal{I}$ between 3D PD and FE domains (see **Figure 2**). In this way, the congruence conditions are satisfied, along with the elimination of the singularity in the global stiffness matrix. The linear system to be solved is expressed as follows:

$$\begin{bmatrix} K & B^T \\ B & 0 \end{bmatrix} \begin{bmatrix} U \\ \lambda \end{bmatrix} = \begin{bmatrix} F \\ 0 \end{bmatrix} \tag{3}$$

where K is the global stiffness matrix, including contributions from both FEM and PD domain, B is the coupling matrix and λ is the Lagrangian's force vector.

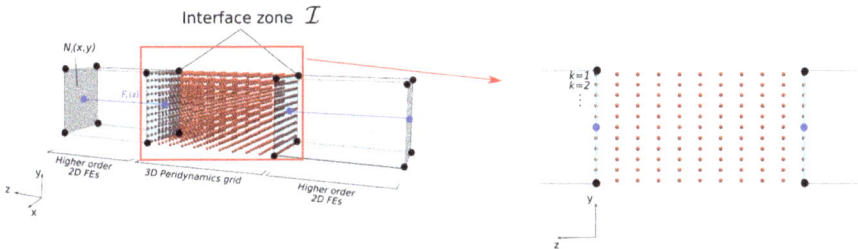

Figure 2. Coupling technique between higher-order 2D FEs and 3D peridynamics grid based on Lagrange multipliers.

Global\local approach and PD\FE coupled models framework
This research combines the G\L approach and the PD\FE coupled models within the same framework. The first step is still represented by the global analysis, performed via finite element analysis in commercial software using 2D classical finite elements. The main novelty lies in the nature of the local domain. In fact, the refined model consists of a 3D peridynamics grid coupled with a FE domain with 2D higher-order finite elements. In this work, a linear static analysis is performed on the local model by application of displacements and rotations from the global nodes as boundary conditions. However, this analysis can be seen as a first step for an iterative progressive failure algorithm, where the PD domain is used for modelling regions where cracks are likely to develop. The implementation of this framework has as objective the reduction of the computational costs of failure analysis while maintaining a high degree of accuracy.

Results
A representative case study is shown in this section. A plate subjected to localized transversal pressure is considered. It is made of an isotropic and homogeneous material, with Young Modulus $E = 200\,GPa$ and $v = 0.25$. The shape of the plate is rectangular, with sides $a = 200\,mm$ and $a = 300\,mm$. The thickness of the plate is equal to $h = 1.25\,mm$. A clamped condition is imposed in one edge, while a localized pressure equal to 10 kPa is applied on a corner element in the opposite edge. The mesh adopted for the global analysis is a 10×15 grid of 2D plate elements. A patch of nine elements is chosen as region to be locally analysed (see Figure 3). Finally, two different local domains have been investigated, one using only higher-order elements and another consisting in a PD\FE coupled model. A 5×5 grid with sixteen-nodes cubic element is adopted

for the in-plane discretization, whereas a single four-node cubic element is used over the thickness. Concerning the PD domain, a grid spacing $\Delta x = 0.125\ mm$ and a horizon $\delta = 3\Delta x$ is chosen.

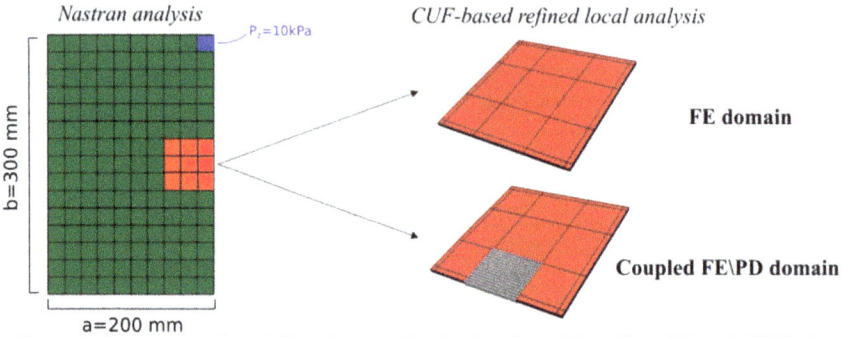

Figure 3. Geometry and modelling features for the global model and both FE and FE\PD local models.

Figure 4 depicts the displacement fields in the full FE and the coupled FE\PD model. A clear correspondence between the two cases is shown.

Moreover, in Figure 5 the distribution of vertical displacement along the blue line in the local domain is illustrated. The solutions from both cases are compared. The solid blue line represents the displacements in the full FE local domain, whereas the dots describe the transversal displacements in the coupled local domain. More specifically, the red dots are the displacement computed in the FE domain, whereas the green ones are those retrieved in the PD region. It is evident that the coupled FE\PD local analysis is able to accurately reproduce the behaviour of the full FE local domain.

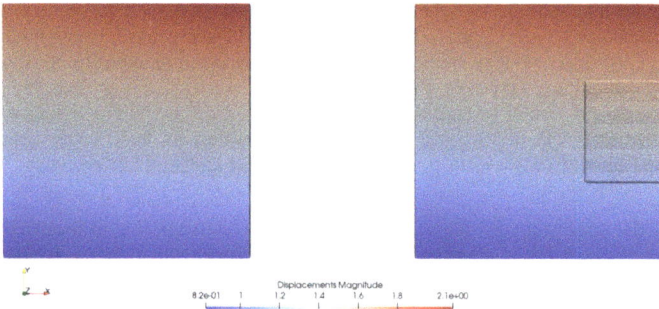

Figure 4. Displacement field in FE and coupled FE\PD local domains.

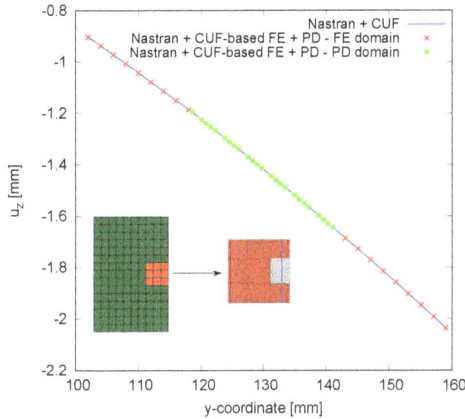

Figure 5. Distribution of transversal displacement along the blue line in the local domain. The solid blue curve represents the results from the full FE domain, whereas the red and green dots are the displacements computed in the coupled local region for FE and PD domains, respectively.

Conclusions

This manuscript has presented the pairing of a global\local approach and a model coupling Finite Elements (FEs) and peridynamics (PD). First, the coupling method between two-dimensional (2D) higher-order finite elements and a 3D peridynamics grid has been implemented for the first time. Then, a global\local analysis on an isotropic plate has been performed. The global analysis is performed via Nastran and by using 2D classical finite elements, whereas the refined local model is represented by a coupled FE\PD domain. Results show that this new coupled local model is able to accurately describe the displacement fields retrieved by a FE-based local analysis.

The presented technique represents a first fundamental step towards building a standalone framework, which will combine the global\local approach with FE\PD coupled models for progressive failure analysis, to reduce the computational costs while maintaining a high degree of solution accuracy.

Acknowledgements

The author would like to acknowledge the DEVISU project which was supported by the Ministero dell'Istruzione, dell'Università e della Ricerca research funding programme PRIN 2017.

References

[1] A. Hillerborg, M. Modéeer, and P. Petersson, Analysis of crack formation and crack growth in concrete by means of fracture mechanics and finite elements.,Cement and Concrete Research, 6 (1976):773-781, https://doi.org/10.1016/0008-8846(76)90007-7

[2] N. Moes, J. Dolbow, and T. Belytschko, A finite element method for crack growth without remeshing, International Journal for Numerical Methods in Engineering, 46 (1999) 131-150. https://doi.org/10.1002/(SICI)1097-0207(19990910)46:1<131::AID-NME726>3.0.CO;2-J

[3] S.A. Silling, Reformulation of elasticity theory for discontinuities and long-range forces, Journal of the Mechanics and Physics of Solids 48 (2000) 175-209. https://doi.org/10.1016/S0022-5096(99)00029-0

[4] A. Pagani, E. Carrera, Coupling 3D peridynamics and high order 1D finite elements based on local elasticity for the linear static analysis of solid beams and thin-walled reinforced structures, International Journal for Numerical Methods in Engineering 121(2020) 5066–5081. https://doi.org/10.1002/nme.6510

[5] E. Carrera, M. Cinefra, M. Petrolo, E. Zappino. Finite Element Analysis of Structures through Unified Formulation. Wiley & Sons, Hoboken, New Jersey. 2014.

[6] A. Pagani, M. Enea, E. Carrera, Quasi-static fracture analysis by coupled three-dimensional peridynamics and high order one-dimensional finite elements based on local elasticity. International Journal for Numerical Methods in Engineering 123(2022) 1098-1113. https://doi.org/10.1002/nme.6890

[7] E. Carrera, A. de Miguel, M. Filippi, I. Kaleel, A. Pagani, M. Petrolo, E. Zappino, Global-local plug-in for high-fidelity composite stress analysis in Femap/NX Nastran, Mechanics of Advanced Materials and Structures, 28 (2021) 1121-1127. https://doi.org/10.1080/15376494.2019.1655689

[8] R. Augello, A. Pagani, E. Carrera, A. Iannotta. Stress and failure onset analysis of thin composite deployables by global/local approach. AIAA Journal, (2022) 1-13. https://doi.org/10.2514/1.J061899

Aerospace Science and Engineering - III Aerospace PhD-Days   Materials Research Forum LLC
Materials Research Proceedings 33 (2023) 347-354   https://doi.org/10.21741/9781644902677-51

Peridynamic simulation of elastic wave propagation by applying the boundary conditions with the surface node method

Francesco Scabbia[1,a,*], Mirco Zaccariotto[1,2], Ugo Galvanetto[1,2], Florin Bobaru[3]

[1] Centro di Ateneo di Studi e Attività Spaziali "Giuseppe Colombo" (CISAS), Università degli Studi di Padova, Padova, 35131, Italy

[2] Industrial Engineering Department (DII), Università degli Studi di Padova, Padova, 35131, Italy

[3] Department of Mechanical and Materials Engineering, University of Nebraska-Lincoln, Lincoln, NE, USA

[a] francesco.scabbia@phd.unipd.it

Keywords: Peridynamics, Wave Propagation, Surface Node Method, Surface Effect, Nonlocal Boundary Conditions

Abstract. Peridynamics is a novel nonlocal theory able to deal with discontinuities, such as crack initiation and propagation. Near the boundaries, due to the incomplete nonlocal region, the peridynamic surface effect is present, and its reduction relies on using a very small horizon, which ends up being expensive computationally. Furthermore, the imposition of nonlocal boundary conditions in a local way is often required. The surface node method has been proposed to solve both the aforementioned issues, providing enhanced accuracy near the boundaries of the body. This method has been verified in the cases of quasi-static elastic problems and diffusion problems evolving over time, but it has never been applied to a elastodynamic problems. In this work, we show the capabilities of the surface node method to solve a peridynamic problem of elastic wave propagation in a homogeneous body. The numerical results converge to the corresponding peridynamic analytical solution under grid refinement and exhibit no unphysical fluctuations near the boundaries throughout the whole timespan of the simulation.

Motivation

When structures are exposed to unfavorable conditions, e.g., high temperature gradients, hostile environmental actions, excessive mechanical loading, or extreme events, cracks may initiate, propagate, and coalesce over time, compromising the functionality of the affected structure. Aircraft integrity assessments often consist of scheduled maintenance to inspect damage propagating in the structures. The first concern is the safety of passengers since in between two scheduled inspections, little can be said about the status of the aircraft so that an unexpected failure could bring catastrophic consequences. The second drawback is related to financial losses, as each airplane has to be grounded for several hours/days during inspections.

Fracture is difficult to predict because of complex multiphysical interactions and multiscale mechanisms that influence the behavior/evolution of cracks and damage. The capability of currently available computational tools based on classical mechanics, such as the Finite Element Method, to describe crack propagation and fragmentation is rather limited. On the other hand, the peridynamic theory allows to model fracture phenomena with ease.

Introduction to the peridynamic theory

Peridynamics (PD) is a nonlocal continuum theory based on integro-differential equations [1,2]. Peridynamics is more general compared to classical continuum models based on partial differential equations since discontinuities in the unknown displacements can arise and evolve in the domain without leading to mathematical inconsistencies or singularities in the problem. Successful

Aerospace Science and Engineering - III Aerospace PhD-Days   Materials Research Forum LLC
Materials Research Proceedings 33 (2023) 347-354   https://doi.org/10.21741/9781644902677-51

applications of this theory are, for instance, initiation and propagation of cracks in solid bodies [3,4], damage mechanisms in corrosion (with autonomously evolving interfaces) [5], etc.

In a PD body B, a point interacts with all the points within a finite distance δ, named *horizon size*. As shown in Fig. 1, the neighborhood H_x is the set of points with which a generic point x interacts:

$$H_x = \{x' \in B : |x' - x| \leq \delta\}. \tag{1}$$

Each interaction between any points x and x' is called *bond*. Thus, the PD equation of motion for a generic point x in a 1D, homogeneous, linear elastic body [6,7] is given as

$$\ddot{u}(x,t) = v^2 \int_{H_x} \frac{u(x',t) - u(x,t)}{\delta(x'-x)^2} \, dx', \tag{2}$$

where u is the displacement, $\ddot{u}$ is the acceleration and v is the wave speed.

Figure 1: Example of the neighborhoods of two points (x_1 and x_2) in a peridynamic body. Note that the neighborhood of point x_2 is incomplete due to the closeness with the boundary of the body, whereas the neighborhood of point x_1 in the bulk of the body is complete. The blue curved lines represent the PD bonds between points x_1, x_2, and their neighboring points, respectively.

However, nonlocal theories are well-known to have issues near the boundaries of the body, for the reason illustrated in Fig. 1. Due to the lack of some bonds, points near the boundaries of the body exhibit an apparent variation in stiffness properties, if one uses the same micromodulus for bonds of points near the boundaries as for bonds of points in the bulk. This phenomenon is called *peridynamic surface effect* and is usually undesired in structural analyses [8-10]. Moreover, nonlocal models require the imposition of nonlocal boundary conditions, i.e., loads and constraints have to be enforced in a layer of finite thickness. This is in contrast with the concept of local boundary conditions that are enforced just at the boundary of the body. Since experiments generally provide measurements at the boundary, imposing local boundary conditions in a nonlocal model is usually desired. In this work, the surface node method (see details in the following) is used to mitigate the PD surface effect and impose local boundary conditions in the peridynamic model. This method has been applied to quasi-static mechanical problems [9-11] and to a diffusion problem evolving over time [12]. Here we extend it to elastodynamic problems in 1D. Further extensions to higher dimensions are immediate.

Discretization in space and time

A peridynamic body is commonly discretized by means of the meshfree method with a uniform grid spacing Δx [13]. Therefore, each node is representative of a cell of length Δx, as shown in Fig. 2. After the discretization in space, the peridynamic equation of motion of a generic node i is written as

$$\ddot{u}(x_i, t) = \frac{v^2}{\delta} \sum_{j \in H_i} \frac{u(x_j,t) - u(x_i,t)}{(x_j - x_i)^2} \beta_{ij} \Delta x, \tag{3}$$

Aerospace Science and Engineering - III Aerospace PhD-Days   Materials Research Forum LLC
Materials Research Proceedings 33 (2023) 347-354   https://doi.org/10.21741/9781644902677-51

where x_i and x_j are respectively the coordinates of node i and any node j within the neighborhood H_i of node i, and β_{ij} is the quadrature coefficient of node j with respect to H_i. β_{ij} is defined as the fraction of cell of node j which actually lies within the neighborhood H_i [14,15]. Hence, $\beta_{ij} = 1$ if the cell of node j lies completely inside H_i and $0 < \beta_{ij} < 1$ if the cell of node j lies partially inside H_i.

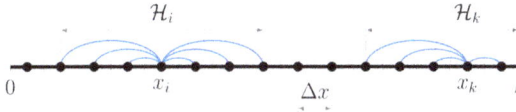

Figure 2: Discretization of the peridynamic body by means of the meshfree method with a uniform grid spacing Δx. The black solid dots and the blue curved lines represent the nodes and the bonds, respectively.

The explicit central difference is employed to integrate in time [13]:

$$\ddot{u}(x_i, t_n) = \frac{u(x_i, t_{n+1}) - 2u(x_i, t_n) + u(x_i, t_{n-1})}{\Delta t^2},$$ (4)

where Δt is the time step size and n stands for the index of the current time step. Therefore, the iterative procedure to obtain the displacement of node i at the time step $n + 1$ is determined by the following formula:

$$u(x_i, t_{n+1}) = 2u(x_i, t_n) - u(x_i, t_{n-1}) + \frac{(v\Delta t)^2}{\delta} \sum_{j \in H_i} \frac{u(x_j, t) - u(x_i, t)}{(x_j - x_i)^2} \beta_{ij} \Delta x.$$ (5)

The initial displacement $u(x_i, t_0)$ and velocity $\dot{u}(x_i, t_0)$ are known at any node. However, to obtain the displacement at the first time step $u(x_i, t_1)$, the knowledge of $u(x_i, t_{-1})$ is required as well. This displacement can be computed as

$$u(x_i, t_{-1}) = u(x_i, t_0) - \dot{u}(x_i, t_0)\Delta t + \ddot{u}(x_i, t_0)\frac{\Delta t^2}{2},$$ (6)

where

$$\ddot{u}(x_i, t_0) = \frac{v^2}{\delta} \sum_{j \in H_i} \frac{u(x_j, t_0) - u(x_i, t_0)}{(x_j - x_i)^2} \beta_{ij} \Delta x.$$ (7)

Review of the Surface Node Method

Let us call *interior nodes* the nodes lying within the peridynamic body. In order to complete the neighborhoods of the interior nodes near the boundaries of the body, two fictitious domains are added at the two boundaries of the body, as shown in Fig. 3. The nodes lying in the fictitious domains are named *fictitious nodes*. Moreover, we introduce the *surface nodes* at the two ends of the body, a new type of nodes that do not have PD bond connections with other nodes and are used to impose the peridynamic boundary conditions in a local way.

Aerospace Science and Engineering - III Aerospace PhD-Days   Materials Research Forum LLC
Materials Research Proceedings 33 (2023) 347-354   https://doi.org/10.21741/9781644902677-51

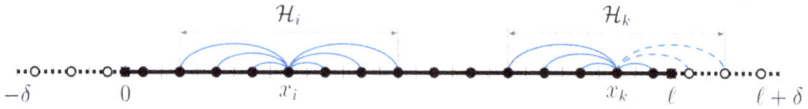

Figure 3: Introduction of the fictitious domain to complete the neighborhoods of the nodes near the boundaries of the body. The solid and empty dots represent the interior and the fictitious nodes, respectively. The solid squares at $x = 0$ and $x = \ell$ represent the surface nodes. The solid and dashed blue lines represent the bonds between nodes.

In this work, we apply only Dirichlet boundary conditions at the surface nodes, i.e., the displacements of the surface nodes are constrained. For the imposition of Neumann boundary conditions, the equations enforced at the surface nodes based on the peridynamic force flux can be found in [9-12]. To determine the displacements of the fictitious nodes, we assume that a fictitious domain deforms as the corresponding closest surface node. In other words, we determine the displacements of a fictitious node f via a Taylor-based extrapolation, truncated at the linear term for simplicity:

$$u(x_f, t_n) = u(x_s, t_n) + (x_f - x_s)\frac{\partial u(x_s, t_n)}{\partial x}, \tag{8}$$

where the derivative of the displacement at the surface node s is computed with the finite difference method as

$$\frac{\partial u(x_s, t_n)}{\partial x} = \frac{u(x_s, t_n) - u(x_p, t_n)}{x_s - x_p}, \tag{9}$$

where p is the index of the interior node closest to the surface node s. Note that, by plugging Eq. 9 into Eq. 8, the displacements of the fictitious nodes are functions of the displacements of the surface node and of its closest interior node. An example of the Taylor-based extrapolation is illustrated in Fig. 4. At this point, since the displacements of the surface nodes are prescribed as boundary conditions, the only unknowns are the displacements of the interior nodes, which can be found by solving, at each time step, the system of equations generated by Eq. 5.

Analytical solution of elastic wave propagation in Peridynamics
Even though Peridynamics handles with ease the treatment of discontinuities thanks to the integro-differential equations, determining analytical solutions to such equations is a much more difficult task to accomplish. For this reason, peridynamic numerical results are often compared with those obtained with classical mechanics. However, due to the different formulations of the theories, analytical solutions to the peridynamic and classical problems may be (and usually are) different from one another, and even more so in dynamics. A recent work [6,7] has shown that the method of separation of variables can be applied to peridynamic models to obtain their analytical solutions. Therefore, we present here one of the examples in [7] with the analytical solution to a peridynamic elastodynamic (elastic wave propagation) problem.

Aerospace Science and Engineering - III Aerospace PhD-Days   Materials Research Forum LLC
Materials Research Proceedings 33 (2023) 347-354   https://doi.org/10.21741/9781644902677-51

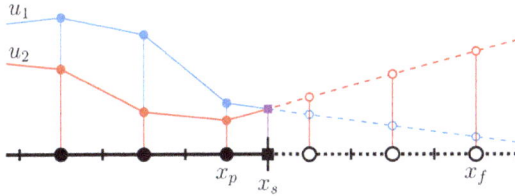

Figure 4: Examples of displacement field (u_1 and u_2 at two different time steps) determined by a linear Taylor-based extrapolation over the fictitious domain when Dirichlet boundary conditions are enforced at the surface node (purple square). The displacements of the fictitious nodes (empty dots) depend on the value of the constraint and the displacement of the interior node closest to the boundary.

The initial boundary value problem of wave propagation in a peridynamic medium is given as

$$\begin{cases} \ddot{u}(x,t) = \frac{v^2}{\delta} \int_{H_x} \frac{u(x',t)-u(x,t)}{(x'-x)^2} \, dx' & \text{for } 0 < x < \ell, \ t > 0, \\ u(0,t) = u(\ell,t) = 0 & \text{for } t > 0, \\ u(x,0) = 0.02e^{-100(x-0.5)^2}, \ \dot{u}(x,0) = 0 & \text{for } 0 < x < \ell, \end{cases} \tag{10}$$

where ℓ is the length of the peridynamic body. Note that the initial displacement field $u(x,0)$ has the shape of a Gaussian function. The analytical solution to this peridynamic problem is computed by means of the method of separation of variables [6,7]:

$$u(x,t) = \sum_{m=1,3,5,\dots}^{\infty} \frac{0.004\sqrt{\pi}}{\ell} \sin\left(\frac{m\pi}{2}\right) e^{\frac{-k_m^2}{400}} \sin(k_m x) \cos\left(v\sqrt{-\psi} \, t\right), \tag{11}$$

where $k_m = \frac{m\pi}{\ell}$ and $\psi = \psi(\delta)$ is the nonlocal factor computed as

$$\psi(\delta) = \frac{2}{\delta^2} [k_m \delta \, \mathrm{Si}(k_m \delta) + \cos(k_m \delta) - 1], \tag{12}$$

where $\mathrm{Si}(\cdot)$ is the sine integral function. The analytical solution in Eq. 11 will be used as reference for the numerical results.

Results and discussion
In this section, we solve numerically the problem in Eq. 10 of the peridynamic wave propagation in a homogeneous, linear elastic body by using the meshfree discretization in space and the explicit central difference method for time integration (Eq. 5). This is the first application of the Surface Node Method to an elastodynamic problem.

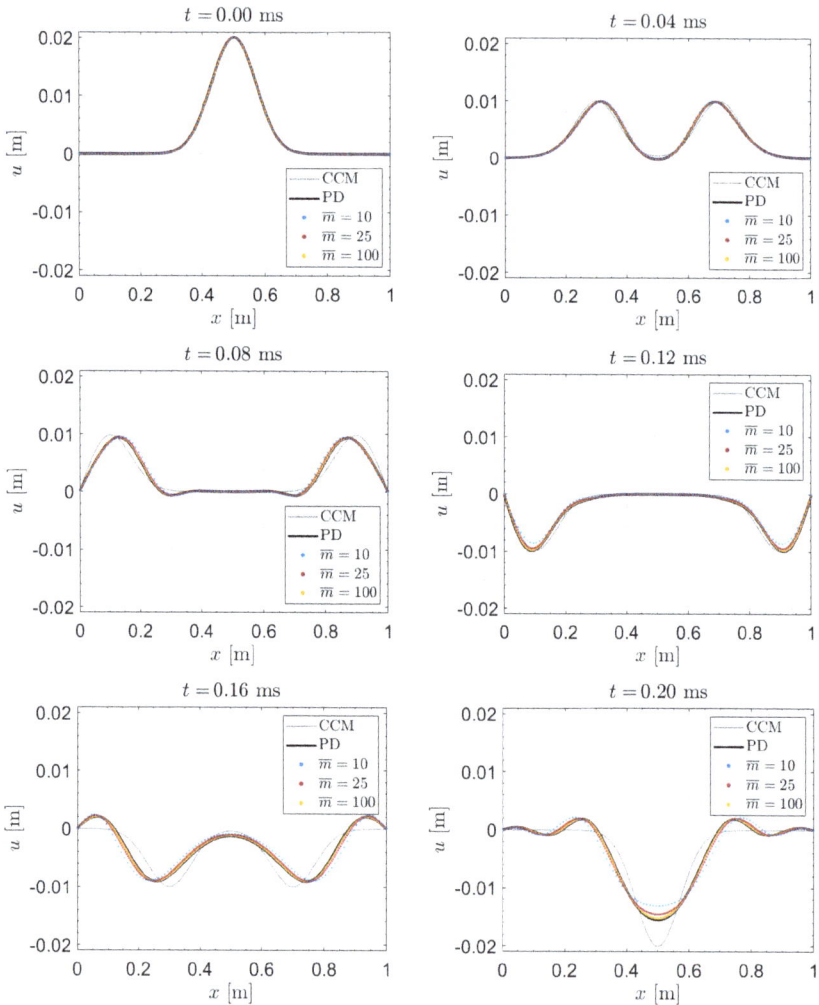

Figure 5: Plots of the propagating wave at different instants of time t. Note that the Classical Continuum Mechanics (CCM) analytical solution is a non-dispersive wave, whereas the peridynamic (PD) analytical solution with $\delta = 0.1\ell$ is a dispersive wave. The peridynamic numerical results (with $\delta = 0.1\ell$) are represented by the colored dots corresponding to the displacements computed at each node. As the grid is refined (namely as the $\bar{m}$-ratio $\bar{m} = \delta/\Delta x$ is increased keeping δ constant), the numerical results converge to the peridynamic analytical solution.

Aerospace Science and Engineering - III Aerospace PhD-Days   Materials Research Forum LLC
Materials Research Proceedings 33 (2023) 347-354   https://doi.org/10.21741/9781644902677-51

The length of the body is assumed to be $\ell = 1$ m and the horizon size $\delta = 0.1\ell$. The wave speed is computed as $v = \sqrt{E/\rho}$, where $E = 200$ GPa is the Young's modulus and $\rho = 8000$ kg/m^3 is the density. Regarding the time integration, we used a time step size of $\Delta t = 0.0002$ ms to discretize a total timespan of $T = 0.2$ ms. The series to compute the analytical solution in Eq. 11 is truncated after the first 40 terms. In order to prove that the numerical model converges to the peridynamic analytical solution, different grid sizes are adopted. Since the horizon size has a fixed value, the grid refinement is equivalent to increasing the value of the $\bar{m}$-ratio, defined as $\bar{m} = \delta/\Delta x$, i.e., increasing the number of nodes lying within a neighborhood. This is why, in Peridynamics, the convergence analysis related to the grid refinement is called $\bar{m}$-convergence. Therefore, we choose $\bar{m} = 10, 25, 100$, corresponding to a grid spacing $\Delta x = 10, 4, 1$ mm and a number of interior nodes $N = 100, 250, 1000$, respectively.

Fig. 5 shows the analytical solutions obtained with Classical Continuum Mechanics (CCM) and Peridynamics (PD), and the peridynamic numerical results. We observe that in this simple dynamic problem the CCM solution cannot serve as reference solution for the numerical PD cases since it is clearly different from the exact PD solution. Moreover, note that elastic waves modeled with CCM are non-dispersive, but this is not the case for waves propagating in a peridynamic medium [1,7]. The dispersion in a peridynamic medium can be reduced by decreasing the horizon size δ, since the PD solution approaches the CCM one as $\delta \to 0$ [7,16].

The peridynamic numerical results are close to the PD analytical solution at any instant of time and their accuracy is improved by increasing the value of the $\bar{m}$-ratio, as expected in a $\bar{m}$-convergence analysis. No unphysical fluctuations or kinks, typically observed in relation to surface effect and/or nonlocal boundary conditions (see for example [8-11]), are exhibited near the boundaries of the body. Therefore, the Surface Node Method has been shown to be accurate and reliable in the solution of dynamic problems.

Conclusions

The peridynamic framework is a promising theory to model fracture phenomena in solid bodies. However, the intrinsic nonlocality of the model leads to the well-known peridynamic surface effect and fluctuations of the solution of an elastodynamic problem near the boundaries of the body. Moreover, there is the need for a method to impose local boundary conditions in these nonlocal models. The surface node method provides an easy and automatic way to considerably reduce the oscillations near the boundaries and impose local boundary conditions in a peridynamic model. In this work, we verified that the surface node method is accurate when applied to solve an elastodynamic problem in a homogeneous, linear elastic body. In particular, we showed that the numerical results exhibit no unphysical fluctuations near the boundaries and converge to the peridynamic analytical solution under grid refinement.

Acknowledgements

The authors would like to acknowledge the support they received from the Italian Ministry of University and Research under the PRIN 2017 research project "DEVISU" (2017ZX9X4K) and from University of Padova under the research project BIRD2020 NR.202824/20.

References

[1] S.A. Silling, Reformulation of elasticity theory for discontinuities and long-range forces, J. Mech. Phys. Solids 48 (2000) 175-209. https://doi.org/10.1016/S0022-5096(99)00029-0

[2] S.A. Silling, M. Epton, O. Weckner, J. Xu, E. Askari, Peridynamic states and constitutive modelling, J. Elast. 88 (2007) 151-184. https://doi.org/10.1007/s10659-007-9125-1

[3] F. Bobaru, G. Zhang, Why do cracks branch? A peridynamic investigation of dynamic brittle fracture, Int. J. Fract. 196 (2015) 59-98. https://doi.org/10.1007/s10704-015-0056-8

[4] M. Zaccariotto, F. Luongo, U. Galvanetto, G. Sarego, Examples of applications of the peridynamic theory to the solution of static equilibrium problems, Aeronaut. J. 119 (2015) 677-700. https://doi.org/10.1017/S0001924000010770

[5] Z. Chen, F. Bobaru, Peridynamic modeling of pitting corrosion damage, J. Mech. Phys. Solids 78 (2015) 352-381. https://doi.org/10.1016/j.jmps.2015.02.015

[6] Z. Chen, X. Peng, S. Jafarzadeh, F. Bobaru, Analytical solutions of peridynamic equations. Part I: transient heat diffusion, J. Peridyn. Nonlocal Model (2022) 303-335. https://doi.org/10.1007/s42102-022-00080-7

[7] Z. Chen, X. Peng, S. Jafarzadeh, F. Bobaru, Analytical solutions of peridynamic equations. Part II: elastic wave propagation, *submitted*.

[8] Q.V. Le, F. Bobaru, Surface corrections for peridynamic models in elasticity and fracture, Comput. Mech. 61 (2018) 499-518. https://doi.org/10.1007/s00466-017-1469-1

[9] F. Scabbia, M. Zaccariotto, U. Galvanetto, A novel and effective way to impose boundary conditions and to mitigate the surface effect in state-based Peridynamics, Int. J. Numer. Methods. Eng. 122 (2021) 5773-5811. https://doi.org/10.1002/nme.6773

[10] F. Scabbia, M. Zaccariotto, U. Galvanetto, A new method based on Taylor expansion and nearest-node strategy to impose Dirichlet and Neumann boundary conditions in ordinary state-based Peridynamics, Comp. Mech. 70 (2022) 1-27. https://doi.org/10.1007/s00466-022-02153-2

[11] F. Scabbia, M. Zaccariotto, U. Galvanetto, A new surface node method to accurately model the mechanical behavior of the boundary in 3D state-based Peridynamics, J. Peridyn. Nonlocal Model (2023) 1-35. https://doi.org/10.1007/s42102-022-00094-1

[12] F. Scabbia, M. Zaccariotto, U. Galvanetto, F. Bobaru, Stability and convergence analyses for peridynamic models with autonomously evolving interfaces, *in preparation*.

[13] S.A. Silling, E. Askari, A meshfree method based on the peridynamic model of solid mechanics, Comput. Struct. 83 (2005) 1526-1535. https://doi.org/10.1016/j.compstruc.2004.11.026

[14] P. Seleson, Improved one-point quadrature algorithms for two-dimensional peridynamic models based on analytical calculations, Comput. Methods Appl. Mech. Eng. 282 (2014) 184-217. https://doi.org/10.1016/j.cma.2014.06.016

[15] F. Scabbia, M. Zaccariotto, U. Galvanetto, Accurate computation of partial volumes in 3D peridynamics, Eng. Comput. (2022) 1-33. https://doi.org/10.1007/s00366-022-01725-3

[16] S.A. Silling, R.B. Lehoucq, Convergence of peridynamics to classical elasticity theory, J. Elast. 93 (2008) 13-37. https://doi.org/10.1007/s10659-008-9163-3

Aerospace Science and Engineering - III Aerospace PhD-Days   Materials Research Forum LLC
Materials Research Proceedings 33 (2023) 355-361   https://doi.org/10.21741/9781644902677-52

Photogrammetric analysis for inspection and damage detection: preliminary assessment and future extension to large-volume structures

Mattia Trombini[1,a] *, Domenico A. Maisano[1,b], Alfonso Pagani[2,c]

[1]Department of Management and Production Engineering, Politecnico di Torino, Italy

[2]Department of Mechanical and Aerospace Engineering, Politecnico di Torino, Italy

[a]mattia.trombini@polito.it, [b]domenico.maisano@polito.it, [c]alfonso.pagani@polito.it

Keywords: Photogrammetry, Geometry Reconstruction, Damage Detection, Visual Inspection, Digital Twin

Abstract. This work aims to provide a preliminary understanding of the advantages of using visual techniques for the inspection and damage detection of large-volume structures. Some preliminary results provide a precise indication of tolerance of photogrammetric reconstructions and some indications to avoid the effect of reflectance of materials. Moreover, the overlap between reference CAD and the reconstructed model provides a picture of the most critical areas to cover with the camera. Finally, some colorband plots exploit the distribution of distances between reference and reconstructed models.

Introduction

Nowadays, the widely spread of composite materials into the aeronautical sector has improved the mechanical performances of components, with a significant reduction of weight [1]. Still, their multi-scale nature requires the development of numerical models for analyzing their mechanical behavior, achieving an accuracy comparable to isotropic materials. Unfortunately, simulating the behavior of composites involves time and effort. One of the most critical aspects of numerical simulation of composites concerns the study of the propagation of the crack within the material, which can have a hard-to-understand progression. Moreover, the constituents have different behavior, creating a gap in the literature regarding the prediction of fatigue life of composites. Some empirical models have been developed [2,3], but designing components in composites using a damage tolerant approach is still complicated. Consequently, the impact of maintenance and health monitoring of composite structures is therefore crucial within the whole aeronautical sector. In order to maintain a high level of flight safety, planes are periodically inspected [4] by human operators. However, the inspection of structures is conducted manually and is time-consuming. Consequently, the recent development of vision-based equipment and evolute image processing technologies has provided a new era of structural monitoring [5].

Among the well-established visual techniques, photogrammetry has a significant role [6]. The main attractive point of using photogrammetry is the capability of reconstructing 3D shape of large-volume structures, e.g., airplanes and large ships and identifying surface defects [7]. In [8], the authors mapped cracked areas by determining the strain field on the surface. Various Image Processing algorithms are applied to these areas: High-Pass Filter, to improve the quality of the image; Otsu's Thresholding, to transform the image into binary code; morphological operators, for noise reduction.

This work presents some preliminary results of using photogrammetry for geometry reconstruction of a 3D printed component with a parabolic shape. This preliminary part consists of the first stages in developing a geometry reconstruction and damage detection system using photogrammetry for operating large-volume structure. At this stage, a relevant aspect is given to

Aerospace Science and Engineering - III Aerospace PhD-Days   Materials Research Forum LLC
Materials Research Proceedings 33 (2023) 355-361   https://doi.org/10.21741/9781644902677-52

the classification of damage issues. As a matter of fact, a structure can contain a non-conformity, defined as a variation from a design specification or a defect, considered as a lack of fulfillment of an operative requirement. Within this context, damage detection stems from localizing and classifying both non-conformities and defects, thus enriching data collection of the structure and achieving better health status and operating life prediction. In order to inspect structures in areas that are difficult to access, the use of a drone assumes an essential function, leading to more precise health monitoring, thus scheduling maintenance activities more accurately. Meanwhile, digital twin (DT) models will be developed to receive real-time data from drone inspection activities and physical sensors, also incorporating high-order structural theories based on Carrera Unified Formulation (CUF) [9] and progressive failure models [10]. Progressive damage models consist of numerical approaches for predicting the initiation and evolution of defects, e.g., cracks and interface debonding within the component. As described in [11], a digital twin is a digital representation of an engineering system that can simulate, monitor, diagnose, predict and optimize the behaviors of its corresponding engineering system in real-time. The simulated data will be coupled with photogrammetric results to perform real-time monitoring and life prediction of structures.

Description of instrumentation
This section describes the tools useful for photogrammetric analysis. First of all, a camera with an appropriate resolution is required: Basler Lens C125-0818-5M-P f8mm - Lens (see Fig. 1). Moreover, using a rotary shelf, shown in Fig. 2a, avoids the excessive movement of the camera, thus maintaining the exact resolution. Above the platform, 6 markers are set and their relative distance permits getting the true coordinates of the cloud points. The position of markers can be detected in Fig. 2b, whereas their coordinates from the origin of the reference system, located in the center of the platform, are given in Table 1. Table 2 contains the name of the employed programs and a brief description of their function.

Figure 1 Camera Basler Lens C125-0818-5M-P f8mm - Lens used for images collection.

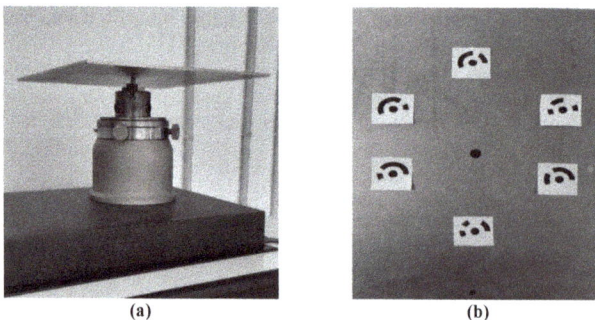

| (a) | (b) |

Figure 2 Rotating shelf (a) with markes (b) above the surface.

Aerospace Science and Engineering - III Aerospace PhD-Days   Materials Research Forum LLC
Materials Research Proceedings 33 (2023) 355-361   https://doi.org/10.21741/9781644902677-52

Table 1 Reference frame of the markers.

Marker	X coord [mm]	Y coord [mm]	Z coord [mm]
Target 4	-0.925	114.288	-0.303
Target 7	84.197	41.484	-0.049
Target 15	86.652	-43.806	0.036
Target 14	6.067	-108.842	0.317
Target 6	-89.377	-47.805	0.112
Target 11	-86.613	44.681	-0.114

Table 2 Software toolkit involved in the photogrammetric process.

Tool	Function
Pylon Viewer	Images takeover and storage
Agisoft Metashape	Images processing and points cloud/mesh generation
MeshLab	Points cloud/mesh view
CloudCompare	Overlapping CAD/reconstructed model and data processing

Preliminary results

The preliminary test case involves the 3D reconstruction of an Onyx-made dish component, designed using Inspire and manufactured via fused deposition modeling (FDM). In order to reduce the reflectance of the dish component, a thin film of anti-reflex spray MR 2000 is applied. This spray is suited for optical 3D-metrology and guarantees an extensive and durable white coating, as shown in Fig. 3. For the sake of clarity, no repetitive or well-establish manual techniques are adopted during the spray application, since the operator generates a random blast over the surface of the component.

(a)   (b)

Figure 3 Representation of the component without surface treatment (a) and covered with a solvent-based anti-reflex spray (b).

The first assessment consists of the geometry reconstruction performed using a deployment of 106 photos taken from 5 different stalls. This procedure aims to verify the capability of the spray to prevent the effect of light. Consequently, any study of convergence on the overlap of images is taken into account. Figure 4 illustrates an example of pictures collected for each camera position.

Figure 4 Five different stalls for images capture.

The images are directly imported into Agisoft Metashape Professional. The elaborating process for points cloud and mesh generation required more than 40 minutes. Table 3 collects the parameters associated with the processing. Considering the model generation, the memory usage can be split as illustrated in Table 4, whereas Table 5 contains the numerical outcomes associated with both points cloud and the mesh.

Table 3 Parameters associated with the generation of points cloud and mesh.

Object	Time [s]	Quality	Memory usage [MB]	File size [MB]
Points cloud	2,598	Medium	1291.48	12.42
Model	2,593	Medium	3481.71	21.49

Table 4 Partition of memory usage during the generation of the mesh.

Step	% of memory usage
Points cloud	4.93%
Model	19.79%
Texturing	75.28%

Table 5 Number of points, faces and vertices in the reconstructed models.

Step	Points cloud	Model
Number of points	923,553	-
Number of faces	-	66,672
Number of vertices	-	33,438

Aerospace Science and Engineering - III Aerospace PhD-Days   Materials Research Forum LLC
Materials Research Proceedings 33 (2023) 355-361   https://doi.org/10.21741/9781644902677-52

The final shape of the points cloud and the model is shown in Fig. 5, inside MeshLab environment. Subsequently, the model and the related file *.stl* are imported within CloudCompare environment.

(a)   **(b)**

Figure 5 Final points cloud (a) and reconstructed mesh (b).

The overlapping between the reference and reconstructed model is performed by selecting 8 reference points, corresponding to the vertices at the edges of the base (see Fig. 6). Figure 7 illustrates the final superposition from four different views, including the colorband containing the distances between the models. Furthermore, histogram in Fig. 8 plots the number of counts with the relative distance, whereas the range of distances between the models is observed in the following relation:

$$-1.607 \text{ mm} \leq \text{distance} \leq 2.140 \text{ mm} \tag{1}$$

Figure 6 Localization of some reference points in the component.

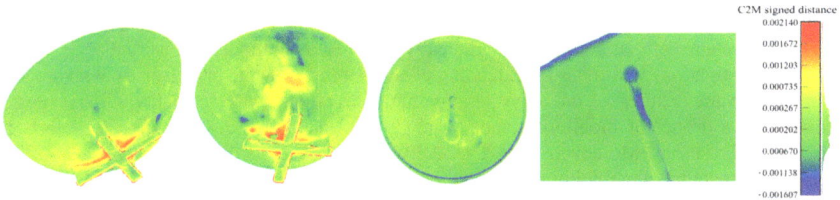

Figure 7 Colorbands concerning the distances between the reconstructed points cloud and the reference geometry. The distances are in given in [m].

Aerospace Science and Engineering - III Aerospace PhD-Days   Materials Research Forum LLC
Materials Research Proceedings 33 (2023) 355-361   https://doi.org/10.21741/9781644902677-52

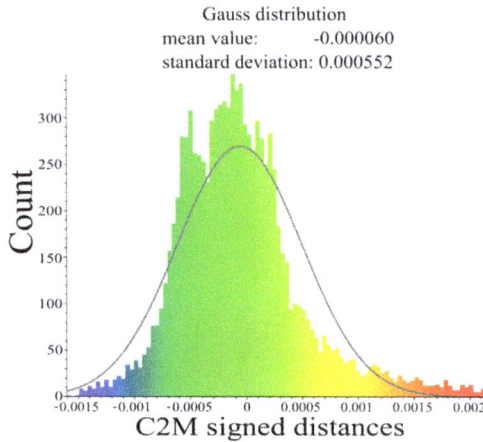

Figure 8 Histogram with counts plot over the signed distances in [m].

The results suggest that:

1. The anti-reflex treatment successfully avoids the effect of reflectance. The number of acquisitions is high (106), but a good distribution of points is detected.
2. The overlapping between reconstructed and CAD models demonstrates a reasonable accuracy. The internal part of the dish correctly represents the reference, whereas the most relevant differences are in the lower part of the component. Furthermore, the middle upright section results less precise than expected.
3. Most distances between the models are held in the range [-0.5 mm;0.3 mm]. In addition, the larger errors in the geometry reconstruction correspond to values around -1.6 cm in case of underestimating and 2.1 cm if the reference CAD is overestimated.

Conclusions and future perspectives

This work initially investigated the employment of photogrammetry for reconstructing component geometry. At first, a description of the equipment used for the experimental campaign is given and some outcomes are shown.

Then, the following considerations are made:

1. The best outcomes demonstrate that the measurement errors are around 0.5 mm. Consequently, there are better solutions than photogrammetry for visualizing small structures with small dimensions and deformations.
2. The initial test cases highlight a set of photogrammetry-related problems. For instance, the reflectance of materials can generate very poor results in model reconstruction.

Future investigations will focus on extending photogrammetry as the primary visual technique to extract relevant features of operative large-volume structures. Also, cameras will become the payload of a drone for monitoring specific areas of structures and localizing the defects and geometric deviations. Moreover, the DT model will be used as a virtualized representation with real-time updating combining photogrammetric results, physical sensors and numerical assessments.

Some thematic areas of research can be identified:

1. The extension of photogrammetry to large-volume structures (planes, ship hulls, spacecraft).

Aerospace Science and Engineering - III Aerospace PhD-Days   Materials Research Forum LLC
Materials Research Proceedings 33 (2023) 355-361   https://doi.org/10.21741/9781644902677-52

2. Using the camera as the payload of a drone and performing damage detection of harsh-to-observe areas, aiming to monitor and classify non-conformities and defects.

3. Coupling photogrammetry and high-order finite elements based on CUF to develop and improve digital twin models for life prediction of components.

Acknowledgments

This work is part of a collaboration between the Department of Management and Production Engineering and the Department of Mechanical and Aerospace Engineering of Politecnico di Torino. The authors would also like to thank Eng. Giacomo Maculotti for providing the instrumentation and training on the photogrammetric procedure adopted in this work.

References

[1] Degenhardt R, Castro SGP, Arbelo MA, Zimmerman R, Khakimova R, Kling A. Future structural stability design for composite space and airframe structures. Thin-Walled Structures, 81:29-38, 2014. https://doi.org/10.1016/j.tws.2014.02.020

[2] Tamadur A, Gathercole N, Reiter H, Harris B. Fatigue life prediction for carbon fibre composites. Adv Compos Lett, 1:23-26, 01 1992. https://doi.org/10.1177/096369359200100106

[3] Mao H, Mahadevan S. Fatigue damage modelling of composite materials. Composite Structures, 58(4):405-410, 2002. https://doi.org/10.1016/S0263-8223(02)00126-5

[4] Jovančević I, Arafat A, Orteu JJ, Sentenac T. Airplane tire inspection by image processing techniques. In 2016 5th Mediterranean Conference on Embedded Computing (MECO), 176-179, 2016. https://doi.org/10.1109/MECO.2016.7525733

[5] Zhao S, Kang F, Li J. Concrete dam damage detection and localization based yolov5s-hsc and photogrammetric 3D reconstruction. Automation in Construction, 143:104555, 2022. https://doi.org/10.1016/j.autcon.2022.104555

[6] Galantucci RA, Fatiguso F. Advanced damage detection techniques in historical buildings using digital photogrammetry and 3D surface analysis. Journal of Cultural Heritage, 36:51-62, 2019. https://doi.org/10.1016/j.culher.2018.09.014

[7] Benzon HH, X. Chen X, Belcher L, Castro O, Branner K, Smit J. An operational image-based digital twin for large-scale structures. Applied Science, 12(7), 2022. https://doi.org/10.3390/app12073216

[8] Valença J, Dias da Costa D, Júlio E, Araújo H, Costa H. Automatic crack monitoring using photogrammetry and image processing. Measurement, 46(1):433-411, 2013. https://doi.org/10.1016/j.measurement.2012.07.019

[9] Carrera E, Cinefra M, Petrolo M, Zappino E. Finite element analysis of structures through unified formulation. John Wiley & Sons, 2014. https://doi.org/10.1002/9781118536643

[10] Nagaraj MH, Reiner J, Vaziri R, Carrera E, Petrolo M. Progressive damage analysis of composite structures using high-order layer-wise elements. Composites Part B: Engineering 190:107921, 2020. https://doi.org/10.1016/j.compositesb.2020.107921

[11] Ye Y, Yang Q, Yang F, Huo Y, Meng S. Digital twin for the structural health management of reusable spacecraft: A case study. Engineering Fracture Mechanics, 234:107076, 2020. https://doi.org/10.1016/j.engfracmech.2020.107076

Aerospace Science and Engineering - III Aerospace PhD-Days   Materials Research Forum LLC
Materials Research Proceedings 33 (2023) 362-368   https://doi.org/10.21741/9781644902677-53

Sensitivity analysis of analytically—corrected acoustic metamaterials into the spacetime domain

Giada Colombo[1,a] *

[1]Roma Tre University, Department of Civil, Computer Science and Aeronautical Technologies Engineering, Via della Vasca Navale 79, Rome, Italy, 00146

[a]giada.colombo@uniroma3.it

Keywords: Metacontinua, Spacetime, Aeroacoustic, Parametric Study

Abstract. The present work is focused on the sensitivity analysis of analytically--corrected acoustic metamaterials with respect to the variation of relevant parameters for aeronautical applications. The performance decay of acoustic metamaterials designed in static conditions when operating in a moving flow can be mitigated using a design process involving coordinate transformations capable of recasting the convective wave equation in the form of the static D'Alembertian. To this aim, a spacetime reformulation of the problem is needed in order to ensure the formal invariance of equations under the action of spacetime coordinate transformations. The considered test case sees a circumferential domain occupied by a metafluid embedded by a conventional medium characterized by a uniform background flow and within which a monopole source is located. All the numerical simulations are done in the frequency domain using the commercial finite element method solver. A parametric study is conducted to analyze the influence of the acoustic source position with respect to the background flow and the variation of the intensity of the latter.

Introduction

The present work's topic is related to the aircraft mobility sustainability task sought by the research efforts of the last decades. Particularly, it arises from the necessity to examine possible strategies to reduce the community noise generated by aircraft mobility. Given the discrepancy between the emission reduction request and the technological advances, a breakthrough is needed. With the advent of the concept of metamaterials, thanks to the work of Veselago, in [1], Pendry and Smith in [2, 3], possibilities have opened up to develop new strategies that could be implemented with current technologies and that could allow achieving the goals set. Firstly though for electromagnetic applications, the metamaterial devices are conceived as the repetition of a sample geometric architecture within the space. The lattice obtained, combined with the materials' properties, showed atypical behaviours not reachable in nature and highly adaptable for satisfying the most diverse application requirements.

Only after the development of the transformation optics techniques [4-6] their enormous potentiality has been actually exploited by releasing the metamaterial properties from the specific design and linking them only to the specific application. As a consequence, their applicability has also been extended to other fields of physics and engineering, acoustics among others.

Following those steps, Norris developed the mathematical model that describes the atypical behavior of acoustic metamaterials obtainable through coordinate transformation addressing the definition of *metacontinuum* able to produce the so called *acoustic mirage* [7, 8]

Both transformation optics techniques and the metacontinuum model of Norris rely on the concept of formal invariance of the governing equation under coordinate transformations provided that unconventional material properties are identified.

Aerospace Science and Engineering - III Aerospace PhD-Days   Materials Research Forum LLC
Materials Research Proceedings 33 (2023) 362-368   https://doi.org/10.21741/9781644902677-53

However, their application to aeronautical contexts has led to a performance decay due to the presence of a background flow that couples the fluid dynamic phenomenon with the acoustic one as seen by Iemma in [9] and Iemma and Palma in [10-12]. In fact, the convection effects result in a wave operator defined through mixed space and time derivatives that make the formal invariance fails. The solution proposed by the authors could be seen as a generalization of the coordinate transformation approach where the formal invariance of the governing equation is restored, describing the propagation phenomenon of the acoustic perturbation in the spacetime domain. In this new manifold, the transformation considered is based on Taylor's or Prandtl-Glauert's transformations and is applied directly to the acoustic metacontinuum model. Hence, analytically adapted acoustic metamaterial properties to convective applications are obtained.

The validation process of the proposed design strategy started in Colombo, Palma, and Iemma in [13] and here is further developed. Specifically, this work introduces a sensitivity analysis of the adapted metamaterial behaviour with respect to the variation of characteristic parameters of aeronautical problems, such as the relative position of the sound source with respect to the background flow and the Mach number intensity. Moreover, since the purpose is to highlight the corrections' behaviour, the acoustic metacontinuum properties are designed to mimic a domain filled with air. Firstly the mathematical model of the generic acoustic metacontinuum is shown, and subsequently, the methodology for its extension for convective applications is presented in the spacetime domain. Then the numerical set-up and the numerical results obtained with the parametric study are discussed.

Methodology
Relying on the definition of the acoustic metacontinuum model of Norris is possible to obtain atypical behaviours thanks to the proper characterization of the medium in terms of inertia and stiffness properties. In the most general case, those arise when anisotropic inertia and bulk modulus are defined. Following Norris' formulation, once the properties are identified, the governing equation that describes the propagation of the acoustic perturbation within such medium assumes the form

$$-\partial_{tt}p + K_{ref}\mathbf{S}:\nabla(\boldsymbol{\rho}^{-1}\mathbf{S}\nabla p) = 0 , \tag{1}$$

where K_{ref}, ρ_{ref} and $c_{ref} = \sqrt{K_{ref}/\rho_{ref}}$ are the referenced bulk modulus, density and speed of sound respectively. In Eq. (1) two second order tensors are present: the $\boldsymbol{\rho}$ and the $\mathbf{S}$ tensors that enclose the anisotropic inertia and the anisotropic stiffness respectively. The actual shape of those quantities depends on the specific application and are generally obtained through coordinate transformation. Since in the tested case the metacontinuum is thought to mimic a domain filled with air, the identity transformation is used and its parameters reduce to the following simple form:

$$K_{ref} = K_0, \qquad \boldsymbol{\rho} = \rho_0\mathbf{I}, \qquad \mathbf{S} = \mathbf{I}, \tag{2}$$

where the subscript "0" refers to the characteristic quantities of a conventional fluid. It is underlined that Eq. (1) is known only in terms of pseudo-pressure and in a quiescent fluid.

Analytical Correction for Acoustic Metacontinua
Here the mathematical tools that allow extending the applicability of acoustic metamaterials to aeronautical context are explained but, firstly, the reformulation of the governing equation into the spacetime domain is needed. In fact, since the wave operator of aeroacoustic phenomenon is defined through mixed space and time derivatives, the acoustic mirage approach is no longer

Aerospace Science and Engineering - III Aerospace PhD-Days   Materials Research Forum LLC
Materials Research Proceedings 33 (2023) 362-368   https://doi.org/10.21741/9781644902677-53

applicable since the formal invariance no longer holds. However, the substitution of a point in the 3D euclidean space with a point-event in a 3D+1 pseudo-Riemannian manifold allows the recovery of the formal invariance under general spacetime coordinate transformation. In this new manifold, any partial differential equation describing the propagation of a scalar field could be recast into the spacetime domain as

$$\partial_\mu(W^{\mu\nu}\partial_\nu\varphi) = 0 \, , \tag{3}$$

where $\partial = (1/c_{ref}\, \partial_t, \partial_{xi})$ is the differential operator in the new manifold and $W^{\mu\nu}$ are the contravariant components of a second order tensor that involves all the information about the phenomenon observed. The tensor $\mathbf{W}$ is linked to the metric tensor by $\mathbf{W} = \sqrt{-g}\,\mathbf{g}^{-1}$. Thus, if coordinate transformations are introduced, they will modify directly the component of such tensor. Specifically, the analytical correction method is based on Taylor's and Prandl-Glauert's coordinate transformations. Both allow the transformation of convective patterns into static ones under the assumption of potential flow. Thus, their definition is directly linked to the characteristics of the background flow. Considering a description of the phenomenon into the spacetime domain, they assume the following forms:

$$\xi_0' = \beta\xi_0 + M_\infty\frac{\xi_1}{\beta}, \quad \xi_1' = \frac{\xi_1}{\beta}, \quad \xi_2' = \xi_2, \quad \xi_3' = \xi_3, \tag{4}$$

for Prantl-Glauert's one, whereas for Taylor's transformation, the form assumed is

$$\xi_0' = \xi_0 + M_\infty\widehat{\boldsymbol{\Phi}}(\boldsymbol{\xi}) = \xi_0 + M_\infty(\xi_1 + \hat{\phi}), \quad \xi_i' = \xi_i, \tag{5}$$

with the aerodynamic potential of the background flow normalized as $\widehat{\boldsymbol{\Phi}} = \Phi/\| \mathbf{U}_\infty \|$.
The application of such coordinate transformations goes under the definition of the transformation matrix $\boldsymbol{\Lambda}$ which has to be inversely applied in order to analytically introduce convection onto static propagation as

$$\boldsymbol{W}_c = \boldsymbol{\Lambda}^{-1}\boldsymbol{W}\boldsymbol{\Lambda}^{-T} \, . \tag{6}$$

Therefore, the transformation in Eq. (6) is thought to be applied to the spacetime reformulation of the governing equation (1), and the adapted metamaterial properties are obtained.

Numerical Results
The numerical results obtained concerned the parametric study of the behaviour of both the corrections proposed in (4) and (5). Numerical simulations have been done for a monochromatic acoustic perturbation exploiting the reformulation of the problem in the frequency domain. Specifically, the interest relies on the capability of the coordinate transformation approach to fictitiously induce flow in a domain where static propagation was originally defined.
    To do so, a domain completely occupied by air is considered and virtually subdivided into two concentric regions, namely an inner circle of radius R_{in} and an annulus with outer radius $R_{out} = 30$ R_{in}. The air in the inner circle is actually modelled by an acoustic metafluid, whose parameters mimic air following Eq. (2) . The simulations are performed using the commercial FEM solver Comsol Multiphysics [14] and the corrected metafluid equations (Eq. (3) for acoustic pressure applying the corrections of Eq. (6)) are implemented using the Partial Differential Equation (PDE) module. The acoustic propagation in the outer *hosting* fluid domain is modelled through the Linearized Euler Equations, with the dedicated module in the mentioned software. The continuity

Aerospace Science and Engineering - III Aerospace PhD-Days   Materials Research Forum LLC
Materials Research Proceedings 33 (2023) 362-368   https://doi.org/10.21741/9781644902677-53

equation for the hosting fluid domain is forced by a monopole source with unitary intensity. Its location varies along an arc of radius $R_{source} = 7.5\ R_{in}$, centered in $\mathbf{x} = (0,0)$, from $\theta = -45°$ (upstream direction) to $\theta = 45°$ (downstream direction), with θ being the angle respect to the vertical axis. Moreover, the hosting fluid is characterized by a background flow; numerical simulations are carried out with a Mach number associated with the free stream ranging from 0.1 to 0.3. To numerically reproduce the free space propagation, the Asymptotic Far Field Radiation condition is imposed on the outer boundary. The Dirichlet and Neumann boundary conditions are imposed as described in [13] at the interface between the two domains for the correct exchange of information between the two modules and to guarantee the acoustic field continuity among the two domains in terms of pressure and normal velocity component.

The underlying hypothesis of the present work is that by applying the above mentioned analytical corrections to the tensor $\mathbf{W}$, the convective effects induced by the presence of a flow are simulated and virtually introduced in the acoustic propagation within the metacontinuum. In this way, the scattering that would appear coupling the static air-metafluid with the flowing air in the hosting domain should be abated, obtaining the propagation pattern of a monopole source in a free field in the presence of aerodynamic convection. The scattering cross section σ is defined as the merit parameter to evaluate the residual scattering happening using the corrected metacontinuum to model the air in the inner circle:

$$\sigma = \int_C \left| L_{p_{metacontinuum}} - L_{p_{idealcase}} \right| dl \, , \tag{7}$$

where $p = 20\ [\mu Pa]$ is the reference pressure for the evaluation of the L_p. Figure 1 shows the values for σ on a circle of radius $R_{SCS} = 1.5\ R_{in}$. The parameter σ is evaluated for each source position $\theta = \pm\ [15°, 25°, 35°, 45°]$ and Mach number associated with the background flow $M_\infty = [0.1, 0.2, 0.3]$.

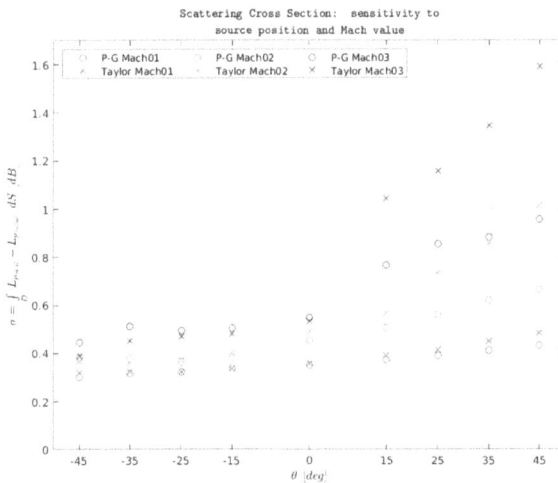

Figure 1. *Point plot of the σ evaluated for $M_\infty = [0.1, 0.2, 0.3]$ and for every source location θ considered*
- Prandtl-Glauert's (o-marks) and Taylor's (x-marks) correction sensitivity comparison

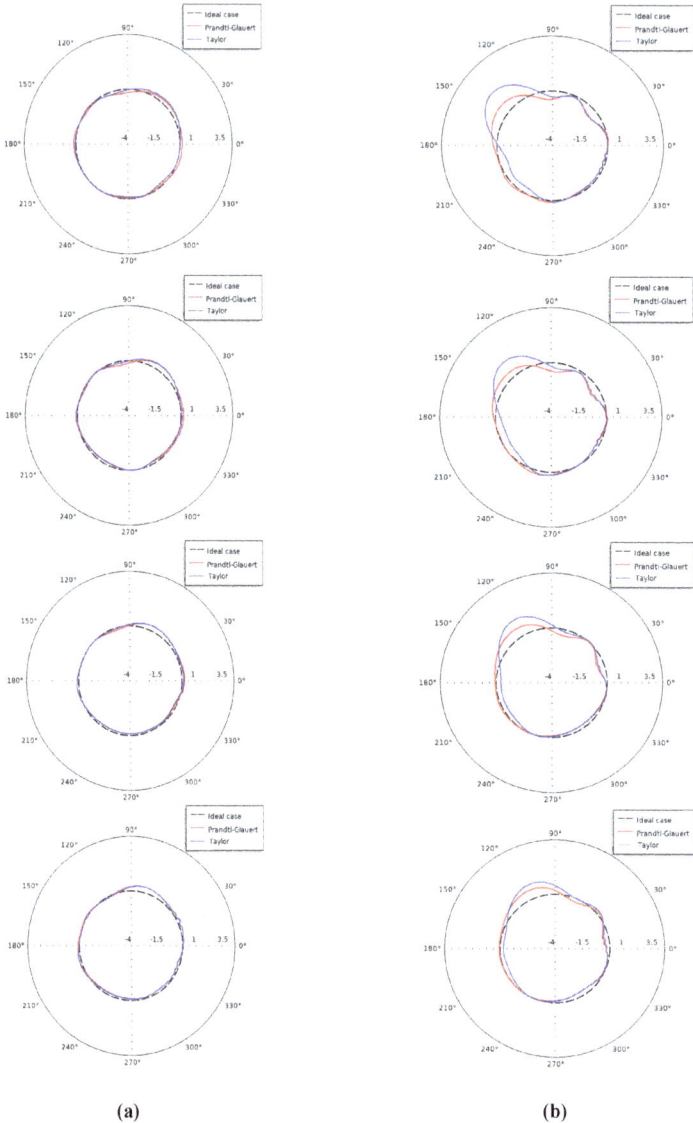

(a) **(b)**

Figure 2. *Polar [dB] diagram of the scattered field for $M_x = 0.3$ at $r = 1.5\ R_{in}$ - Prandtl-Glauert's (red) and Taylor's (blue) correction sensitivity to the variation of the source location - left column 2a relative to $\theta = -[45°, 35°, 25°, 15°]$ values; right column 2b relative to $\theta = [45°, 35°, 25°, 15°]$ values.*

Aerospace Science and Engineering - III Aerospace PhD-Days   Materials Research Forum LLC
Materials Research Proceedings 33 (2023) 362-368   https://doi.org/10.21741/9781644902677-53

A trend in the σ appears from Fig. 1: the more the source location moves downstream, the higher the values of the SCS. The effect of Mach is more evident for the downstream θ positions, increasing the values of σ. This is also in agreement with previous findings in Iemma and Palma [11]. At least for Taylor's coordinate transformation, the authors addressed the effect of the neglected terms as a pattern field connected to a dipole propagation in the case of an acoustic disturbance impinging an impermeable body. The study highlights the presence of sectors of high intensity where the correction loses part of its efficiency. In the case considered, it seems that the sector where the misbehaviour of the corrections arises coincides with the downstream location of the acoustic source. Thus, it is plausible that the same considerations done by the authors for their application of Taylor's transformation could also be extended here. Even with limited amplitude compared to the Taylor one, the performances of the analytic corrections seem to depend on the source position also for the PG transformation. The causes are not completely clear to the authors. In order to verify these effects, the insertion loss (IL) is evaluated on R_{SCS} and shown in a polar plot. In Fig. 2 the results obtained with the application of each analytical correction for every θ angle and with the Mach number set to its higher value are shown. The scattering directivity is increased when the source is moved towards downstream positions. Moreover, some small ripples appear in the IL, suggesting their origin to be sought in possible numerical issues.

Conclusions

The parametric study conducted has brought more insights into the coordinate transformations' behaviour. The qualitative analysis of their sensitivity to the variation of the characteristic aeronautical parameters considered has led to the identification of an influence zone where the corrections are losing part of their efficacy. The fictitious introduction of flow in the metafluid domain seems more efficient whenever the source location is in the upstream zone. This condition guarantees a low value of the merit parameter σ independently from the Mach number value for both coordinate transformations. The present analysis confirms previous works' findings and represents a step forward in defining more suitable strategies for using adapted meta-devices in aeronautical contexts. However, a higher comprehension of the coordinate transformations behaviour requires the analytical approximation's effect quantification and its influence on the performances of convective-designed metamaterials.

References

[1] Veselago, V. G., "The Electrodynamics of Substances with Simultaneously Negative Values of ε and μ," Physics-Uspekhi, Vol. 10, 1968, pp. 509–514. https://doi.org/10.1070/PU1968v010n04ABEH003699

[2] Pendry, J. B., "Negative Refraction Makes Perfect Lens," The American Physical Society, Vol. 85, No. 18, 2000. https://doi.org/10.1103/PhysRevLett.85.3966

[3] Smith, D. R., "Composite medium with simultaneously negative permeability and permittivity," The American Physical Society, Vol. 84, No. 18, 1999. https://doi.org/10.1103/PhysRevLett.84.4184

[4] Pendry, J. B., Schurig, D., and Smith, D. R., "Controlling Electromagnetic Fields," Science, Vol. 312, 2006, pp. 1780–1782. https://doi.org/10.1126/science.1125907

[5] Leonhardt, U., "Optical conformal mapping," Science, Vol. 23, 2006. https://doi.org/10.1126/science.1126493

[6] Leonhardt, U., and Philbin, T. G., "General relativity in electrical engineering," New Journal of Physics, Vol. 8, 2006. https://doi.org/10.1088/1367-2630/8/10/247

[7] Norris, A., "Acoustic cloaking theory." Proceedings of the Royal Society A: Mathematical, Physical and Engineering Sciences, Vol. 464, 2008, pp. 2411–2434. https://doi.org/10.1098/rspa.2008.0076

[8] Norris, A., "Acoustic metafluids." The Journal of the Acoustical Society of America, Vol. 125, 2009, pp. 839–849. https://doi.org/10.1121/1.3050288

[9] Iemma, U., "Theoretical and Numerical Modeling of Acoustic Metamaterials for Aeroacoustic Applications," Aerospace, Vol. 3, No. 2, 2016. https://doi.org/10.3390/aerospace3020015

[10]   Iemma, U., and Palma, G., "On the Use of the Analogue Transformation Acoustics in Aeroacoustics," Mathematical Problems in Engineering, Vol. vol. 2017, 2017, p. 16 pages. https://doi.org/https://doi.org/10.1155/2017/8981731

[11]   Iemma, U., and Palma, G., "Convective correction of metafluid devices based on Taylor transformation," Journal of Sound and Vibration, Vol. 443, 2018, pp. 238–252. https://doi.org/10.1016/j.jsv.2018.11.047

[12]   Iemma, U., and Palma, G., "Design of metacontinua in the aeroacustic spacetime", Scientific Reports, Vol. 10, 2020. https://doi.org/10.1038/s41598-020-74304-5

[13]   Colombo, G., Palma, G., and Iemma, U., "Validation of analytic convective corrections for metacontinua in the aeroacoustic spacetime," Proceedings of the 28th International Congress on Sound and Vibration, ICSV, 2022

[14]   "COMSOL Multiphysics® v. 5.4", 2018. URL http://www.comsol.com

Aerospace Science and Engineering - III Aerospace PhD-Days
Materials Research Proceedings 33 (2023) 369-375

Materials Research Forum LLC
https://doi.org/10.21741/9781644902677-54

Simulation of flow field characteristics in gap between high-speed rocket sled slipper and track

Tianjiao Dang [1,2,a], Zhen Liu [1,b,*], Pierangelo Masarati [2,c]

[1]State Key Laboratory for Strength and Vibration of Mechanical Structures, Xi'an Jiaotong University, Xi'an, Shaanxi, 710049, P.R. China

[2]Department of Aerospace Science and Technology, Politecnico di Milano, via La Masa 34, Milano, 20156, Italy

[a] tianjiao.dang@polimi.it, [b,*] liuz@mail.xjtu.edu.cn, [c] pierangelo.masarati@polimi.it

Keywords: Gap Flow Field, Computational Fluid Dynamics, Aerodynamics, Supersonic

Abstract. Accurate simulation of flow field characteristics within the gap between the slipper and track is essential for the prevention of aerodynamic heat damage to high-speed rocket sleds. A three-dimensional structured mesh was utilized to establish the flow field model in the slipper-track gap, while computational fluid dynamics was employed for simulating the flow field movement. The results revealed that the bow shock wave at the head of the rocket sled has a significant influence on the flow field characteristics within the gap. Specifically, the velocity of the mainstream in the gap initially exhibited an increasing trend, followed by a decreasing trend, and then a re-increase. Conversely, the mainstream temperature displayed a decreasing trend initially, followed by an increasing trend, and then a decrease once more. The air compression within the slipper-track gap resulted in remarkably high temperatures, with the maximum temperature reaching 1160 K at $Ma = 4$ in the immediate vicinity of the slipper. The current investigation provides valuable insights that can guide future research on the structural characteristics of slippers in high-temperature environments.

Introduction

The high-speed rocket sled is a remarkable and precise piece of ground testing equipment that was developed in the mid to late 20th century. It can solve a range of performance test issues encountered during aircraft aerodynamic experiments in high-speed and high-load environments [1-4]. In contrast to wind tunnel tests that employ scaled-down models, the rocket sled is tested using a full-scale model of the aircraft. It is extensively employed in the aviation, aerospace, and defense industries to research and develop a variety of products, such as aviation lifesaving seat ejection tests, missile guidance, and aircraft rain erosion. During the test, the rocket engine propels the sled along the track at high speed, while the test product's performance parameters affixed to the sled are measured with precision.

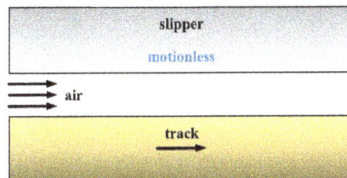

Fig. 1. Gap between slipper and track

Aerospace Science and Engineering - III Aerospace PhD-Days   Materials Research Forum LLC
Materials Research Proceedings 33 (2023) 369-375   https://doi.org/10.21741/9781644902677-54

The *slipper* is a crucial component of the rocket sled that makes contacts with the track, preventing the sled from flying off during movement. During the design phase, a narrow gap is intentionally left between the slipper and the track to avoid it getting stuck during operation. However, due to the high speed of the slipper and the stationary nature of the track, there exists a significant velocity difference between the upper and lower surfaces of the gap, as illustrated in Fig. 1. This results in a high-gradient flow field within the gap, which can generate extreme temperatures. Such high-speed and high-gradient flow fields can cause damage to the slipper and alter its material properties. Thus, it is crucial to accurately simulate the flow field characteristics in the gap to ensure the thermal protection of the rocket sled slipper.

Two main methods, theoretical analysis and numerical simulation, have been employed to investigate the characteristics of the high-speed rocket sled slipper-track gap flow field. Notably, Korkegi and Briggs [5] have made significant contributions to the study of slipper-track gap flow. Given the high velocity of fluid in the slipper-track gap of high-speed rocket sleds, the fluid is compressible. The authors [6] identified the presence of compressible turbulence in the gap and developed a theoretical model to derive the velocity and temperature curves of two-dimensional compressible flow. When the fluid flows close to a solid surface, the velocity of the fluid decreases due to drag, resulting in the formation of a boundary layer. The authors [7] argued that supersonic flow entering the initial stage of the gap would create shock waves, followed by the formation of boundary layers on the surface of both the slipper and the track. These two boundary layers converge at a specific point within the gap, as deduced by the authors. In a subsequent paper, the same authors [8] demonstrated that a decrease in gap height results in a corresponding pressure decrease, and vice versa.

With the increase in computer processing speed, computational fluid dynamics (CFD) method was applied to the numerical simulation of the slipper-track gap flow field of high-speed rocket sled. The Air Force Institute of Technology (AFIT) at the Air University in the United States has conducted extensive research in this field. Lofthouse [9] utilized the CFD method to compute the three-dimensional steady inviscid flow field of the slipper-track gap of high-speed rocket sleds at varying speeds, thereby predicting the pressure gradient within the narrow gap. However, the accuracy of the results was limited to first order due to the use of an unstructured mesh. Moreover, since the flow field was assumed to be inviscid, the temperature within the narrow gap could not be accurately calculated, although its influence on pressure calculation was minimal. Alban [10] conducted a comprehensive investigation of the aerodynamic heating within the slipper-track gap and improved the accuracy of temperature predictions using a one-dimensional finite difference model. Subsequently, Alban replaced the one-dimensional finite difference model with a two-dimensional finite element model to simulate the temperature distribution within the gap with greater precision.

The study of the flow field in the slipper-track gap of high-speed rocket sled has made significant progress, yet there is still much room for improvement. The current studies have shown high accuracy in one and two dimensions but have used an unstructured mesh in the three-dimensional model, leading to reduced computational accuracy. To address this issue, this study seeks to enhance the accuracy of the three-dimensional model using a structured mesh. This will enable a more precise calculation of the flow field in the slipper-track gap of high-speed rocket sled and allow for exploration of the movement of the flow field within the gap. Ultimately, the results of this study will provide a foundation for further research on heat transfer in the slipper-track gap.

This paper is organized as follows: Section 1 provides an overview of the study's background, significance, and current status. Section 2 outlines the theoretical framework and modeling approach. Section 3 presents the calculation results and analysis. Finally, Section 4 summarizes the key findings and conclusions.

Aerospace Science and Engineering - III Aerospace PhD-Days   Materials Research Forum LLC
Materials Research Proceedings 33 (2023) 369-375   https://doi.org/10.21741/9781644902677-54

Methods

In this study, the numerical method was utilized to calculate the flow field in the slipper-track gap of high-speed rocket sled. The finite volume method was employed to solve the three-dimensional steady compressible Navier-Stokes equations. The governing equation is expressed as

$$\frac{d}{dt} \iiint Q d\Omega + \oiint \vec{H}^I \cdot \vec{n} dS = \oiint \vec{H}^V \cdot \vec{n} dS, \tag{1}$$

where Q is the state variable, including the density, momentum component, and total energy of the fluid. $\vec{H}^I$ is the inviscid flux vector on the boundary surface of the control body, $\vec{H}^V$ is the viscous flux vector, Ω is the volume of the control body, S is the area of the boundary surface of the control body, and $\vec{n}$ is the normal vector of the boundary surface.

The solution format and turbulence model are critical factors that can significantly impact the accuracy of numerical calculations. To minimize the impact of non-physical dissipation on accuracy, inviscid fluxes were discretized using Roe's Flux Difference Splitting scheme [11], and the entropy method was used to correct non-physical solutions. The Roe format has proven to be one of the most effective formats in practical applications due to its exceptional ability to resolve viscous and shock wave discontinuities. For the spatial discretization, the second-order upwind scheme was utilized. In this study, the two-equation $k-\omega$ SST turbulence model was employed to calculate turbulence. This model combines the advantages of $k-\varepsilon$ model and $k-\omega$ model. The equations of this model are

$$\frac{\partial}{\partial t}(\rho k) + \frac{\partial}{\partial x_j}(\rho k u_j) = \frac{\partial}{\partial x_j}\left[\left(\mu + \frac{\mu_t}{\sigma_k}\right)\frac{\partial k}{\partial x_j}\right] + G_k - Y_k + S_k \tag{2}$$

and

$$\frac{\partial}{\partial t}(\rho \omega) + \frac{\partial}{\partial x_j}(\rho \omega u_j) = \frac{\partial}{\partial x_j}\left[\left(\mu + \frac{\mu_t}{\sigma_\omega}\right)\frac{\partial \omega}{\partial x_j}\right] + G_\omega - Y_\omega + D_\omega + S_k, \tag{3}$$

where G_k and G_ω are the generation terms of k and ω. Y_k and Y_ω are the dissipation terms of k and ω. D_ω is the cross-diffusion term, and S_k is the self-defined source term.

In this study, a three-dimensional model of the rocket sled and track was developed to investigate the flow field in the slipper-track gap. The model consisted of a rocket sled, a slipper, and a track, with a narrow gap between the slipper and track. To numerically solve the flow field, the finite volume method was employed and the domain was discretized into a structured mesh, as illustrated in Fig. 2. The discretization of the mesh and calculation were based on the commercial software ANSYS ICEM CFD [12] and Fluent [13], respectively. The resulting cell number was approximately 9 million. In the areas where the flow field parameters vary greatly, such as the sled head, windward side of the slipper, and boundary layer, a denser grid was employed to better capture the characteristics of the flow field near the structure surface.

The flow field was set as a cuboid. The inlet and far field of the flow field were set as far-field boundaries; the outlet was set as a supersonic outlet boundary; the sled and the slipper were set as non-slip adiabatic walls. The track and the ground were set as translational walls, whose translational speed was consistent with the incoming flow speed. A free incoming flow static pressure $p_0 = 101,325$ Pa and static temperature $T_0 = 288.15$ K were defined. Sea level pressure and temperature parameters were used, and the ideal gas model was applied. The slipper-track gap flow field characteristics of the rocket sled at Mach numbers of 2 and 4 were computed. After 25,000 steps, the residual curve converged below 10^{-4} to reach the steady state

Aerospace Science and Engineering - III Aerospace PhD-Days   Materials Research Forum LLC
Materials Research Proceedings 33 (2023) 369-375   https://doi.org/10.21741/9781644902677-54

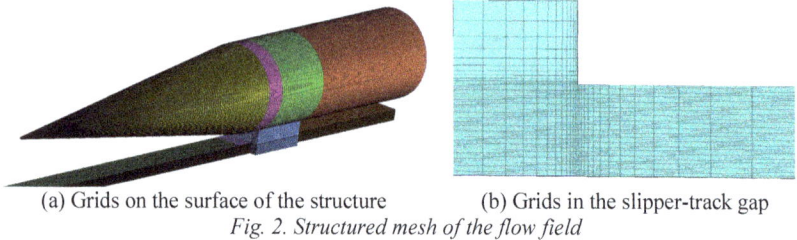

(a) Grids on the surface of the structure   (b) Grids in the slipper-track gap

Fig. 2. Structured mesh of the flow field

Results and discussion

Fig. 3 illustrates the pressure distribution at $Ma = 2$. The figure showcases several features of the flow field that merit discussion. Specifically, Fig. 3(a) reveals the formation of bow shock waves at the fore of the rocket sled. These bow shock waves are comprised of a central normal shock wave, surrounded by oblique shock waves on either side. Fig. 3(b) further highlights the emergence of an annular high-pressure region at the outlet of the flow field and on the ground. Furthermore, the slipper-track gap flow field characteristics are influenced by the shock wave reflecting off the track and the ground. Finally, Fig. 3(c) also indicates the presence of a distinct normal shock wave at the gap's entrance, which is consistent with the theory advanced in Reference [7].

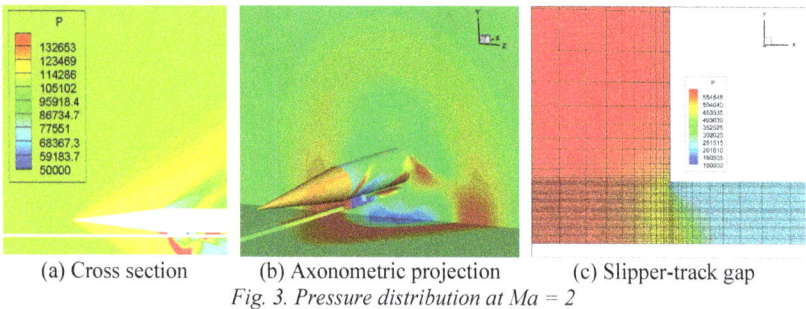

(a) Cross section   (b) Axonometric projection   (c) Slipper-track gap

Fig. 3. Pressure distribution at Ma = 2

In this study, we analyzed the flow field at $Ma = 4$ with the aim of determining the distribution of physical quantities and the variation rules within the slipper-track gap. To achieve this objective, we selected ten observation points along the X-axis of the gap, as illustrated in Fig. 4. Observation point 1 is located at the leading edge of the gap, while observation point 10 is situated at the trailing edge. To account for the significant variation of the flow field at the leading and trailing edges and the comparatively small variation in the middle section, we adopted a dense-sparse-dense distribution mode for the observation points.

Fig. 4. Observation points at the slipper-track gap

In Fig. 5, we present the physical quantity profiles at different locations within the slipper-track gap at $Ma = 4$. Fig. 5(a) shows the velocity profiles at ten observation points. The velocity curves exhibit a similar trend along the gap flow direction. At the gap inlet, the mainstream velocity is about 600 m/s. Subsequently, the velocity gradually increases from the inlet to the stable section before slightly decreasing again. Upon reaching the stable stage, the velocity near the upper and lower walls rises due to the increase in fluid viscosity, while the mainstream velocity slightly

Aerospace Science and Engineering - III Aerospace PhD-Days   Materials Research Forum LLC
Materials Research Proceedings 33 (2023) 369-375   https://doi.org/10.21741/9781644902677-54

decreases. At the outlet section, the mainstream velocity experiences a sharp increase, reaching approximately 900 m/s. Fig. 5(b) presents the temperature profiles of ten observation points. At the gap inlet, there is a high-temperature zone in the upper region, whereas the temperature near the lower wall is lower. While the temperature in the inlet section remains consistent in shape, it slightly decreases. The temperature then increases from the inlet section to the stable section. The temperature profile gradually transitions to a lower temperature in the mainstream area and a higher temperature at the upper and lower walls, with the temperature gradient near the lower wall being zero. The temperature profile in the stable stage remains constant, with the mainstream temperature slightly increasing to 1110 K. In the outlet section, the temperature experiences a significant decrease, with the upper part of the gap outlet being a low-temperature zone and the temperature near the lower wall being higher. The mainstream temperature at the outlet is approximately 870 K. To summarize, the narrow gap's temperature is exceptionally high, which could modify the slipper's material properties.

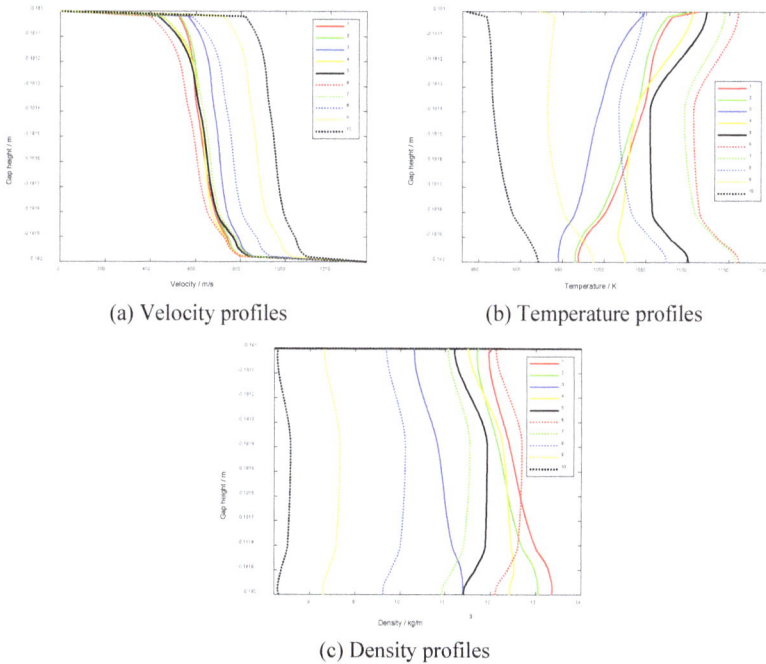

(a) Velocity profiles   (b) Temperature profiles

(c) Density profiles

Fig. 5. Physical quantity profiles at different positions in the gap

In Fig. 5(c), we present the density profiles of ten observation points at $Ma = 4$. The compression effect is more pronounced in the lower part of the fluid near the gap inlet. Therefore, the air density in the inlet area increases from the upper wall to the lower wall. The density profile retains the same shape, but there is a slight decrease. As one moves from the inlet section to the stable section, the density increases, and the mainstream density reaches 13 kg/m^3. The density profile gradually shifts such that the mainstream density is higher, while the density near the upper and lower walls is lower. The density profile remains constant until the outlet section, where there

Aerospace Science and Engineering - III Aerospace PhD-Days   Materials Research Forum LLC
Materials Research Proceedings 33 (2023) 369-375   https://doi.org/10.21741/9781644902677-54

is an abrupt drop in density, and the mainstream density is about 7.5 kg/m^3. Thus, the high air density in the narrow gap suggests significant air compression.

Conclusions

The high-speed rocket sled is a crucial tool in the aviation and aerospace. This study utilized a three-dimensional structured mesh to establish a relatively accurate model of the flow field in the slipper-track gap of a high-speed rocket sled. The simulation provided insight into the movement law of the flow field within the gap.

At $Ma = 2$, the pressure distribution within the gap revealed that the bow shock wave at the head of the rocket sled impacted the flow field characteristics. Moreover, an obvious normal shock wave was detected at the entrance of the gap. At $Ma = 4$, the shapes of velocity curves along the gap flow direction were nearly identical. Initially, the mainstream velocity increased, then decreased, and finally increased again. In contrast, the mainstream temperature first decreased, then increased, and eventually decreased once more. The temperature near the upper wall of the gap inlet was higher, whereas the temperature at the lower part of the outlet was elevated. The air density of the lower portion at the gap inlet was the largest, and it gradually became the largest in the mainstream area.

In conclusion, the air in the slipper-track gap undergoes significant compression, resulting in very high temperatures. At $Ma = 4$, the highest temperature near the slipper can reach 1160 K, which poses a potential risk of damage to the slipper. This study provides a foundation for further research on the structural characteristics of slippers in high-temperature environments.

Acknowledgments

The first author acknowledges the financial support by the China Scholarship Council (No. 202106280004).

References

[1] Dang T, Liu Z, and Zhou X, et al, Dynamic response of a hypersonic rocket sled considering friction and wear, Journal of Spacecraft and Rockets. 59 (2022) 1289-1303. https://doi.org/10.2514/1.A35267

[2] Cinnamon John D., and Anthony N. Palazotto, Analysis and simulation of hypervelocity gouging impacts for a high speed sled test, International Journal of Impact Engineering. 36 (2009) 254-262. https://doi.org/10.1016/j.ijimpeng.2007.11.009

[3] Dang T, Li B, and Hu D, et al, Aerodynamic design optimization of a hypersonic rocket sled deflector using the free-form deformation technique, Proceedings of the Institution of Mechanical Engineers, Part G: Journal of Aerospace Engineering. 235 (2021) 2240-2248. https://doi.org/10.1177/09544100221075071

[4] Szmerekovsky Andrew G., and Anthony N. Palazotto, Structural dynamic considerations for a hydrocode analysis of hypervelocity test sled impacts, AIAA Journal. 44 (2006) 1350-1359. https://doi.org/10.2514/1.13803

[5] Korkegi Robert H., and Ronald A. Briggs, On compressible turbulent-plane Couette flow, AIAA Journal. 6 (1968) 742-744. https://doi.org/10.2514/3.4583

[6] Korkegi Robert H., and Ronald A. Briggs, Compressible turbulent plane Couette flow with variable wall temperature, Aerospace research labs Wright-Patterson AFB Ohio, 1970. https://apps.dtic.mil/sti/citations/AD0708162

[7] Korkegi Robert H., and Ronald A. Briggs, The hypersonic slipper bearing-A test track problem, Journal of Spacecraft and Rockets. 6 (1969) 210-212. https://doi.org/10.2514/3.29570

[8] Korkegi R. H., and R. A. Briggs, Compressible turbulent plane Couette flow with variable heat transfer based on von Karman model, AIAA Journal. 8 (1970) 817-819. https://doi.org/10.2514/3.5767

[9] Lofthouse A., M. Hughson, and A. Palazotto, Hypersonic test sled external flow field investigation using computational fluid dynamics, 40th AIAA Aerospace Sciences Meeting & Exhibit, 2002. https://doi.org/10.2514/6.2002-306

[10] Alban Christopher J, Thermal and melt wear characterization of materials in sliding contact at high speed, Air Force Institute of Technology, Wright-Patterson AFB Ohio, Graduate school of engineering and management, 2014. https://apps.dtic.mil/sti/citations/ADA599170

[11] Roe Philip L, Approximate Riemann solvers, parameter vectors, and difference schemes, Journal of computational physics. 43 (1981) 357-372. https://doi.org/10.1016/0021-9991(81)90128-5

[12] Ansys® ICEM CFD, Release 2020 R1, help system, ANSYS, Inc.

[13] Ansys® Fluent, Release 2020 R1, help system, ANSYS, Inc.

Aerospace Science and Engineering - III Aerospace PhD-Days   Materials Research Forum LLC
Materials Research Proceedings 33 (2023) 376-381   https://doi.org/10.21741/9781644902677-55

Improving satellite pose estimation across domain gap with generative adversarial networks

Alessandro Lotti[1,a] *

[1]Department of Industrial Engineering, Alma Mater Studiorum Università di Bologna
Via Fontanelle 40, 47121, Forlì, Italy

[a]alessandro.lotti4@unibo.it

Keywords: Pose Estimation, Computer Vision, Vision-Based Navigation, Domain Gap, Generative Adversarial Networks

Abstract. Pose estimation from a monocular camera is a critical technology for in-orbit servicing missions. However, collecting large image datasets in space for training neural networks is impractical, resulting in the use of synthetic images. Unfortunately, these often fail to accurately replicate real image features, leading to a significant domain gap. This work explores the use of generative adversarial networks as a solution for bridging this gap from the data level by making synthetic images more closely resemble real ones. A generative model is trained on a small subset of unpaired synthetic and real pictures from the SPEED+ dataset. The entire synthetic dataset is then augmented using the generator, and employed to train a regression model, based on the MetaFormer architecture, which locates a set of landmarks. By comparing the model's pose estimation accuracy on real images with and without generator preprocessing, it is observed that the augmentation effectively reduces the median pose estimation error by a factor 1.4 to 5. This compelling result validates the efficacy of these tools and justifies further research in their utilization.

Introduction

Accurately navigating an active probe around a target spacecraft is crucial for many space missions, such as satellite servicing and debris removal. In this context, the ability to operate autonomously is essential to ensure that the servicer can make timely maneuvers and respond quickly to changing conditions, especially to avoid collisions. Therefore, the chaser shall be able to measure and control its state relative to the client throughout the rendezvous.

In this scenario, monocular cameras have emerged as an appealing sensor solution because of their low power consumption and small form factor. To estimate the relative chaser-to-target pose (i.e. position and attitude), a set of 2D landmarks extracted from the image is matched with the 3D model of the spacecraft using Perspective-n-Point solvers [1], as illustrated in Fig. 1. Image processing (IP) is a critical aspect of this software routine and recently many authors proposed to leverage neural networks (NNs) for feature extraction. Unfortunately, these models require large-scale datasets for training which are highly impractical to collect and label in orbit with accurate pose information. To overcome this challenge, researchers have turned to synthetic datasets generated through 3D computer graphics software [2,3,4]. On one side, these allow for rapid generation of thousands of images but, on the other side, they struggle to accurately reproduce the visual characteristics and large diversity of spaceborne pictures. This discrepancy between synthetic and real-world images is known as domain gap, and is a major obstacle to the adoption of NNs for vision-based navigation. Indeed, NNs tend to develop over-reliance on synthetic features, leading to inaccurate predictions when applied to real-world scenarios.

Aerospace Science and Engineering - III Aerospace PhD-Days   Materials Research Forum LLC
Materials Research Proceedings 33 (2023) 376-381   https://doi.org/10.21741/9781644902677-55

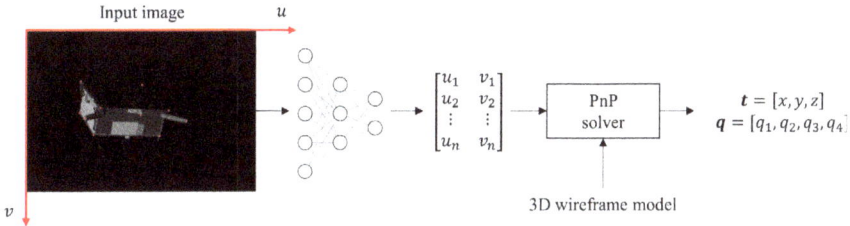

Fig. 1 - Schematization of the pose estimation process.

While domain gap is actively researched within the computer vision community (see [5] for a throughout survey), few works addressed this issue in the framework of satellite pose estimation. According to the results of the ESA sponsored "Satellite Pose Estimation Competition 2021" [6], the use of adversarial training proved to be an effective technique. Although training the NN to also deceive a discriminator can help to learn domain invariant features, this method demands a large number of real pictures. Obtaining such images, however, can be costly and require specialized robotic facilities.

Generative adversarial networks (GANs) represent a possible approach to address this challenge by narrowing domain gap at the data level. Indeed, they could be employed to enhance synthetic datasets, making them more closely resemble real-world images.

GAN models typically comprise two neural networks, a generator and a discriminator, that are trained together to learn a mapping from a source domain to a target domain. The purpose is to transfer the visual characteristic of real pictures to synthetic images which should become indistinguishable from the real ones. GANs provide a versatile and cost-effective way to address the scarcity of spaceborne images as the same generator could in principle be applied to any synthetic dataset regardless of the depicted target.

This work demonstrates how the regression error on real images can be greatly reduced by simply preprocessing synthetic images through a trained GAN.

Methods

The SPEED+ dataset [7] is adopted for this study. This includes synthetic grayscale images (47966 for training and 11994 for validation) and real pictures of a spacecraft mockup captured in a laboratory environment at a maximum distance of 10 m from the camera. Fig. 2 displays synthetic and real samples, which feature two different illumination conditions, namely *lightbox* (6740) and *sunlamp* (2791). Throughout the workflow, input images are resized from the original 1200x1920 px resolution to 320x512 px to reduce the computational time while preserving the original aspect ratio. This project employs a state-of-the-art image-to-image (I2I) translation algorithm based on the contrastive learning method proposed in [8], which does not require paired samples from the source and target domains during training. The generator network is built using a Residual Network [9] with 9 residual blocks. The I2I model is trained on randomly selected subsets of 1000 synthetic and real images for 200 epochs, with a batch size of 1. The learning rate is linearly decreased from 2e-4 to 0 starting from epoch 100. Training is repeated separately for *sunlamp* and *lightbox* domains. Later, all synthetic images from the SPEED+ train partition are processed through the generator obtaining two sets of synthetic-enhanced pictures.

Aerospace Science and Engineering - III Aerospace PhD-Days   Materials Research Forum LLC
Materials Research Proceedings 33 (2023) 376-381   https://doi.org/10.21741/9781644902677-55

Fig. 2 - Images from the SPEED+ dataset [7] depicting the satellite in similar poses: synthetic (left), lightbox (center), sunlamp (right).

During translation, for each synthetic image a random sample from the 1000 real pictures is selected to provide a context for the I2I translation.

A collection of original and enhanced images is illustrated in Figure 3, showcasing the ability of the generator to capture the appearance of the target domain while preserving most of the content of the original image.

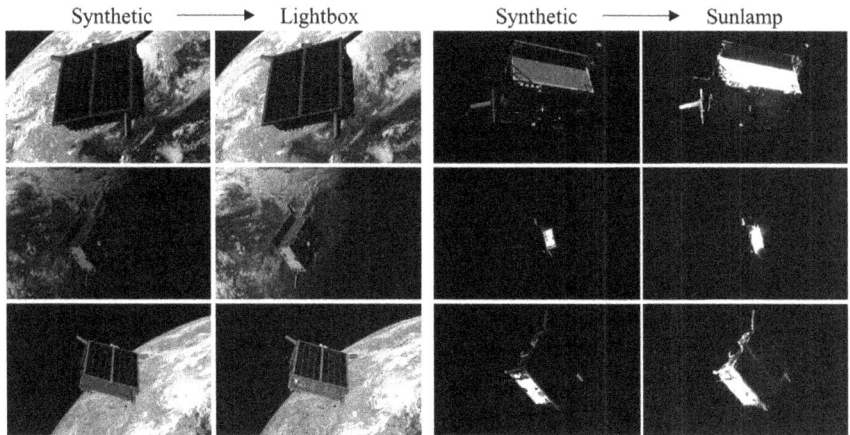

Fig. 3 - Synthetic vs GAN-enhanced images.

To assess the effectiveness of the learned mapping, a pose estimation pipeline according to the scheme illustrated in Fig. 1 has been set up. In this framework a NN is trained to predict the location of 11 landmarks on the enhanced-synthetic datasets and tested on actual *sunlamp* and *lightbox* images. To this end a *ConvFormerS18* model [10] is employed as the backbone of a 2-layer regression head, comprising a depthwise separable convolution, with ReLu activation, followed by a fully connected layer. The ConvFormer belongs to the family of MetaFormer models [11], inspired by Transformers [12], which showcased improved robustness to domain gap over convolutional neural networks [13]. The architecture leverages depthwise separable convolutions as token mixers, which are well-suited for embedded processors. The network is trained using mean absolute error loss for 40 epochs with a batch size of 48 images. The fitting is carried out on Google Colab's Tensor Processing Units with Adam optimizer and cosine decay learning rate, starting from 7.5e-5. The backbone is initialized with Imagenet weights [14] and common data augmentations are applied, these include random image rotation, brightness and contrast adjustments, blurring and gaussian noise.

At inference time, the regressed landmarks are fed into an EPnP solver [15] together with the 3D satellite model and camera parameters.

Results and Discussion

The pose regression error is then evaluated through the following metrics:

$$e_t = |t_{BC} - \hat{t}_{BC}|_2 \tag{1}$$

$$\bar{e}_t = \frac{e_t}{|t_{BC}|_2} \tag{2}$$

$$e_q = 2\,arccos(|q^T \hat{q}|) \tag{3}$$

$$E = \bar{e}_t + e_q \tag{4}$$

Where t_{BC}, q and $\hat{t}_{BC}$, $\hat{q}$ represent respectively the ground truth and estimated position vectors and attitude quaternions aligning the chaser-mounted camera frame (C) and the target body frame (B). Table 1 compares the performance of the NN trained on synthetic enhanced images with that of the model trained without GAN preprocessing. The symbol $\langle\ \rangle$ denotes the median operator. A solution is defined to be of high quality (HQ) if both the normalized position and rotation errors fall below specific thresholds. For the rotation error, the threshold is 5° when the satellite is within 5 meters, and 10° when it is farther away. Meanwhile, the limit for normalized position error is fixed at 0.1. Notably, GAN preprocessing allowed to decrease the median pose error of a factor 5 and 1.4 on *sunlamp* and *lightbox* respectively. The enhancement is further evidenced by the increase in the percentage of HQ solutions.

Table 1 - Pose estimation errors obtained by training the NN with and without GAN preprocessing.

GAN preprocessing	Lightbox				Sunlamp			
	$\langle e_t \rangle$ [m]	$\langle e_q \rangle$ [deg]	$\langle E \rangle$	HQ	$\langle e_t \rangle$ [m]	$\langle e_q \rangle$ [deg]	$\langle E \rangle$	HQ
With	**0.369**	**11.2**	**0.288**	**38.4%**	**0.131**	**5.08**	**0.116**	**67.6%**
Without	0.488	15.8	0.403	30.6%	0.580	25.5	0.597	15.8%

Conclusions and Future Work

Overall, the results of this preliminary study demonstrate that GAN preprocessing is effective in reducing the domain gap at the data level, with the added benefit of requiring only a small fraction of the dataset size for training. However, further investigations are needed to explain the discrepancy in the advantages achieved on the two domains, as illustrated in Table 1. Additionally, the reliability of the MetaFormer architecture for the IP step has also been demonstrated by its ability to achieve competitive pose estimation errors despite having only around 24 million parameters.

As a next step, the potential of fusing *lightbox* and *sunlamp* characteristics in a single generator will be investigated to produce even more realistic images that capture the visual features of both domains. The ability of the GAN to translate images depicting different targets from those featured in the training data, without compromising their content, will also be explored. Furthermore, the benefits of combining GAN preprocessing with other domain generalization methods, such as the extensive augmentations and multi-task learning solutions proposed in the literature [16], will be assessed.

Aerospace Science and Engineering - III Aerospace PhD-Days   Materials Research Forum LLC
Materials Research Proceedings 33 (2023) 376-381   https://doi.org/10.21741/9781644902677-55

References

[1] B. Chen, J. Cao, A. Parra, T.-J. Chin, Satellite Pose Estimation with Deep Landmark Regression and Nonlinear Pose Refinement, Proceedings of the IEEE/CVF International Conference on Computer Vision Workshops, IEEE, New York, 2019, pp. 2816–2824. https://doi.org/10.1109/ICCVW.2019.00343

[2] S., Sharma, S. D'Amico, Pose Estimation for Non-Cooperative Rendezvous Using Neural Networks, 2019 AAS/AIAA Astrodynamics Specialist Conference, American Astronautical Society Paper 19-350, Springfield, VA, 2019, pp. 1–20.

[3] P. F. Proenca, Y. Gao, Deep Learning for Spacecraft Pose Estimation from Photorealistic Rendering, 2020 IEEE International Conference on Robotics and Automation, IEEE, New York, 2020, pp. 6007–6013. https://doi.org/10.1109/ICRA40945.2020.9197244

[4] A. Lotti, D. Modenini, P. Tortora, M. Saponara, and M. A. Perino, Deep Learning for Real-Time Satellite Pose Estimation on Tensor Processing Units, Journal of Spacecraft and Rockets, 2023, pp. 1–5. https://doi.org/10.2514/1.A35496

[5] J. Wang, C. Lan, C. Liu, Y. Ouyang,T. Qin, W. Lu, Y. Chen, W. Zeng, P.S. Yu, Generalizing to Unseen Domains: A Survey on Domain Generalization, IEEE Transactions on Knowledge and Data Engineering. https://doi.org/10.1109/TKDE.2022.3178128

[6] T. H. Park, M. Märtens, M. Jawaid, Z. Wang, B. Chen, T.-J. Chin, D. Izzo, S. D'Amico, Satellite Pose Estimation Competition 2021: Results and Analyses, Acta Astronautica, vol. 204, 2023, pp. 640–665. https://doi.org/10.1016/j.actaastro.2023.01.002

[7] T. H. Park, M. Martens, G. Lecuyer, D. Izzo, and S. D'Amico, SPEED+: Next-Generation Dataset for Spacecraft Pose Estimation across Domain Gap, IEEE Aerospace Conference (AERO), Big Sky, MT, USA, 2022, pp. 1–15. https://doi.org/10.1109/AERO53065.2022.9843439

[8] T. Park, A. A. Efros, R. Zhang, and J.-Y. Zhu, Contrastive Learning for Unpaired Image-to-Image Translation, European Conference on Computer Vision (ECCV), Springer, 2020, pp. 319–345.

[9] J. Johnson, A. Alahi, and L. Fei-Fei, Perceptual Losses for Real-Time Style Transfer and Super-Resolution, European Conference on Computer Vision (ECCV), Springer, 2016, pp 694-711.

[10] W. Yu, C. Si, P. Zhou, M. Luo, Y. Zhou, J. Feng, S. Yan, X. Wang, MetaFormer Baselines for Vision. arXiv, 2022. https://doi.org/10.48550/arXiv.2210.13452

[11] W. Yu, M. Luo, P. Zhou, C. Si, Y. Zhou, X. Wang, J. Feng, S. Yan, MetaFormer Is Actually What You Need for Vision, Proceedings of the IEEE/CVF Conference on Computer Vision and Pattern Recognition (CVPR), 2022, pp 10819-10829.

[12] A. Vaswani, N. Shazeer, N. Parmar, J. Uszkoreit, L. Jones, A. N. Gomez, Ł. Kaiser, I. Polosukhin, Attention is all you need, Advances in Neural Information Processing Systems, 2017, pp 5998–6008.

[13] C. Zhang, M. Zhang, S. Zhang, D. Jin, Q. Zhou, Z. Cai, H. Zhao, X. Liu, Z. Liu, Delving Deep into the Generalization of Vision Transformers under Distribution Shifts, Proceedings of the IEEE/CVF Conference on Computer Vision and Pattern Recognition (CVPR), 2022, pp 7277-7286.

Aerospace Science and Engineering - III Aerospace PhD-Days   Materials Research Forum LLC
Materials Research Proceedings 33 (2023) 376-381   https://doi.org/10.21741/9781644902677-55

[14] A. Krizhevsky, I. Sutskever, G. E. Hinton, ImageNet Classification with Deep Convolutional Neural Networks, Advances in Neural Information Processing Systems, Vol. 25, No. 2, 2012, pp. 1106–1114. https://doi.org/10.1145/3065386

[15] V. Lepetit, F. Moreno-Noguer, and P. Fua, EPnP: An Accurate O(n) Solution to the PnP Problem, International Journal of Computer Vision, Vol. 81, No. 2, 2009, pp. 155–166. https://doi.org/10.1007/s11263-008-0152-6

[16] T.H. Park, S. D'Amico, Robust Multi-Task Learning and Online Refinement for Space-craft Pose Estimation across Domain Gap, 2022. ArXiv: abs/2203.04275

Aerospace Science and Engineering - III Aerospace PhD-Days
Materials Research Proceedings 33 (2023) 382-387

Materials Research Forum LLC
https://doi.org/10.21741/9781644902677-56

Towards a CO₂ emission standard for supersonic transport: A Mach 2 concept case study

Oscar Gori[1,a*]

[1]Politecnico di Torino, Torino, Italy

[a]oscar.gori@polito.it

Keywords: Supersonic Aircraft, Environmental Sustainability, CO₂ Emissions Standards

Abstract. This paper reports the work performed in the field of the H2020 MORE&LESS Project to contribute to shaping global environmental regulations for supersonic aviation. The existing CO_2 emission standard, which is defined for subsonic aircraft, is analyzed. However, the applicability of this standard to supersonic concepts needs to be assessed. A case study of a Mach 2 concept, based on the Concorde configuration, is considered to explore modifications needed for the CO_2 metric value calculation. The results provide insights into the CO_2 metric value for a supersonic aircraft and its comparison with subsonic aircraft limits.

Introduction

During the last decades, the aerospace community has witnessed a renewed interest in high-speed civil passenger transport, with numerous projects underway to design civil passenger aircraft which fly faster than the speed of sound. However, it is crucial to consider the environmental impact of such concepts and prioritize environmental sustainability and social acceptance. To this end, the H2020 MORE&LESS Project (MDO and REgulations for Low boom and Environmentally Sustainable Supersonic aviation) was initiated in January 2021 with funding from the European Commission [1]. The project, which is expected to run for four years, aims to support Europe in shaping global environmental regulations for future supersonic aviation.

However, the current environmental regulations apply only to subsonic aircraft. A CO_2 emission standard exists and it is defined in the ICAO Annex 16 Volume III [2]. The CO_2 metric value is used to measure the fuel burn performance of an aircraft and it is designed to be common across different aircraft categories, regardless of their purpose or capabilities [3]. The CO_2 emission standard relies on three key factors linked to aircraft technology and design: cruise fuel burn performance, aircraft size and aircraft weight. It has been designed to ensure that effective enhancements measured through the system will lead to a corresponding reduction in CO_2 emissions during regular aircraft operations.

The work towards a potential CO_2 emission standard for supersonic transport (SST) is presented in this paper. One of the case studies of the More&Less project is considered for this analysis: a Mach 2 concept, derived from the Concorde configuration. More details on this case study are provided in the next sections.

The CO_2 metric value is currently defined for subsonic aircraft only. Nevertheless, with the growing interest in supersonic aviation, there arises a necessity for a new metric value to be established specifically for this type of aircraft. The initial step to take is to evaluate whether any elements of the existing requirements are directly applicable to civil supersonic concepts. If not, it is crucial to determine which modifications are required.

Methodology

The CO_2 emission standard value is a fuel-efficiency standard, and it is based on 2 parameters:

1. Specific Air Range (SAR) during cruise flight.

2. Reference Geometric Factor (RGF), a measure of cabin size.

When an aircraft type undergoes CO_2 certification, a CO_2 metric value is calculated based on these parameters, and then compared to a limit that is dependent on the Maximum Take-Off Mass (MTOM) of the aircraft. The CO_2 metric value can be evaluated as:

$$CO_2 \; metric \; value = \frac{\left(\frac{1}{SAR}\right)_{avg}}{RGF^{0.24}} \,. \tag{1}$$

Where $(1/SAR)_{avg}$ is the average of the reciprocal of the specific air range (SAR) [kg/km], which is evaluated at three cruise flight reference points, and the reference geometric factor (RGF) is a dimensionless measure of the cabin size. SAR values are evaluated for 3 reference cruise conditions, which are defined as a function of the MTOM:

1. High gross mass: $0.92 \cdot$ MTOM.
2. Low gross mass: $(0.45 \cdot \text{MTOM}) + (0.63 \cdot \text{MTOM}^{0.924})$.
3. Mid gross mass: average of high and low gross masses.

For a generic subsonic mission, the three points should be located at beginning, end and mid of cruise, respectively, as can be seen in Fig. 1.

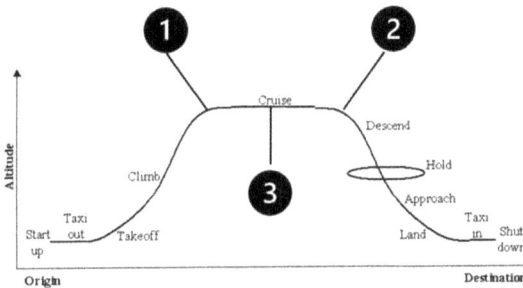

Fig. 1 – Mass points location for a generic subsonic mission.

The applicability of these correlations is limited to subsonic aircraft, as they are linked to the fuel utilization percentage during ascent and cruising stages, in relation to the maximum take-off weight. Consequently, these magnitudes may vary when applied to a supersonic mission. Then, an additional evaluation is carried out to explore potential modifications in the definition of mass-points, which considers the supersonic cruise conditions. Eventually, the CO_2 metric value can be evaluated for these mass-points.

Case study

The case study considered for the analysis is a Mach 2 aircraft powered with biofuel. The Concorde aircraft has been used as reference configuration, targeting a minimization of noise and pollutant emissions as well as the lowest environmental impact at local, regional and global level. The aircraft characterization is an ongoing work and includes different aspects: vehicle design, aerodynamic characterization, propulsive characterization and mission simulation. The GTO mass of the aircraft is about 177 tons, with a maximum payload of 15280 kg and a fuel mass of 82180 kg. [4] The aircraft main geometric parameters are reported in Table 1, while an image of the aircraft is shown in Fig. 2.

Table 1 – Mach 2 case study main parameters.

Parameter	Value
Fuselage length [m]	62.25
Wing span [m]	25.60
Fuselage width [m]	2.90
Wing surface [m^2]	327.00
MTOM [Mg]	176.85
Passengers [-]	~120
Range [km]	6500

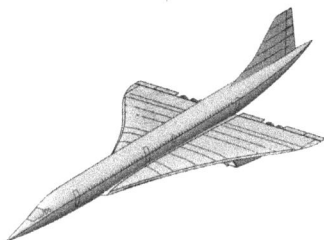

Fig. 2 – Mach 2 case study.

Results

First, in accordance with the standards for subsonic aircraft, Specific Air Range (SAR) values are calculated at the three specific reference mass points, which are dependent on the Maximum Take-Off Mass (MTOM). Although they are intended to represent various segments of the subsonic mission's cruise phase, they could be not situated in cruise conditions for a supersonic mission. The results obtained for the Mach 2 case study are reported in Table 2. The location of the mass points along the mission altitude profile is shown in Fig. 3. As expected, it can be seen that they do not represent the cruise conditions, but they are shifted towards the beginning of the mission. This is due to the fact that the percentage of fuel consumed to complete the climb is higher with respect to subsonic aircraft.

Table 2 – Mass-points and SAR evaluation.

	Mass [Mg]	Altitude [km]	SAR [km/kg]
1. High-mass point	162.70	08.84	0.047
2. Low-mass point	124.06	16.04	0.108
3. Mid-mass point	143.38	15.00	0.073

Materials Research Forum LLC
https://doi.org/10.21741/9781644902677-56

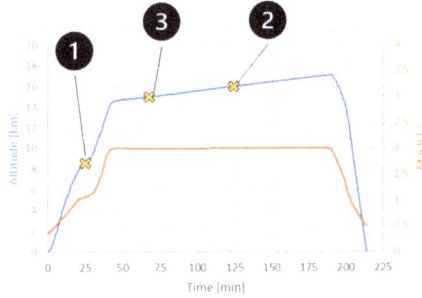

Fig. 3 – Altitude and Mach profile vs Time with mass points location.

Then, the actual cruise conditions are considered. The mass of the aircraft is evaluated at the beginning (BOC), end (EOC) and mid (MOC) of cruise and SAR is then computed accordingly. The results are reported in Table 3, while the modified position of the mass points is shown in Fig. 4.

Table 3 – Mass and SAR evaluated at real cruise conditions.

	Mass [Mg]	Altitude [km]	SAR [km/kg]
BOC	153.70	14.60	0.087
EOC	105.40	17.00	0.128
MOC	129.55	15.80	0.104

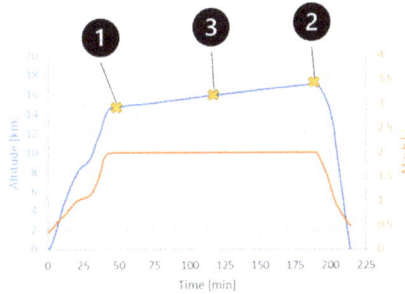

Fig. 4 – Altitude and Mach profile vs Time with mass points evaluated at real cruise conditions.

Eventually, the high-mass and low-mass points can be expressed as a function of MTOM for the actual cruise condition, as reported in Table 4. These results indicate that a representative high-mass point (at initial cruise conditions) shifts to a lower fraction of MTOM for SSTs compared to subsonic aircraft. Initial analyses with higher-Mach concepts in More&Less confirm that this trend increases with design Mach number, as more fuel is burned in the climb and acceleration segments.

Aerospace Science and Engineering - III Aerospace PhD-Days   Materials Research Forum LLC
Materials Research Proceedings 33 (2023) 382-387   https://doi.org/10.21741/9781644902677-56

Table 4 – Mass points for the different cases considered.

	High-mass point	Low-mass point
Subsonic aircrafts	$0.92 \cdot MTOM$	$(0.45 \cdot MTOM) + (0.63 \cdot MTOM^{0.924})$
Mach 2 Case Study	$0.87 \cdot MTOM$	$0.59 \cdot MTOM$

Moreover, SAR can be also evaluated during the entire mission, including climb, cruise and descent phases. An overview of the altitude and instantaneous SAR as function of ground distance is reported in Fig. 5 and Fig. 6. As expected, SAR reaches its highest level during cruise conditions, except for the final stage of the mission when the aircraft descends toward the destination airport and the thrust is kept to lower values.

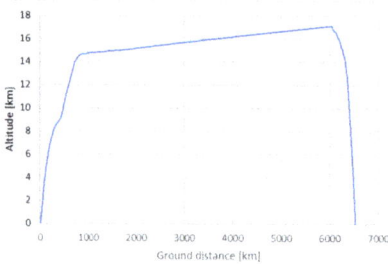

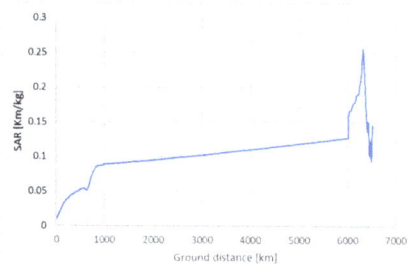

Fig. 5 – Altitude profile vs ground distance   *Fig. 6 – Instantaneous SAR vs ground distance*

Eventually, the CO_2 metric value can be evaluated considering the mass points at the representative cruise conditions:

$$CO_2 \, MV = \frac{\left(\frac{1}{SAR}\right)_{avg}}{RGF^{0.24}} = 3.1 \, kg/km$$

The CO_2 metric value can be compared to the CO_2 limits for subsonic aircrafts, as defined in Annex 16 Vol III (Fig. 7). Some additional points are also included in the plot, representing two supersonic concepts from NASA and DLR. However, these concepts are quite different from the Mach 2 case, since they are much smaller and they are designed to fly at lower Mach numbers (1.4 and 1.6, respectively). For that reason, a direct comparison is not possible.

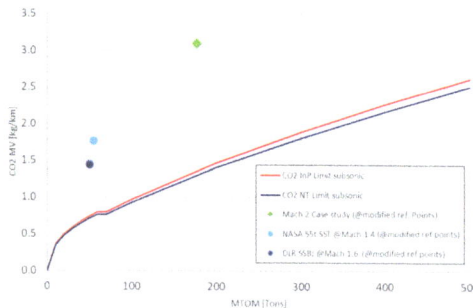

Fig. 7 – CO_2 metric values vs MTOM

Aerospace Science and Engineering - III Aerospace PhD-Days   Materials Research Forum LLC
Materials Research Proceedings 33 (2023) 382-387   https://doi.org/10.21741/9781644902677-56

Conclusions and future steps

The current CO_2 standard is defined for subsonic aircrafts, and it cannot be directly applied to civil supersonic aircraft. The More&Less project aims to contribute to the assessment of the metric value by considering a Mach 2 case study as a baseline. Specific Air Range (SAR) and Reference Geometric Factor (RGF) are evaluated, to understand to what extent the present requirements are applicable to civil supersonic aircraft, and which modifications are needed. One crucial change is the re-definition of reference mass-points, to cover the cruise phase when evaluating SAR. However, further investigation on SAR and RGF and their impact on CO_2 MV is also necessary. Different configurations will be explored at various Mach numbers, such as Mach 1.5 and 1.7, to increase the number of available data. The results from these configurations can be exploited to further contribute to the assessment of CO_2 MV for civil supersonic aircraft.

References

[1] "MDO and REgulations for Low-boom and Environmentally Sustainable Supersonic aviation (MORE&LESS Project)," [Online]. Available: https://cordis.europa.eu/project/id/101006856. [Accessed 07 03 2023].

[2] "ICAO Annex 16 Vol III".

[3] "ICAO AIRCRAFT CO2 EMISSIONS STANDARD," [Online]. Available: https://www.icao.int/environmental-protection/Documents/CO2%20Metric%20System%20-%20Information%20Sheet.pdf. [Accessed 07 03 2023].

[4] O. Gori, "Simplified models for the preliminary aerodynamic characterization of high-speed vehicles to support mission analysis," in *33rd Congress of the International Council of the Aeronautical Sciences*, Stockholm, Sweden, 2022.

Aerospace Science and Engineering - III Aerospace PhD-Days   Materials Research Forum LLC
Materials Research Proceedings 33 (2023) 388-396   https://doi.org/10.21741/9781644902677-57

High-fidelity simulation of shock-wave/boundary layer interactions

A. Ceci

Sapienza University of Rome, Department of Mechanical and Aerospace Engineering, via Eudossiana 18, Rome 00184

alessandro.ceci@uniroma1.it

Keywords: Compressible Boundary Layers, Shock Waves, Turbulence Simulation

Abstract. We perform direct numerical simulations of impinging shock-boundary layer interaction on a flat plate, in which the shock is not orthogonal to the boundary layer flow. The analysis relies on an idealized configuration, where a spanwise flow component is used to introduce the effect of the sweep angle between a statistically two-dimensional boundary layer and the shock. A quantitative comparison is carried out between the swept case and the corresponding unswept one, and the effect of the domain spanwise width is examined. The analysis reveals that, while the time-averaged swept flow characteristics are basically unaffected by the choice of the domain width, the spectral dynamics of the flow dramatically changes with it. For very narrow domains, a pure two-dimensional, low-frequency component can be detected, which resembles the low-frequency oscillation of the unswept case. The present work is also devoted to compare the performance of Digital Filtering (DF) and Recycling-Rescaling methods (RR) in reaching an equilibrium state for the Direct Numerical Simulation (DNS) of a turbulent boundary layer. We performed two sets of DNS of supersonic and hypersonic boundary layers, based on previous numerical studies. It is found that, overall, the RR method is the most appropriate choice, to quickly reach a correct trend of the wall pressure fluctuations, whereas the DF method is more capable in obtain small deviations of the skin friction coefficient with respect to the benchmark.

Introduction

Shock-wave/turbulent-boundary-layer interactions (SBLIs) are encountered in a multitude of aeronautics and aerospace applications. Interactions of this kind frequently occur in external flows, owing to aerodynamic interference between aircraft appendices/boosters and main body, as well as internal flows, for instance air intakes. More generally, SBLIs are found whenever a shock wave sweeps across a turbulent boundary layer developing on a solid surface. The presence of SBLI may lead to significant drawback on aerodynamic performance of aircraft, yielding loss of efficiency of the aerodynamic surfaces, unwanted wall pressure fluctuations possibly leading to structural vibrations, and localized heat transfer peaks, especially when extensive flow separation occurs [3],[9],[13]. The present work focuses on the study of these interactions when the boundary layer upstream of the shock impingement is in the turbulent regime, using direct numerical simulations.

Most available works on turbulent SBLI [19],[21],[23],[24],[25] focus on simplified configurations in which the shock is perfectly orthogonal to the main stream. A peculiar aspect of this kind of interactions is the coupling of the separation flow with the turbulent structures in the upcoming boundary layer, which locally modifies the shear layer embedded in the separation bubble, generating low-frequency oscillation of the separated region [19],[25]. This mechanism yields intense low-frequency tones in the temporal spectrum of the wall pressure fluctuations, whose prediction is important for the safety and integrity of aircraft structures.

Engineering applications, however, often feature more geometrically complex interactions, in which the shock impingement line is not orthogonal to the incoming flow. This is the case of fully three-dimensional SBLIs, whose prototypes are flows over swept compression ramps

Aerospace Science and Engineering - III Aerospace PhD-Days   Materials Research Forum LLC
Materials Research Proceedings 33 (2023) 388-396   https://doi.org/10.21741/9781644902677-57

[10],[28],[33],[34] and around fins [14],[27]. These interactions are made more complicated from the presence of a cross-flow with respect to the shock impingement lines, and by the fact that the separation bubble (when present) is of open type as contrasted to the case of 3D interactions in which separation streamlines must be closed. It is now established that, depending on relative strength of the incoming shock, the near-wall interaction region may have either cylindrical symmetry (namely, the size does not vary along the span) in the case of weak-to-mild interactions, or conical symmetry with length scale linearly increasing along the span in the case of strong interactions [29],[32]. Shock skewing is known to modify the frequency and the intensity of the wall pressure fluctuations, depending on the flow sweep angle [10] also by introducing cross-flow instabilities within the separation bubble [33]. Owing to greater geometrical complexity and difficulty to get converged statistics, three-dimensional SBLIs have been addressed in a much more limited number of studies than their two-dimensional counterparts ([1],[2],[12],[39]).

Researchers have recently attempted to overcome the complexity of three-dimensional SBLI investigations by considering simple geometries able to mimic at least part of the typical phenomena arising in the aforementioned applications. The arguably simplest configuration which can be used at this aim is the interaction between a swept oblique shock wave and a fully developed flat plate boundary layer. Both experimental [18] and numerical works [15] have been recently published using this approach: the former study used a swept shock generator to introduce a shock wave that is not orthogonal to the boundary layer mean flow direction, whereas the aforementioned numerical investigation employs a swept inflow condition for the boundary layer to generate a three-dimensional interaction.

A critical ingredient in the numerical setup of DNS of spatially developing flows is the choice of the inflow conditions. In fact, it is now known that both inflow mean velocity profile [1] and the velocity fluctuations [7] may affect the statistical properties of DNS. The data scatter between simulations with the same free-stream properties but different inflow strategies is the source of large uncertainties in the evaluation of the main quantities of engineering interest (as the skin friction coefficient distribution), which is still a major modeling bottleneck in hypersonic research.

From a computational perspective, a conservative approach to achieve an equilibrium state is to use very long computational domains [30],[31]. Experimental studies have also highlighted the need of taking flow measurements sufficiently far from the wind tunnel inlet section [11]. Recent works [35] have revealed that the streamwise length necessary to reach fully developed turbulence increases monotonically as the free-stream Mach number increases. Extending the analysis of Schlatter et al. [26], those authors considered fulfillment of the Von Karman equation, namely balance of friction and streamwise momentum flux as a quantitative criterion for the evaluation of the inflow length. Although their analysis was restricted to free-stream Mach number M_0 between 0.3 and 2.5, it may be expected that increasing trend is also valid at higher M_0, thus prompting new methods specifically tailored to minimize the development length.The current state-of-the-art modeling of inflow fluctuations is generally based on two classes of numerical methods: recycling-rescaling methods (RR) [17],[37] and digital filtering (DF) methods [16]. Quantitative evaluation of the performance of those methods is still lacking in supersonic/hypersonic flow, which based on the previous observations would be of great value. In this respect, introducing quantitative criteria for estimating the development length is a mandatory prerequisite.

We have carried out numerical simulations of supersonic SBLI in presence of crossflow, using the 2D/3C (two-dimensional, three-component) numerical approach previously employed by the research group for supersonic flow simulations and discussed by Di Renzo et al. [8]. A canonical two-dimensional SBLI, swept by an angle γ_0, is introduced into the computational domain, whose spanwise ends are orthogonal to the shock impingement line. Periodic boundary conditions are applied at the spanwise boundaries, which makes the configuration representative of a 3D swept SBLI (SSBLI) flow with cylindrical symmetry [32]. This configuration allows comparison both

Aerospace Science and Engineering - III Aerospace PhD-Days   Materials Research Forum LLC
Materials Research Proceedings 33 (2023) 388-396   https://doi.org/10.21741/9781644902677-57

with standard hypersonic boundary layers and with swept ones, at reduced computational cost as compared to fully three-dimensional simulations.

Regarding the assessment of inflow condition for supersonic boundary layers, a series of DNS have been carried out to analyze the spatial development of turbulence, using the data of Pirozzoli & Bernardini [22] as a benchmark. As previously mentioned, two goals are pursued: first, identifying suitable criteria to evaluate turbulence development towards an equilibrium state, and second look for modifications of the standard RR and DF techniques, in the attempt of overcoming their weaknesses.

Methodology

The analysis presented in this abstract relies on the numerical solution of the conservation equations for mass, momentum, and energy of a compressible fluid, which read:

$$\frac{\partial \rho}{\partial t} + \nabla \cdot (\rho \boldsymbol{u}) = 0$$
$$\frac{\partial (\rho \boldsymbol{u})}{\partial t} + \nabla \cdot (\rho \boldsymbol{u} \otimes \boldsymbol{u}) = -\nabla P + \nabla \cdot \underline{\underline{\tau}}$$
$$\frac{\partial (\rho e_0)}{\partial t} + \nabla \cdot (\rho e_0 \boldsymbol{u}) = \nabla \cdot \left(\underline{\underline{\tau}} \boldsymbol{u} + \lambda \nabla T + P \boldsymbol{u} \right).$$

In this formulation, ρ, P, and T are the density, pressure, and temperature of the gas, respectively, and $\boldsymbol{u}$ is the flow velocity vector field. The total energy per unit of mass of the gas is defined as $e_0 = e + u^2/2$, where $e = 1/(\gamma_g - 1) RT$ is the internal energy of the mixture per unit of mass, $\gamma_g = 1.4$ is the ratio of the heat capacities of the gas, R is the gas constant. The system of equations is complemented with the ideal gas equation of state

$$P = \rho RT$$

The local shear stress of the fluid is computed with the relation

$$\underline{\underline{\tau}} = \mu \left[\nabla \boldsymbol{u} + \nabla u^T + 2(\nabla \cdot \boldsymbol{u}) \underline{\underline{I}}/3 \right]$$

where μ is the local dynamic viscosity of the mixture, evaluated using a power law of the type $\mu/\mu_0 = (T/T_0)^{0.76}$, whereas the thermal conductivity λ is computed using a constant Prandtl number $Pr = 0.72$.

The Navier-Stokes equations are solved using the in-house high-fidelity code STREAmS [4] for direct numerical simulations of compressible wall-bounded flows. The convective fluxes are discretized by means of a hybrid scheme which combines the energy-preserving properties of a sixth order skew-symmetric central difference scheme [20] with the shock-capturing properties of a fifth order weighted essentially non-oscillatory (WENO) scheme. The switch between the two methods is controlled by a modified Ducros sensor

$$\theta = \frac{(\nabla \cdot \boldsymbol{u})^2}{(\nabla \cdot \boldsymbol{u})^2 + (\nabla \times \boldsymbol{u})^2 + (u_0/\delta_0)^2}$$

which is activated when $\Theta > 0.5$ in any point of the WENO stencil. The diffusive fluxes are discretized with sixth order central formulas, and the time advancement is carried out with a low-storage third order Runge-Kutta scheme [36].

Aerospace Science and Engineering - III Aerospace PhD-Days   Materials Research Forum LLC
Materials Research Proceedings 33 (2023) 388-396   https://doi.org/10.21741/9781644902677-57

SBLI setup

A schematic of the flow under scrutiny for the swept SBLI numerical campaign is provided in Figure 1. A turbulent boundary layer with thickness δ_0 is injected at the left boundary of the computational domain ($x/\delta_0 = 0$), and it develops over the bottom wall swept by an angle γ_0 with respect to the positive x direction. An oblique shock spanning the z direction impinges the boundary layer with an angle β with respect to the flat plate. The velocity inflow condition is obtained as a combination of a Van Driest-transformed incompressible Musker profile for the time-averaged field and velocity fluctuations obtained from a plane in x_r using a recycling-rescaling approach [17] suitable for compressible flows. The temperature fluctuation field is obtained from the streamwise velocity one using the strong Reynolds analogy (SRA).

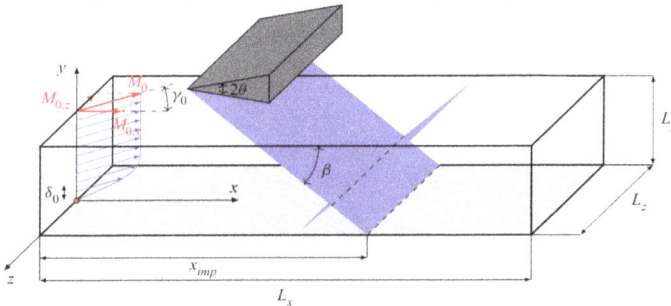

Figure 1. Schematic of SBLI setup: γ_0 is the incoming flow skew angle; δ_0 is the incoming boundary layer thickness; β is the shock angle, θ is the flow deflection angle and x_{imp} is the nominal location of the shock impingement (Figure from Ceci et al. [6], Creative Common license http://creativecommons.org/licenses/by/4.0/)

The bottom wall is assumed to be isothermal, and the wall temperature is set to its nominal turbulent recovery value for the incoming boundary layer (i.e. $T_w/T_r=1$, where $T_r/T_0 = 1 + r\ (\gamma_g - 1)/2\ M_0^2$ is the recovery to free-stream temperature ratio, $r = Pr^{1/3}$ is the recovery factor and $Pr = 0.72$ the Prandtl number). Periodicity of the flow is assumed in the z direction. Non-reflecting boundary conditions are imposed at the inlet and outlet boundaries to minimize numerical feedback. Nonreflecting boundary conditions are also used at the top boundary, except for a narrow zone where the incoming shock is injected into the computational domain by hard enforcement of the Rankine-Hugoniot jump relations. Various values of the inflow skew angle γ_0 and shock generator θ are considered. Extensive flow separation occurs for all the cases under investigation. Table 1 contains the key computational parameters of the simulation campaign.

Table 1. Summary of the SBLI flow cases

Case Label	M_0	M_{0x}	γ_0 (deg.)	θ (deg.)	$Re_{\delta 0}$	$(L_x \times L_y \times L_z)/\delta_0$	x_r/δ_0	x_{imp}/δ_0	T_w/T_r
G00_T08	2.28	2.28	0	8	15800	$96 \times 20 \times 96$	30	64	1
G00_T10	2.28	2.28	0	10.4	15800	$96 \times 20 \times 96$	30	64	1
G07_T10	2.3	2.28	7.5	10.4	15800	$96 \times 20 \times 96$	30	64	1
G15_T10	2.36	2.28	15	10.4	16200	$96 \times 20 \times 96$	30	64	1
G30_T08	2.63	2.28	30	8	19000	$96 \times 20 \times 96$	30	64	1
G30_T09	2.63	2.28	30	9.2	19000	$96 \times 20 \times 96$	30	64	1
G30_T10	2.63	2.28	30	10.4	19000	$96 \times 20 \times 96$	30	64	1
G45_T10	3.22	2.28	45	10.4	27500	$96 \times 20 \times 96$	30	64	1

Aerospace Science and Engineering - III Aerospace PhD-Days   Materials Research Forum LLC
Materials Research Proceedings 33 (2023) 388-396   https://doi.org/10.21741/9781644902677-57

Turbulent boundary layer (TBL) setup

The performance evaluation of standard turbulent inflow conditions for the DNS of supersonic/hypersonic turbulent boundary layers and the development of new inflow methods have been carried out with reference to the computational case of Pirozzoli & Bernardini [22] and Zhang et al. [38]. The first studied a spatially developing, supersonic zero-pressure-gradient (ZPG) boundary layer on a flat plate, with free-stream Mach number $M_0=2$ and nominally adiabatic wall conditions ($T_w/T_r=1$); the latter a spatially developing, ZPG hypersonic boundary layer at $M_0=5.84$ and cooled walls ($T_w/T_r=0.25$).

A series of direct numerical simulations has been performed, using several inflow conditions based both on recycling-rescaling (RR) and digital filtering (DF). A schematic representation of the recycling-rescaling setup is shown in Figure 2.

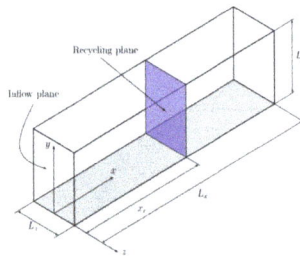

Figure 2. Set-up for recycling/rescaling: x_r denotes the position of the recycling plane, and L_x, L_y and L_z denote the size of the computational box (Figure from Ceci et al. [5], Creative Common license http://creativecommons.org/licenses/by/4.0/)

Standard recycling-rescaling and digital filtering inflow generations have been already implemented in the publicly available version of the STREAmS solver. Those routines have been modified throughout the course of the present project to test two novel inflow conditions. The flow is assumed to be periodic in the z direction, and non-reflecting boundary conditions are imposed on both the top and right boundaries. Time-averaging has been obtained by collecting instantaneous data for at least 800 convective time units δ_0/U_0; spanwise-averaging of the velocity and pressure fields is also performed. The list of DNS and the key computational setup are reported in Table 2.

Table 2. Summary of TBL flow cases

Flow Case	M_0	$Re_{\delta 0}$	$(L_x \times L_y \times L_z)/\delta_0$	x_r/δ_0	T_w/T_r
M2-RR	2	12662	$106 \times 8.3 \times 9.6$	53	1
M2-DF	2	12662	$159 \times 8.3 \times 9.6$	-	1
M2-L1	2	4479	$310 \times 26 \times 32$	53	1
M2-L2	2	8230	$310 \times 26 \times 26$	53	1
M2-L3	2	12662	$318 \times 16.6 \times 19.2$	53	1
M5.84-RR	5.84	23152	$150 \times 10 \times 9$	53	0.25
M5.84-DF	5.84	23152	$150 \times 10 \times 9$	-	0.25
M5.84-L1	5.84	10650	$300 \times 20 \times 18$	53	0.25
M5.84-L2	5.84	16788	$300 \times 20 \times 18$	53	0.25
M5.84-L3	5.84	23152	$300 \times 20 \times 18$	53	0.25

Aerospace Science and Engineering - III Aerospace PhD-Days   Materials Research Forum LLC
Materials Research Proceedings 33 (2023) 388-396   https://doi.org/10.21741/9781644902677-57

Results of the supersonic SBLI study

We have developed a simple model to characterize low-frequency unsteadiness in swept SBLIs, which is robustly supported from analysis of DNS data. We provide a scaling law for the spanwise undulation of the separation line and for the convection velocity of pressure disturbances, which concur to predict growth of the typical pressure oscillation frequency with the skew angle, consistent with trends observed in DNS. The proposed behavior of pressure fluctuations along the shock foot is described as

$$ St_L = \left| St_{L,0} \pm \frac{\eta \, tan(\gamma_0)}{\alpha} \right| $$

with $\alpha \approx 2$ and $\eta \approx 0.7$, as obtained from the SBLI study, and $St_{L,0} \approx 0.04$.

Quantitative comparison of the numerically computed peak frequencies with the above prediction is presented in Figure 3. The prediction is clearly quite good, perhaps with exception of the single data point corresponding to $\gamma_0 = 30°$, $\theta = 8°$, which has a small separation bubble. Overall, the agreement becomes more satisfactory as the sweep angle increases [6].

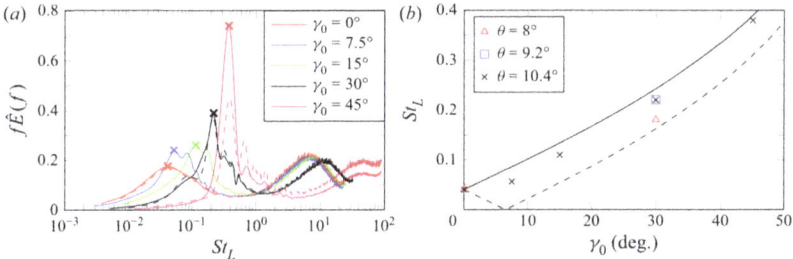

Figure 3: (a) Pre-multiplied normalized frequency spectra of wall pressure at the mean separation line for various sweep angles and for fixed shock strength ($\theta = 10.4°$). Peaks are marked with crosses. Solid lines denote PSD obtained with the full-time window, whereas dashed lines denote PSD obtained with 50 % shorter time windows. (b) Peak frequency as a function of sweep angle: the solid and dashed lines denote the proposed prediction (Figure from Ceci et al. [6], Creative Common license http://creativecommons.org/licenses/by/4.0/).

Results of the TBL numerical tripping

Two reference flow cases have been selected, one representative of supersonic adiabatic boundary layers and the other of hypersonic cooled boundary layers. For both flow cases, two series of DNS have been carried out, one on relatively short domains, which serve to quantify effects of inflow seeding (RR- or DF-type), as compared to benchmark simulations, carried out in very long domains, which are verified to be yield to a healthy state of developed turbulence. The supersonic data set includes six DNS in short domains and three DNS in long domains, while the hypersonic data set includes four DNS in short domains, and three DNS in long domains.

We have derived a procedure to assess the TBL equilibrium conditions in numerical simulation by monitoring the deviation of significant metrics from a reference trend. Such metrics are the friction coefficient, wall pressure root mean square, Reynolds stress peaks and Stanton number [5]. In this respect, no single criterion can be used to define the inflow length for arbitrary flow conditions, but rather different metrics suggest different inflow adaptation lengths, which can also change as a result of the flow conditions. We have found that the friction coefficient is particularly sensitive to inflow seeding, and it can bear memory of inflow seeding quite far from the inflow plane [5].

Aerospace Science and Engineering - III Aerospace PhD-Days   Materials Research Forum LLC
Materials Research Proceedings 33 (2023) 388-396   https://doi.org/10.21741/9781644902677-57

References

[1]Adler, M. C. & Gaitonde, D. V. 2018 Dynamic linear response of a shock/turbulent-boundary-layer interaction using constrained perturbations. Journal of Fluid Mechanics 840. https://doi.org/10.1017/jfm.2018.70

[2]Adler, M. C. & Gaitonde, D. V. 2019 Flow similarity in strong swept-shock/ turbulent-boundary-layer interactions. AIAA Journal 57. https://doi.org/10.2514/1.J057534

[3]Babinsky, H. & Harvey, J. K. 2011 Shock wave-boundary-layer interactions. Cambridge University Press. https://doi.org/10.1017/CBO9780511842757

[4]Bernardini, M., Modesti, D., Salvadore, F. & Pirozzoli, S. 2021 Streams: A high-fidelity accelerated solver for direct numerical simulation of compressible turbulent flows. Computer Physics Communications 263, 107906. https://doi.org/10.1016/j.cpc.2021.107906

[5]Ceci , A., Palumbo , A., Larsson , J. & Pirozzoli , S. 2022 Numerical tripping of high-speed turbulent boundary layers. Theoretical and Computational Fluid Dynamics 36. https://doi.org/10.1007/s00162-022-00623-0

[6]Ceci , A., Palumbo , A., Larsson , J. & Pirozzoli , S. 2023 On low-frequency unsteadiness in swept shock wave-boundary layer interactions. Journal of Fluid Mechanics 956. https://doi.org/10.1017/jfm.2023.2

[7]Dhamankar, N. S., Blaisdell, G. A. & Lyrintzis, A. S. 2018 Overview of turbulent inflow boundary conditions for large-eddy simulations. AIAA Journal 56. https://doi.org/10.2514/1.J055528

[8]Di Renzo, M., Oberoi, N., Larsson, J. & Pirozzoli, S. 2021 Crossflow effects on shock wave/turbulent boundary layer interactions. Theoretical and Computational Fluid Dynamics. https://doi.org/10.1007/s00162-021-00574-y

[9]Dolling, D. S. 2001 Fifty years of shock-wave/boundary-layer interaction research: What next? AIAA Journal 39. https://doi.org/10.2514/3.14896

[10]Erengil, M. E. & Dolling, D. S. 1993 Effects of sweepback on unsteady separation in mach 5 compression ramp interactions. AIAA Journal 31. https://doi.org/10.2514/3.60176

[11]Erm, L. P. & Joubert, P. N. 1991 Low-reynolds-number turbulent boundary layers. Journal of Fluid Mechanics 230. https://doi.org/10.1017/S0022112091000691

[12]Fang, J., Yao, Y., Zheltovodov, A. A. & Lu, L. 2017 Investigation of Three-Dimensional ShockWave/Turbulent-Boundary-Layer Interaction Initiated by a Single Fin. AIAA Journal 55. https://doi.org/10.2514/1.J055283

[13]Gaitonde, D. V. 2015 Progress in shock wave/boundary layer interactions. Progress in Aerospace Sciences 72. https://doi.org/10.1016/j.paerosci.2014.09.002

[14]Gaitonde, D. V., Shang, J. S., Garrison, T. J., Zheltovodov, A. A. & Maksimov, A. I. 1999 Three-dimensional turbulent interactions caused by asymmetric crossing-shock configurations. AIAA journal 37. https://doi.org/10.2514/2.660

[15]Gross, A. & Fasel, H. F. 2016 Numerical investigation of shock boundary-layer interactions. In 54th AIAA Aerospace Sciences Meeting. https://doi.org/10.2514/6.2016-0347

[16]Klein, M., Sadiki, A. & Janicka, J. 2003 A digital filter based generation of inflow data for spatially developing direct numerical or large eddy simulations. Journal of Computational Physics 186. https://doi.org/10.1016/S0021-9991(03)00090-1

[17]Lund, T. S., Wu, X. & Squires, K. D. 1998 Generation of turbulent inflow data for spatially-developing boundary layer simulations. Journal of computational physics 140. https://doi.org/10.1006/jcph.1998.5882

[18]Padmanabhan, S., Maldonado, J. C., Threadgill, J. A. & Little, J. C. 2021 Experimental study of swept impinging oblique shock/boundary-layer interactions. AIAA journal 59. https://doi.org/10.2514/1.J058910

[19]Piponniau, S., Dussauge, J. P., Debiève, J. F. & Dupont, P. 2009 A simple model for low-frequency unsteadiness in shock-induced separation. Journal of Fluid Mechanics 629. https://doi.org/10.1017/S0022112009006417

[20]Pirozzoli, S. 2010 Generalized conservative approximations of split convective derivative operators. Journal of Computational Physics 229. https://doi.org/10.1016/j.jcp.2010.06.006

[21]Pirozzoli, S. & Bernardini, M. 2011a Direct numerical simulation database for impinging shock wave/turbulent boundary-layer interaction. AIAA Journal 49. https://doi.org/10.2514/1.J050901

[22]Pirozzoli, S. & Bernardini, M. 2011b Turbulence in supersonic boundary layers at moderate reynolds number. Journal of Fluid Mechanics 688. https://doi.org/10.1017/jfm.2011.368

[23]Pirozzoli, S., Bernardini, M. & Grasso, F. 2010 Direct numerical simulation of transonic shock/boundary layer interaction under conditions of incipient separation. Journal of Fluid Mechanics 657. https://doi.org/10.1017/S0022112010001710

[24]Pirozzoli, S. & Grasso, F. 2006 Direct numerical simulation of impinging shock wave/turbulent boundary layer interaction at M=2.25. Physics of Fluids 18. https://doi.org/10.1063/1.2216989

[25]Priebe, S. & Pino Martin, M. 2012 Low-frequency unsteadiness in shock wave-turbulent boundary layer interaction. Journal of Fluid Mechanics 699. https://doi.org/10.1017/jfm.2011.560

[26]Schlatter, P., Li, Q., Brethouwer, G., Johansson, A. V. & Henningson, D. S. 2010 Simulations of spatially evolving turbulent boundary layers up to Reθ= 4300. International Journal of Heat and Fluid Flow 31. https://doi.org/10.1016/j.ijheatfluidflow.2009.12.011

[27]Schmisseur, J. D. & Dolling, D. S. 1994 Fluctuating wall pressures near separation in highly swept turbulent interactions. AIAA Journal 32. https://doi.org/10.2514/3.12114

[28]Settles, G. S., Perkins, J. J. & Bogdonoff, S. M. 1980 Investigation of three-dimensional shock/boundary-layer interactions at swept compression corners. AIAA Journal 18. https://doi.org/10.2514/3.50819

[29]Settles, G. S. & Teng, H. Y. 1984 Cylindrical and conical flow regimes of three-dimensional shock/boundary-layer interactions. AIAA Journal 22. https://doi.org/10.2514/3.8367

[30]Sillero, J. A., Jiménez, J. & Moser, R. D. 2013 One-point statistics for turbulent wall-bounded flows at reynolds numbers up to δ+ 2000. Physics of Fluids 25, 105102. https://doi.org/10.1063/1.4823831

[31]Simens, M. P., Jim´enez, J., Hoyas, S. & Mizuno, Y. 2009 A high-resolution code for turbulent boundary layers. Journal of Computational Physics 228. https://doi.org/10.1016/j.jcp.2009.02.031

[32]Threadgill, J. A. & Little, J. C. 2020 An inviscid analysis of swept oblique shock reflections. Journal of Fluid Mechanics 890. https://doi.org/10.1017/jfm.2020.117

Aerospace Science and Engineering - III Aerospace PhD-Days   Materials Research Forum LLC
Materials Research Proceedings 33 (2023) 388-396   https://doi.org/10.21741/9781644902677-57

[33]Vanstone, L. & Clemens, N. T. 2018 Pod analysis of unsteadiness mechanisms within a swept compression ramp shock-wave boundary-layer interaction at Mach 2. In AIAA Aerospace Sciences Meeting, 2018. https://doi.org/10.2514/6.2018-2073

[34]Vanstone, L., Musta, M. N., Seckin, S. & Clemens, N. T. 2018 Experimental study of the mean structure and quasi-conical scaling of a swept-compression-ramp interaction at Mach 2. Journal of Fluid Mechanics 841. https://doi.org/10.1017/jfm.2018.8

[35]Wenzel, C., Selent, B., Kloker, M. & Rist, U. 2018 DNS of compressible turbulent boundary layers and assessment of data/scaling-law quality. Journal of Fluid Mechanics 842, 428-468. https://doi.org/10.1017/jfm.2018.179

[36]Wray, A. A. 1990 Minimal storage time advancement schemes for spectral methods. Tech. Rep.NASA Ames Research Center, Moffett Field, CA.

[37]Xu, S. & Martin, M. P. 2004 Assessment of inflow boundary conditions for compressible turbulent boundary layers. Physics of Fluids 16. https://doi.org/10.1063/1.1758218

[38]Zhang, C., Duan, L., Choudhari, M.M. 2018 Direct numerical simulation database for supersonic and hypersonic turbulent boundary layers. AIAA J. 56(11), 4297-4311 https://doi.org/10.2514/1.J057296

[39]Zuo, F. Y., Memmolo, A., Huang, G. P. & Pirozzoli, S. 2019 Direct numerical simulation of conical shock wave-turbulent boundary layer interaction. Journal of Fluid Mechanics 877. https://doi.org/10.1017/jfm.2019.558

Keyword Index

e